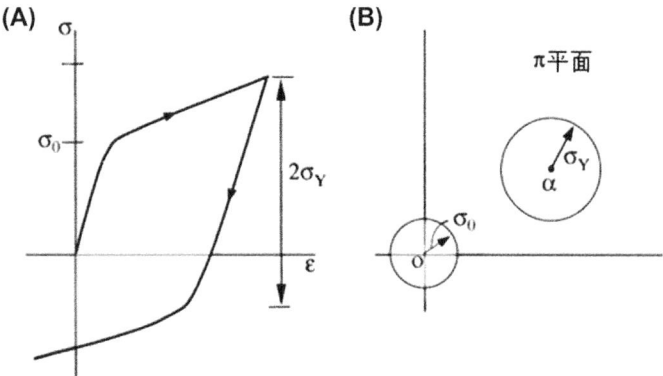

Figure 2.2

Combined isotropic-kinematic hardening illustration: (A) Bauschinger effect, and (B) translation and expansion of yield surface.

in compression is reduced after an initial yield in tension. A way to conceptualize this behavior is to observe that the center of the yield surface moves in a direction of the plastic flow. The schematic in Fig. 2.2(B) is for multiaxial stress states. The expansion of a circular yield surface corresponds to isotropic hardening and translation of its center corresponds to kinematic hardening. This means that the isotropic hardening corresponding to the yield surface expands but its center does not move, whereas the kinematic hardening corresponding to the center of yield surface moves but its surface does not expand. If the yield surface is in expansion and its center is also in motion, combination hardening occurs, as shown in Fig. 2.2(B).

Prager (1945) and Ziegler (1959) introduced a simple kinematic hardening plasticity model to consider this phenomenon. In the kinematic hardening model, another internal variable, α, called back stress, is introduced into both the plastic flow relation and the yield condition:

Plastic flow law:

$$\dot{\varepsilon}^p = \dot{\lambda}\frac{\partial\psi}{\partial\sigma}, \quad \psi = |\sigma - \alpha| \tag{2.2.15}$$

Yield condition:

$$f = |\sigma - \alpha| - \sigma_Y(\bar{\varepsilon}) \tag{2.2.16}$$

Note that $\partial\psi/\partial\sigma = \partial f/\partial\sigma = \text{sign}(\sigma - \alpha)$ and $\dot{\bar{\varepsilon}} = \dot{\lambda}$ from Eq. (2.2.8). An evolution equation is required for the internal variable α (back stress). The simplest form is called linear kinematic hardening and is specified as

$$\dot{\alpha} = \kappa\dot{\varepsilon}^p \tag{2.2.17}$$

Dislocation Mechanism-Based Crystal Plasticity

Dislocation Mechanism-Based Crystal Plasticity

Theory and Computation at the Micron and Submicron Scale

Zhuo Zhuang

Zhanli Liu

Yinan Cui

ACADEMIC PRESS

An imprint of Elsevier

Library of Congress Cataloging-in-Publication Data
A catalog record for this book is available from the Library of Congress

British Library Cataloguing-in-Publication Data
A catalogue record for this book is available from the British Library

ISBN: 978-0-12-814591-3

For information on all Academic Press publications visit our
website at https://www.elsevier.com/books-and-journals

Working together
to grow libraries in
developing countries

www.elsevier.com • www.bookaid.org

Publisher: Matthew Deans
Acquisition Editor: Glyn Jones
Editorial Project Manager: Naomi Robertson
Production Project Manager: Sruthi Satheesh
Cover Designer: Vicky Pearson

Typeset by TNQ Technologies

Contents

Preface

This book aims to provide a comprehensive introduction to the theoretical models and computational methods of dislocation mechanism-based crystal plasticity at the micron and submicron scales (100 nm—100 μm). The popular notion that smaller is stronger has accentuated the need to introduce a size effect. Contrary to the prevalent notion that prestrain leads to hardening and annealing leads to softening, a small-scaled material exhibits softening when it is prestrained and anneal hardening when the temperature is increased. It is also interesting that a variety of atypical mechanical behaviors are observed in the crystal when the material size is reduced from the macroscopic to the micron to submicron scales: for example, size-dependent yield stress and intermittent plastic flow during compression tests of micron pillars. The widely observed intermittent plastic flow at small scales prompts a paradigm shift away from traditional well-defined continuous and determined plastic flow behavior. These unconventional features have attracted a great deal of academic attention worldwide in the engineering mechanics' and materials' community. Much research work has been conducted. This book attempts to provide insights into understanding micron and submicron plasticity and to present a theoretical framework and simulation efforts.

This book is structured in two parts. Part I focuses on continuum dislocation mechanism-based crystal plasticity. In contrast, Part II focuses on discrete dislocation mechanism-based crystal plasticity. Part I, including Chapters 2—6, presents the continua dislocation mechanism-based crystal plasticity theory and computations, which include the fundamental concept of conventional crystal plasticity theory without a size effect at macroscale; the strain gradient crystal plasticity theory based on the Taylor law dislocation mechanism considering a size effect at micron scale; the developed dislocation-based crystal plasticity model within the framework of continuum mechanics theories by introducing the dislocation mechanism revealed by experiments and dislocation dynamic simulations at the submicron scale; and the phase-field theory of crystal plasticity. Part II, including Chapters 7—12, describes the discrete dislocation mechanism-based theory and computations at the submicron scale, which includes the single-crystal plasticity theory and the discrete-continuous model of crystal plasticity by coupling three-dimensional discrete dislocation dynamics and the finite element method. Three kinds of plastic deformation mechanisms for submicron pillars are systematically presented: single-arm dislocation source-controlled plastic flow in micropillars, confined plasticity in coated micropillars, and dislocation

starvation under low-amplitude cyclic loading in micropillars. Flow stress is deduced according to the single-arm dislocation source mechanism, which is a significant achievement in crystal plasticity theoretical work at a submicron scale. Two other interesting issues regarding crystal plasticity related to discrete dislocation evolution are described. One is dislocation nucleation and multiplication at a high strain rate, with examples of compression micropillars. The other is the temperature effect for the dislocation annihilation mechanism considering a void diffusion-based dislocation climb model and helical dislocations based on a coupled glide-climb model.

The plasticity theory formula, computational modeling, and experiment data of crystalline materials at the micron and submicron scales are discussed in this book. The focus is mainly on dislocation motion-controlled plastic behavior in crystals. Understanding the properties of crystalline materials to be captured by an appropriate constitutive model is the key to modeling large plastic deformation. Many books describe crystal plasticity theory models, computation methods, and experiment data at the macroscale. However, few books are dedicated to describing crystal plasticity covering the micron and submicron scales; thus, it may be that this is the first book on both continua and discrete dislocation mechanism-based crystal plasticity theories and computations at these scales.

There are more than 600 related articles listed in references, more than 25 of which were published by our group. This book originated from doctoral thesis research work during 2004–2018 at the Department of Engineering Mechanics, Tsinghua University in Beijing. The contributors were Yu Guo, Xiaoming Liu, Zhanli Liu, Junfeng Nie, Zhaohui Zhang, Xuechuan Zhao, Yuan Gao, Yinan Cui, Peng Lin, Liyuan Wang, Jianqiao Hu, and Fengxian Liu. Specially, Prof. Zhanli Liu and Dr. Yinan Cui are coauthors of this book.

We sincerely appreciate Prof. Keh-Chih Hwang, who gave us a lot of encouragement and fruitful discussion to understand theory models and constitutive relations in solid mechanics.

We would like to thank Mr. Lei Shi and Ms. Ya Qi at Tsinghua University Press for helping publish this book.

Zhuo Zhuang, Zhanli Liu, Yinan Cui
Department of Engineering Mechanics, School of Aerospace Engineering, Tsinghua University,
Beijing, China

plastic deformation mechanisms for submicron pillars are systematically presented: single-arm dislocation, source-controlled plastic flow in micropillars; confined plasticity in coated micropillars; and dislocation starvation under low-amplitude cyclic loading in micropillars. Flow stress is deduced according to the source mechanism of single-arm dislocation, which is a significant achievement of work on crystal plasticity theory at a submicron scale. Two other interesting issues of crystal plasticity related to the evolution of discrete dislocation are described. One is dislocation nucleation and starvation at a high strain rate, with examples of compression micropillars. The other is the temperature effect for the mechanism of dislocation annihilation considering a diffusion model-based dislocation climb model and helical dislocations based on a coupled glide-climb model.

The plasticity theory formula, computation modeling, and experimental data of crystalline materials on the micron and submicron scales are discussed in this book. The focus is mainly on dislocation motion-controlled plastic behavior in crystals. Understanding the properties of crystals captured by an appropriate constitutive model is the key to modeling large plastic deformation.

1.2 Polycrystalline and Single-Crystal Plasticity

It is an experimental fact that the stress-strain behavior of a large polycrystalline aggregate is different from that of a single crystal, which will define the local response of a material point in crystalline plasticity models, as can be seen in Fig. 1.1. The local plastic response of a single crystal is generally characterized by three distinct stages: easy glide, hardening, and recovery (Fig. 1.1B), which are governed by the dislocation motion. In general, these three stages cannot be discerned at the macroscale (Fig. 1.1A), which may be thought of as a homogenization of all local responses in a polycrystalline aggregate.

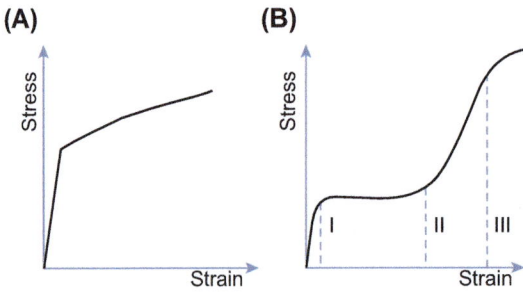

Figure 1.1
Stress and strain in (A) a polycrystalline aggregate, and (B) a single crystal.

Another way in which a large polycrystalline aggregate is different from that of a single crystal is the local and global rates of deformation, in which the global rate is the appropriate volume average of local rates. When dislocation evolution is stable, the local plastic response will be averaged over a uniformly deforming specimen. When dislocation evolution becomes unstable at some material points, only part of the crystal becomes plastically active and localized deformation for the single crystal will ensue (Estrin and Kubin, 1986). For specimens composed of a multitude of randomly oriented crystals, their plastic response may appear isotropic. However, single crystals are strongly anisotropic in their plasticity. This discrepancy has been explained in terms of the preference of dislocations to nucleate in only specific planes of crystals and to propagate along only certain directions contained in those planes, which is a main concern of this book.

Macroscopic plastic deformation may appear homogeneous and isotropic in some cases whereas the origins of plasticity on a microscale are neither homogeneous nor isotropic. The inhomogeneity of plastic deformation is an important aspect of large deformation mechanics that phenomenological constitutive laws are unable to capture properly. For instance, in applications of metal forming such as deep drawing, effects such as springback, and earing are widely observed. It is important to reproduce the evolving plastic inhomogeneity but it remains difficult for phenomenological plasticity models to capture this. On the other hand, in its atomic origins, plasticity involves the passage of atoms between positions of equilibrium in the crystal (Bulatov and Cai, 2006; Hull and Bacon, 2011), so that it is necessary to keep track of patterns of atomic movement to describe them. To model plastic deformation efficiently, however, because of the large scale of engineering applications, it is infeasible to retain an atomic resolution of crystals (for example, molecule dynamics [MD]), so a continuum approach becomes necessary. To this end, the theory of single-crystal plasticity (Asaro and Rice, 1977) was developed, which offers a unique theoretical insight into the microstructural mechanisms that govern the mechanical behavior of crystals in general, and metals and alloys in particular.

Crystalline plasticity recognizes that plastic deformation occurs along preferred planes and directions in crystals, so that only specific orientations at a material point are used to construct the plastic deformation tensors. In this manner, plastic anisotropy is automatically built into the constitutive model. The possibility of inhomogeneous plastic deformation was shown to depend on the evolution of densities of dislocations (Estrin and Kubin, 1986), which are in fact the bearers of plasticity in crystals. Thus, an explicit connection to dislocation densities as internal variables needs to be made in crystal plasticity formulations, so as to reproduce the mechanisms of glide, hardening, and recovery that physically govern plasticity and its patterns of inhomogeneity.

1.3 Size Effect on Crystal Plasticity at Micron and Submicron Scales
1.3.1 Size Effect Observed in Material Experiments

For conventional crystal plasticity, there is no size effect in the constitutive relation and the stress-strain curve is smooth and continuous. However, during the last decade of the 20th century, through experiments with copper wire torsion, for a size effect at a micron scale, when diameter $d = 170$ μm reduced to $d = 12$ μm, the torsion strength increased three times, as shown in Fig. 1.2A (Fleck et al., 1994). In contrast, there was no size effect for stress-strain curves under uniaxial tension without a strain gradient even though the diameters varied from $d = 170$ μm to $d = 12$ μm, as shown in Fig. 1.2B. During a thin-beam bending test, for a size effect when thickness $h = 100$ μm reduced to $h = 12.5$ μm, the bending strength also increased significantly (Stölken and Evans, 1998). This unconventional crystal plasticity behavior created a challenge for the mechanician. Afterward, the continua dislocation mechanism of Taylor law-based strain gradient plasticity theory and other theories at a microscale was established, in which the material characteristic size was involved in the constitutive relations. This work is summarized in Chapter 3.

The question is whether, if the material size were reduced to a submicron scale under tension or compression without a strain gradient, a size effect would exist. If there is a size effect, what would the critical size be?

In the first decade of the 21st century, size effect was disclosed through uniaxial compression experimental micropillars when the diameter was reduced from bulk to a few

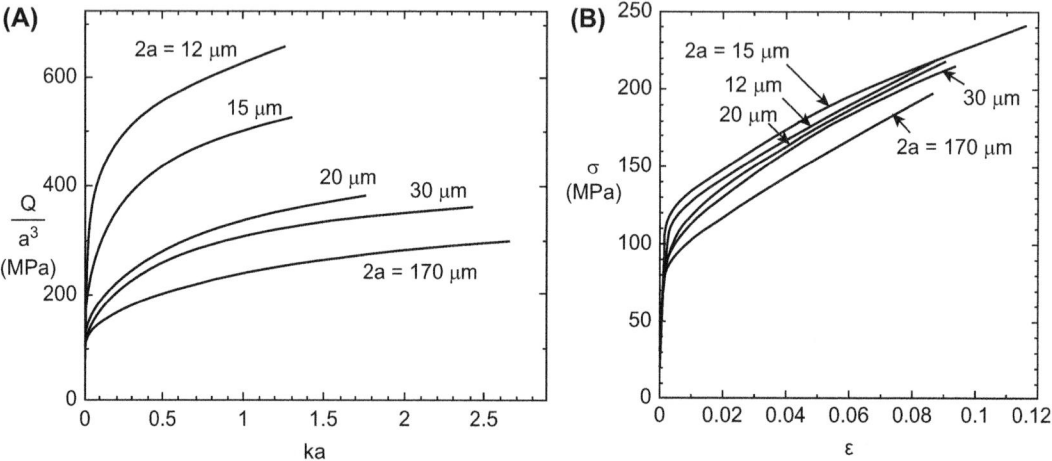

Figure 1.2

Fleck et al. (1994) thin copper wire test: (A) torsion; and (B) tension, where Q-torsion, k-torsion angle per unit length. *Reprinted from Fleck, N.A., Muller, G.M., Ashby, M.F., Hutchinson, J.W., 1994. Strain gradient plasticity - theory and experiment. Acta Metallurgica et Materialia 42, 475–487. Copyright 1994, with permission from Elsevier.*

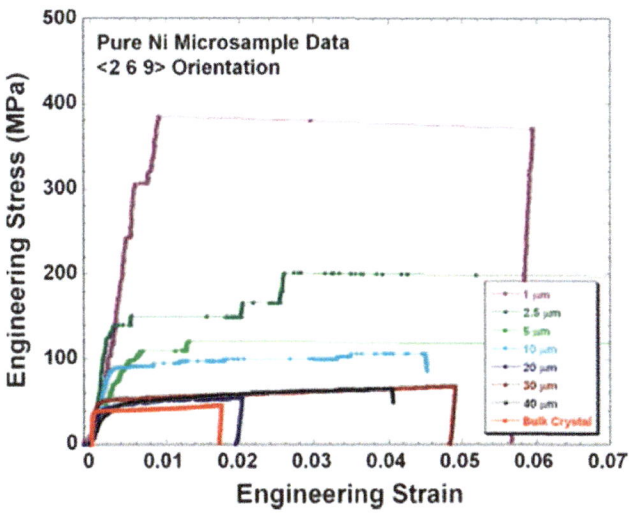

Figure 1.3

Uniaxial compression experiments on Ni micropillars when the diameter reduced from micro to submicro scale. *Reprinted from Dimiduk, D.M., Uchic, M.D., Parthasarathy, T.A., 2005. Size-affected single-slip behavior of pure nickel microcrystals. Acta Materialia 53, 4065—4077. Copyright 2005, with permission from Elsevier.*

micrometers (Dimiduk et al., 2005). As shown in Figs. 1.3 and 1.4, when the material size was reduced to the submicron scale under tension or compression without a strain gradient, there was a size effect of smaller and harder. In addition, the stress-strain curves were not smooth and continuous but exhibited significant intermittency. These behaviors brought a new challenge to the mechanician because the continua dislocation mechanism of Taylor law-based strain gradient plasticity theory and other theories could not be used at a submicron scale. How to understand the origin of the size effect and intermittent plasticity and how to develop a crystal plasticity theory based on these new features emerged as two main problems.

When the material size is reduced to the micron and submicron scale, simply using the dislocation density as the internal variable is insufficient to capture the new plastic behavior. Considering that the submicron sample size is close to the characteristic length of the internal microstructure, the discrete nature of dislocations becomes especially important. A conventional simulation and analysis method based on continuum mechanics is no longer applicable. This creates the need to capture the discrete nature of dislocation and reveal the new dislocation mechanism at small scales, which is important not only for the reliable design and improved performance of submicron devices but also to shed light on developing theories of submicron crystal plasticity.

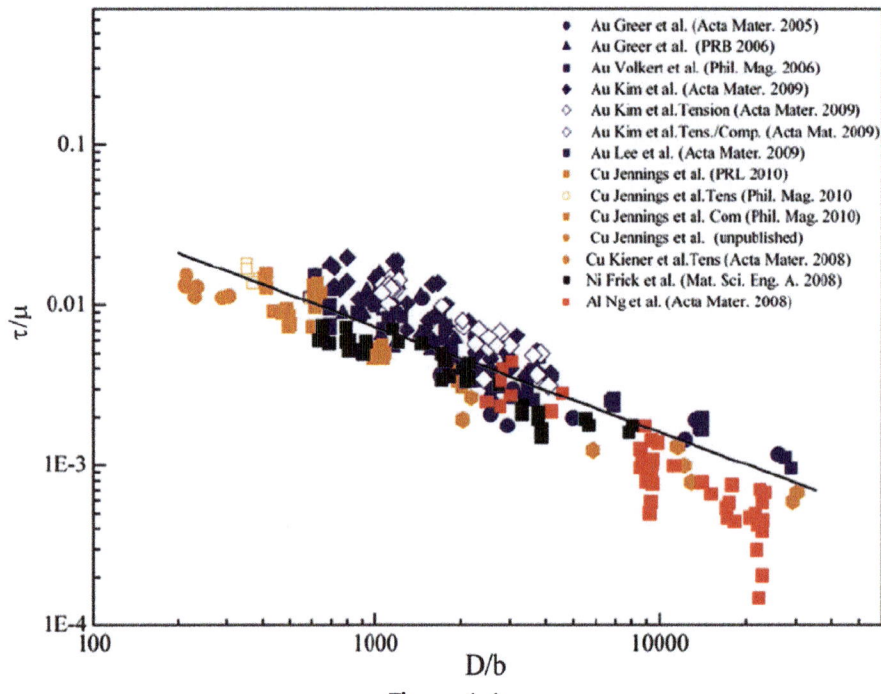

Figure 1.4

The reduced size effect makes the pillars harder under uniaxial compression. *Reprinted from Greer, J.R., De Hosson, J.T.M., 2011. Plasticity in small-sized metallic systems: intrinsic versus extrinsic size effect. Progress in Materials Science 56, 654–724. Copyright 2011, with permission from Elsevier.*

To achieve this goal, the discrete dislocation-dominated mechanism-based crystal plasticity theory and computational method were developed on a submicron scale, in which the material characteristic size and stable discrete dislocation source length were considered in the flow stress equations. This background was the main motivation that has stimulated our research work for several decades. The theory and method are summarized in Chapters 6–11.

The following section further reviews size-dependent plastic flow on the micron and submicron scales to familiarize readers.

1.3.2 Size Effect of Yield Stress

The size effect of strength is generally described by the equation for dislocation-controlled plasticity:

$$\tau = \tau_0 + K\sqrt{pL^{-2n} + q} \tag{1.3.1}$$

Table 1.1: Mechanism Model of Size Effect for Dislocation-Controlled Plasticity (Cui, 2016)

Source	Typical Examples	Model	Descriptions
Deformation constraint	**Nano Indentation**	Pileup hardening by geometrically necessary dislocations, strain gradient plasticity theory	$\tau = \alpha\mu b \sqrt{\left(\frac{3\tan^2\theta}{2b}\right)L^{-1} + \rho_{SSD}}$ (Nix and Gao, 1998) where α is constant and ρ_{SSD} is statistically stored dislocation density
Microstructure constraint	**Grain boundary**	Dislocation pileup model, Hall-Petch effect	$\tau = \tau_0 + K_{HP} L^{-1/2}$ (Hall, 1951; Petch, 1953), where K_{HP} is constant; when $L <$ critical grain size (~ 20 nm), it ceases to hold (El-Sherik et al., 1992; Schuh et al., 2002)
	Precipitation	Dislocation bowing out model, Orowan mechanism	$\tau = \alpha'\mu bL^{-1}$ α' depends on defect strength, when dislocation cannot pass $\alpha' = 1$ (Kelly and Nicholson, 1972)
Geometrical constraint	**Passivated film** $L=H/\sin\varphi$ **Free film** $L=H/\sin\varphi$	Dislocation channel model, similar to Orowan mechanism	Passivated film: $\tau = \mu_{eff}bL^{-1}$ Top free film: $\tau = \mu_{eff}bL^{-1/2}$ μ_{eff} is effective shear modulus, H is the film thickness, and φ is angle between slip plane and surface normal (Nix, 1989)
	Micropillar compression	Dislocation starvation model, single-arm source model, etc.	$\tau = \tau_0 + K_\rho L^{n_p}$ For face-centered cubic crystal, $n_p = 0.61-0.97$ (Uchic et al., 2009b)

Here, τ is the critical resolved shear stress, μ is the shear modulus, and b is the Burgers vector magnitude.

where τ is the resolved shear stress, τ_0 is the corresponding shear stress for bulk material, L is the internal characteristic length, and K, p, q, and n are constants, as summarized in Table 1.1.

Although these size effects can be described by this universal relation, the underlying mechanisms are different. Generally, the size effect falls roughly into three classes (Arzt, 1998; Gil Sevillano et al., 2001): the first is induced by the inhomogeneity of the microstructure or the boundary condition. For example, during nanoindentation (Bushby and Dunstan, 2004) or a thin metal wire rotation test (Fleck et al., 1994; Liu et al., 2013a), numerous geometrically necessary dislocations (GND) need to be stored to accommodate

the plastic deformation gradient, which leads to significant hardening. This kind of size effect is well-captured by the strain gradient plasticity theory, as described in Chapter 3. The second results from the resistant effect on the dislocation glide by the internal microstructure. The most widely observed microstructure constraint includes grain boundary and second phase particles. The finer the grain or the more dispersed the particle, the more significant the hardening effect. These can be predicted by the Hall-Petch effect and Orowan mechanism, respectively. The third size effect emerges when the geometrical size of the sample is close to the characteristic length of the dislocation structure. Related research was first carried out on thin films. The researchers proposed that the dislocation channel model would predict the source operation stress in passivated or free film (Tu and Rosenberg, 1982). It can be seen from the equations in Table 1.1 that the operation stress was similar to the Orowan equation describing participation hardening. The main difference is that the characteristic length changed from the participation distance to the effective length of the film. In addition, to consider the influence of the misfit dislocation, the concept of effective shear modulus is introduced (Nix, 1989). Actually, the size effect of the film also originated from the resistance effect to dislocation motion by an interface or passivated layer. On the other hand, a micropillar compression test exhibited a size effect with no geometrical constraint (Uchic et al., 2004). Here, uniform loading avoided introducing a strain gradient and dislocation glide was not constraint by the precipitation and interface. The mechanism of this kind of size effect quickly received a great deal of attention, as described in detail in Chapters 6−11.

Some researchers stated that this size effect was related to external conditions. For example, focused ion beam fabrication introduces surface defects owing to gallium ion bombardment on the surface of the sample. This leads to a high initial dislocation density. At the same time, a slight taper is created (Bei et al., 2007; Shim et al., 2009). Furthermore, both deformation of the substrate during uniaxial compression (Ouyang et al., 2009) and lateral constraint between the pillar top surface and the compression tip significantly influence the recorded stress-strain curve. However, a size effect is observed with different fabrication methods (Jennings et al., 2010) and under different substrate and lateral constraint conditions (Gao et al., 2010; Shade et al., 2009), which implies that these external factors are not the dominant reasons for the size effect. Until now, an interesting but unanswered question has been how to depict the influences of these external factors quantitatively.

To reveal the underlying dislocation mechanism of the size effect for submicron crystals, several models are proposed. Among them, two are most widely accepted. One is the dislocation starvation model proposed by Greer (2006), and Greer et al. (2005). It assumes that it is easy for dislocations to glide out from the free surface in small-scale samples, and the dislocation annihilation rate is even higher than their multiplication rate. Therefore, the dislocation density gradually decreases during deformation and even

reaches a starvation state. Then, further plastic deformation requires an increase in the external load to trigger the nucleation of new dislocations. Through in situ transmission electron microscopy (TEM), Shan et al. (2008) observed the dislocation surface annihilation process during a uniaxial compression test, verifying the dislocation starvation hypothesis. Afterwards, Liu et al. (2009a,b) captured the dislocation starvation trend and the accompanying hardening behavior using 3D DDD simulations. Benzerga (2009) further pointed out that dislocation starvation tends to occur in samples with a low initial dislocation density. Based on that phenomenological consideration, Nix and Lee (2010) developed a dislocation nucleation-dominated plastic model to predict the size effect.

The other typical model is single-arm source (SAS). Both DDD simulations (Akarapu et al., 2010; El-Awady et al., 2009a; Tang et al., 2008) and TEM observations (Oh et al., 2009) revealed that typical dislocation sources in submicron crystals exhibited new features. A phase diagram of plastic deformation mechanisms in a face-centered cubic (FCC) single crystal is illustrated in Fig. 1.5. The Frank-Read source with two pinning points in bulk samples becomes an SAS with one pinning point as a result of the surface truncation effect, which is the schematic figure of the SAS evolutionary process. Rao et al. (2007) discussed the anisotropic effect of the line tension force on the operation stress of SAS based on 3D DDD simulations. Parthasarathy et al. (2007) proposed an SAS model based on statistical considerations, which could be used to predict the initial yield strength and size effect of an FCC crystal according to the strength of the weakest SAS. Lee and Nix (2012) found that this SAS model was also suitable for describing a body-centered cubic crystal and discussed the dependence of the power law exponent for the size effect (see n_p in Table 1.1) on material properties according to this model.

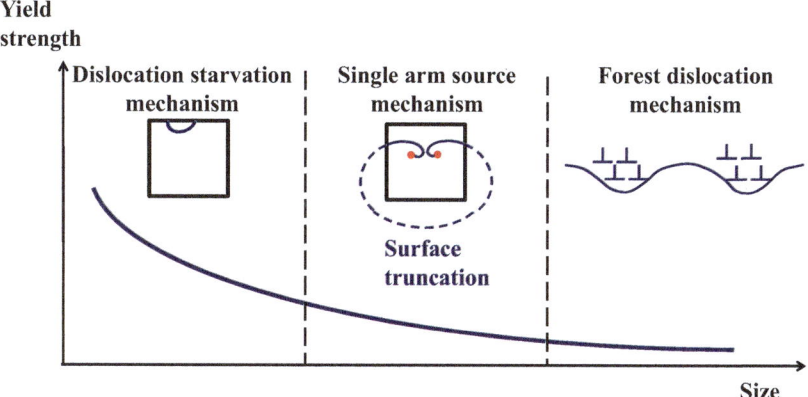

Figure 1.5
Phase diagram showing plastic deformation mechanisms in a face-centered cubic single crystal (Cui, 2016).

of a single-crystal Si beam using a depth-sensing nano-indenter with a harmonic force. Kiener et al. (2010) analyzed the relation between the Bauschinger effect and the dislocation pileup configuration for a cyclic bending test of a single-crystal Cu microbeam. In most available experimental (Kiener et al., 2010; Wang et al., 2008a; Zhang et al., 2003, 2006) and simulation studies (Brinckmann and Van der Giessen, 2004; Déprés et al., 2004; Shin et al., 2005) on the cyclic response of submicron crystals, there were internal microstructure (such as grain boundary) or strain gradients that introduced resistance or tramping effects on dislocation motion. Hence, it is difficult to isolate the effect of the external geometrical size on cyclic behavior. Namazu and Isono (2009) investigated the cyclic tension behavior of single-crystal Si, but the minimum characteristic sample size reached only 76 μm. Obviously, if the sample size were further reduced, the mean free path for dislocation gliding would become shorter. Meanwhile, the attractive image force would become more significant, leading to easier surface annihilation. How these features influence cyclic response on a small scale is an interesting problem.

Kiener and Minor (2011b) carried out multiple tension tests on a single-crystal sample with a minimum characteristic length of about 100−200 nm. The experimental results showed that the initial high dislocation density gradually decreased during multiple cycle tests (Fig. 1.7). This is distinctly different from the dislocation accumulation process in bulk materials. Moreover, Wang et al. (2015) carried out a low-amplitude cyclic tension

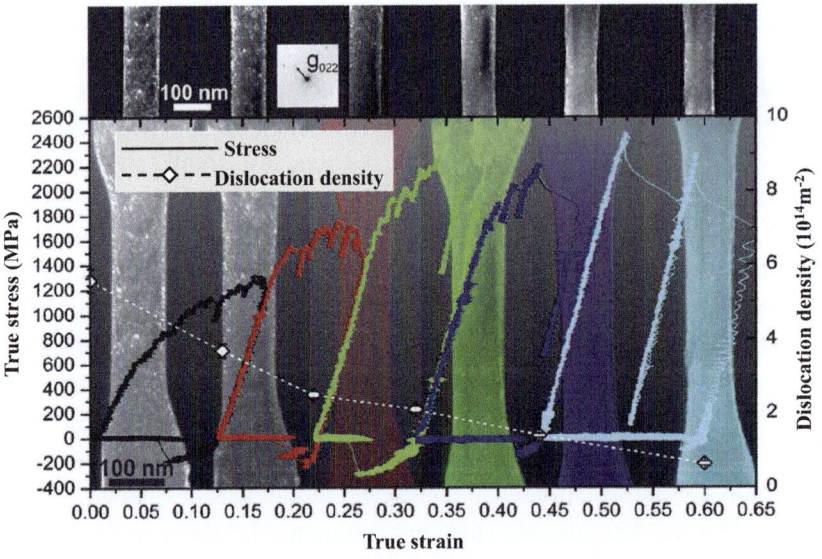

Figure 1.7

True stress versus true strain curve for Cu with a diameter of 133 nm under multiple tension tests, and the corresponding dislocation density evolution. *Reprinted from Kiener, D., Minor, A.M., 2011b. Source truncation and exhaustion: insights from quantitative in-situ TEM tensile testing. Nano Letters. Copyright 2011, with permission from ACS, https://pubs.acs.org/doi/10.1021/nl201890s.*

test on a single-crystal Al pillar with a characteristic length of 387 nm. The dislocation density gradually decreased and even reached a starvation state as cyclic loading proceeded. Furthermore, after cyclic tension, the dislocation-free sample exhibited relative high flow stress and clear necking features. This finding is important for the application of microdevices because it provides a new mechanical-based method to adjust the internal dislocation density, and the dislocation density has a pronounced effect on the mechanical, thermal, and electrical properties. The change in the shape of the sample was small because of the low cyclic loading stress, which means that this finding provides the possibility of obtaining a dislocation-free sample with little change in the fabrication geometry.

Several key questions must be answered before this method can be applied practically. First, how does low cyclic stress contribute to the decline in dislocation density? How will cyclic loading affect the dislocation annihilation and multiplication process? In addition, both the line tension model (Dupuy and Fivel, 2002) and atomic-level analysis (Rodney and Phillips, 1999) indicate that failure of the dislocation junction often requires an adequately high level of stress to be applied. How can dislocation junctions be destroyed without high stress under cyclic loading? In addition, how can the critical conditions for mechanical annealing be estimated? These important problems are investigated in Chapter 9.

1.3.5 Size Effect of Deformation Morphology of Compressed Micropillars

The size-dependent strength and intermittent flow of crystal micropillars were observed under uniaxial compression. Greer (2006) pointed out that small pillars under compression lacked stage II work hardening, as shown in Fig. 1.1, because small samples tended to deform in an individual slip system instead of activating multiple slip systems. After deformation, large pillars had barrel shapes with almost homogeneous deformation whereas small ones had severe shear bands (Dimiduk et al., 2005), as shown in Fig. 1.8A. Even loaded in a high-symmetry orientation <100> (Kiener and Minor, 2011a), the pillars revealed a transition from multiple slip systems to a single slip system with a decreasing diameter of 30 to 1.0 μm, as shown in Fig. 1.8B. This diameter reduction reduced the size scale from a micron to submicron scale. The dislocation slipping mechanism was only along the single slip system, which is obviously different from slipping along double-slip systems with symmetry deformation. The theoretical model was developed by Lin et al. (2016). It tends to deform with double slips and symmetry deformation for large pillars $d > 10$ μm; it deforms with a single slip and form a shear band for small pillars 0.2 μm $< d < 10$ μm because the second slip system operation is restricted. The computation results are plotted in Fig. 1.9. This theoretical model and the computation results are described in Chapter 4.

(A)

(B)

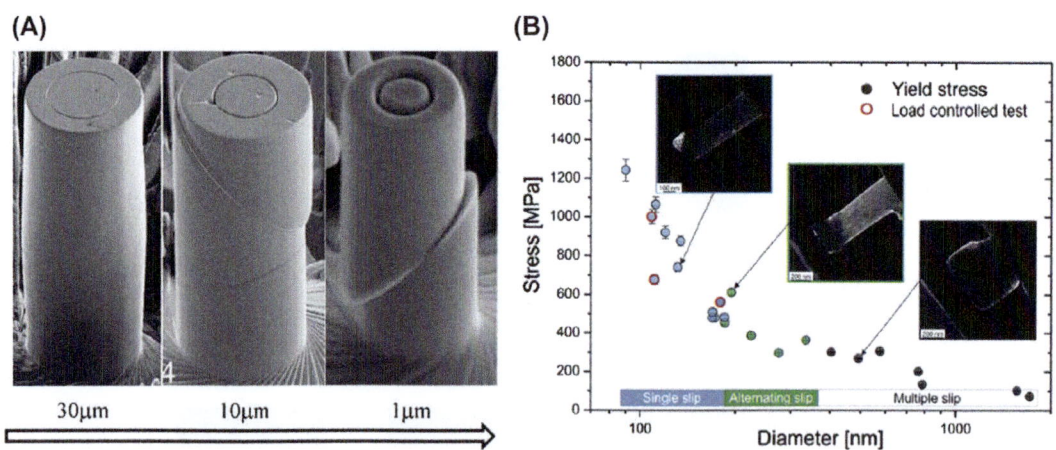

Figure 1.8

(A) Transition from barrel shape to shear band with decreasing pillar size, from experiments of Dimiduk et al. (2005); (B) In addition to a size effect on strength, a change in the deformation morphology is shown, as indicated by *differently colored symbols* and *color bars* indicating the regimes, from the experiments of Kiener and Minor (2011b). *(A) Reprinted from Dimiduk, D.M., Uchic, M.D., Parthasarathy, T.A., 2005. Size-affected single-slip behavior of pure nickel microcrystals. Acta Materialia 53, 4065–4077. Copyright 2005, with permission from Elsevier. (B) Reprinted from Kiener, D., Minor, A.M., 2011a. Source-controlled yield and hardening of Cu (100) studied by in situ transmission electron microscopy. Acta Materialia 59 (4), 1328–1337. Copyright 2011, with permission from Elsevier.*

1.4 Method to Bridge Size Effect

1.4.1 Supersurface From Macro to Micron

As discussed in Section 1.3, there are many features of size effect when the material size is reduced from a micron to submicron scale. How can the behavior be linked as "smaller is harder"? The MD simulation results (Guo et al., 2007) are illustrated in a log-log plot of normalized resolved yield stress (NRYS) as a function of the length scale parameter, as shown in Fig. 1.10, which are compared with other atomistic simulations (Horstemeyer et al., 2001) and experimental data (Fleck et al., 1994; Edington, 1969; Greer et al., 2005; McElhaney et al., 1998; Michalske and Houston, 1998), as well as a power law given by Horstemeyer et al. (2001). As is shown in Fig. 1.10 (also called the supersurface), the simulation results coincide with the power law (the dashed line) that predicts simple shear and indentation responses well but exhibits some departures in the torsion and compression data. The power law takes the form

$$y^* = 3.2 \times 10^{-5} x^{-0.38} \tag{1.4.1}$$

where y^* refers to NRYS at the relatively low strain rate and x is a length scale parameter, defined as the ratio of the volume to the surface area. This power law bridges the spatial scales from nano-dimensions to microns but it does not accurately predict the yield stress of

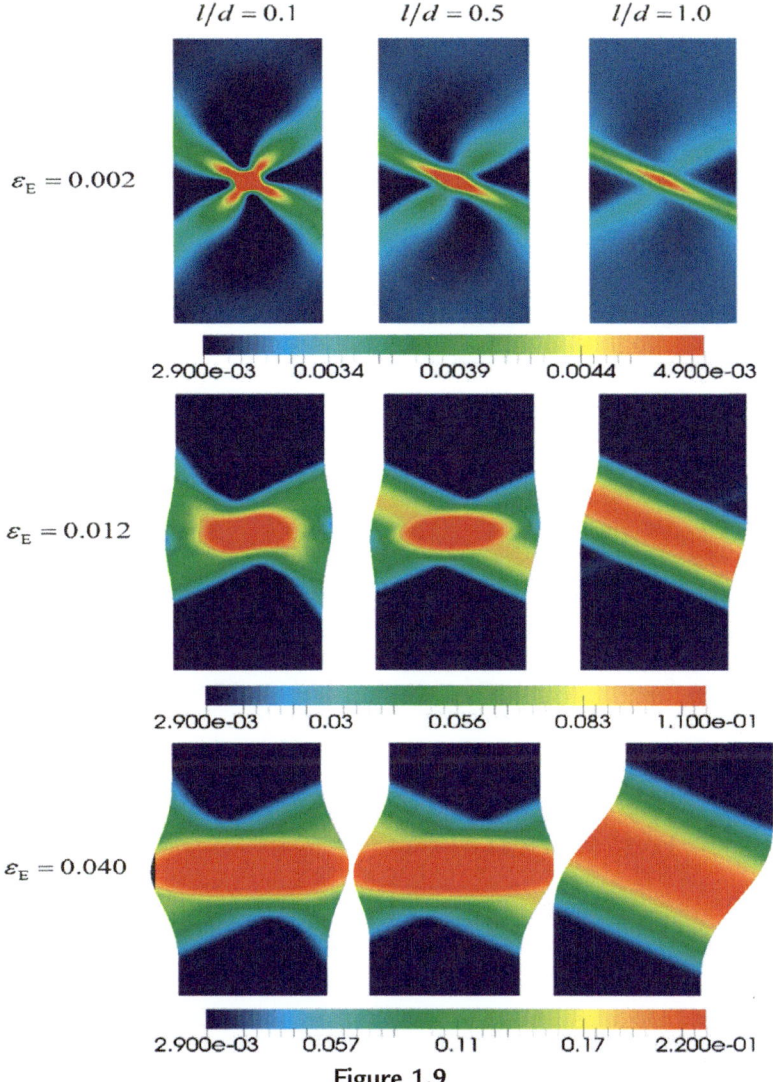

Figure 1.9

The same pillars during loading are plotted in the same column whereas different sizes of pillars at the same engineering strain are plotted in the same row. The model tends to deform with double slips and symmetry deformation for large pillars $d > 10$ µm, and it deforms with a single slip and forms a shear band for small pillars 0.2 µm $< d < 10$ µm because the second slip system operation is restricted. *Reprinted from Lin, P., Liu, Z., Zhuang, Z., 2016. Numerical study of the size-dependent deformation morphology in micropillar compressions by a dislocation-based crystal plasticity model. International Journal of Plasticity 87, 32—47. Copyright 2016, with permission from Elsevier.*

the specimens at a larger scale (millimeters, for example), as demonstrated in Fig. 1.10. By checking Eq. (1.4.1), it can be found that the NRYS tends to 0 as the sample length is increased to infinity. Hence, an attempt is made here to modify the power law so that the mechanical properties across nanoscale, microscale, and macroscale may be better represented.

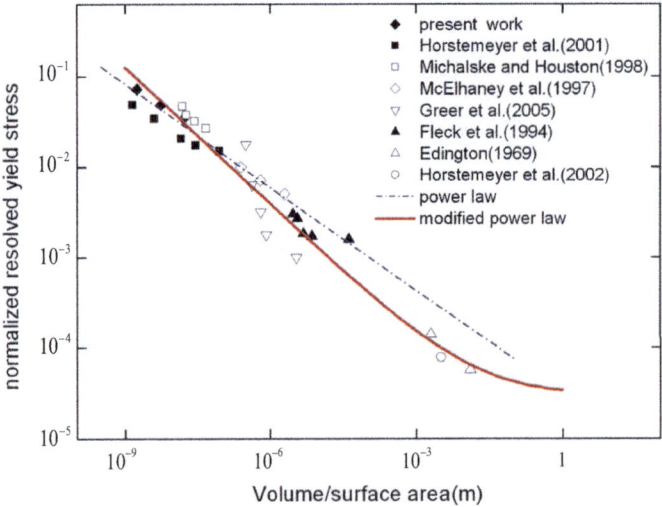

Figure 1.10

Yield stress normalized by the elastic modulus and resolved on a (111) slip plane versus the ratio of the volume to the surface area for copper, nickel, and gold, with various experiments and atomistic simulations. *Reprinted from Guo, Y., Zhuang, Z., Li, X., Chen, Z., 2007. An investigation of the combined size and rate effects on the mechanical responses of FCC metals. International Journal of Solids and Structures 44, 1180–1195. Copyright 2007, with permission from Elsevier.*

The advantage of this simple log-log plot is that it clearly expresses the size effect of yield stress across the length scale. The drawback of this rough curve is that it misses some theoretical and numerical information and lacks a physical mechanism because almost all of the results are collected from the experiments at micron to submicron scales and an MD simulation at nanoscale, but singular information comes from the theoretical analysis. To remedy this problem, related theoretical models were developed and numerical simulations were carried out, some of which were performed by our group, as will be described in detail later in this book.

1.4.2 Nonlocal Crystal Plasticity

As summarized in Section 1.3, both mechanical behavior and the dislocation features exhibit a significant dependence on size. Conventional continuum mechanics ignore the long-range stress field and nonlocal effect induced by the dislocation microstructure. There is no internal characteristic length, so it cannot describe scale-dependent plastic behavior, as described in Chapter 2. To consider the size effect, some researchers introduced the effect of strain gradient on hardening through the Taylor model and described the strain gradient by GND (Gao et al., 1999; Han et al., 2005a,b; Huang et al., 2000), as described in Chapter 3. When the characteristic length of the inhomogeneous deformation field is much larger than the internal length scale, the classical J_2 plastic flow theory is obtained.

This theory is easily implemented into finite element programming, which considers the slip resistance induced by GND and captures isotropic hardening well but is limited by having to deal with back stress and boundary conditions. To capture mixed hardening, some researchers directly incorporated the back stress induced by GND into the crystal slip model (Evers et al., 2004b; Geers et al., 2007). Others introduced high-order stress, which was work-conjugated to the strain gradient, then derived mechanical equilibrium equations based on the principle of virtual work, and determined the constitutive equation according to the second law of thermodynamics (Borg, 2007; Gurtin, 2002, 2010). In this framework, the variables may include not only the displacement but also the plastic deformation, rigid movement, rotation, or GND (Evers et al., 2004b) etc. Although the theoretical backgrounds and mathematical formats for different theoretical models are different, they are somehow equivalent (Evans and Hutchinson, 2009; Kuroda and Tvergaard, 2008a,b). This nonlocal (gradient) plastic theory predicts the size effect well during nano-indentation, the bending of thin beam, the growth of microvoids, microrotation (Huang et al., 2000), the tension of polycrystals (Borg, 2007), and the tension of thin film (Liu et al., 2011), as well as crack tip toughness (Qu et al., 2004; Wei, 2006) and interface yielding problems (Aifantis and Ngan, 2007).

Because crystal plasticity theory is based on the statistically averaged description of dislocation, ignoring the discrete nature of the dislocation structure, only the deterministic and continuous stress-strain curve can be predicted. However, when the crystal size is smaller than 20 μm, the dislocation behavior and mechanical response become stochastic (Dimiduk et al., 2005) (see Fig. 1.10). Considering this, Zaiser and Aifantis (2006) took the local fluctuation effect into account and combined the deterministic strain gradient theory and stochastic microstructure effect, which to some extent captures the spatially and temporally heterogeneous plastic feature. Zhang and Aifantis (2011) assumed that the single crystal is composed of several slip layers, and used gradient plastic theory to explain the origin of strain burst by accounting for interfaces between the slip layers. In further work (Zhang and Shang, 2014), they further introduced the microboundary condition and revealed the relation between strain burst and negative second-order work: $d^2 W_2 = \int_V \widehat{\sigma} : \varepsilon dV \leq 0$, where $\widehat{\sigma}$ is the Jaumann stress rate, ε is strain, and V is the instantaneous volume.

Although nonlocal plastic theory has an important role in understanding submicron plasticity, there are still several unresolved problems: (1) The characteristic material length is generally obtained by dimensional analysis or fitting experimental data, without clear physical meaning. How the relation between this characteristic length and the internal dislocation mechanism can be determined requires further investigation. (2) Whether the contribution of GND should be considered in slip resistance is still under debate. Some researchers introduced it by the Taylor relation (Fleck et al., 1994; Han et al., 2005a). However, others believed that the back stress term is already considered the contribution

of GND; its role is double counted if it is also considered in the Taylor relation (Guruprasad and Benzerga, 2008a) (Mayeur and McDowell, 2013). (3) For submicron crystals, compared with the effect of the strain gradient on GND accumulation, size-dependent new dislocation mechanisms have a more important role. For example, to explain the size effect during uniaxial compression described in Section 1.3.2, the new dislocation mechanism at submicron scales must be taken into account. Because the experimental investigations cannot provide all information about the underlying microstructure mechanism, the development of nonlocal plastic theory requires more information obtained from smaller-scaled simulations (LeSar, 2014).

1.4.3 Discrete Dislocation Dynamics Simulation Method

To gain insight into the atomic-scaled dislocation mechanism, substantial MD simulations were conducted that gave many valuable results (Xu et al., 2013; Zhou et al., 1998; Zuo and Ngan, 2006). However, because of the limitation of computational ability, the spatial and temporal scales that can be accessed by this method are limited. Meanwhile, the high strain rate that is used to reduce computation time influences the dislocation evolution features (Needleman, 2000). The discrete dislocation dynamics (DDD) simulation method, on the other hand, can be used to study problems with larger spatial and temporal scales and is an ideal method to reveal plastic mechanisms on a submicron scale.

The basic idea of the DDD simulation method is schematically shown in Fig. 1.11. Dislocations are carriers of plastic deformation; that is, plasticity is accommodated by

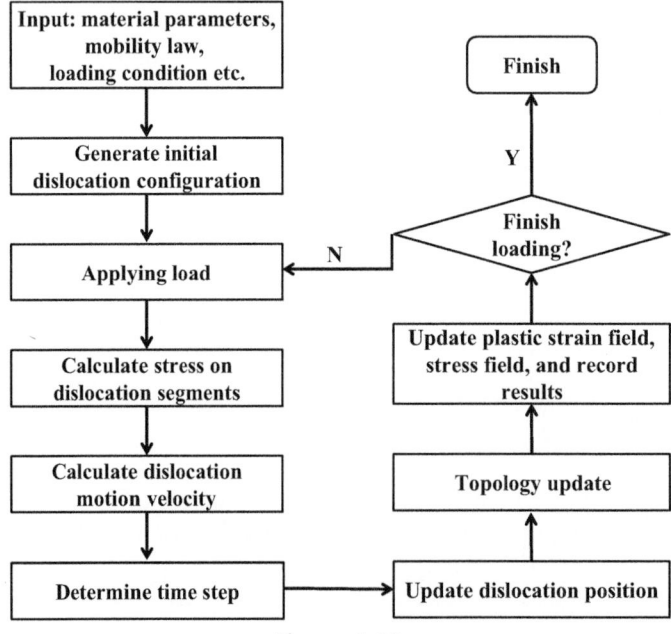

Figure 1.11
Calculation procedure of discrete dislocation dynamics.

the motion of dislocations. The DDD simulation method deals directly with discrete dislocation lines and updates the dislocation positions according to the instantaneous stress field. Here, the stress field includes the external loading effect calculated by solving the boundary value problem, the interaction stress induced by other dislocations or other kinds of defects, and the image force induced by the free surface or interface, etc. Then, according to the mobility law, the dislocation motion velocity is evaluated, which is used to update the dislocation position. The dislocation mobility law depends on the crystal structure and material properties and is generally obtained from molecular dynamics simulation results, theoretical analysis, or by fitting experimental data. After the dislocation slip area is known, the corresponding plastic strain can be calculated based on the Orowan relation. Then, the elastic strain can be calculated by subtracting the total strain with the plastic strain. Afterward, the stress field is updated according to the calculated elastic strain. In addition to calculating the dislocation velocity, the topology needs to be updated to deal with short-range interactions between dislocations, such as junction formation and mutual annihilation between close opposite-sign dislocations. More details about the DDD method are given in Section 7.1. To enhance the capability of DDD simulation methods further, DDD is sometimes coupled with finite element methods to give it the ability to treat with complex boundary conditions, capture the finite strain effect, and accurately calculate the image force. Related efforts are reviewed in Section 7.2.

Advances in developing DDD have made it a rigorous and effective tool for revealing microstructural mechanisms, predicting mechanical responses, and shedding light on developing crystal plasticity theory. For example, DDD simulations provide a great deal of physical insight into the dislocation mechanism at submicron scales. The dislocation starvation mechanism and source truncation mechanism described in Section 1.3.2 are captured (Liu et al., 2009a,b; Rao et al., 2008). The relation between strain burst and dislocation configuration evolution is studied (Cui et al., 2014, 2016b; Rao et al., 2008; Tang et al., 2008). Chapters 6–9 present these typical applications of DDD to illustrate how to use DDD to disclose the underlying physics of submicron plasticity and develop prediction theoretical models. Chapter 10 demonstrates how to use DDD to tackle high-strain rate problems. In Chapter 11, a high-temperature version of DDD is presented that takes into account both dislocation glide and climb. The mystery of high-temperature anneal hardening at small scales is disclosed.

1.5 Content of This Book

This book is separated into two parts. Part I focuses on the continuum-based mechanism of crystal plasticity, including Chapters 2–6. Part II focuses on the discrete dislocation-based mechanism of crystal plasticity, including Chapters 7–12.

In Chapter 2, the fundamental concept of plasticity constitutive law is briefly introduced, which includes the one-dimensional plasticity of material constitution and multiaxial stress state; the von Mises J_2 flow theory, which is frequently used for the plastic deformation of rate-independent and rate-dependent materials; the Mohr-Coulomb and Drucker-Prager models, which introduce nonassociative plasticity; the constitution of porous elastic-plastic solids combining the Gurson model, which is used to describe metal material deformation, damage, and failure; and the co-rotational stress formulation to solve the stress objectivity problem during a large rotation. All of these material constitutions and theoretical models are described on a macroscale. The purpose is to set up the essential concept for understanding material plastic behavior and to guide the reader through the following chapters on crystal plasticity on the micron and submicron scales.

In Chapter 3, many experiments have shown that materials display a strong size effect when the characteristic length scale is on the order of microns. At a material point, stress depends not only on the strain at the same point but also on the strain gradient at nearby points. Therefore, it is necessary to establish a micron-level or nonlocal continuum theory. Three strain gradient plasticity theories are described: couple stresses, stretch and rotation gradient, and mechanism-based strain gradient deformation.

Chapter 4 introduces conventional single crystal plasticity theory and progress. First, the conventional constitutive model for single crystal is described in detail, including the kinematics of single-crystal plasticity, dislocation density evolution, and stress update,. Then, a new dislocation-based crystal plasticity model is developed by the continuum description of the collective behavior of dislocations. Compared with conventional crystal plasticity models, a microforce balance equation is involved by combining the equation of dislocations motion and Orowan's law. The microforce balance equation contains back stress to consider the short-range interactions of dislocations. As an application, a strong size effect and prominent Bauschinger effect in the thin films with a passivation layer during loading and unloading are numerically studied using the proposed crystal plasticity model.

Size-dependent phenomena appear when the sample dimension is of a small scale. At these scales, dislocation-surface and dislocation-dislocation interactions have an important role in the plastic deformation of crystals. These phenomena can be considered in two ways. First, for the overall response, the size-dependent flow stress and serrated stress-strain curves of crystal micropillars are observed in uniaxial compression tests. Second, for the local deformation of the pillar, for example, the deformation morphology and microstructure evolution also dependent on size. In Chapter 5, the size effect and strain burst in compressed single-crystal micropillars and the size-dependent deformation morphology of compressed single-crystal micropillars are successfully reproduced by applying the models of dislocation-based crystal plasticity.

Chapter 6 develops a microscale crystal plasticity model based on phase field theory. The stress-free inelastic strain induced by dislocations is directly considered to be the plastic strain, based on crystal plasticity theory in the microscale crystal plasticity model. The plastic slip associated with each slip system is described by a phase field to model a single dislocation. The elastic strain energy is expressed as a function of elastic strain through the crystal plasticity constitutive model. Based on a thermodynamically consistent framework that differentiates between energetic and dissipative mechanisms during plastic deformation, the coupled balances of quasistatic stress equilibrium and plastic slip evolution are derived from the principle of virtual power. Then, the boundary value problem is solved directly by FEM. One advantage of the proposed model is that it can be used for complex structures or boundary conditions in which the analytical Green's function solution is unavailable. Another advantage is that the elastic modulus mismatch in heteroepitaxial structures is easily treated without additional complications.

In Chapter 7, the discrete-continuous model of crystal plasticity is proposed on a submicron scale. The basic theory and algorithmic details of the 3D DDD method are first described. The coupling framework between 3D DDD and FEM is introduced. Existing problems of the coupling method are addressed. Then, the methods are improved and verified by quantitatively comparing them with typical cases. As examples, typical dislocation behavior and stress distribution in heteroepitaxial film and the plastic behavior of micropillars are studied.

In Chapter 8, the investigation focuses on the strain burst, the size effect of flow stress, and the dislocation density evolution dominated by the SAS mechanism. A discrete dislocation simulation is carried out on the compression test of a single-crystal micropillar. The most important achievement is that the critical resolved shear stress is proposed, which is only a function of the Burgers vector, the shear modulus, and a length of the statistically average effective SAS. The corresponding theoretical model is further built to predict the stress-strain curve and dislocation density evolution.

The discrete dislocation simulation is carried out in Chapter 9 to study the compression behavior of a coated micropillar. The features and operation stress of the dislocation source in a coated pillar are revealed. The dislocation density evolution and the relation between the back stress and trapped dislocation density are disclosed, which are further used preliminarily to discuss the connection between discrete dislocation simulation results and nonlocal crystal plasticity theory. A theoretical model that considers the stochastic distribution of SAS is established to predict the upper and lower bounds of flow stress in the coated micropillar. The coating failure mechanism is also preliminarily discussed.

In Chapter 10, the discrete dislocation simulations and theoretical analyses are carried out to investigate a low-amplitude cyclic loading test on micropillars. The evolution of the dislocation configuration and density are discussed. The stability and different failure

mechanisms of the dislocation junction under different loading modes are revealed. Furthermore, the dislocation starvation model under low-amplitude cyclic loading is proposed. By introducing this mechanics into dislocation density evolution equations, the critical conditions are compared for mechanical annealing under cyclic and monotonic loading conditions.

Chapter 11 investigates DDD simulations of the strain rate effect on the deformation of finite-sized single-crystal coppers under uniaxial compression, tension, and hydrostatic compression. Although the rate-dependent stress-strain relationship is not observed under hydrostatic compression, the evolution of dislocation density exhibits a significant rate dependence. With an increase in the strain rate, the yield stress increases rapidly. A critical strain rate exists in each single-crystal copper block for the given size and dislocation sources, below which the yield stress is relatively insensitive to the strain rate. The dislocation patterning changes from nonuniform to uniform under a high strain rate. Bandlike dislocation walls and their shielding effect on other dislocations are observed in the shocked single-crystal copper. Both fast homogeneous nucleation and avalanche-like dislocation multiplication become involved and lead to the softening of shear stress. Through a comparison of dynamic behavior under different impact speeds, a threshold speed of around 1000 m/s for the dislocation dominant mechanism is proposed, beyond which effects of other defects such as stacking faults and twinning would be prominent.

The coupled dislocation glide-climb model is developed in Chapter 12 to study the intrinsic formation mechanism of helical dislocations and high-temperature anneal hardening in the submicron pillars. Both thermally activated dislocation glide and climb are dealt with in the framework of 3D DDD. The climb rate is determined by the vacancy volumetric flux across the dislocation core, which is obtained by solving vacancy diffusion equations analytically with an assumption of steady-state diffusion or using FEM. An adaptive coupled scheme is proposed to bridge the great timescale separation between glide and climb, which is accurate and efficient for solving the glide-climb coupling process. Through the dislocation glide-climb simulation, the mechanism of formation of helical dislocation and influential factors in helical configuration may be studied. The remarkable softening effect after prestraining and the hardening effect after annealing are observed in the submicron pillars. The simulation results agree well with the experiment data.

Continuum Dislocation Mechanism-Based Crystal Plasticity

Fundamental Conventional Concept of Plasticity Constitution

Chapter Outline

2.1 Introduction

In the mechanics description of material behavior, the response of a material is characterized by a constitutive equation that gives stress as a function of the deformation history of the material. The fundamental concept of plasticity constitutions is briefly introduced, which includes the one-dimensional plasticity of the material constitution and multiaxial stress state; the von Mises J_2 flow theory, which is frequently used for the plastic deformation of rate-independent and rate-dependent materials; the Mohr-Coulomb and Drucker-Prager models, which introduce nonassociative plasticity; the constitution of porous elastic-plastic solids combining the Gurson model, which is used to describe metal material deformation, damage, and failure; and the corotational stress formulation to solve the stress objectivity problem during a large rotation. All of these material constitutions

Dislocation Mechanism-Based Crystal Plasticity. https://doi.org/10.1016/B978-0-12-814591-3.00002-9

and theoretical models are described at the macroscale. The purpose is to set up the essential concept for understanding material plastic behavior and to help the reader comprehend the following chapters on crystal plasticity at the micron and submicron scales.

2.2 One-Dimensional Plasticity

Materials for which permanent strains are developed upon unloading are called plastic materials. Many materials exhibit elastic behavior up to a stress called the yield strength, such as metals, polymers, soils, and concretes. Once a loading is beyond the initial yield strength, plastic strains occur. Elastic-plastic materials are further classified as rate-independent materials, in which the stress is independent of the strain rate, and rate-dependent materials, in which the stress depends on the strain rate. The latter are also classified as rate-sensitive materials.

The major ingredients of plastic theory are: (1) decomposition of each increment of strain into an elastic reversible part $d\varepsilon^e$ and a plastic irreversible part $d\varepsilon^p$; (2) a flow rule governing the plastic flow and determining the plastic strain increments; (3) a yield function $f(\sigma, q_\alpha)$ governing the onset and continuance of plastic deformation, where q_α is a set of internal variables; and (4) the evolution equations of internal variables governing the evolution of yield functions, including the strain-hardening or strain-softening relations.

Elastic-plastic laws are path-dependent and dissipative. A large part of the work is expended in material plastically deforming, which is irreversibly converted to the other forms of energy, particularly heat. Stress depends on an entire history of the deformation and cannot be expressed as a single-valued function of the strain; rather, it can only be specified as a relation between rates of stress and strain.

2.2.1 Isotropic Hardening

A typical stress-strain curve for a metal under uniaxial stress is illustrated in Fig. 2.1. From initial loading until the initial yield stress is attained, the material exhibits elastic behavior, which is usually assumed to be linear elastic. This regime is followed by an elastic-plastic regime in which the permanent irreversible plastic strains are induced upon further loading. The stress reversal is called unloading. During unloading, the stress-strain response is assumed to be governed by the elastic law. The increments in strain are assumed to be decomposed into elastic and plastic parts. Thus, we can write:

$$d\varepsilon = d\varepsilon^e + d\varepsilon^p \qquad (2.2.1)$$

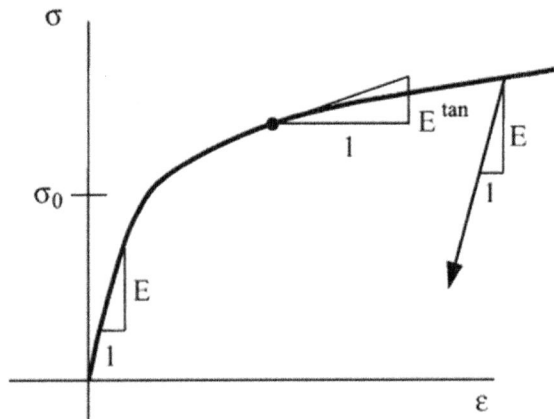

Figure 2.1
Typical stress-strain curve for a metal under uniaxial stress.

Dividing both sides of this equation by a differential time increment dt obtains the rate form

$$\dot{\varepsilon} = \dot{\varepsilon}^e + \dot{\varepsilon}^p \tag{2.2.2}$$

The stress increment (rate) is always related by the elastic modulus to the increment (rate) of elastic strain:

$$d\sigma = Ed\varepsilon^e, \quad \dot{\sigma} = E\dot{\varepsilon}^e \tag{2.2.3}$$

In the nonlinear elastic-plastic regime, the stress-strain relation is given by

$$d\sigma = Ed\varepsilon^e = E^{\text{tan}}d\varepsilon, \quad \dot{\sigma} = E\dot{\varepsilon}^e = E^{\text{tan}}\dot{\varepsilon} \tag{2.2.4}$$

where the elastic-plastic tangent modulus, E^{tan}, is the slope of the stress-strain curve (Fig. 2.1).

These relations are homogeneous in the rates of stress and strain. Thus, if time is scaled by an arbitrary factor, the constitutive relations remain unchanged. Therefore, the material response is rate-independent even though it is expressed in terms of a strain rate. Consequently, the rate form of constitutive relations will be used. For the notation of increment relations, the rate form may be inconvenient, especially for large-strain formulations.

The plastic strain rate is given by a flow rule, which is often specified in terms of the plastic flow potential ψ:

$$\dot{\varepsilon}^p = \dot{\lambda}\frac{\partial\psi}{\partial\sigma} \tag{2.2.5}$$

where λ is called a plastic rate parameter. An example of the flow potential is

$$\psi = |\sigma| = \overline{\sigma} = \sigma \, \text{sign}(\sigma), \quad \frac{\partial \psi}{\partial \sigma} = \text{sign}(\sigma) \tag{2.2.6}$$

where $\overline{\sigma}$ is called the effective stress, which is defined by stress with a sign function.

The yield condition is

$$f = \overline{\sigma} - \sigma_Y(\overline{\varepsilon}) = 0 \tag{2.2.7}$$

where $\sigma_Y(\overline{\varepsilon})$ is the yield strength under uniaxial tension and $\overline{\varepsilon}$ is an effective plastic strain.

The increase in the yield strength after the initial yield is called work hardening or strain hardening. Material-hardening behavior is generally a function of the history of plastic deformation. In metal plasticity, the history of plastic deformation is often characterized by the effective plastic strain, which is given by

$$\overline{\varepsilon} = \int \dot{\overline{\varepsilon}} \, dt, \quad \dot{\overline{\varepsilon}} = \sqrt{\dot{\varepsilon}^p \dot{\varepsilon}^p} \tag{2.2.8}$$

where $\dot{\overline{\varepsilon}}$ is an effective plastic strain rate. The effective plastic strain $\overline{\varepsilon}$ is an example of an internal variable used to characterize the material inelastic response. An alternative internal variable for hardening is the plastic work (Hill, 1950), which is expressed by $W^p = \int \sigma \dot{\varepsilon}^p dt$.

The yield behavior given by Eq. (2.2.7) is called isotropic hardening: the yield strengths in tension and compression are always equal and are given by $\sigma_Y(\overline{\varepsilon})$. A typical hardening curve is shown in Fig. 2.1. The slope of this curve beyond σ_0 is the plastic modulus, $E^{\text{tan}} = H$, i.e., $H = d\sigma_Y(\overline{\varepsilon})/d\overline{\varepsilon}$. In contrast, the slope of this curve below σ_0 for a pure linear part is called Young's modulus, $E^{\text{tan}} = E$. There is a small nonlinear elastic part between them for metal, which is ignored in analysis here.

For a particular model, combining Eqs. (2.2.5), (2.2.6) and (2.2.8) yields $\lambda = \dot{\overline{\varepsilon}}$. Therefore, the plastic strain rate (2.2.5) can be written as

$$\dot{\varepsilon}^p = \dot{\overline{\varepsilon}} \, \text{sign}(\sigma) = \dot{\overline{\varepsilon}} \frac{\partial f}{\partial \sigma} \tag{2.2.9}$$

where the result $\partial f/\partial \sigma = \partial \overline{\sigma}/\partial \sigma = \text{sign}(\sigma)$ is used. We may find that $\partial f/\partial \sigma = \partial \psi/\partial \sigma$. The plastic flow for which $\partial f/\partial \sigma \sim \partial \psi/\partial \sigma$ is called associative; otherwise, plastic flow is called nonassociative. For associative plasticity, plastic flow is in a direction normal to the yield surface. This distinction is important in multiaxial plasticity models and will be elaborated upon in Section 2.3.

Plastic deformation occurs only when the yield condition $f = 0$ in Eq. (2.2.7) is satisfied. During plastic loading, the stress must remain on the yield surface, so $\dot{f} = 0$. Enforcement of this leads to the consistency condition

$$\dot{f} = \dot{\bar{\sigma}} - \dot{\sigma}_Y(\bar{\varepsilon}) = 0 \tag{2.2.10}$$

which gives

$$\dot{\bar{\sigma}} = \frac{\mathrm{d}\sigma_Y(\bar{\varepsilon})}{\mathrm{d}\bar{\varepsilon}}\dot{\bar{\varepsilon}} = H\dot{\bar{\varepsilon}} \tag{2.2.11}$$

where $H = \mathrm{d}\sigma_Y(\bar{\varepsilon})/\mathrm{d}\bar{\varepsilon}$ is called the plastic modulus. Using Eqs. (2.2.2), (2.2.4), (2.2.11) and (2.2.5) it follows that

$$\frac{1}{E^{\text{tan}}} = \frac{1}{E} + \frac{1}{H} \quad \text{or} \quad E^{\text{tan}} = \frac{EH}{E+H} = E - \frac{E^2}{E+H} \tag{2.2.12}$$

Consider a plastic switch parameter β with $\beta = 1$ corresponding to plastic loading and $\beta = 0$ corresponding to a purely elastic response (loading or unloading). Then, the tangent modulus is

$$E^{\text{tan}} = E - \beta \frac{E^2}{E+H} \tag{2.2.13}$$

The loading-unloading conditions can also be stated as

$$\dot{\lambda} \geq 0, \quad f \leq 0, \quad \dot{\lambda}f = 0 \tag{2.2.14}$$

These are sometimes referred to as Kuhn-Tucker conditions. The first condition indicates that the plastic rate parameter is nonnegative whereas the second condition indicates that the stress state must lie on or within the yield surface. The last condition ensures that the stress lies on the yield surface during plastic loading, $\dot{\lambda} > 0$, which can also be stated in a rate form through what is known as the consistency condition, $\dot{f} = 0$. For plastic loading ($\dot{\lambda} > 0$), the stress state must remain on the yield surface, $f = 0$; therefore, $\dot{f} = 0$. For elastic loading or unloading $\dot{\lambda} = 0$, i.e., there is no plastic flow. An extension of the isotropic hardening model to a case of kinematic hardening is given in following section. More general constitutive relations use additional internal variables.

2.2.2 Kinematic Hardening

In cyclic loading, an isotropic hardening model provides a poor representation of the stress-strain response for many metals. For example, Fig. 2.2(A) illustrates a phenomenon observed in cyclic plasticity known as the Bauschinger effect, in which the yield strength

Similar Eq. (3.2.51), it is assumed that the relation of total equivalent strain E_{SG} and total equivalent stress Σ is given:

$$\left(\frac{E_{SG}}{E_{SG0}}\right) = \left(\frac{\Sigma}{\Sigma_0}\right)^n \tag{3.3.49}$$

where E_{SG0} and Σ_0 are material constants. As in Eqs. (3.2.52) and (3.2.53), w and ϕ can be determined, respectively:

$$w = \frac{1}{2}K\varepsilon_V^2 + \frac{n}{n+1}\Sigma_0 E_{SG0}\left(\frac{E_{SG}}{E_{SG0}}\right)^{\frac{n+1}{n}} \tag{3.3.50}$$

$$\phi = \frac{1}{2K}\sigma_m^2 + \frac{1}{n+1}\Sigma_0 E_{SG0}\left(\frac{\Sigma}{\Sigma_0}\right)^{n+1} \tag{3.3.51}$$

where σ_m is an average stress.

The field equation, also known as the principle and boundary condition of SG strain gradient theory, may be found in Fleck and Hutchinson (1997). Some applications were also obtained using this theory. For example, research was carried out on hardening and porous instability that resulted in rigid particle inclusions in a metal matrix and in approximation solutions for mode I and mode II fractures.

The FEM is also used to analyze the crack tip field. Because of a large tension gradient at the crack tip location that is almost incompressible, Wei and Hutchinson (1998) developed a new FEM to demonstrate an accurate strain gradient effect at the crack tip. The results solved by this kind of element under the $n = 1$ critical condition were in good agreement with elastic analysis solutions obtained by Shi et al., (2000), who used the Wiener-Hopf method. This agreement is not only for the crack tip location but for the whole field solution. These results indicated that the FEM developed by Wei and Hutchinson (1998) can be used to exhibit the strain gradient effect at the crack tip, including rotational gradient and tensional gradient. The SG theory used by Wei and Hutchinson (1997) for mode I fracture steady-state propagation obtained a much higher stress field at the crack tip than results obtained using CS theory. Another increment-type theory of SG was proposed by Fleck and Hutchinson (1997).

3.4 Microscale Mechanism-Based Strain Gradient Plasticity Theory

This section focuses on MSG deformation plasticity theory. Microindentation and fracture are studied. The results of microindentation using MSG theory agree well with the experimental data, with an indent depth of one-tenth of a micron to several microns. This indicates that MSG theory can characterize plastic behavior accurately at the micron and submicron scales.

The consistency condition is given by differencing Eq. (2.2.16):

$$\dot{f} = (\dot{\sigma} - \dot{\alpha})\text{sign}(\sigma - \alpha) - H\dot{\bar{\varepsilon}} = 0 \qquad (2.2.18)$$

and thus

$$\dot{\bar{\varepsilon}} = \frac{1}{H}(\dot{\sigma} - \dot{\alpha})\text{sign}(\sigma - \alpha) \qquad (2.2.19)$$

From Eqs. (2.2.3), (2.2.15) and (2.2.19), then we have

$$\dot{\sigma} = E(\dot{\varepsilon} - \dot{\varepsilon}^p) = E\big(\dot{\varepsilon} - \dot{\bar{\varepsilon}}\,\text{sign}(\sigma - \alpha)\big) \qquad (2.2.20)$$

Subtracting $\dot{\alpha}$,

$$
\begin{aligned}
(\dot{\sigma} - \dot{\alpha}) &= E\big(\dot{\varepsilon} - \dot{\bar{\varepsilon}}\,\text{sign}(\sigma - \alpha)\big) - \dot{\alpha} \\
&= E\big(\dot{\varepsilon} - \dot{\bar{\varepsilon}}\,\text{sign}(\sigma - \alpha)\big) - \kappa\dot{\bar{\varepsilon}}\,\text{sign}(\sigma - \alpha)
\end{aligned}
\qquad (2.2.21)
$$

where Eqs. (2.2.15) and (2.2.17) are used. From Eqs. (2.2.19) and (2.2.21), an expression for the effective plastic strain rate is obtained:

$$\dot{\bar{\varepsilon}} = \frac{1}{H}\Big[E\dot{\varepsilon}\,\text{sign}(\sigma - \alpha) - E\dot{\bar{\varepsilon}} - \kappa\dot{\bar{\varepsilon}}\Big] = \frac{E\dot{\varepsilon}\,\text{sign}(\sigma - \alpha)}{E - (H + \kappa)} \qquad (2.2.22)$$

Substituting this expression into (2.2.20) gives $\dot{\sigma} = E^{\tan}\dot{\varepsilon}$ for plastic loading, where

$$E^{\tan} = \frac{E(H + \kappa)}{(H + \kappa) + E} = E - \frac{E^2}{(H + \kappa) + E} \qquad (2.2.23)$$

For elastic loading or unloading, the tangent modulus is simply the elastic modulus, $E^{\tan} = E$. Alternatively, the loading-unloading conditions can be written in terms of the Kuhn-Tucker conditions (Eq. 2.2.14).

2.2.3 Rate-Dependent Plasticity

In rate-dependent plasticity, the material plastic response depends on the loading rate. The elastic response in a rate form was given previously by

$$\dot{\sigma} = E\dot{\varepsilon}^e = E(\dot{\varepsilon} - \dot{\varepsilon}^p) \qquad (2.2.24)$$

In contrast to rate-independent plasticity, in which the yield condition cannot be exceeded, the yield condition must be met or exceeded for plastic deformation to occur. The plastic strain rate combining isotropic and kinematic hardening is given by

$$\dot{\varepsilon}^p = \dot{\lambda}\frac{\partial\psi}{\partial\sigma}, \quad \dot{\bar{\varepsilon}} = \dot{\lambda}, \quad \sigma' = \sigma - \alpha, \quad \psi = |\sigma'| \qquad (2.2.25)$$

For a rate-dependent material, one way to describe plastic response is by means of an overstress model:

$$\dot{\bar{\varepsilon}} = \frac{\varphi(\sigma, \bar{\varepsilon}, \alpha)}{\eta} \tag{2.2.26}$$

where φ is the overstress and η is the viscosity. In an overstress model, the effective plastic strain rate depends on how much the yield stress is exceeded. Instead of consistency conditions (2.2.18) and (2.2.19) to obtain $\dot{\bar{\varepsilon}}$, the effective plastic strain rate is given by an empirical law. For example, the overstress model of Perzyna (1971) is

$$\phi = \sigma_Y \left\langle \frac{|\sigma - \alpha|}{\sigma_Y} - 1 \right\rangle^n \tag{2.2.27}$$

where n is called the rate-sensitivity exponent. Plastic strain occurs when the yield condition $|\sigma - \alpha| - \sigma_Y(\bar{\varepsilon}) = 0$ is exceeded. This is incorporated using the Macaulay bracket $\langle f \rangle = f$ if $f > 0$ and $\langle f \rangle = 0$ if $f \le 0$. Using Eqs. (2.2.23) and (2.2.25)–(2.2.26), an expression for the stress rate is given by

$$\dot{\sigma} = E \left(\dot{\varepsilon} - \frac{\varphi(\sigma, \bar{\varepsilon}, \alpha)}{\eta} \operatorname{sign}(\sigma') \right) \tag{2.2.28}$$

This is a differential equation for stress evolution. Comparing this equation with Eq. (2.2.5), Eq. (2.2.28) is inhomogeneous in the rates and therefore the material response is rate-dependent. More elaborate models with additional internal variables are given in Lubliner (1990) and Khan and Huang (1995).

In rate-dependent plasticity, an alternative form of the effective plastic strain rate is given by Peirce et al. (1984):

$$\dot{\bar{\varepsilon}} = \dot{\varepsilon}_0 \left(\frac{|\sigma - \alpha|}{\sigma_Y(\bar{\varepsilon})} \right)^{1/m} \tag{2.2.29}$$

This model does not include an explicit yield surface. For plastic straining at the rate $\dot{\varepsilon}_0$, the reference response $|\sigma - \alpha| = \sigma_Y$ is obtained. For rates that exceed $\dot{\varepsilon}_0$, the stress is elevated above the reference stress, σ_Y, whereas for lower rates the stress falls below the reference stress. A case of interest is the near rate-independent limit when the rate-sensitivity exponent $m \to 0$. It can be found from Eq. (2.2.29) (with $m \to 0$) that for $|\sigma - \alpha| < \sigma_Y$, the effective plastic strain rate is negligible, whereas for a finite plastic strain rate, $|\sigma - \alpha|$ is approximately equal to the reference stress, σ_Y. In this way, the model exhibits yielding together with the near-elastic unloading and rate-independent response.

2.3 Multiaxial Plasticity

The one-dimensional plastic constitutive relations presented in Section 2.2 are now extended to multiaxial plasticity. We begin with a general treatment of hypoelastic-plastic constitutive relations for large strain. These formulations are typically based on an additive decomposition of the rate-of-deformation tensor into elastic and plastic parts. The elastic part is taken as a hypoelastic response. The results are described from a modification of rate-independence for the case of rate-dependent plasticity, namely viscoplasticity, as discussed in Section 2.4. A discussion follows of large deformation plasticity formulations based on a multiplicative decomposition of the deformation gradient into elastic and plastic parts. Elastic-plastic behavior is based on a hyperelastic representation of the elastic response. As a special case, the reduction of a general large strain formulation to the case of small strain is also given.

2.3.1 Hypoelastic-Plastic Materials

When elastic strains are small compared with the plastic strains, hypoelastic-plastic models are generally used. The energy is not conserved in a closed deformation cycle for hypoelastic materials. However, for small elastic strains, the energy error is insignificant and hypoelastic descriptions of the elastic response are often adequate. In these constitutive models, the additive decomposition of the rate-of-deformation tensor $\mathbf{D}$ into elastic and plastic parts is assumed:

$$\mathbf{D} = \mathbf{D}^e + \mathbf{D}^p \tag{2.3.1}$$

where $\mathbf{D}$ is defined as the symmetric part of velocity gradient. The velocity gradient is defined by $\mathbf{L} = \frac{\partial \mathbf{v}}{\partial \mathbf{x}} = (\nabla \mathbf{v})^T = (\text{grad } \mathbf{v})^T$ or $L_{ij} = \frac{\partial v_i}{\partial x_j}$, which is additively decomposed into the symmetric and skew-symmetric parts as

$$L = \frac{1}{2}\left(L + L^T\right) + \frac{1}{2}\left(L - L^T\right) = D + W \quad \text{or} \quad L_{ij} = \frac{1}{2}\left(L_{ij} + L_{ji}\right) + \frac{1}{2}\left(L_{ij} - L_{ji}\right), \tag{2.3.2}$$

respectively. The elastic response is hypoelastic. A suitable objective rate of stress is related to the elastic part of the rate-of-deformation tensor. Considered in Section 2.2.1, the stress rate is related to the total rate-of-deformation tensor in pure hypoelasticity. The choice of objective stress rate in the constitutive response depends on several factors. One example of stress rates is the Jaumann rate of Cauchy stress, which is given by

$$\sigma^{\nabla J} = \dot{\sigma} - \mathbf{W} \cdot \sigma - \sigma \cdot \mathbf{W}^T \quad \sigma^{\nabla J} = \mathbf{C}^{\sigma J} : \mathbf{D} \tag{2.3.3}$$

where $\mathbf{W}$ is the spin tensor in Eq. (2.3.2), which is the skew-symmetric part of the velocity gradient. The Jaumann rate of Cauchy stress leads to nonsymmetric tangent stiffness

matrices, whereas the Jaumann rate of Kirchhoff stress $\tau = J\sigma$ can lead to symmetric stiffness matrices. We will identify the circumstances under which these choices are appropriate and what advantages may be gained by choosing one rate over another. The choice of objective stress rate should not be confused with the expression of a given constitutive relation in terms of different stress rates. The latter is accomplished simply by using the appropriate transformation between rates.

Based on Cauchy stress with elastic response specified in terms of the Jaumann rate, we present a model in which the elastic response is specified by applying the hypoelastic law to the elastic part of the rate-of-deformation:

$$\sigma^{\nabla J} = \mathbf{C}^{\sigma J} : \mathbf{D}^e = \mathbf{C}^{\sigma J} : (\mathbf{D} - \mathbf{D}^p) \tag{2.3.4}$$

If the elastic moduli, $\mathbf{C}^{\sigma J}$, are taken as constant, they must be isotropic to satisfy the principle of material frame indifference.

The rate of plastic flow is given by

$$\mathbf{D}^p = \dot{\lambda}\mathbf{r}(\boldsymbol{\sigma}, \mathbf{q}), \quad D_{ij}^p = \dot{\lambda}r_{ij}(\boldsymbol{\sigma}, \mathbf{q}) \tag{2.3.5}$$

where $\dot{\lambda}$ is a scalar plastic flow rate; $\mathbf{r}(\boldsymbol{\sigma}, \mathbf{q})$ is plastic flow direction, which is often specified as $\mathbf{r} = \partial\psi/\partial\boldsymbol{\sigma}$, where ψ is called the plastic flow potential. To avoid confusing the plasticity parameter with the Lamé constant, the Lamé constant is subsequently denoted as λ^e. The plastic flow direction depends on the Cauchy stress $\boldsymbol{\sigma}$ and on a set of internal variables denoted collectively as $\mathbf{q}$. The examples of scalar internal variables are the accumulated effective plastic strain and the void volume fraction. The back stress in kinematic hardening models is an example of an internal variable that is a second-order tensor.

The evolution equations for internal variables are required for most plasticity models, which can be specified as

$$\dot{\mathbf{q}} = \dot{\lambda}\mathbf{h}(\boldsymbol{\sigma}, \mathbf{q}), \quad \dot{q}_\alpha = \dot{\lambda}h_\alpha(\boldsymbol{\sigma}, \mathbf{q}) \tag{2.3.6}$$

where α ranges over the number of internal variables. Here, the internal variables are a collection of scalars and the material time derivative is an objective rate. The plastic parameter λ or some function of it may be one of the internal variables. The evolution equation for the plastic parameter is obtained through the consistency condition given subsequently. The yield condition is

$$f(\boldsymbol{\sigma}, \mathbf{q}) = 0 \tag{2.3.7}$$

and loading-unloading conditions are given by

$$\dot{\lambda} \geq 0, \quad f \leq 0, \quad \dot{\lambda}f = 0 \tag{2.3.8}$$

During plastic loading ($\dot{\lambda} > 0$), stress is required to remain on the yield surface, $f = 0$. This can be also stated in terms of the consistency condition $\dot{f} = 0$, which can be expanded through the chain rule to give

$$\dot{f} = f_{\boldsymbol{\sigma}} : \dot{\boldsymbol{\sigma}} + f_{\mathbf{q}} \cdot \dot{\mathbf{q}} = 0, \quad \dot{f} = (f_{\sigma})_{ij} : \dot{\sigma}_{ij} + (f_q)_{\alpha} \cdot \dot{q}_{\alpha} = 0 \tag{2.3.9}$$

where the notations $f_{\boldsymbol{\sigma}} = \partial f / \partial \boldsymbol{\sigma}$ and $f_{\mathbf{q}} = \partial f / \partial \mathbf{q}$ are adopted.

The consistency condition involves the normal to the yield surface $f_{\boldsymbol{\sigma}}$. If the plastic flow direction is proportional to the normal of the yield surface, i.e., $\mathbf{r} \sim f_{\boldsymbol{\sigma}}$, plastic flow is said to be associative; otherwise, it is said to be nonassociative. When the flow direction is given by the derivative of a plastic flow potential, the condition for associative plasticity is $\psi_{\boldsymbol{\sigma}} \sim f_{\boldsymbol{\sigma}}$. For many materials, an appropriate choice for the plastic potential is $\psi = f$, which gives rise to an associative flow rule. When the yield surface is convex, and if the strain hardening is positive, Drucker has proved that associative plasticity models are stable for small strains.

Considering the Jaumann rate of Cauchy stress, if $f_{\boldsymbol{\sigma}}$ and $\boldsymbol{\sigma}$ commute, i.e.,

$$f_{\boldsymbol{\sigma}} \cdot \boldsymbol{\sigma} = \boldsymbol{\sigma} \cdot f_{\boldsymbol{\sigma}} \tag{2.3.10}$$

it follows (Prager, 1961):

$$f_{\boldsymbol{\sigma}} : \dot{\boldsymbol{\sigma}} = f_{\boldsymbol{\sigma}} : \boldsymbol{\sigma}^{\nabla J} \tag{2.3.11}$$

In continua mechanics, the derivatives of principal invariants of a second-order tensor with respect to the tensor itself are often required in constitutive equations. From Eq. (2.3.10), when f is a function of the invariants of the stress, $f_{\boldsymbol{\sigma}}$ and $\boldsymbol{\sigma}$ commute and thus Eqs. (2.3.10) and (2.3.11) hold. The stress objectivity requires yield functions of the form Eq. (2.3.7) to be isotropic functions of the stress, and hence functions of the principal invariants of stress. For example, the von Mises yield function depends on the second invariant of the deviatoric stress, $I_2(\sigma^{\mathrm{dev}}) \equiv -J_2 = -\frac{1}{2}\sigma^{\mathrm{dev}} : \sigma^{\mathrm{dev}}$. Substituting Eq. (2.3.11) into Eq. (2.3.9) gives

$$\dot{f} = f_{\boldsymbol{\sigma}} : \dot{\boldsymbol{\sigma}}^{\nabla J} + f_{\mathbf{q}} \cdot \dot{\mathbf{q}} = 0 \tag{2.3.12}$$

By using the hypoelastic relation (2.3.4), the plastic flow relation (2.3.5) and evolution Eq. (2.3.6) in Eq. (2.3.12), we have

$$0 = f_{\boldsymbol{\sigma}} : \mathbf{C}_{el}^{\sigma J} : (\mathbf{D} - \mathbf{D}^p) + f_{\mathbf{q}} \cdot \dot{\mathbf{q}} = f_{\boldsymbol{\sigma}} : \mathbf{C}_{el}^{\sigma J} : (\mathbf{D} - \dot{\lambda}\,\mathbf{r}) + f_{\mathbf{q}} \cdot \dot{\lambda}\,\mathbf{h} \tag{2.3.13}$$

which can be solved for $\dot{\lambda}$ to obtain

$$\dot{\lambda} = \frac{f_{\boldsymbol{\sigma}} : \mathbf{C}_{el}^{\sigma J} : \mathbf{D}}{-f_{\mathbf{q}} \cdot \mathbf{h} + f_{\boldsymbol{\sigma}} : \mathbf{C}_{el}^{\sigma J} : \mathbf{r}} \tag{2.3.14}$$

Eq. (2.3.11) also holds for the other spin-based rates of stress but not for the Truesdell rate. When the elastic response is specified in terms of the spin-based rate, the simplification (2.3.13) occurs. With the Truesdell rate, it is needed to account for the additional terms.

Substituting Eq. (2.3.14) together with the plastic flow rule (2.3.5) into Eq. (2.3.4), we obtain a relation between the Jaumann rate of Cauchy stress and the total rate-of-deformation tensor:

$$\boldsymbol{\sigma}^{\nabla J} = \mathbf{C}_{el}^{\sigma J} : (\mathbf{D} - \dot{\lambda}\,\mathbf{r}) = \mathbf{C}_{el}^{\sigma J} : \left(\mathbf{D} - \frac{f_{\boldsymbol{\sigma}} : \mathbf{C}_{el}^{\sigma J} : \mathbf{D}}{-f_{\mathbf{q}} \cdot \mathbf{h} + f_{\boldsymbol{\sigma}} : \mathbf{C}_{el}^{\sigma J} : \mathbf{r}}\,\mathbf{r} \right) = \mathbf{C}^{\sigma J} : \mathbf{D} \qquad (2.3.15)$$

The fourth-order tensor $\mathbf{C}^{\sigma J}$ is called the continuum elastic-plastic tangent modulus, which is obtained by rearranging the expression in Eq. (2.3.15):

$$\mathbf{C}^{\sigma J} = \mathbf{C}_{el}^{\sigma J} - \frac{\left(\mathbf{C}_{el}^{\sigma J} : \mathbf{r} \right) \otimes \left(f_{\boldsymbol{\sigma}} : \mathbf{C}_{el}^{\sigma J} \right)}{-f_{\mathbf{q}} \cdot \mathbf{h} + f_{\boldsymbol{\sigma}} : \mathbf{C}_{el}^{\sigma J} : \mathbf{r}}$$

$$C_{ijkl}^{\sigma J} = \left(C_{el}^{\sigma J} \right)_{ijkl} - \frac{\left(C_{el}^{\sigma J} \right)_{ijmn} : r_{mn} (f_{\sigma})_{pq} \left(C_{el}^{\sigma J} \right)_{pqkl}}{-(f_q)_\alpha \cdot h_\alpha + (f_\sigma)_{rs} \left(C_{el}^{\sigma J} \right)_{rstu} r_{tu}} \qquad (2.3.16)$$

The symbol $\otimes$ denotes the tensor or open product, which is defined in the vector product by $\mathbf{a} \otimes \mathbf{b}$; in indicial notation $\mathbf{a} \otimes \mathbf{b} \rightarrow a_i b_j$; and in matrix notation $\mathbf{a} \otimes \mathbf{b} \rightarrow \{a\}\{b\}^T$, respectively.

The elastic-plastic tangent modulus consists of the elastic tangent modulus and a term attributable to plastic flow. When written in Voigt matrix form, the plastic flow contribution is a rank 1 matrix and is often referred to as a rank 1 correction (to the elastic moduli). From symmetry of the stress rate and rate-of-deformation, the elastic-plastic tangent modulus $\mathbf{C}^{\sigma J}$ has both minor symmetries. It has major symmetry, $C_{ijkl}^{\sigma J} = C_{klij}^{\sigma J}$, when $\mathbf{C}_{el}^{\sigma J} : \mathbf{r} \sim f_{\boldsymbol{\sigma}} : \mathbf{C}_{el}^{\sigma J}$ or, alternatively, when plastic flow is associative, i.e., $\mathbf{r} \sim f_{\boldsymbol{\sigma}}$ (the major symmetry of elastic moduli is assumed).

The elastic-plastic tangent modulus $\mathbf{C}^{\sigma J}$ in Eq. (2.3.16) has major symmetry when plastic flow is associative. However, the corresponding modulus $\mathbf{C}^{\sigma T}$ for a Truesdell rate of Cauchy stress is not symmetric. This lack of symmetry is a result of the plastic flow equations based on Cauchy stress. If the plasticity equations are formulated in terms of the Kirchhoff stress, and if plastic flow is associative, $\mathbf{C}^{\sigma T}$ will have major symmetry.

The Cauchy stress is preferred in the plastic yield function and flow rule because it is a true stress. For plastic constitutive relations, plastic flow is essentially isotropic and volume remains unchanged. So, we have $J \approx 1$ (elastic strains are small) and Kirchhoff

stress is virtually indistinguishable from the Cauchy stress. This is the case for a broad class of metals described by classical J_2 flow theory, in which experiments show that plastic strains produce little or no volume change.

For dilatant materials and porous plastic solids, as in the Gurson model described in Section 2.6, large dilations accompany plastic deformations and the assumption $J \approx 1$ is no longer valid. In this situation, the yield function is preferably expressed in terms of the Cauchy stress and the resulting tangent stiffness is not symmetric. The Kirchhoff stress formulation is analogous to the Cauchy stress formulation and may be obtained from formulations described in this section by replacing the Cauchy stress with the Kirchhoff stress. The specific case of J_2 plastic flow theory will be illustrated in Section 2.4.

2.3.2 Small Strain Plasticity

The general formulations for rate-independent large-deformation plasticity represented in the previous discussion can be readily reduced to small strain formulations. There is no distinction between stress measures required. Thus, Cauchy stress $\boldsymbol{\sigma}$ is used. Because objectivity requirements are not relevant in the small strain setting, the material time derivative is related to the stress rate, and strain rate $\dot{\boldsymbol{\varepsilon}}$ replaces the rate-of-deformation D. These small strain formulations given subsequently are also valid for anisotropic elastic moduli, **C**, and yield function f.

The additive decomposition of the strain rate into elastic and plastic parts gives

$$\dot{\boldsymbol{\varepsilon}} = \dot{\boldsymbol{\varepsilon}}^e + \dot{\boldsymbol{\varepsilon}}^p \tag{2.3.17}$$

The relation between the stress rate and elastic strain rate is

$$\dot{\boldsymbol{\sigma}} = \mathbf{C} : \dot{\boldsymbol{\varepsilon}}^e = \mathbf{C} : (\dot{\boldsymbol{\varepsilon}} - \dot{\boldsymbol{\varepsilon}}^p) \tag{2.3.18}$$

The plastic flow rule and evolution equations are given by

$$\dot{\boldsymbol{\varepsilon}}^p = \dot{\lambda} \mathbf{r}(\boldsymbol{\sigma}, \mathbf{q}) \quad \dot{\mathbf{q}} = \dot{\lambda} \mathbf{h} \tag{2.3.19}$$

The yield condition is similar Eq. (2.3.7):

$$f(\boldsymbol{\sigma}, \mathbf{q}) = 0 \tag{2.3.20}$$

and loading-unloading conditions are given by Eq. (2.3.8):

$$\dot{\lambda} \geq 0, \quad f \leq 0, \quad \dot{\lambda} f = 0 \tag{2.3.21}$$

The plastic rate parameter from the consistency condition is given by

$$\dot{\lambda} = \frac{f_\sigma : \mathbf{C} : \dot{\boldsymbol{\varepsilon}}}{-f_\mathbf{q} \cdot \mathbf{h} + f_\sigma : \mathbf{C} : \mathbf{r}} \tag{2.3.22}$$

Then, the relation between stress rate and strain rate is

$$\dot{\sigma} = \mathbf{C}^{ep} : \dot{\varepsilon} \tag{2.3.23}$$

Because the objectivity requirements are not relevant in the small strain setting, the continuum elastic-plastic tangent modulus is obtained subsequently:

$$\mathbf{C}^{ep} = \mathbf{C} - \frac{(\mathbf{C} : \mathbf{r}) \otimes (f_\sigma : \mathbf{C})}{-f_\mathbf{q} \cdot \mathbf{h} + f_\sigma : \mathbf{C} : \mathbf{r}} \tag{2.3.24}$$

which is symmetric if plastic flow is associative ($\mathbf{C}{:}\mathbf{r} \sim f_\sigma{:}\mathbf{C}$).

2.4 J_2 Flow Theory Plasticity

A special case of the general model presented in Section 2.3 is the J_2 flow model based on the von Mises yield surface. This model is especially useful for metal plasticity, for which it was developed, and is described in detail in this section. The key assumption of the J_2 model is that plastic flow in metals is unaffected by pressure, which was experimentally proved by Bridgman (1949). The yield condition and plastic flow direction are based on the deviatoric part of the stress tensor. von Mises effective stress is used to generalize the observed behavior in uniaxial stress or in shear stress to multiaxial stress states. In this section, we begin with a general treatment of Kirchhoff stress formulations of J_2 flow theory plasticity. These formulations are typically based on an additive decomposition of the rate-of-deformation tensor into elastic and plastic parts. The elastic part is taken as a hypoelastic response. The results are described from modifying rate-independent for rate-dependent plasticity, namely viscoplasticity. They then extend the isotropic hardening formulation to combine kinematic hardening. The viscoplasticity constitutive relations can be extended to multidimensions in a similar manner by generalizing the one-dimensional rate-dependent plasticity equations of Section 2.2. In rate-independent plasticity, the plastic rate parameter is obtained from the consistency condition, whereas in rate-dependent plasticity, this parameter is obtained as an empirical function of stress and internal variables. It is typically given by an overstress function.

2.4.1 Kirchhoff Stress Formulation of J_2 Flow Theory Plasticity

In the constitutive models of J_2 flow theory plasticity, the additive decomposition of the rate-of-deformation tensor $\mathbf{D}$ into elastic and plastic parts is assumed in Eq. (2.3.1):

$$\mathbf{D} = \mathbf{D}^e + \mathbf{D}^p$$

The stress rate relation is

$$\tau^{\nabla J} = \mathbf{C}_{el}^{\tau J} : \mathbf{D}^e = \mathbf{C}_{el}^{\tau J} : (\mathbf{D} - \mathbf{D}^p) \tag{2.4.1}$$

The plastic flow rule and evolution equations are given by

$$\mathbf{D}^p = \dot{\lambda}\mathbf{r}(\tau, \mathbf{q}), \quad \mathbf{r} = \frac{3}{2\bar{\sigma}}\tau^{\text{dev}} \quad \tau^{\text{dev}} = \tau - \frac{1}{3}\text{trace}(\tau)\mathbf{I}, \quad \bar{\sigma} = \sqrt{\frac{3}{2}\tau^{\text{dev}} : \tau^{\text{dev}}}$$

$$\dot{q}_1 = \dot{\lambda}h_1 \quad q_1 = \bar{\varepsilon} = \int \dot{\bar{\varepsilon}}\, dt \quad \dot{\lambda} = \dot{\bar{\varepsilon}} \quad h_1 = 1 \tag{2.4.2}$$

where the only internal variable is the accumulated effective plastic strain $q_1 \equiv \bar{\varepsilon}$. τ^{dev} is the deviatoric part of the Kirchhoff stress and $\bar{\sigma}$ is the von Mises effective stress. Note that $\bar{\sigma}$ and $\dot{\bar{\varepsilon}}$ are plastic work rate conjugates: $\boldsymbol{\sigma} : \mathbf{D}^p = \bar{\sigma}\dot{\bar{\varepsilon}}$. For the case of uniaxial stress, $\bar{\sigma} = \sigma$.

The yield condition is

$$f(\boldsymbol{\tau}, \mathbf{q}) = \bar{\sigma} - \sigma_Y(\bar{\varepsilon}) = 0 \tag{2.4.3}$$

where $\sigma_Y(\bar{\varepsilon})$ is the yield stress under uniaxial tension. If $\frac{\partial f}{\partial \tau} = \frac{3}{2\bar{\sigma}}\tau^{\text{dev}} = \mathbf{r}$, it is an associative plasticity and $\frac{\partial f}{\partial q_1} = -\frac{d}{d\bar{\varepsilon}}\sigma_Y(\bar{\varepsilon}) = -H(\bar{\varepsilon})$, where $H(\bar{\varepsilon})$ is the plastic modulus.

The loading-unloading conditions are given by Eq. (2.3.8):

$$\dot{\lambda} \geq 0, \quad f \leq 0, \quad \dot{\lambda}f = 0 \tag{2.4.4}$$

The plastic rate parameter from the consistency condition is given by

$$\dot{\lambda} = \frac{f_\tau : \mathbf{C}_{el}^{\tau J} : \mathbf{D}}{-f_{\mathbf{q}} \cdot \mathbf{h} + f_\tau : \mathbf{C}_{el}^{\tau J} : \mathbf{r}} = \dot{\bar{\varepsilon}} = \frac{\mathbf{r} : \mathbf{C}_{el}^{\tau J} : \mathbf{D}}{H + \mathbf{r} : \mathbf{C}_{el}^{\tau J} : \mathbf{r}} \tag{2.4.5}$$

Based on the Cauchy stress with elastic response specified in terms of the Jaumann rate, it turns to Kirchhoff stress with a Jaumann rate $\tau^{\nabla J}$. The relation between the stress rate and total rate-of-deformation is

$$\boldsymbol{\tau}^{\nabla J} = \mathbf{C}^{\tau J} : \mathbf{D} \quad \tau_{ij}^{\nabla J} = C_{ijkl}^{\tau J} D_{kl} \tag{2.4.6}$$

The continuum elastic-plastic tangent modulus is given by

$$\mathbf{C}^{\tau J} = \mathbf{C}_{el}^{\tau J} - \frac{\left(\mathbf{C}_{el}^{\tau J} : \mathbf{r}\right) \otimes \left(f_\tau : \mathbf{C}_{el}^{\tau J}\right)}{-f_{\mathbf{q}} \cdot \mathbf{h} + f_\tau : \mathbf{C}_{el}^{\tau J} : \mathbf{r}} = \mathbf{C}_{el}^{\tau J} - \frac{\left(\mathbf{C}_{el}^{\tau J} : \mathbf{r}\right) \otimes \left(\mathbf{r} : \mathbf{C}_{el}^{\tau J}\right)}{H + \mathbf{r} : \mathbf{C}_{el}^{\tau J} : \mathbf{r}} \tag{2.4.7}$$

where the elastic modulus is expressed in terms of bulk and deviatoric parts:

$$\mathbf{C}_{el}^{\tau J} = K\mathbf{I} \otimes \mathbf{I} + 2\mu \mathbf{I}^{\text{dev}}, \quad \mathbf{I}^{\text{dev}} = \mathbf{I} - \frac{1}{3}\mathbf{I} \otimes \mathbf{I} \tag{2.4.8}$$

Noting that $\mathbf{r}$ is deviatoric, it follows that

$$\mathbf{C}_{el}^{\tau J} : \mathbf{r} = 2\mu\mathbf{r}, \quad \mathbf{r} : \mathbf{C}_{el}^{\tau J} : \mathbf{r} = 3\mu \tag{2.4.9}$$

The elastic-plastic modulus is given by

$$\mathbf{C}^{\tau J} = K\mathbf{I} \otimes \mathbf{I} + 2\mu\left(\mathbf{I}^{\text{dev}} - \gamma\hat{\mathbf{n}} \otimes \hat{\mathbf{n}}\right) = \lambda^e\mathbf{I} \otimes \mathbf{I} + 2\eta\mathbf{I} - 2\eta\gamma\hat{\mathbf{n}} \otimes \hat{\mathbf{n}} \qquad (2.4.10)$$

where $\gamma = \frac{1}{1+(H/3\mu)}$ and $\hat{\mathbf{n}} = \sqrt{\frac{2}{3}}\mathbf{r}$. Here, the Lamé constant is denoted by λ^e to avoid confusion with the plasticity parameter λ. For elastic loading or unloading, the same tangent modulus $\mathbf{C}^{\tau J} = \mathbf{C}_{el}^{\tau J}$ is used.

The overall tangent modulus that relates the Truesdell rate of Cauchy stress and the rate-of deformation tensor, $\boldsymbol{\sigma}^{\nabla T} = \mathbf{C}^{\sigma T}{:}\mathbf{D}$, is given by

$$\mathbf{C}^{\sigma T} = J^{-1}\mathbf{C}^{\tau J} - \mathbf{C}' \qquad (2.4.11)$$

which has major and both minor symmetries, where $C' = \frac{1}{2}(\delta_{ik}\sigma_{jl} + \delta_{il}\sigma_{jk} + \delta_{jk}\sigma_{il} + \delta_{jl}\sigma_{ik})$. For plane strain, the tangent modulus is written in matrix form using Voigt notation as

$$\left[C_{ab}^{\sigma T}\right] = J^{-1}\begin{bmatrix} \lambda^e + 2\mu & \lambda^e & 0 \\ \lambda^e & \lambda^e + 2\mu & 0 \\ 0 & 0 & \mu \end{bmatrix} - 2\mu\gamma J^{-1}\begin{bmatrix} \hat{n}_1\hat{n}_1 & \hat{n}_1\hat{n}_2 & \hat{n}_1\hat{n}_3 \\ \hat{n}_2\hat{n}_1 & \hat{n}_2\hat{n}_2 & \hat{n}_2\hat{n}_3 \\ \hat{n}_3\hat{n}_1 & \hat{n}_3\hat{n}_2 & \hat{n}_3\hat{n}_3 \end{bmatrix}$$
$$- \frac{1}{2}\begin{bmatrix} 4\sigma_1 & 0 & 2\sigma_3 \\ 0 & 4\sigma_2 & 2\sigma_3 \\ 2\sigma_3 & 2\sigma_3 & \sigma_1 + \sigma_2 \end{bmatrix} \qquad (2.4.12)$$

where $\hat{n}_1 = \hat{n}_{11}, \hat{n}_2 = \hat{n}_{22}, \hat{n}_3 = \hat{n}_{12}$, and $\sigma_1 = \sigma_{11}, \sigma_2 = \sigma_{22}, \sigma_3 = \sigma_{12}$.

2.4.2 Extension to Kinematic Hardening

The isotropic hardening formulation presented earlier can be extended to combine kinematic hardening following the same procedure outlined in Section 2.2. In multiaxial large strain kinematic hardening models, the objective rate of back stress tensor $\boldsymbol{\alpha}$ is required. To generalize the one-dimensional kinematic hardening model presented in Section 2.2, the overstress tensor $\sum = \boldsymbol{\tau} - \boldsymbol{\alpha}$ is introduced, where $\boldsymbol{\alpha}$ is the center of the yield surface. The evolution of the back stress tensor is given in terms of the Jaumann rate, $\boldsymbol{\alpha}^{\nabla J} = \kappa\mathbf{D}^p$, where κ is the kinematic hardening modulus. In simple shear at large deformations, Nagtegaal and DeJong (1981) showed that the Jaumann rate in the back stress evolution law results in nonphysical stress oscillations. This model is acceptable when strains are less than about 0.4, which is explained here with that caveat.

For the J_2 flow theory hypoelastic-plastic constitutive model with combined isotropic kinematic hardening, the plastic flow equations and tangent modulus are summarized here. The plastic flow rule and evolution equations are given by

$$\mathbf{D}^p = \dot{\lambda}\mathbf{r}(\boldsymbol{\Sigma},\mathbf{q}), \quad \mathbf{r} = \frac{3}{2\bar{\sigma}}\Sigma^{\text{dev}}, \quad \boldsymbol{\Sigma} = \boldsymbol{\tau} - \boldsymbol{\alpha}, \quad \Sigma^{\text{dev}} = \tau^{\text{dev}} - \boldsymbol{\alpha},$$

$$\tau^{\text{dev}} = \tau - \frac{1}{3}\text{trace}(\tau)\text{I}, \quad \bar{\sigma} = \sqrt{\frac{3}{2}\sum^{\text{dev}} : \sum^{\text{dev}}} \tag{2.4.13}$$

$$\dot{q}_1 = \dot{\lambda}h_1 \quad q_1 = \bar{\varepsilon} = \int \dot{\bar{\varepsilon}}\, dt \quad \dot{\lambda} = \dot{\bar{\varepsilon}} \quad h_1 = 1, \quad \alpha^{\nabla J} = \kappa\mathbf{D}^p = \kappa\dot{\lambda}\mathbf{r}$$

where κ is the kinematic hardening modulus and the internal variables are accumulated effective plastic stress $\bar{\varepsilon}$ and back stress tensor $\boldsymbol{\alpha}$.

The yield condition is

$$f\left(\sum, \mathbf{q}\right) = \bar{\sigma} - \sigma_Y(\bar{\varepsilon}) = 0 \tag{2.4.14}$$

and

$$\frac{\partial f}{\partial \sum} = \frac{3}{2\bar{\sigma}}\sum^{\text{dev}} = \mathbf{r}\,(\text{associative plasticity}), \quad \frac{\partial f}{\partial q_1} = -\frac{d}{d\bar{\varepsilon}}\sigma_Y(\bar{\varepsilon}) = -H(\bar{\varepsilon}) \tag{2.4.15}$$

where $\sigma_Y(\bar{\varepsilon})$ is the yield stress in uniaxial tension and $H(\bar{\varepsilon})$ is a plastic modulus.

The loading-unloading conditions are given by Eq. (2.4.4):

$$\dot{\lambda} \geq 0, \quad f \leq 0, \quad \dot{\lambda}f = 0$$

The plastic rate parameter from the consistency condition is given by

$$\dot{\lambda} = \frac{f_{\sum} : \mathbf{C}_{el}^{\tau J} : \mathbf{D}}{-f_\mathbf{q}\cdot\mathbf{h} + f_{\sum} : \kappa\mathbf{r} + f_{\sum} : \mathbf{C}_{el}^{\tau J} : \mathbf{r}} = \dot{\bar{\varepsilon}} = \frac{\mathbf{r} : \mathbf{C}_{el}^{\tau J} : \mathbf{D}}{H + \kappa' + \mathbf{r} : \mathbf{C}_{el}^{\tau J} : \mathbf{r}} \tag{2.4.16}$$

where $\kappa' = \frac{3}{2}\kappa$. Note that there is some difference from Eq. (2.4.5).

The relation between the stress rate and the total rate-of-deformation is given in Eq. (2.4.6):

$$\tau^{\nabla J} = \mathbf{C}^{\tau J} : \mathbf{D} \quad \tau_{ij}^{\nabla J} = C_{ijkl}^{\tau J}D_{kl} \tag{2.4.17}$$

The continuum elastic-plastic tangent modulus is given by

$$\mathbf{C}^{\tau J} = \mathbf{C}_{el}^{\tau J} - \frac{\left(\mathbf{C}_{el}^{\tau J} : \mathbf{r}\right) \otimes \left(f_{\sum} : \mathbf{C}_{el}^{\tau J}\right)}{-f_\mathbf{q}\cdot\mathbf{h} + f_{\sum} : \kappa\mathbf{r} + f_{\sum} : \mathbf{C}_{el}^{\tau J} : \mathbf{r}} = \mathbf{C}_{el}^{\tau J} - \frac{\left(\mathbf{C}_{el}^{\tau J} : \mathbf{r}\right) \otimes \left(\mathbf{r} : \mathbf{C}_{el}^{\tau J}\right)}{H + \kappa' + \mathbf{r} : \mathbf{C}_{el}^{\tau J} : \mathbf{r}} \tag{2.4.18}$$

The elastic-plastic tangent modulus is also given:

$$\mathbf{C}^{\tau J} = K\mathbf{I}\otimes\mathbf{I} + 2\mu\left(\mathbf{I}^{\text{dev}} - \gamma\hat{\mathbf{n}}\otimes\hat{\mathbf{n}}\right) = \lambda^e\mathbf{I}\otimes\mathbf{I} + 2\eta I - 2\eta\gamma\hat{\mathbf{n}}\otimes\hat{\mathbf{n}} \tag{2.4.19}$$

$$\gamma = \frac{1}{1 + ((H + \kappa')/3\mu)}, \quad \hat{\mathbf{n}} = \sqrt{\frac{2}{3}}\mathbf{r}$$

For elastic loading or unloading, there is $\mathbf{C}^{\tau J} = \mathbf{C}_{el}^{\tau J}$. The overall tangent modulus is obtained in an manner analogous to Eqs. (2.4.11) and (2.4.12).

Johnson and Bammann (1984) showed that the Green-Naghdi rate of stress and back stress could eliminate nonphysical oscillations. A hypoelastic-plastic formulation based on the Green-Naghdi rate is given in Section 2.7.

2.4.3 Large Strain Viscoplasticity

Rate-dependent plasticity, i.e., viscoplasticity, constitutive relations can be extended to multidimension in a similar manner by generalizing the one-dimensional rate-dependent plasticity equations of Section 2.2. In rate-independent plasticity, the plastic rate parameter is obtained from the consistency condition, whereas in rate-dependent plasticity, this parameter is obtained as an empirical function of stress and internal variables. It is typically given by an overstress function, as will be seen next. Therefore, we have the same form of the plastic flow rule and evolution equations for the internal variables:

$$\mathbf{D}^p = \dot{\lambda}\mathbf{r}(\boldsymbol{\sigma}, \mathbf{q}), \quad \dot{\mathbf{q}} = \dot{\lambda}\mathbf{h} \tag{2.4.20}$$

where the plastic rate parameter is given by

$$\dot{\lambda} = \frac{\phi(\boldsymbol{\sigma}, \mathbf{q})}{\eta} \tag{2.4.21}$$

where ϕ is an overstress function. Note that ϕ has dimensions of stress, which can be thought of as the driving force for the plastic strain rate. η is the viscosity and has dimensions of stress × time.

Owing to Perzyna (1971), a typical example of Eq. (2.4.21) of the overstress function for J_2 plasticity flow is

$$\phi = \sigma_Y(\bar{\varepsilon})\left\langle \frac{\bar{\sigma}}{\sigma_Y(\bar{\varepsilon})} - 1 \right\rangle^n \tag{2.4.22}$$

where $\langle \bullet \rangle$ are Macaulay brackets, $\bar{\sigma}$ is the von Mises effective stress Eq. (2.4.2), n is a rate-sensitivity exponent, and $\sigma_Y(\bar{\varepsilon})$ is the yield stress under uniaxial tension. For the J_2 plasticity flow theory, Peirce et al. (1984) gave an alternative viscoplastic model as

$$\dot{\lambda} = \dot{\bar{\varepsilon}} = \dot{\varepsilon}_0 \left(\frac{\bar{\sigma}}{\sigma_Y(\bar{\varepsilon})} \right)^{1/m} \tag{2.4.23}$$

where m is the rate-sensitivity exponent and $\dot{\varepsilon}_0$ is a reference strain rate. This model does not use an explicit yield function. However, because $m \to 0$, rate-independent plasticity with yield stress $\sigma_Y(\bar{\varepsilon})$ is approached. These constitutive equations for rate-dependent plasticity, namely viscoplasticity, are summarized subsequently.

The additive decomposition of rate-of-deformation tensor into elastic and plastic parts is given by

$$\mathbf{D} = \mathbf{D}^e + \mathbf{D}^p \tag{2.4.24}$$

The relation between the stress rate and rate-of-deformation is

$$\boldsymbol{\sigma}^{\nabla J} = \mathbf{C}_{el}^{\sigma J} : \mathbf{D}^e = \mathbf{C}_{el}^{\sigma J} : (\mathbf{D} - \mathbf{D}^p) \tag{2.4.25}$$

The plastic flow rule and evolution equations are given by

$$\mathbf{D}^p = \dot{\lambda}\mathbf{r}(\boldsymbol{\sigma}, \mathbf{q}), \quad \dot{\lambda} = \frac{\phi(\boldsymbol{\sigma}, \mathbf{q})}{\eta}, \quad \dot{\mathbf{q}} = \dot{\lambda}\mathbf{h}(\boldsymbol{\sigma}, \mathbf{q}) \tag{2.4.26}$$

The relation between the stress rate and total rate-of-deformation is

$$\boldsymbol{\sigma}^{\nabla J} = \mathbf{C}_{el}^{\sigma J} : \mathbf{D} - \frac{\phi}{\eta}\mathbf{C}_{el}^{\sigma J} : \mathbf{r} \tag{2.4.27}$$

2.5 Rock-Soil Constitutive Model

For materials such as soil and rock, frictional and dilatational effects are significant. The J_2 flow models presented earlier are inappropriate for these materials. Instead, yield functions representing material friction behaviors were developed. In these materials, plastic behavior depends on pressure, in contrast to von Mises plasticity, which is independent of pressure. Furthermore, for frictional materials, associative plasticity laws are no longer inappropriate. There is another contrast between associative plasticity for J_2 flow theory and nonassociative plasticity for rock-soil models.

2.5.1 Mohr-Coulomb Constitutive Model

To illustrate the frictional behavior of materials, consider a block under a normal force N and a tangential force Q, as shown in Fig. 2.3. The block rests on a rough surface with the static friction coefficient μ. If Coulomb's law is assumed to hold, the maximum frictional resistance is given by $F_{\max} = \mu N$. The onset of sliding occurs when the yield condition is satisfied by

$$f = Q - \mu N = 0 \tag{2.5.1}$$

The yield surface Eq. (2.5.1) is plotted in Fig. 2.3. The direction of sliding (plastic flow) is horizontal, which is in the direction of Q and not normal to the yield surface. This is an example of nonassociative plastic flow behavior. The Mohr-Coulomb criterion has a generalization of this behavior to continua and to multiaxial states of stress and strain. It is widely used in modeling the behavior of granular materials (soils) and rocks.

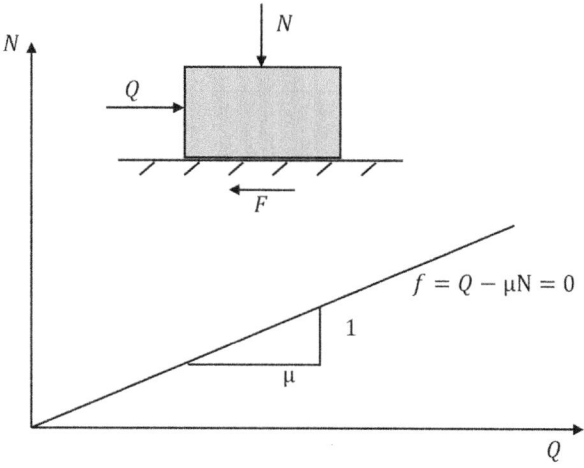

Figure 2.3
Yield surface for friction sliding.

The Mohr-Coulomb criterion is based on the concept that yielding in the material occurs on any plane when a critical combination of shear stress and mean normal stress is reached. The criterion is stated as

$$\tau = c - \mu\sigma \tag{2.5.2}$$

where τ is a magnitude of the shear stress, σ is a normal stress on the plane, and c is the cohesion. The internal friction angle ϕ is defined through $\mu = \tan\phi$. Eq. (2.5.2) is represented by two straight lines (depending on the friction sliding directions) in the Mohr plane, which are the envelopes of the Mohr circles and are called Mohr failure or rupture envelopes. In more general forms, these lines are curves (Khan and Huang, 1995). If all three Mohr circles associated with the principal stresses lie between the failure envelopes, no yielding occurs. When the yield surface is tangential to one of the Mohr's circles, yielding occurs. For example, in three dimensions, the stress state at yield is depicted in Fig. 2.4(A), where it is assumed that the principal stresses $\sigma_1 > \sigma_2 > \sigma_3$. The stress state is given by

$$\tau = \frac{1}{2}(\sigma_1 - \sigma_3)\cos\phi \ \text{ and } \ \sigma = \frac{1}{2}(\sigma_1 + \sigma_3) + \frac{1}{2}(\sigma_1 - \sigma_3)\sin\phi \tag{2.5.3}$$

The yield criterion (2.5.2) is therefore given by

$$f(\boldsymbol{\sigma}) = \sigma_1 - \sigma_3 + (\sigma_1 + \sigma_3)\sin\phi - 2c\cos\phi = 0 \tag{2.5.4}$$

This equation is a conical surface in principal stress space. The intersection of the yield surface with π-plane ($\sigma_{kk} = 0$) is shown in Fig. 2.4(B) and is seen to be an irregular hexagon. Considering the special case $\phi = 0$ and letting $c = k$ denote the yield strength in shear, Eq. (2.5.4) reduces to $\sigma_1 - \sigma_3 - 2k = 0$, which is the Tresca criterion (Hill, 1950).

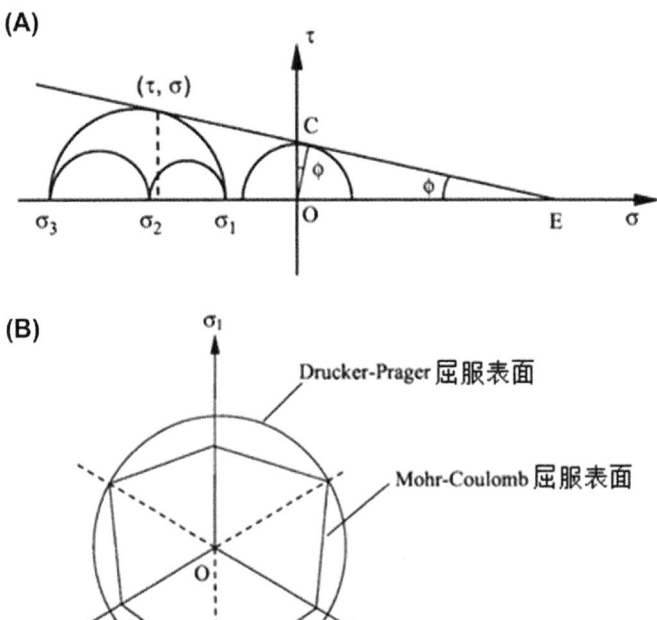

(A)

(B)

Figure 2.4
(A) Mohr-Coulomb yield behavior, and (B) Drucker-Prager and Mohr-Coulomb yield surfaces.

In the materials of soil and rock, plastic behavior depends on pressure, in contrast to von Mises plasticity for metals, which is independent of the pressure. There is significant strength behavior for the compression to be larger than tension on the Mohr-Coulomb yield surface, as shown in Fig. 2.4(B), which has an irregular hexagon. This is in contrast to the Tresca yield surface, as shown in Fig. 2.5(A) for three dimensions on a π plane, and in Fig. 2.5(B), on a two-dimensional $\sigma_1 - \sigma_2$ plane, respectively, which has a regular hexagon.

2.5.2 Drucker-Prager Constitutive Model

The straight-line segments on the Tresca and Mohr-Coulomb yield surfaces make these surfaces convenient for analytical solutions of plasticity problems, as shown in Figs. 2.4(B) and 2.5. However, from the standpoint of computation and program implementation, the corners make the constitutive equations difficult. For example, in computing the normal to the yield surface, it is hard to choose the normal corresponding to the yield function at the corner. It is fortunate that the von Mises and Drucker-Prager yield criteria avoid this problem associated with corners by modifying the von Mises yield criterion (2.2.7) to incorporate the effects of pressure:

$$f = \bar{\sigma} - \alpha \boldsymbol{\sigma} : \mathbf{I} - Y = 0 \qquad (2.5.5)$$

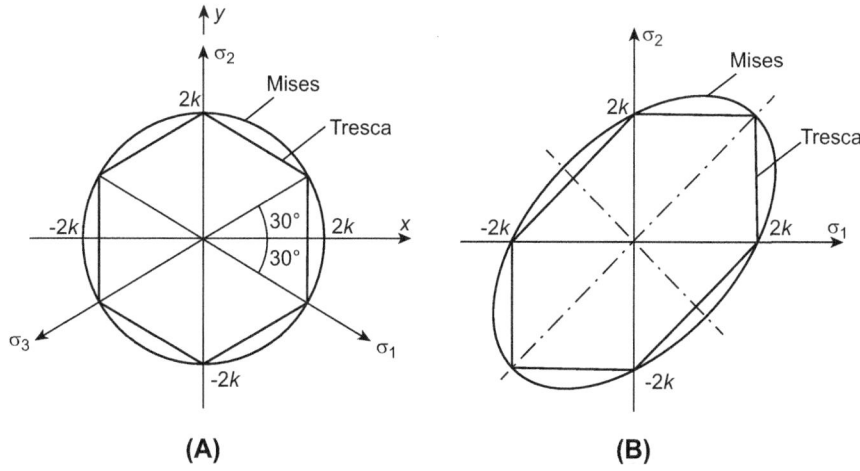

Figure 2.5

Tresca and Mises yield surfaces: (A) on a three-dimensional π plane, and (B) on a two-dimensional $\sigma_1 - \sigma_2$ plane.

which is the equation of a smooth circular cone. In Eq. (2.5.4), $\bar{\sigma}$ is the effective Cauchy stress. By choosing the constants α and Y as

$$\alpha = \frac{2 \sin \phi}{3 \pm \sin \phi}, \quad Y = \frac{6c \cos \phi}{3 \pm \sin \phi} \tag{2.5.6}$$

the Drucker-Prager yield surface passes through the inner or outer apexes of the Mohr-Coulomb yield surface (with the plus sign corresponding to the inner apexes and the minus sign corresponding to the outer apexes).

The elastic response is given by a hypoelastic relation for the Jaumann rate of Cauchy stress. Both associative and nonassociative models can be developed. The associative plastic flow rule is $\mathbf{D}^p = \dot{\lambda}\mathbf{r}(\boldsymbol{\sigma},\mathbf{q})$, where

$$\mathbf{r} = \frac{\partial f}{\partial \boldsymbol{\sigma}} = \frac{3}{2\bar{\sigma}}\sigma^{\text{dev}} - \alpha\mathbf{I} \tag{2.5.7}$$

Many nonassociative rules can be developed. One of them is

$$\mathbf{r} = \frac{\partial \psi}{\partial \boldsymbol{\sigma}} = \frac{3}{2\bar{\sigma}}\sigma^{\text{dev}}, \quad \psi = \bar{\sigma} \tag{2.5.8}$$

It can be seen from this that the volumetric plastic flow is nonzero and material dilates under compression for the associative law (2.5.7). This contradicts the observed behavior of granular materials. In the nonassociative law (2.5.8), plastic flow is isotropic.

The complete formulation of the model is given in Section 2.3.1 with the definitions of yield condition (2.5.5) and flow rule (2.5.7) or (2.5.8). Because the model is based on the Cauchy stress, the overall tangent modulus is not symmetric.

2.6 Gurson Model for Porous Elastic-Plastic Solids

The key assumption of the J_2 model is that plastic flow in metals is unaffected by pressure. However, for porous elastic-plastic solids, one needs to consider the volume change affected by pressure. For dilatant materials and porous plastic solids, as in the Gurson model, large dilations accompany plastic deformations and the assumption $J \approx 1$ is no longer valid. In this situation, the yield function is preferably expressed in terms of the Cauchy stress and the resulting tangent stiffness is not symmetric. The Gurson constitutive model (Gurson, 1977) was developed to simulate progressive microrupture through void nucleation and growth. It is extensively used to model the ductile rupture of metals (Tvergaard and Needleman, 1984).

Different versions of the Gurson model can be formulated. For example, Narasimhan et al. (1992) considered fracture initiation in ductile steel using a small strain rate-independent plasticity version of the model. Here, we present a large deformation, hypoelastic, rate-independent plasticity version. The rate-dependent formulation is provided by Pan et al. (1983).

The material consists of a matrix and voids with volume fraction f_v. In this section, Φ is used to denote the yield function. The void volume fraction and the accumulated plastic strain in the matrix material are internal variables in the model. For the constitutive model, it is an additive decomposition of the rate-of-deformation tensor into elastic and plastic parts. The Jaumann rate of Cauchy stress is adopted in the hypoelastic stress rate relation. The moduli are usually taken to be constant and isotropic. The plastic flow equations are based on Cauchy stress and the von Mises-type yield condition is used.

The yield function Φ also acts as a potential for plastic flow, so that this theory is associative. The yield condition is given by

$$\Phi = \frac{\sigma_e^2}{\overline{\sigma}^2} + 2f^* \beta_1 \cosh\left(\frac{\beta_2 \boldsymbol{\sigma} : \mathbf{I}}{2\overline{\sigma}}\right) - 1 - (\beta_1 f^*)^2 = 0 \tag{2.6.1}$$

where $\sigma_e = \sqrt{\frac{3}{2}\sigma^{\text{dev}} : \sigma^{\text{dev}}}$ is the effective macroscopic Cauchy stress, in which $\sigma^{\text{dev}} = \sigma - \frac{1}{3}\text{trace}(\sigma)\mathbf{I}$ is the deviatoric Cauchy stress, $\overline{\sigma}$ is the effective stress in the matrix material, and f^* is a function of the void volume fraction given subsequently.

The model was originally introduced by Gurson (1977) for rate-independent plasticity with the values of β_1 and β_2 set to unity. Later, the parameters β_1 and β_2 were introduced by Tvergaard (1981) to model behavior more accurately at a low void volume fraction. Parameter f^* was introduced by Tvergaard and Needleman (1984) to simulate the rapid loss of strength in the final stages of void coalescence. In Gurson's original model,

parameter f^* is simply void volume fraction f_v. In the Tvergaard-Needleman approach, a modification is invoked when f_v reaches a critical value f_c and is given by

$$f^* = \begin{cases} f_v & f_v \leq f_c \\ f_c + (f_u - f_c)(f_v - f_c)/(f_f - f_c) & f_v > f_c \end{cases} \tag{2.6.2}$$

where $f_u = 1/\beta_1$ and $f^*(f_f) = f_u$. Note that f_f is the void volume fraction at which the material has completely lost stress-carrying capacity.

The plastic flow direction is given by the associative rule $\mathbf{D}^P = \dot{\lambda}\mathbf{r}$, where

$$\mathbf{r} = \frac{\partial \Phi}{\partial \boldsymbol{\sigma}} = \frac{3}{\bar{\sigma}^2}\boldsymbol{\sigma}^{\text{dev}} + (f^*\beta_1\beta_2/\bar{\sigma})\sinh\left(\frac{\beta_2\boldsymbol{\sigma}:\mathbf{I}}{2\bar{\sigma}}\right)\mathbf{I} \tag{2.6.3}$$

It also requires evolution equations for internal variables $q_1 = f_v$ and $q_2 = \bar{\varepsilon}$. The increase in voids in the material is the result of the growth of existing voids and the nucleation of new voids, which can be written as

$$\dot{f}_v = \dot{f}_{\text{growth}} + \dot{f}_{\text{nucleation}} \tag{2.6.4}$$

In an incompressible matrix, where the small contribution from elastic strain is neglected, by using the kinematic of void growth and the macroscopic plastic flow rule, the expression for void growth is obtained as

$$\dot{f}_{\text{growth}} = (1 - f_v)\text{trace}(\mathbf{D}^P) = \dot{\lambda}(1 - f_v)\text{trace}(\mathbf{r}) \tag{2.6.5}$$

Nucleation is typically regarded as strain or stress controlled, which is neglected here for simplicity.

During plastic loading, the effective stress in the matrix material must be lie on the matrix yield surface $\bar{\sigma} - \sigma_Y(\bar{\varepsilon}) = 0$. The consistency condition in the matrix material is obtained by differentiating this expression, to give

$$\dot{\bar{\sigma}} = H(\bar{\varepsilon})\dot{\bar{\varepsilon}} \tag{2.6.6}$$

where $H(\bar{\varepsilon}) = d\sigma_Y(\bar{\varepsilon})/d\bar{\varepsilon}$ is the matrix plastic modulus. It follows Eq. (2.6.6) as

$$\frac{\partial}{\partial q_2} \equiv \frac{\partial}{\partial \bar{\varepsilon}} = H\frac{\partial}{\partial \bar{\sigma}} \tag{2.6.7}$$

which is used subsequently to obtain the derivatives of the yield function.

An evolution expression for the accumulated effective plastic strain is obtained by equating macroscopic and microscopic rates of plastic work, i.e.,

$$\boldsymbol{\sigma}:\mathbf{D}^P = (1 - f_v)\bar{\sigma}\dot{\bar{\varepsilon}} \tag{2.6.8}$$

From this equation, the relation between the effective plastic strain rate and plastic rate-of-deformation is obtained as

$$\dot{\bar{\varepsilon}} = \frac{\boldsymbol{\sigma} : \mathbf{D}^p}{(1 - f_v)\bar{\sigma}} = \dot{\lambda}\frac{\boldsymbol{\sigma} : \mathbf{r}}{(1 - f_v)\bar{\sigma}} \tag{2.6.9}$$

The equations for the rate-independent Gurson model are summarized next.

The additive decomposition of rate-of-deformation tensor into elastic and plastic parts is given by

$$\mathbf{D} = \mathbf{D}^e + \mathbf{D}^p \tag{2.6.10}$$

The relation between the stress rate and rate-of-deformation is

$$\boldsymbol{\sigma}^{\nabla J} = \mathbf{C}_{el}^{\sigma J} : \mathbf{D}^e = \mathbf{C}_{el}^{\sigma J} : (\mathbf{D} - \mathbf{D}^p) \tag{2.6.11}$$

The plastic flow rule and evolution equations are given by

$$\mathbf{D}^p = \dot{\lambda}\mathbf{r}(\boldsymbol{\sigma}, \mathbf{q}), \quad \mathbf{r} = \frac{\partial \Phi}{\partial \boldsymbol{\sigma}}, \quad \dot{\mathbf{q}} = \dot{\lambda}\mathbf{h}, \quad q_1 = f_v, \quad q_2 = \bar{\varepsilon}$$

$$h_1 = (1 - f_v)\mathrm{trace}(\mathbf{r}), \quad h_2 = \frac{\boldsymbol{\sigma} : \mathbf{r}}{(1 - f_v)\bar{\sigma}}, \quad \dot{\bar{\sigma}} = H\dot{\bar{\varepsilon}} \tag{2.6.12}$$

The yield condition is seen in Eq. (2.6.1):

$$\Phi = \frac{\sigma_e^2}{\bar{\sigma}^2} + 2f^* \beta_1 \cosh\left(\frac{\beta_2 \boldsymbol{\sigma} : \mathbf{I}}{2\bar{\sigma}}\right) - 1 - (\beta_1 f^*)^2 = 0 \tag{2.6.13}$$

The loading-unloading conditions are given by

$$\dot{\lambda} \geq 0, \quad \Phi \leq 0, \quad \dot{\lambda}\Phi = 0 \tag{2.6.14}$$

The plastic rate parameter from consistency condition $\dot{\Phi} = 0$ is obtained as

$$\dot{\lambda} = \frac{\Phi_{\boldsymbol{\sigma}} : \mathbf{C}_{el}^{\sigma J} : \mathbf{D}}{-\Phi_{\mathbf{q}} \cdot \mathbf{h} + \Phi_{\boldsymbol{\sigma}} : \mathbf{C}_{el}^{\sigma J} : \mathbf{r}} = \frac{\mathbf{r} : \mathbf{C}_{el}^{\sigma J} : \mathbf{D}}{-\Phi_{\mathbf{q}} \cdot \mathbf{h} + \mathbf{r} : \mathbf{C}_{el}^{\sigma J} : \mathbf{r}} \tag{2.6.15}$$

Note that from Eq. (2.6.7) that $\partial\Phi/\partial q_2 = H\partial\Phi/\partial\bar{\sigma}$.

The relation between the stress rate and total rate-of-deformation is

$$\boldsymbol{\sigma}^{\nabla J} = \mathbf{C}^{\sigma J} : \mathbf{D} \quad \sigma_{ij}^{\nabla J} = C_{ijkl}^{\sigma J}D_{kl} \tag{2.6.16}$$

where the continuum elastic-plastic tangent modulus is given by

$$\mathbf{C}^{\sigma J} = \mathbf{C}_{el}^{\sigma J} - \frac{\left(\mathbf{C}_{el}^{\sigma J} : \mathbf{r}\right) \otimes \left(\Phi_{\boldsymbol{\sigma}} : \mathbf{C}_{el}^{\sigma J}\right)}{-\Phi_{\mathbf{q}} \cdot \mathbf{h} + \Phi_{\boldsymbol{\sigma}} : \mathbf{C}_{el}^{\sigma J} : \mathbf{r}} = \mathbf{C}_{el}^{\sigma J} - \frac{\left(\mathbf{C}_{el}^{\sigma J} : \mathbf{r}\right) \otimes \left(\mathbf{r} : \mathbf{C}_{el}^{\sigma J}\right)}{-\Phi_{\mathbf{q}} \cdot \mathbf{h} + \mathbf{r} : \mathbf{C}_{el}^{\sigma J} : \mathbf{r}} \tag{2.6.17}$$

which has a major symmetry. For elastic loading or unloading, $\mathbf{C}^{\sigma J} = \mathbf{C}^{\sigma J}_{el}$. The overall tangent modulus is given by $\mathbf{C}^{\sigma T} = \mathbf{C}^{\sigma J} - \mathbf{C}^*$, which does not possess major symmetry because $\mathbf{C}^*$ is not symmetric, since

$$\mathbf{C}^* = \mathbf{C}' - \sigma \otimes I, \quad \mathbf{C}' : D = D \cdot \sigma + \sigma \cdot D \tag{2.6.18}$$

where $\mathbf{C}'$ is referenced in Eqs. (2.4.11) and (2.4.12).

When $f_v = 0$, which means that there is no void in the material, the rate-independent Gurson model is reduced to rate-independent J_2 plastic flow theory.

2.7 Corotational Stress Formulation

The previously described hypoelastic-plastic formulations are typically used with constant elastic moduli. Objectivity requires these moduli to be isotropic, and as discussed earlier, the yield function is restricted to be an isotropic function of the stress. The corotational stress formulation presented in this section is not restricted to be an isotropic response; as will be seen, the tangent modulus is not symmetric even through the model is based on the Kirchhoff stress.

The corotational Kirchhoff stress tensor, $\widehat{\tau}$, is defined by

$$\widehat{\tau} = \mathbf{R}^T \cdot \tau \cdot \mathbf{R} = J\widehat{\sigma} \tag{2.7.1}$$

where $\widehat{\sigma}$ is the corotational Cauchy stress and R is the rotation tensor, also called a rotation matrix, which is used to express rigid body rotation. For two dimensions,

$$R(t) = \begin{bmatrix} \cos \omega t & -\sin \omega t \\ \sin \omega t & \cos \omega t \end{bmatrix} \tag{2.7.2}$$

The $\theta = \omega t$ is used here to express motion as a function of time. The relation between the stress rate and the elastic strain rate is written as

$$\dot{\widehat{\tau}} = \widehat{C}^\tau_{el} : \widehat{D}^e \tag{2.7.3}$$

where $\widehat{D}^e$ is the elastic part of the corotational rate-of-deformation tensor $\widehat{D} = R^T \cdot D \cdot R$ or $\widehat{D}_{ij} = R^T_{ik} D_{kl} R_{lj}$. The plasticity equations and elastic-plastic tangent modulus are analogous to those in Section 2.3.1.

The corotational Kirchhoff stress formulations of a hypoelastic-plastic constitutive model are given next, for review.

The additive decomposition of a corotational rate-of-deformation tensor into elastic and plastic parts is given by

$$\widehat{\mathbf{D}} = \widehat{\mathbf{D}}^e + \widehat{\mathbf{D}}^p \tag{2.7.4}$$

The relation between the stress rate and the rate-of-deformation is

$$\dot{\widehat{\tau}} = \widehat{\mathbf{C}}_{el}^{\tau} : \widehat{\mathbf{D}}^{e} = \widehat{\mathbf{C}}_{el}^{\tau} : \left(\widehat{\mathbf{D}} - \widehat{\mathbf{D}}^{p} \right) \tag{2.7.5}$$

The plastic flow rule and evolution equations are given by

$$\widehat{\mathbf{D}}^{p} = \dot{\lambda}\widehat{\mathbf{r}}(\widehat{\tau}, \widehat{\mathbf{q}}) \quad \dot{\widehat{\mathbf{q}}} = \dot{\lambda}\widehat{\mathbf{h}}(\widehat{\tau}, \widehat{\mathbf{q}}) \tag{2.7.6}$$

The yield condition is

$$\widehat{f}(\widehat{\tau}, \widehat{\mathbf{q}}) = 0 \tag{2.7.7}$$

The loading-unloading conditions are given by

$$\dot{\lambda} \geq 0, \quad \widehat{f} \leq 0, \quad \dot{\lambda}\widehat{f} = 0 \tag{2.7.8}$$

The plastic rate parameter from the consistency condition is obtained as

$$\dot{\lambda} = \frac{\widehat{f}_{\widehat{\tau}} : \widehat{\mathbf{C}}_{el}^{\tau} : \widehat{\mathbf{D}}}{-\widehat{f}_{\mathbf{q}} \cdot \widehat{\mathbf{h}} + \widehat{f}_{\widehat{\tau}} : \widehat{\mathbf{C}}_{el}^{\tau} : \widehat{\mathbf{r}}} \tag{2.7.9}$$

A relation between the material time derivative of Kirchhoff stress and the corotational rate-of-deformation tensor is denoted by

$$\dot{\widehat{\tau}} = \widehat{\mathbf{C}}^{\tau} : \widehat{\mathbf{D}} \quad \dot{\widehat{\tau}}_{ij} = \widehat{C}_{ijkl}^{\tau} \widehat{D}_{kl} \tag{2.7.10}$$

For only elastic loading or unloading,

$$\widehat{\mathbf{C}}^{\tau} = \widehat{\mathbf{C}}_{el}^{\tau}$$

For plastic loading, the continuum elastic-plastic tangent modulus is given by

$$\widehat{\mathbf{C}}^{\tau} = \widehat{\mathbf{C}}_{el}^{\tau} - \frac{\left(\widehat{\mathbf{C}}_{el}^{\tau} : \widehat{\mathbf{r}} \right) \otimes \left(\widehat{f}_{\widehat{\tau}} : \widehat{\mathbf{C}}_{el}^{\tau} \right)}{-\widehat{f}_{\widehat{\mathbf{q}}} \cdot \mathbf{h} + \widehat{f}_{\widehat{\tau}} : \widehat{\mathbf{C}}_{el}^{\tau} : \widehat{\mathbf{r}}} \tag{2.7.11}$$

$$\widehat{C}_{ijkl}^{\tau} = \left(\widehat{C}_{el}^{\tau} \right)_{ijkl} - \frac{\left(\widehat{C}_{el}^{\tau} \right)_{ijmn} \widehat{r}_{mn} \left(\widehat{f}_{\widehat{\tau}} \right)_{pq} \left(\widehat{C}_{el}^{\tau} \right)_{pqkl}}{-\left(\widehat{f}_{\widehat{\mathbf{q}}} \right)_{\alpha} \widehat{h}_{\alpha} + \left(\widehat{f}_{\widehat{\tau}} \right)_{rs} \left(\widehat{C}_{el}^{\tau} \right)_{rstu} \widehat{r}_{tu}}$$

Now, note that the Green-Naghdi rate of Kirchhoff stress is given by

$$\tau^{\nabla G} = \dot{\tau} - \Omega \cdot \tau - \tau \cdot \Omega^{T} = \mathbf{R} \cdot \dot{\widehat{\tau}} \cdot \mathbf{R}^{T} \tag{2.7.12}$$

Then from Eq. (2.7.10), we obtain

$$\tau^{\nabla G} = \mathbf{C}^{\tau G} : \mathbf{D} \quad C_{ijkl}^{\tau G} = R_{im} R_{jn} R_{kp} R_{lq} \widehat{C}_{mnpq}^{\tau} \tag{2.7.13}$$

where $\mathbf{D} = \mathbf{R} \cdot \hat{\mathbf{D}} \cdot \mathbf{R}^T$. The overall tangent modulus (Eq. 2.6.18), which is not symmetric owing to C^{spin}, is given by

$$\mathbf{C}^{\sigma T} = J^{-1} \mathbf{C}^{\tau G} - \mathbf{C}' - \mathbf{C}^{spin} \qquad (2.7.14)$$

where $\mathbf{C}^{spin}: D = (W - \Omega) \cdot \sigma + \sigma \cdot (W - \Omega)^T$, in which $\Omega = \dot{R} \cdot R^T$ is called the angular velocity tensor with skew-symmetric. $\Omega = -\Omega^T$ is associated with the rotation R and W is the rate-of-rotation or spin tensor, which is the skew-symmetric part of the velocity gradient. In the absence of deformation, the spin tensor and angular velocity tensor are equal: $W = \Omega$.

An advantage of the corotational stress formulation is that frame invariance requirements do not limit the model to isotropic elastic moduli or isotropic yield behavior, as was the case for the previously described Jaumann rate of Cauchy stress or Kirchhoff stress formulations. The rotated stress is insensitive to rigid rotations of the current configuration:

$$\hat{\tau}^* = \mathbf{R}^{*T} \cdot \tau^* \cdot \mathbf{R}^* = \mathbf{R}^T \cdot \mathbf{Q}^T \cdot \mathbf{Q} \cdot \tau \cdot \mathbf{Q}^T \cdot \mathbf{Q} \cdot \mathbf{R} = \mathbf{R}^T \cdot \tau \cdot \mathbf{R} = \hat{\tau} \qquad (2.7.15)$$

where $\mathbf{R}^* = \mathbf{Q} \cdot \mathbf{R}$ and $\tau^* = \mathbf{Q} \cdot \tau \cdot \mathbf{Q}^T$, in which $\mathbf{Q}$ is a rotation tensor or matrix, which is used here to express rotation as a function of time. Thus, the elastic moduli $\hat{\mathbf{C}}^\tau$ may be anisotropic and the yield function f may be an arbitrary function of corotational stress $\hat{\tau}$.

2.8 Summary

The essential constitutive theories of plasticity are described in this chapter. The purpose is to provide referenced knowledge points so that the following chapters of this book may easily be read. Although many theory formulations are described, they are only a tip of the iceberg for overall material plasticity constitutive models. For example, the special case of the single crystal plasticity theory is discussed in Appendix 1. We also do not introduce hyperelastic-plastic constitutive models because this book focuses on crystal plasticity especially at the micron scale, as described in Chapter 3, and the submicron scale, as discussed in Chapters 4–12. The constitutions at a macroscale are time- but not size-dependent. In Chapter 3, strain gradient plasticity theories demonstrate a size effect because the constitutions are related to the material characteristic size at a micron scale. For more knowledge about conventional plasticity mechanics, the readers may study the listed references.

Strain Gradient Plasticity Theory at the Microscale

3.1 Size Dependence of Material Behavior at the Microscale

Many experiments have shown that materials have a strong size effect when the characteristic length scale is on the order of microns: The smaller, the stronger. Conventional plasticity theories cannot predict this size dependence of material behavior at the micron scale because their constitutive models possess no internal length scale, as described in Chapter 2. The fundamental assumption in conventional plasticity theory is that stress at a material point depends on the strain or deformation history at the same point. Consequently, the response of a material is characterized by a constitutive equation

that gives the stress as a function of the deformation history of the material. In contrast, at the micron scale, the stress at a material point depends not on only the strain at the same point but also on the strain gradient at nearby points. Therefore, it is important to establish a micron-level or nonlocal continuum theory (for example, strain gradient plasticity theory) that considers the intrinsic material length and deformation gradient.

In the 1990s, mounting experimental evidence showed that materials displayed strong size effects when the characteristic length scale associated with nonuniform plastic deformation was on the order of micrometers. Four examples are provided: (1) In the torsion test of thin copper wire, Fleck et al. (1994) observed that torsion strength increased three times when the diameters of copper wire was reduced from d = 170 μm to d = 12 μm, as shown in Fig. 3.1A. In contrast, there was no size effect for the stress-strain curves under uniaxial tension without a strain gradient even though the diameters varied from d = 170 μm to d = 12 μm, as shown in Fig. 3.1B. (2) For the thin beam bending test, Stolken and Evans (1998) found that the bending strength increased significantly when the thickness of h = 100 μm was reduced to h = 12.5 μm. (3) The size effect was also observed in the microscopic hardness test on the order of microns. The hardness of metal increased one time when the indentation depth was reduced from h = 10 μm to h = 1 μm (Nix, 1989; De Guzman et al., 1993; Stelmashenko et al., 1993; Ma and Clarke, 1995, Poole et al., 1996, McElhaney et al., 1998). (4) The size effect was observed in particle-reinforced metal-matrix composites. For silicon carbide particle-reinforced aluminum-silicon matrix composites, Lloyd (1994) observed that the strength of composites increased dramatically when the particle diameters were reduced from

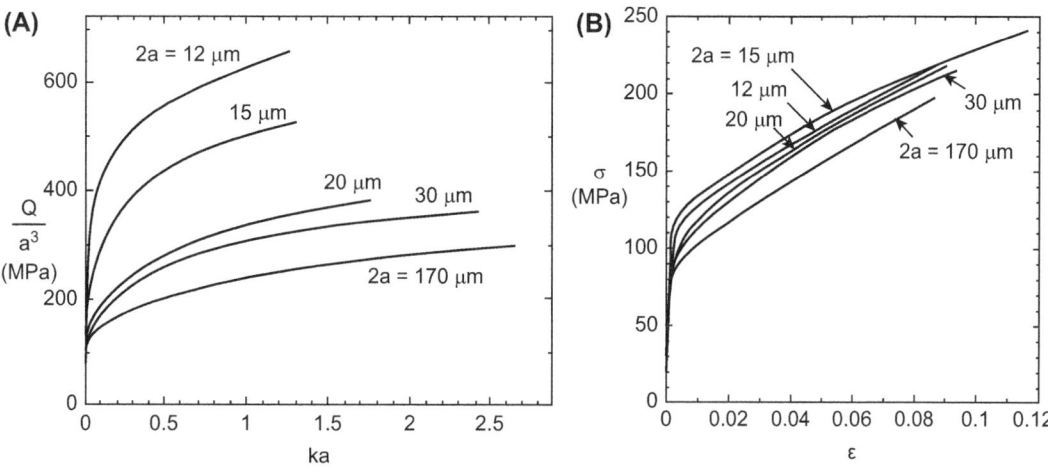

Figure 3.1
The thin copper wire test of Fleck et al. (1994): (A) torsion; and (B) tension, for Q-torsion and k-torsion angle per unit length. *Reprinted from Fleck, N.A., Muller, G.M., Ashby, M.F., Hutchinson, J.W., 1994. Strain gradient plasticity - theory and experiment. Acta Metallurgica et Materialia 42, 475–487. Copyright 1994, with permission from Elsevier.*

d = 16 μm to d = 7.5 μm under the condition of keeping the particle volume fraction at 15%. Motivated by these experiments the strain gradient plasticity theories have been developed, which are introduced in this chapter.

Conventional plasticity theories cannot predict this size dependence of material behavior at the micron scale because their constitutive models possess no internal length scale. However, in engineering practice, there is a strong need to deal with designing and manufacturing devices at the microscale: for example, thin film at micrometers or even smaller thicknesses, sensor systems, microelectronic devices or packages at the size of micrometers, composite materials including particles or fibers at the microscale, and so on. Current analysis and design methods such as the finite element method (FEM) and computer-aided engineering were developed based on conventional plasticity theory, which cannot be used at this size scale. On the other hand, it is still hard to handle these structures at the microscale using quantum mechanics and molecular dynamics because of the limited length and time scales. Therefore, under the continuum mechanics framework at the microscale, the objective of establishing a constitutive model considering the size effect is an essential bridge to linking conventional plasticity theory and microscale simulations.

Another objective in establishing a constitutive model under the continuum mechanics framework at the microscale is to set up a relation between macroscopic fracture behavior and the atom fracture process for toughness materials. In a series of experiments, Elssner et al. (1994) measured macroscopic fracture toughness and atom separation at the interface of single-crystal niobium-sapphire. A specially designed four-point bending test specimen was used to measure macroscopic fracture to obtain the interface toughness. Atomic separation was determined through microvoid balance configuration at the interface. Although niobium is a tough material with many dislocations, the bimaterial interface still maintained atomic-scale sharp cracks. The crack tips were not blunt. The separation force of the atomic lattice or strong interface needs to be about $0.03E$ or $10\sigma_y$ (E is the Young's modulus and σ_y is the tension yield stress). However, according to conventional plasticity theory, Hutchinson (1997) pointed out that the maximum stress can reach four to five times σ_y. Obviously, this stress is far from results in the data observed in the test by Elssner et al. (1994), which cannot separate atoms at the crack tip. Consequently, the effect of the strain gradient may explain this phenomenon.

Theoretical and computational work has been developed in this area in parallel. It is mainly divided into couple stress (CS) theory, stretch and rotation gradient (SG) theory, and mechanism-based strain gradient (MSG) deformation plasticity theory. These theories are described in detail in this chapter; other strain gradient theories are not introduced here. For example, Aifantis and colleagues (Aifantis, 1984; Zbib and Aifantis, 1989; Muhlhaus and Aifantis, 1991) expressed the strain gradient in terms of the first Laplace

operator and second Laplace operator of equivalent strain. However, the work conjugation term of strain gradient was not defined in their theories. Considering the possible configuration form of strain gradient plasticity, Acharya and Bassani (1995) proposed a kind of flow theory that should maintain the fundamental form of conventional plasticity and ensure the limitation of thermal mechanics. As a kind of internal variable, the strain gradient was expressed as the increase in the current tangential hardening modulus. However, because there was no systemic method with which to construct the tangential modulus, this method was not specified.

In the following sections, CS theory is introduced in Section 3.2. SG theory is expressed in Section 3.3. Section 3.4 focuses on MSG deformation plasticity theory. Finally, a summary is provided in Section 3.5.

3.2 Couple Stress Theory

Dislocation theory states that the plastic hardening of a material originates from geometrically necessary dislocation (GND) and statistically stored dislocation (SSD). According to this theory, Fleck and Hutchinson (1993) and Fleck et al. (1994) developed one type of strain gradient plasticity theory, which is extended by the conventional J_2 flow theory or deformation theory. They introduced CS to consider the effect of rotation gradient. By using CS theory, size effects at the micron scale were successfully predicted for thin copper wire torsion (Fleck et al., 1994), thin beam bending (Stolken and Evans, 1998), and particle-reinforced aluminum and magnesium matrix composites (Lloyd, 1994). In this constitutive relation, material constant l (length dimension) is introduced to balance the dimension of the strain and strain gradient. This length constant is regarded as an intrinsic length of material that depends on the micron structures in the material. It is estimated as $l = 4$ μm for copper and $l = 6$ μm for nickel, respectively. When characteristic length L in an inhomogeneous deformation field is much larger than intrinsic length l of the material, the strain gradient effect is too small to be neglected. The reason is that the contribution of the strain gradient term is much smaller than that of strain term ε, i.e., $l\mathrm{d}\varepsilon/\mathrm{d}x \sim (l/L)\varepsilon \ll \varepsilon$. Thus, CS theory is reduced to conventional J_2 flow theory, as mentioned in Section 2.4. However, when characteristic length L in an inhomogeneous deformation field belongs to the same order with intrinsic length l of the material, as mentioned in the previous experiments, the strain gradient effect is significant.

3.2.1 Couple Stresses

In conventional plasticity theory, there are no CSs, so the Cauchy stress is a symmetric tensor. However, in the CS theory, Cauchy stress is not a symmetric tensor because there are also CSs. The Cauchy stress tensor is denoted by t and CS tensor is denoted by m.

The force and moment of force acting on a unit area of the plane with unit normal vector *n* are given by, respectively,

$$T = t \cdot n, \quad q = m \cdot n \tag{3.2.1}$$

It is assumed that there is no body force and body moment of force; then, the equilibrium equations of force and moment of force are given by, respectively,

$$t \cdot \nabla = 0, \quad t_{ij,j} = 0 \tag{3.2.2}$$

and

$$m \cdot \nabla = e : t, \quad m_{ip,p} = e_{irs} t_{rs} \tag{3.2.3}$$

where *e* is the substitution tensor of the third order and is denoted by

$$e = e_{ijk} e_i e_j e_k = e'_{ijk} e'_i e'_j e'_k \tag{3.2.4}$$

By using $e \sim \delta$ equality in the tensor analysis,

$$e_{ijk} e_{irs} = \delta_{jr} \delta_{ks} - \delta_{js} \delta_{kr} \tag{3.2.5}$$

It can be obtained as

$$e \cdot m \cdot \nabla = t - t^T, \quad e_{ijk} m_{ip,p} = t_{jk} - t_{kj} \tag{3.2.6}$$

This equation can be proved through the substitution relations of the third-order tensor. It must maintain the right-hand coordinate after transformation if the coordinate is on the right hand previously. Otherwise, a negative sign should be used. In the conventional theory, there are no CSs, i.e., *m* = 0, so the Cauchy stress *t* is a symmetric tensor. However, in general, Cauchy stress is not a symmetric tensor: for example, CSs exist.

The additive decomposition of Cauchy stress *t* into symmetric part σ and skew-symmetric part τ is assumed:

$$t = \sigma + \tau, \quad t_{ij} = \sigma_{ij} + \tau \tag{3.2.7}$$

where

$$\sigma = \frac{1}{2}(t + t^T), \quad \sigma_{ij} = \frac{1}{2}(t_{ij} + t_{ji}) \tag{3.2.8a}$$

$$\tau = \frac{1}{2}(t - t^T), \quad \tau_{ij} = \frac{1}{2}(t_{ij} - t_{ji}) \tag{3.2.8b}$$

Then, substituting Eq. (3.2.8b) into Eq. (3.2.6), it becomes

$$\frac{1}{2} e \cdot m \cdot \nabla = \tau, \quad \frac{1}{2} e_{ijk} m_{ip,p} = \tau_{jk} \tag{3.2.9}$$

This equation indicates that antisymmetric stress τ is associated with the divergence of CSs m; that is,

$$\frac{1}{2}m_{1p,p} = \tau_{23} = -\tau_{32}$$

$$\frac{1}{2}m_{2p,p} = \tau_{31} = -\tau_{13} \qquad (3.2.10)$$

$$\frac{1}{2}m_{3p,p} = \tau_{12} = -\tau_{21}$$

Substituting an equation, $t = \sigma + \tau$, and τ of Eq. (3.2.9) into the equilibrium equations of force (3.2.2) gives

$$\sigma \cdot \nabla = -\frac{1}{2}[e \cdot (m \cdot \nabla)] \cdot \nabla \qquad (3.2.11)$$

Writing m as the additive decomposition of the deviatoric part and hydrostatic part $(\operatorname{tr} m)1/3$ demonstrates that the right-hand side terms in Eq. (3.2.11) are not related to the hydrostatic part:

$$\sigma \cdot \nabla = -\frac{1}{2}[e \cdot (m' \cdot \nabla)] \cdot \nabla \qquad (3.2.12)$$

This was indicated by Fleck and Hutchinson (1993). Early work by Koiter (1964) proved that the hydrostatic part of m did not appear in all of the field equations and in constitutive equations, as given later in Eq. (3.2.34). Thus, it is assumed that the hydrostatic part of m is 0 and m is a tensor without trace:

$$m = m', \quad \operatorname{tr} m = 0 \qquad (3.2.13)$$

In general, m is a nonsymmetric tensor.

3.2.2 Rotation and Rotation Gradient

Displacement is denoted by u. The displacement difference between two neighbor points, x and $x + dx$, is given by

$$du = (u\nabla) \cdot dx \qquad (3.2.14)$$

The additive decomposition of displacement gradient $u\nabla$ into symmetric tensor ε and antisymmetric tensor ω gives

$$u\nabla = \varepsilon + \omega \quad u_{i,j} = \varepsilon_{ij} + \omega_{ij} \qquad (3.2.15)$$

where ε is the strain tensor and ω is the spin tensor:

$$\varepsilon = \frac{1}{2}(u\nabla + \nabla u), \quad \varepsilon_{ij} = \frac{1}{2}\left(u_{i,j} + u_{j,i}\right) \qquad (3.2.16a)$$

$$\omega = \frac{1}{2}(u\nabla - \nabla u), \quad \omega_{ij} = \frac{1}{2}\left(u_{i,j} - u_{j,i}\right) \qquad (3.2.16b)$$

Substituting Eq. (3.2.16) into Eq. (3.2.14) gives

$$\mathrm{d}\boldsymbol{u} = \boldsymbol{\varepsilon} \cdot \mathrm{d}x + \boldsymbol{\omega} \cdot \mathrm{d}x \tag{3.2.17}$$

Tensor analysis states that the scalar product of the antisymmetric or skew-symmetric tensor $\boldsymbol{\omega}$ and vector $\mathrm{d}x$ can be expressed as the vector product of anticouple vector $\boldsymbol{\theta}$ and vector $\mathrm{d}x$:

$$\boldsymbol{\omega} \cdot \mathrm{d}x = \boldsymbol{\theta} \times \mathrm{d}x \tag{3.2.18}$$

where $\boldsymbol{\theta}$ denotes the rotational vector, which has the relation to skew-symmetric tensor $\boldsymbol{\omega}$:

$$\boldsymbol{\theta} = -\frac{1}{2}\boldsymbol{e} : \boldsymbol{\omega}, \quad \boldsymbol{\omega} = -\boldsymbol{e} \cdot \boldsymbol{\theta} = -\boldsymbol{\theta} \cdot \boldsymbol{e} \tag{3.2.19a}$$

$$\theta_i = -\frac{1}{2}e_{ijk}\omega_{jk}, \quad \omega_{ij} = -e_{ijk}\theta_k \tag{3.2.19b}$$

In two dimensions, a skew-symmetric tensor has a single independent component and its axial vector is perpendicular to the two-dimensional plane of model, so

$$\boldsymbol{\omega} = \begin{bmatrix} 0 & \omega_{12} \\ -\omega_{12} & 0 \end{bmatrix} = \begin{bmatrix} 0 & -\theta_3 \\ \theta_3 & 0 \end{bmatrix} \tag{3.2.20}$$

In three dimensions, a skew-symmetric tensor has three independent components that are related to three components of its axial vector by Eq. (3.2.19b), giving

$$\boldsymbol{\omega} = \begin{bmatrix} 0 & \omega_{12} & \omega_{13} \\ -\omega_{12} & 0 & \omega_{23} \\ -\omega_{13} & -\omega_{23} & 0 \end{bmatrix} = \begin{bmatrix} 0 & -\theta_3 & \theta_2 \\ \theta_3 & 0 & -\theta_1 \\ -\theta_2 & \theta_1 & 0 \end{bmatrix} \tag{3.2.21}$$

The angular velocity matrix is sometimes defined as being a negative of this equation. Thus, Eq. (3.2.18) is also described for rigid body rotation in dynamic textbooks.

Substituting Eq. (3.2.16b) into Eq. (3.2.19a) gives

$$\boldsymbol{\theta} = -\frac{1}{2}\boldsymbol{e} : \frac{1}{2}(\boldsymbol{u}\nabla - \nabla\boldsymbol{u}) = -\frac{1}{2}\boldsymbol{e} : (\boldsymbol{u}\nabla) = \frac{1}{2}\boldsymbol{e} : [\nabla\boldsymbol{u}] = -\frac{1}{2}\boldsymbol{u} \times \nabla = \frac{1}{2}\nabla \times \boldsymbol{u} \tag{3.2.22}$$

and

$$\theta_i = -\frac{1}{2}e_{ijk}u_{j,k} = \frac{1}{2}e_{ijk}\partial_j u_k$$

where $u_{k,j} = \partial_j u_k$.

It can be seen again from Eq. (3.2.22) that the geometric meaning of $\boldsymbol{\theta}$ is a rotational vector. So, Eq. (3.2.17) can be written as

$$\mathrm{d}\boldsymbol{u} = \boldsymbol{\varepsilon} \cdot \mathrm{d}x + \boldsymbol{\theta} \times \mathrm{d}x \tag{3.2.23}$$

The first term of the right-hand side indicates displacement near point x resulting from deformation. The second term indicates displacement resulting from rotation.

Second-order tensor χ is defined by Fleck and Hutchinson (1993) as

$$\chi = \theta \nabla \quad \chi_{ij} = \theta_{i,j} \tag{3.2.24}$$

where χ is a gradient of rotational vector θ, which is called the rotation or spin gradient. If the body acts as a rigid rotation, the same θ occurs at any point of the body. The gradient of θ is 0. Consequently, χ is a quantity characterizing deformation behavior at any point of the body. Strain tensor ε in Eq. (3.2.16) can be expressed by the first derivative of displacement with respect to the coordinate, whereas spin gradient χ can be expressed by the second derivative of displacement with respect to the coordinate. Substituting Eq. (3.2.22) into Eq. (3.2.24) gives

$$\chi = -\frac{1}{2}e : (u\nabla)\nabla = -\frac{1}{2}(u \times \nabla)\nabla \tag{3.2.25}$$

$$\chi_{ij} = -\frac{1}{2}e_{ist}u_{s,tj} = \frac{1}{2}e_{its}u_{s,tj}$$

By using the equality $u_{s,tj} = u_{s,jt}$, it is easy to prove that χ_{ij} can be expressed by the derivative of strain component ε_{ij} with respect to the coordinate (strain gradient):

$$\chi_{ij} = -e_{ist}\varepsilon_{js,t} = e_{its}\varepsilon_{js,t} \tag{3.2.26}$$

Eq. (3.2.26) is called the $\varepsilon \sim \chi$ compatibility equation, in which χ is a nonsymmetric tensor and it can be proved that its trace is 0. Thus, owing to Eq. (3.2.25), it gives

$$\text{tr}\chi = \chi_{ii} = -\frac{1}{2}e_{ist}u_{s,ti} = 0 \tag{3.2.27}$$

There is $\text{tr } \chi = \theta \cdot \nabla$ according to Eq. (3.2.24). Therefore, $\theta = \frac{1}{2}\nabla \times u$ is the spin vector of u based on Eq. (3.2.22) and the divergence $\theta \cdot \nabla = \frac{1}{2}(\nabla \times u) \cdot \nabla$ is obviously equal to 0, which results in Eq. (3.2.27). Thus, ε and χ are used to describe deformation behavior. ε has six independent components and χ has eight independent components.

3.2.3 Virtual Work Principle

The virtual work principle supposes $t = \sigma + \tau$ and m as a set of equilibrium internal forces in the body, which are satisfied with Eqs. (3.2.2) and (3.2.3). ε and χ are a set of strain caused by displacement u in the body, which are satisfied with Eqs. (3.2.15) and (3.2.25). It is assumed that there are no relations between (t, m) and (ε, χ); thus, they may have the virtual work given by

$$\int_A (\mathbf{T} \cdot u + q \cdot \theta)\mathrm{d}A = \int_\Omega (\sigma \cdot \varepsilon + m \cdot \chi)\mathrm{d}\Omega \tag{3.2.28}$$

where Ω is a domain and A is a surface of the body, T is a drag force, and q is a drag moment of force, as given in Eq. (3.2.1). Deformation energy and complementary energy are calculated by a unit volume instead of a unit quality.

In the following, Eq. (3.2.28) is proved started from the left-hand side:

$$\int_A (\boldsymbol{T} \cdot \boldsymbol{u} + \boldsymbol{q} \cdot \boldsymbol{\theta}) \mathrm{d}A = \int_A (\boldsymbol{u} \cdot \boldsymbol{t} \cdot \boldsymbol{n} + \boldsymbol{\theta} \cdot \boldsymbol{m} \cdot \boldsymbol{n}) \mathrm{d}A$$

$$= \int_\Omega [(\boldsymbol{u} \cdot \boldsymbol{t} + \boldsymbol{\theta} \cdot \boldsymbol{m}) \cdot \nabla] \mathrm{d}\Omega \qquad (3.2.29)$$

$$= \int_\Omega [(\boldsymbol{u}\nabla): \boldsymbol{t} + \boldsymbol{u} \cdot (\boldsymbol{t} \cdot \nabla) + (\boldsymbol{\theta} \, \nabla): \boldsymbol{m} + \boldsymbol{\theta} \cdot (\boldsymbol{m} \cdot \nabla)] \mathrm{d}\Omega$$

where the second term in the integral is 0 according to Eq. (3.2.2). By using Eqs. (3.2.22) and (3.2.6), as well as the indicial rotation character of e, i.e., $e_{ijk} = e_{jki} = e_{kij}$, the fourth term in the integral can be changed by

$$\boldsymbol{\theta} \cdot (\boldsymbol{m} \cdot \nabla) = -\frac{1}{2}[e: (\boldsymbol{u}\nabla)] \cdot (\boldsymbol{m} \cdot \nabla) = -\frac{1}{2}(\boldsymbol{u}\nabla): e \cdot (\boldsymbol{m} \cdot \nabla) = -(\boldsymbol{u}\nabla): \boldsymbol{\tau} \qquad (3.2.30)$$

Substituting Eq. (3.2.30) into the last row of Eq. (3.2.29) and considering $t = \sigma + \tau$, it is obtained as

$$\int_A (\boldsymbol{T} \cdot \boldsymbol{u} + \boldsymbol{q} \cdot \boldsymbol{\theta}) \mathrm{d}A = \int_\Omega [(\boldsymbol{u}\nabla): \boldsymbol{\sigma} + (\boldsymbol{\theta} \, \nabla): \boldsymbol{m}] \mathrm{d}\Omega \qquad (3.2.31)$$

Noting that σ is the symmetric stress and Eq. (3.2.24), we immediately have

$$\int_A (\boldsymbol{T} \cdot \boldsymbol{u} + \boldsymbol{q} \cdot \boldsymbol{\theta}) \mathrm{d}A = \int_\Omega [\boldsymbol{\varepsilon}: \boldsymbol{\sigma} + \boldsymbol{\chi}: \boldsymbol{m}] \mathrm{d}\Omega \qquad (3.2.32)$$

This is now Eq. (3.2.28) and the proof is finished.

By the way, χ is not a trace tensor (see Eq. 3.2.27); i.e., tr $\chi = 0$. Therefore, the hydrostatic part of m does not have a part in $m \cdot \chi$, which is on the right-hand side terms in Eq. (3.2.28). That is $m{:}\chi = m'{:}\chi$. It is why m has been supposed to be a no-trace tensor in Eq. (3.2.13).

Eq. (3.2.28) indicates that there is an equal quantity of external work and internal work. Thus, it is stated that symmetric Cauchy stress σ and strain ε are conjugated in work and CS m and spin gradient χ are also conjugated in work.

3.2.4 Constitutive Relation of Couple Stress Strain Gradient Plasticity Theory

In this section, we will deduce the constitutive relation of the CS strain gradient plasticity theory. It supposes that the deformation energy of a body in a unit volume is w, which is a function of strain tensor ε and rotation gradient χ:

$$w = w(\boldsymbol{\varepsilon}, \, \boldsymbol{\chi}) \qquad (3.2.33)$$

Then, the nonlinear elastic constitutive in terms of tensor notation or indicial notation is given by

$$\boldsymbol{\sigma} = \frac{\partial w}{\partial \boldsymbol{\varepsilon}}, \quad \boldsymbol{m} = \frac{\partial w}{\partial \boldsymbol{\chi}} \tag{3.2.34a}$$

$$\sigma_{ij} = \frac{\partial w}{\partial \varepsilon_{ij}}, \quad m_{ij} = \frac{\partial w}{\partial \chi_{ij}} \tag{3.2.34b}$$

As discussed previously, rotation gradient $\boldsymbol{\chi}$ is a nontrace tensor (Eq. 3.2.27); thus, its nine components are incompletely independent. Therefore, $\boldsymbol{m} = \frac{\partial w}{\partial \boldsymbol{\chi}}$ must also be stipulated as a nontrace tensor, which is indicated in Eq. (3.2.13).

The complementary energy of body in a unit volume is defined as

$$\phi = \phi(\boldsymbol{\sigma}, \boldsymbol{m}) = \boldsymbol{\sigma} : \boldsymbol{\varepsilon} + \boldsymbol{m} : \boldsymbol{\chi} - w(\boldsymbol{\varepsilon}, \boldsymbol{\chi}) \tag{3.2.35}$$

If $\boldsymbol{\sigma}$ and $\boldsymbol{m}$ are taken as the independent variables, Eq. (3.2.35) is the Legendre transformation of Eq. (3.2.34). $\boldsymbol{\varepsilon}$ and $\boldsymbol{\chi}$ are functions of $\boldsymbol{\sigma}$ and $\boldsymbol{m}$ determined by Eq. (3.2.34). According to Eqs. (3.2.34) and (3.2.35), they can be deduced through

$$\boldsymbol{\varepsilon} = \frac{\partial \phi}{\partial \boldsymbol{\sigma}}, \quad \boldsymbol{\chi} = \frac{\partial \phi}{\partial \boldsymbol{m}} \tag{3.2.36a}$$

$$\varepsilon_{ij} = \frac{\partial \phi}{\partial \sigma_{ij}}, \quad \chi_{ij} = \frac{\partial \phi}{\partial m_{ij}} \tag{3.2.36b}$$

In Eq. (3.2.36), because $\boldsymbol{m}$ is a nontrace tensor (see Eq. 3.2.13), $\boldsymbol{\chi} = \frac{\partial \phi}{\partial \boldsymbol{m}}$ must also be stipulated as a nontrace tensor (see Eq. 3.2.27).

This assumes that no unloading is occurring; thus, Eq. (3.2.34) or Eq. (3.2.36) is just the constitutive relation of strain gradient plasticity (total strain theory). If the expression is obtained in detail, the constitutive relation must be determined, which is further discussed in Section 3.2.6.

3.2.5 Principles of Minimum Potential Energy and Minimum Complementary Energy

Fleck and Hutchinson (1993) proved the following two principles. Suppose the body boundary in v domain is divided into two parts: displacement boundary a_u and force boundary a_T. The boundary conditions of displacement and the rotation angle at a_u are given by

$$\boldsymbol{u} = \boldsymbol{u}^0, \quad \boldsymbol{\theta} = \boldsymbol{\theta}^0 \quad \text{at} \ a_u \tag{3.2.37}$$

$$u = u_i^0, \quad \theta = \theta_i^0$$

whereas the boundary conditions of force and couples at a_T are given by

$$\boldsymbol{T} = \boldsymbol{t} \cdot \boldsymbol{n} = \boldsymbol{T}^0, \quad \boldsymbol{q} = \boldsymbol{m} \cdot \boldsymbol{n} = \boldsymbol{q}^0 \quad \text{at} \ a_T \tag{3.2.38}$$

$$T_i = t_{ij} n_j = T_i^0, \quad q_i = m_{ij} n_j = q_i^0$$

where the known quantities are denoted as "0" at the upper right corner and n is outer normal unit vector.

A set of calculated solutions is $(u, \theta, \chi, \sigma, m)$, which should satisfy balance Eqs. (3.2.2) and (3.2.3), geometrical relation Eqs. (3.2.16) for (u and ε), (3.2.22) for (u and θ), and (3.2.24) for (χ and θ), as well as boundary conditions (3.2.37) and (3.2.38). This solution is called the real solution.

There is supposed a displacement field, u, which can be used to calculate θ according to Eq. (3.2.22). The u and θ satisfy boundary conditions (3.2.37) at a_u. Then, the u is called the geometrical allowable displacement. θ can be used to calculate χ according to Eq. (3.2.24). Finally, the potential energy is calculated by

$$P(u) = \int_v w(\varepsilon \cdot \chi) \mathrm{d}v - \int_{a_T} (T^0 \colon u + q^0 \colon \theta) \mathrm{d}a \tag{3.2.39}$$

where $P(u)$ can be regarded as a function of geometrical allowable displacement u.

1. The minimum potential energy principle
   If w is a monotonically convex function of $(\varepsilon \cdot \chi)$, the potential energy $P(u)$ of the real solution is the absolute minimum in the all-geometrical allowable displacement field u.

   It is supposed that there is a stress field (σ, m) that satisfies balance Eqs. (3.2.2) and (3.2.3) and boundary condition (3.2.38). Then, field (σ, m) is called the static allowable stress field. The definition of complementary energy is given by

$$C(\sigma \cdot m) = \int_v \phi(\sigma \cdot m) \mathrm{d}v - \int_{a_u} (T \colon u^0 + q \colon \theta^0) \mathrm{d}a \tag{3.2.40}$$

   where (T and q) at a_u are calculated through Eq. (3.2.1). $C(\sigma \cdot m)$ can be regarded as a function of static allowable stress $(\sigma \cdot m)$.

2. The minimum complementary energy principle
   If ϕ is a monotonically convex function of $(\sigma \cdot m)$, the complementary energy $C(\sigma \cdot m)$ of the real solution is the absolute minimum in the all-static allowable stress field $(\sigma \cdot m)$.

   The minimum potential energy principle and minimum complementary energy principle guarantee unique real solutions. A detailed discussion of linear situations can be found in Koiter (1964).

3.2.6 Equivalent Stress and Equivalent Strain

In conventional plastic deformation theory, the definition of Mises equivalent strain ε_{eq} is an invariable given as

$$\varepsilon_{eq} = \left(\frac{2}{3} \varepsilon' \colon \varepsilon'\right)^{\frac{1}{2}} = \left(\frac{2}{3} \varepsilon'_{ij} \colon \varepsilon'_{ij}\right)^{\frac{1}{2}} \tag{3.2.41}$$

Fleck and Hutchinson (1993) use ε_{eq} to denote the contribution from statistically stored dislocation to strain energy density w. After that, Fleck et al. (1994) suggested using an invariable expression, χ_{eq}, which is like Eq. (3.2.41), to denote the contribution from geometrically necessary dislocation to strain energy density w. However, the deviator ε' of strain tensor appears in ε_{eq} defined in Eq. (3.2.41). Thus, volume strain tr ε is taken out. How to take the contribution of volume strain from rotation gradient χ remains problematic (see Section 3.3). For the moment, we keep away from this problem. Fleck and Hutchinson (1993) suppose that the material is incompressible. Thus, copying Eq. (3.2.41), the definition of χ_{eq} is given by

$$\chi_{eq} = \left(\frac{2}{3}\boldsymbol{\chi}:\boldsymbol{\chi}\right)^{\frac{1}{2}} = \left(\frac{2}{3}\chi_{ij}\chi_{ij}\right)^{\frac{1}{2}} \tag{3.2.42}$$

Because χ is a symmetric tensor, there is another invariable:

$$\boldsymbol{\chi}:\boldsymbol{\chi}^{T} = \boldsymbol{\chi}:\boldsymbol{\chi} = \chi_{ij}\chi_{ji} \tag{3.2.43}$$

However, Fleck and Hutchinson (1993) only use Eq. (3.2.42) instead of using Eq. (3.2.43), for mathematic simplicity. Combination Eqs. (3.2.41) and (3.2.42), it is defined as

$$E_{eq}^{2} = \varepsilon_{eq}^{2} + l^{2}\chi_{eq}^{2} \tag{3.2.44}$$

where l is a material parameter with a length dimension, which is required to balance the dimension in Eq. (3.2.44). E_{eq} is called the total equivalent strain.

It is supposed that strain energy w at a per-unit volume is a function of total equivalent strain E_{eq}, $w = w(E_{eq})$. Thus, total equivalent stress Σ is defined as

$$\Sigma = \frac{dw(E_{eq})}{dE_{eq}} \tag{3.2.45}$$

Substituting $w(E_{eq})$ into w in Eq. (3.2.34), the solutions are given by

$$\boldsymbol{\sigma}' = \frac{2}{3}\frac{\Sigma}{E_{eq}}\boldsymbol{\varepsilon}, \quad \mathbf{m} = \frac{2}{3}l^{2}\frac{\Sigma}{E_{eq}}\boldsymbol{\chi} \tag{3.2.46a}$$

$$\sigma_{ij}' = \frac{2}{3}\frac{\Sigma}{E_{eq}}\varepsilon_{ij}, \quad m_{ij} = \frac{2}{3}l^{2}\frac{\Sigma}{E_{eq}}\chi_{ij} \tag{3.2.46b}$$

It is assumed the material is uncompressed here. That is, $\varepsilon' = \varepsilon$. The stress calculated through $\partial w/\partial\varepsilon$ in Eq. (3.2.34a) is only partial stress σ'. For the uncompressed material, hydrostatic pressure tr $\sigma/3$ is not determined by a constitutive relation.

In conventional plasticity theory, the Mises equivalent stress is defined by

$$\sigma_{eq} = \left(\frac{3}{2}\boldsymbol{\sigma}':\boldsymbol{\sigma}'\right)^{\frac{1}{2}} \tag{3.2.47}$$

After Eq. (3.2.47), the equivalent CS can be determined by

$$m_{eq} = \left(\frac{3}{2} m : m \right)^{\frac{1}{2}} \tag{3.2.48}$$

Substituting Eqs. (3.2.46a) and (3.2.46b) into Eqs. (3.2.47) and (3.2.48), respectively, we may obtain

$$\sigma_{eq} = \frac{\Sigma}{E_{eq}} \varepsilon_{eq}, \quad m_{eq} = l^2 \frac{\Sigma}{E_{eq}} \chi_{eq} \tag{3.2.49}$$

Substituted Eq. (3.2.44) into Eq. (3.2.49) results in

$$\Sigma^2 = \sigma_{eq}^2 + l^2 m_{eq}^2 \tag{3.2.50}$$

Consequently, total equivalent stress Σ combines equivalent stress σ_{eq} and equivalent CS m_{eq}. Supposing total equivalent stress Σ has a power function relation with total equivalent strain E_{eq}:

$$\frac{E_{eq}}{E_{eq0}} = \left(\frac{\Sigma}{\Sigma_0} \right)^n \tag{3.2.51}$$

where Σ_0 and E_{eq0} are material constants. Thus, we may obtain the equations

$$w = w(E_{eq}) = \frac{n}{n+1} \Sigma_0 E_{eq0} \left(\frac{E_{eq}}{E_{eq0}} \right)^{\frac{n+1}{n}} \tag{3.2.52}$$

$$\phi = \phi(\Sigma) = \frac{1}{n+1} \Sigma_0 E_{eq0} \left(\frac{\Sigma}{\Sigma_0} \right)^{n+1} \tag{3.2.53}$$

CS gradient theory has been adopted for many applications. Fleck and Hutchinson (1993) used this theory to calculate material macrohardening composited with sparse rigid particle inclusions and macrosoftening resulting in sparse cavities. This theory has also been used to solve fractured problems, which can be found in Huang et al. (1997). Xia and Hutchinson (1996), Huang and Hu (1995) and Huang et al. (1997) obtained asymptotic solutions considering the strain gradient effect at the crack tip for an elastic-power law hardening material.

Because it is hard to arrive at an analytic solution in strain gradient theory, FEM becomes an important tool. The main difficulty of FEM is that the solution depends on the selection of elements. The urgent question is what element is suitable for solving this problem, which may be compared by the analysis solution, while the analytic solution can only be used for the strain gradient elastic situation. Some people have obtained a strain gradient effect for an elastic material based on CS stress theory. Sternberg and Muki (1967) studied the mode I fracture stress field for finite-length cracks in an infinite domain for a

strain gradient elastic material. Atkinson and Leppington (1977) adopted the Wiener-Hopf method to study the entire field solution for a semiinfinite-length crack applied at normal surface force with exponent decay along the distance. Zhang et al. (1998) obtained the entire field solution for mode III fractures caused by a K_{III} far field in an infinite elastic material. Huang et al. (1999) obtained the entire field solutions for mode I and mode II fractures under a K far field in an infinite elastic material. It is noteworthy that the elastic release solution itself is also very important, except for being used to examine the reliability of FEM calculation results. Chen et al. (1998) proved that the continuum description of a honeycomb structure is equal to an elastic material with a strain gradient effect. Intrinsic material length l is equal to the cell size of a honeycomb. Meanwhile, according to a simple failure criterion of a cell wall, they found that the fracture toughness of a honeycomb elastic material is proportional to $\sigma_c h/\sqrt{l}$, in which σ_c is the tension strength of the cell wall and h and l are wall thickness and length of the cell, respectively. Fleck and Shu (1995) found the same strain gradient effect for the fiber-reinforced composite material, in which the material intrinsic length is equal to the fiber thickness. Xia and Hutchinson (1996) and Huang et al. (1997) also carried out FEM analyses under conditions of full plastic and small domain yield.

To a certain extent, the CS theory of strain gradient plasticity was successfully estimated for thin copper wires torsion (Fleck and Hutchinson, 1993), thin beam bending (Stolken and Evans, 1998), and the size effect of stress analysis at the crack tip field, as mentioned previously. However, Shu and Fleck (1998) applied this theory to an indentation problem. The results did not agree well compared with data given by micro- and nanoscale indentation experiments, which observed increases in hardness of 200% or even 300%. According to the analytic and numerical results given by Xia and Hutchinson (1996), as well as Huang and Hu (1995) and Huang et al. (1997), the maximum shear stress was 2.5 times the conventional plasticity solution (Hutchinson-Rice-Rosengren [HRR] solution) for mode II fractures on the crack surface and r-distance away from the crack tip as plastic hardening exponent $n = 5$. However, similar results were not observed for a mode I fracture. The reason is that only a rotational gradient in the second gradient of displacement is involved in strain gradient CS theory. However, crack tip stress for a mode I fracture is in the rotational field. The rotational gradient becomes a low-order term that does not contribute to force on the crack surface. For this reason, Fleck and Hutchinson (1997) proposed another theory called strain gradient plasticity (SG) theory. In this theory, besides considering rotational gradient, the tension gradient is involved. It will be described in the following section.

3.3 Stretch and Rotation Gradient Theory

SG theory is mentioned incidentally in some research.

3.3.1 Strain Gradient Tensor

Fleck and Hutchinson (1997) used second-order gradient $\boldsymbol{\eta}$ (the third-order tensor) of displacement $\boldsymbol{u}$ to describe the deformation state besides strain tensor $\boldsymbol{\varepsilon}$. It is defined as

$$\boldsymbol{\eta} = \nabla\nabla\mathbf{u}, \quad \eta_{ijk} = \partial_i \partial_j u_k = u_{k,ij} \tag{3.3.1}$$

where $\boldsymbol{\eta}$ is called the strain gradient tensor. It is apparent that $\boldsymbol{\eta}$ is symmetric for the first and second indications; that is, $\eta_{ijk} = \eta_{jik}$. From Eq. (3.2.25), we know that $\boldsymbol{\eta}$ has a relation to rotational gradient $\boldsymbol{\chi}$:

$$\chi_{ij} = -\frac{1}{2} e_{ist} \eta_{jts} = \frac{1}{2} e_{its} \eta_{jts} \tag{3.3.2}$$

or

$$\boldsymbol{\chi} = \frac{1}{2} \boldsymbol{e} : \boldsymbol{\eta}$$

where symbol (:) indicates the rear two transfection indicators. From Eq. (3.3.2), it is easy to prove the relation:

$$\eta_{ijk} - \eta_{ikj} = 2 e_{jkp} \chi_{pi} \tag{3.3.3}$$

Toupin (1962) and Mindlin (1965) discussed the situation of a linear elastic solid. Plasticity theory supposes that there is no volume variation during plastic deformation. Thus, to measure plastic deformation, it should remove the contribution of volume deformation in $\boldsymbol{\varepsilon}$ and $\boldsymbol{\eta}$. After removing the volume variation from the hydrostatic part, tr $\boldsymbol{\varepsilon}/3$, in strain tensor $\boldsymbol{\varepsilon}$, we may obtain the strain partial tensor $\boldsymbol{\varepsilon}'$. However, the question is how to decompose $\boldsymbol{\eta}$ into hydrostatic and partial parts. If the material is incompressible, there is obviously

$$\eta_{ikk} = \eta_{kik} = 0 \tag{3.3.4}$$

Eq. (3.3.4) is called an incompressible condition. From this discussion, we may decompose $\boldsymbol{\eta}$ into hydrostatic part $\boldsymbol{\eta}^{\text{Hy}}$ and partial part $\boldsymbol{\eta}'$:

$$\boldsymbol{\eta} = \boldsymbol{\eta}^{\text{Hy}} + \boldsymbol{\eta}', \quad \eta_{ijk} = \eta_{ijk}^{\text{Hy}} + \eta_{ijk}' \tag{3.3.5}$$

Three conditions must be satisfied in Eq. (3.3.5): (1) both $\boldsymbol{\eta}^{\text{Hy}}$ and $\boldsymbol{\eta}'$ are symmetric for the first and second indications: that is, $\eta_{ijk}^{\text{Hy}} = \eta_{jik}^{\text{Hy}}$, $\eta_{ijk}' = \eta_{jik}'$; (2) $\boldsymbol{\eta}'$ satisfies the incompressible condition: that is, $\eta_{ikk}' = 0$ and thus $\eta_{ikk}^{\text{Hy}} = \eta_{jkk}$; and (3) $\boldsymbol{\eta}^{\text{Hy}}$ and $\boldsymbol{\eta}'$ satisfy orthogonality: that is, $\boldsymbol{\eta}^{\text{Hy}} : \boldsymbol{\eta}' = \eta_{ijk}^{\text{Hy}} \eta_{ijk}' = 0$. To satisfy Conditions (1) and (2), suppose

$$\eta_{ijk}^{\text{Hy}} = \alpha \delta_{ij} \eta_{kpp} + \frac{1}{4} (1 - \alpha) \left(\delta_{ik} \eta_{jpp} + \delta_{jk} \eta_{ipp} \right) \tag{3.3.6}$$

where α is an undecided parameter. To satisfy Condition (3), it is necessary for $\alpha = 0$. Finally, the hydrostatic part η^{Hy} is obtained as

$$\eta_{ijk}^{Hy} = \frac{1}{4}\left(\delta_{ik}\eta_{jpp} + \delta_{jk}\eta_{ipp}\right) \tag{3.3.7}$$

It is similar to Eq. (3.2.44) for CS theory. Total equivalent strain E_{SG} is combines all invariables from Mises equivalent strain (3.2.41) and η', given as

$$E_{SG}^2 = \frac{2}{3}\varepsilon'_{ij}\varepsilon'_{ij} + c_1\eta'_{iik}\eta'_{jjk} + c_2\eta'_{ijk}\eta'_{ijk} + c_3\eta'_{ijk}\eta'_{kji} \tag{3.3.8}$$

The relation between χ and η is given in Eq. (3.3.2) and η' is defined in Eq. (3.3.5). Thus, χ' can also be defined according to Eq. (3.3.2):

$$\chi' = \frac{1}{2}e:\eta' \quad \chi'_{ij} = \frac{1}{2}e_{its}\eta'_{jts} \tag{3.3.9}$$

where symbol (:) indicates the rear two transfection indicators. η has been defined as the double gradient of displacement by Eq. (3.3.1). Now, η' defined by Eq. (3.3.5) is no longer the double gradient of some vector. Similarly, χ has been defined as the gradient of rotational vector θ by Eq. (3.2.24). Now, χ' defined by Eq. (3.3.9) is not the gradient of some vector. As we have deduced Eq. (3.3.3) from Eq. (3.3.2), so we may deduce Eq. (3.3.10) from Eq. (3.3.9):

$$\eta'_{ijk} - \eta'_{ikj} = 2e_{jkp}\chi'_{pi} \tag{3.3.10}$$

3.3.2 Decomposition of Strain Gradient Partial Tensor η′ and Total Equivalent Strain E$_{SG}$

Smyshlyaev and Fleck (1996) decomposed η' into three orthogonal parts:

$$\eta' = \eta'^{(1)} + \eta'^{(2)} + \eta'^{(3)} \tag{3.3.11}$$

and the different decomposition part of any two of η and $\widetilde{\eta}$ also satisfies orthogonal condition:

$$\eta'^{(i)}:\widetilde{\eta}^{(j)} = 0 \quad \text{when} \quad i \neq j \tag{3.3.12}$$

The decomposition process is given subsequently.

First, the complete symmetry η'^{Sy} is generated from η', which is determined to be a component of η' by transforming and averaging indications:

$$\eta_{ijk}^{'Sy} = \frac{1}{3}\left(\eta'_{ijk} + \eta'_{jki} + \eta'_{kij}\right) \tag{3.3.13}$$

It can be proved that η'^{Sy} not only has symmetry for the first and second indications, for example $\eta_{ijk}'^{Sy} = \eta_{jik}'^{Sy}$, but also has the character of rotation symmetry, that is $\eta_{ijk}'^{Sy} = \eta_{jki}'^{Sy} = \eta_{kij}'^{Sy}$. Thus, the value of $\eta_{ijk}'^{Sy}$ depends only on the combination (i, j, k) but is not related to the order of (i, j, k). Following this discussion, we define another parameter, η'^{A}:

$$\boldsymbol{\eta}'^{A} = \boldsymbol{\eta}' - \boldsymbol{\eta}'^{Sy}, \quad \eta_{ijk}'^{A} = \eta_{ijk}' + \eta_{ijk}'^{Sy} \tag{3.3.14}$$

Based on Eqs. (3.3.13) and (3.3.14), it is easy to prove the character of η'^{A} as

$$\eta_{ijk}'^{A} + \eta_{jki}'^{A} + \eta_{kij}'^{A} = 0 \tag{3.3.15}$$

and for any of $\boldsymbol{\eta}$ and $\tilde{\boldsymbol{\eta}}$, the $\boldsymbol{\eta}'^{Sy}$ and $\boldsymbol{\eta}'^{A}$ are orthogonal each other. This is given by

$$\boldsymbol{\eta}'^{Sy} : \tilde{\boldsymbol{\eta}}'^{A} = \eta_{ijk}'^{Sy} \eta_{ijk}'^{A} = 0 \tag{3.3.16}$$

Substituting Eq. (3.3.13) into Eq. (3.3.14) and using Eq. (3.3.10), the component of η'^{A} can be expressed through the component of $\boldsymbol{\chi}'$:

$$\eta_{ijk}'^{A} = \frac{2}{3} e_{ikp} \chi_{pj}' + \frac{2}{3} e_{jkp} \chi_{pi}' \tag{3.3.17}$$

Now, second-order tensor $\boldsymbol{\chi}'$ can be decomposed into symmetric part $\boldsymbol{\chi}'^{s}$ and asymmetric part $\boldsymbol{\chi}'^{a}$. The lowercase alphabet used here indicates that superscript "s" and "a" differ from their capital forms in Eq. (3.3.14):

$$\boldsymbol{\chi}' = \boldsymbol{\chi}'^{s} + \boldsymbol{\chi}'^{a} \tag{3.3.18}$$

$$\boldsymbol{\chi}'^{s} = \frac{1}{2} \left(\boldsymbol{\chi}' + \boldsymbol{\chi}'^{T} \right), \quad \chi_{ij}'^{s} = \frac{1}{2} \left(\chi_{ij}' + \chi_{ji}' \right)$$

$$\boldsymbol{\chi}'^{a} = \frac{1}{2} \left(\boldsymbol{\chi}' - \boldsymbol{\chi}'^{T} \right), \quad \chi_{ij}'^{a} = \frac{1}{2} \left(\chi_{ij}' - \chi_{ji}' \right)$$

S and S^* are used to express two invariables of $\boldsymbol{\chi}'$:

$$\begin{aligned} S &= \boldsymbol{\chi}' : \boldsymbol{\chi}' = \chi_{ij}' \chi_{ij}' \\ S^* &= \boldsymbol{\chi}' : \boldsymbol{\chi}'^{T} = \boldsymbol{\chi}' \cdot \cdot \boldsymbol{\chi}' = \chi_{ij}' \chi_{ij}' \end{aligned} \tag{3.3.19}$$

Then, it can be proved that

$$\boldsymbol{\chi}'^{s} : \boldsymbol{\chi}'^{a} = \chi_{ij}'^{s} \chi_{ij}'^{a} = 0$$

$$\boldsymbol{\chi}'^{s} : \boldsymbol{\chi}'^{s} = \chi_{ij}'^{s} \chi_{ij}'^{s} = \frac{1}{2} (S + S^*) \tag{3.3.20}$$

$$\boldsymbol{\chi}'^{a} : \boldsymbol{\chi}'^{a} = \chi_{ij}'^{a} \chi_{ij}'^{a} = \frac{1}{2} (S - S^*)$$

Substituting decomposition Eq. (3.3.18) into Eq. (3.3.17), the decomposition equation of corresponding η'^A can be obtained by

$$\eta'^A = \eta'^{As} + \eta'^{Aa} \tag{3.3.21}$$

where similar to Eq. (3.3.17), we have

$$\eta'^{As}_{ijk} = \frac{2}{3}e_{ikp}\chi'^s_{pj} + \frac{2}{3}e_{jkp}\chi'^s_{pi} \tag{3.3.22a}$$

$$\eta'^{Aa}_{ijk} = \frac{2}{3}e_{ikp}\chi'^a_{pj} + \frac{2}{3}e_{jkp}\chi'^a_{pi} \tag{3.3.22b}$$

Using Eq. (3.3.9), Eq. (3.3.22) can be expressed by the component of η':

$$
\begin{aligned}
\eta'^{As}_{ijk} &= \frac{1}{6}\left[e_{ikp}e_{jlm}\eta'_{lpm} + e_{jkp}e_{ilm}\eta'_{lpm} + 2\eta'_{ijk} - \eta'_{jki} - \eta'_{kij} \right] \\
\eta'^{Aa}_{ijk} &= \frac{1}{6}\left[-e_{ikp}e_{jlm}\eta'_{lpm} - e_{jkp}e_{ilm}\eta'_{lpm} + 2\eta'_{ijk} - \eta'_{jki} - \eta'_{kij} \right]
\end{aligned} \tag{3.3.23}
$$

According to Eqs. (3.3.14) and (3.3.21), we have decomposed η' into three parts:

$$\eta' = \eta'^{Sy} + \eta'^{As} + \eta'^{Aa} \tag{3.3.24}$$

where η' satisfies the incompressible condition as described in Eq. (3.3.5): that is, $\eta'_{ikk} = 0$. It can be also proved that η'^{As} has the following characters by using Eq. (3.3.22a):

$$\eta'^{(2)}_{ikk} = \eta'^{As}_{ikk} = 0 \tag{3.3.25a}$$

$$\eta'^{(2)}_{iik} = \eta'^{As}_{iik} = 0 \tag{3.3.25b}$$

$$\eta'^{(2)}_{ijk} + \eta'^{(2)}_{jki} + \eta'^{(2)}_{kij} = \eta'^{As}_{ijk} + \eta'^{As}_{jki} + \eta'^{As}_{kij} = 0 \tag{3.3.25c}$$

and η'^{As} also satisfies the incompressible condition.

Substituting three parts of η' decomposed by Eq. (3.3.11) into the total equivalent strain E_{SG} defined by Eq. (3.3.8), E^2_{SG} can be written as

$$
\begin{aligned}
E^2_{SG} = &\frac{2}{3}\boldsymbol{\varepsilon'}:\boldsymbol{\varepsilon'} + c_1\frac{5}{2}\boldsymbol{\eta}'^{(3)}:\boldsymbol{\eta}'^{(3)} + c_2\left(\boldsymbol{\eta}'^{(1)}:\boldsymbol{\eta}'^{(1)} + \boldsymbol{\eta}'^{(2)}:\boldsymbol{\eta}'^{(2)} + \boldsymbol{\eta}'^{(3)}:\boldsymbol{\eta}'^{(3)} \right) + \\
&c_3\left(\boldsymbol{\eta}'^{(1)}:\boldsymbol{\eta}'^{(1)} - \frac{1}{2}\boldsymbol{\eta}'^{(2)}:\boldsymbol{\eta}'^{(2)} - \frac{1}{4}\boldsymbol{\eta}'^{(3)}:\boldsymbol{\eta}'^{(3)} \right)
\end{aligned} \tag{3.3.26}
$$

Reorganization of the previous equation gives

$$E^2_{SG} = \frac{2}{3}\boldsymbol{\varepsilon'}:\boldsymbol{\varepsilon'} + l^2_1\boldsymbol{\eta}'^{(1)}:\boldsymbol{\eta}'^{(1)} + l^2_2\boldsymbol{\eta}'^{(2)}:\boldsymbol{\eta}'^{(2)} + l^2_3\boldsymbol{\eta}'^{(3)}:\boldsymbol{\eta}'^{(3)} \tag{3.3.27}$$

where

$$l_1^2 = c_2 + c_3, \quad l_2^2 = c_2 - \frac{1}{2}c_3, \quad l_3^2 = \frac{5}{2}c_1 + c_2 - \frac{1}{4}c_3 \tag{3.3.28}$$

Otherwise, c_1, c_2, c_3 can be indicated by l_1^2, l_2^2, l_3^2

$$c_1 = -\frac{1}{15}l_1^2 - \frac{1}{3}l_2^2 + \frac{2}{5}l_3^2$$

$$c_2 = \frac{1}{3}l_1^2 + \frac{2}{3}l_2^2 \tag{3.3.29}$$

$$c_3 = \frac{2}{3}l_1^2 - \frac{2}{3}l_2^2$$

Through the invariables of $\boldsymbol{\chi}'$ in Eq. (3.3.19), E_{SG}^2 in Eq. (3.3.27) can be expressed as

$$E_{SG}^2 = \frac{2}{3}\boldsymbol{\varepsilon}':\boldsymbol{\varepsilon}' + l_1^2\boldsymbol{\eta}'^{(1)}:\boldsymbol{\eta}'^{(1)} + \left(\frac{4}{3}l_2^2 + \frac{8}{5}l_3^2\right)\boldsymbol{\chi}':\boldsymbol{\chi}' + \left(\frac{4}{3}l_2^2 - \frac{8}{5}l_3^2\right)\boldsymbol{\chi}':\boldsymbol{\chi}'^T \tag{3.3.30}$$

where, l_1^2, l_2^2, l_3^2 are three material constants with a length dimension. If l_2 and l_3 satisfy the relation

$$l_3 = \sqrt{\frac{5}{6}}l_2 \tag{3.3.31}$$

only one invariable $\boldsymbol{\chi}':\boldsymbol{\chi}'$ of $\boldsymbol{\chi}'$ appears at the right-hand side of Eq. (3.3.30) whereas another invariable, does not appear. For the next special material constants,

$$l_1 = 0, \quad l_2 = \frac{1}{2}l, \quad l_3 = \sqrt{\frac{5}{24}}l \tag{3.3.32}$$

then, E_{SG}^2 in Eq. (3.3.30) can be transited into total equivalent strain E_{eq}^2 of CS theory in Eq. (3.2.44). Consequently, CS theory can be regarded as a special case of SG theory with the material constant combination given in Eq. (3.3.32).

Through the thin beam bending test (Stolken and Evans, 1998), thin copper wire torsion test (Fleck et al., 1994), and micron indentation test (Nix, 1989; De Guzman et al., 1993; Stelmashenko et al., 1993; Ma and Clarke, 1995; Poole et al., 1996; McElhaney et al., 1998), Begley and Hutchinson (1997) determined these material constants:

$$l_1 = \frac{1}{16}l \sim \frac{1}{8}l, \quad l_2 = \frac{1}{2}l, \quad l_3 = \sqrt{\frac{5}{24}}l \tag{3.3.33}$$

For example, $l = 4$ µm, $l_1 = 0.25$ µm ~ 0.5 µm, $l_2 = 2$ µm, and $l_3 = 1.8$ µm for copper and $l = 6$ µm, $l_1 = 0.38$ µm ~ 0.75 µm, $l_2 = 3$ µm, and $l_3 = 2.7$ µm for nickel.

The corresponding material constants are given in Eq. (3.3.32) based on rotational gradient theory. According to these material lengths, only a 10%−20% increase in strength is predicted during nano- or microindentation tests, which is a far cry from the increase in strength of 200%−300% observed in the experiments.

Besides considering strain partial tensor $\boldsymbol{\varepsilon}'$ in CS theory, rotational gradient $\boldsymbol{\chi}'$ is considered, but not all of it, because the invariable $\boldsymbol{\chi}':\boldsymbol{\chi}'^{T}$ is ignored. In this section, the invariables $\boldsymbol{\chi}':\boldsymbol{\chi}'$ and $\boldsymbol{\chi}':\boldsymbol{\chi}'^{T}$ are simultaneously considered in SG theory. In addition, $\boldsymbol{\eta}'^{(1)}$ is considered. If the material is incompressible and rotation or the rotational gradient is 0, the total equivalent strains can be given by CS or SG theory, respectively:

$$E_{eq}^{2} = \frac{2}{3}\boldsymbol{\varepsilon}':\boldsymbol{\varepsilon}', \quad E_{SG}^{2} = \frac{2}{3}\boldsymbol{\varepsilon}':\boldsymbol{\varepsilon}' + l_{1}^{2}\boldsymbol{\eta}':\boldsymbol{\eta}', \tag{3.3.34}$$

The elongation gradient of the material must be reflected in E_{SG}^{2}, which is called elongational and rotational gradient theory.

3.3.3 Constitutive Relation of Stretch and Rotation Gradient Strain Gradient Plastic Theory

Stretch and rotation gradient strain gradient plastic theory supposes that strain energy w at a per-unit volume of the body is composed of two parts:

$$w = w_{V}(\varepsilon_{V}) + w(E_{SG}) \tag{3.3.35}$$

where w_{V} is the strain energy resulting in volume change. For an isotropic elastic, ε_{V} is defined in terms of volume deformation; then, we have

$$w_{V} = \frac{1}{2}K\varepsilon_{V}^{2}, \quad \varepsilon_{V} = \operatorname{tr}\boldsymbol{\varepsilon} = \varepsilon_{ii} \tag{3.3.36}$$

where $K = \frac{E}{3(1-2\nu)}$ is the volume modulus, E is the Young's modulus, and ν is Poisson's ratio. Similar to Eq. (3.2.45), total equivalent stress is defined as

$$\Sigma = \frac{\partial w(E_{SG})}{\partial E_{SG}} \tag{3.3.37}$$

As a result of Eq. (3.3.27), E_{SG} depends on $\boldsymbol{\varepsilon}'$, $\boldsymbol{\eta}'^{(1)}$, $\boldsymbol{\eta}'^{(2)}$ and $\boldsymbol{\eta}'^{(3)}$, whereas $\boldsymbol{\eta}'^{(1)}$, $\boldsymbol{\eta}'^{(2)}$ and $\boldsymbol{\eta}'^{(3)}$ are three orthogonal parts decomposed from $\boldsymbol{\eta}'$, as given in Eq. (3.3.11). All of them can only be linearly described by $\boldsymbol{\eta}'$. Regard w in Eq. (3.3.35) as a function of $\boldsymbol{\varepsilon}$ and $\boldsymbol{\eta}'$, for example, $w(\varepsilon, \boldsymbol{\eta}')$; then, the nonlinear elastic relations are given by

$$\boldsymbol{\sigma} = \frac{\partial w(\boldsymbol{\varepsilon}, \boldsymbol{\eta}')}{\partial \boldsymbol{\varepsilon}}, \quad \boldsymbol{\tau}' = \frac{\partial w(\boldsymbol{\varepsilon}, \boldsymbol{\eta}')}{\partial \boldsymbol{\eta}'} \tag{3.3.38}$$

where σ, τ' are work conjunction stresses corresponding to ε, η'. Because ε is symmetric and η' is symmetric to the first and second indications, $\eta'_{ijk} = \eta'_{jik}$, and also satisfies incompressible condition $\eta'_{ikk} = 0$, as given in Eq. (3.3.4), σ should be stipulated as symmetric and τ' is symmetric to the first and second indications, $\tau'_{ijk} = \tau'_{jik}$, and also satisfies incompressible condition $\tau'_{ikk} = 0$. To calculate σ, τ' in Eq. (3.3.38), we need only to calculate dw. Because of Eq. (3.3.35), using of Eqs. (3.3.36) and (3.3.37), we can write

$$dw = dw_V + dw(E_{SG}) = K\varepsilon_V \mathbf{1}: d\boldsymbol{\varepsilon} + \Sigma dE_{SG} \tag{3.3.39}$$

where dE_{SG} can be deduced from Eq. (3.3.27):

$$E_{SG}dE_{SG} = \frac{2}{3}\boldsymbol{\varepsilon}': d\boldsymbol{\varepsilon}' + l_1^2 \boldsymbol{\eta}'^{(1)}:d\boldsymbol{\eta}'^{(1)} + l_2^2 \boldsymbol{\eta}'^{(2)}:d\boldsymbol{\eta}'^{(2)} + l_3^2 \boldsymbol{\eta}'^{(3)}:d\boldsymbol{\eta}'^{(3)} \tag{3.3.40}$$

Owing to the orthogonal condition in Eq. (3.3.12), we obtain $\widetilde{\boldsymbol{\eta}}'$ as d$\boldsymbol{\eta}'$:

$$\boldsymbol{\eta}'^{(i)}:d\boldsymbol{\eta}'^{(j)} = 0 \quad \text{when } i \neq j \tag{3.3.41}$$

Rewriting Eq. (3.3.40) as given subsequently

$$E_{SG}dE_{SG} = \frac{2}{3}\boldsymbol{\varepsilon}': d\boldsymbol{\varepsilon}' + \left[l_1^2 \boldsymbol{\eta}'^{(1)} + l_2^2 \boldsymbol{\eta}'^{(2)} + l_3^2 \boldsymbol{\eta}'^{(3)}\right]:d\boldsymbol{\eta}' \tag{3.3.42}$$

Substituting dE_{SG} of Eq. (3.3.42) into Eq. (3.3.39), obtains

$$dw = \left(K\varepsilon_V \mathbf{1} + \frac{2}{3}\frac{\Sigma}{E_{SG}}\boldsymbol{\varepsilon}'\right): d\boldsymbol{\varepsilon} + \frac{\Sigma}{E_{SG}}\left[l_1^2 \boldsymbol{\eta}'^{(1)} + l_2^2 \boldsymbol{\eta}'^{(2)} + l_3^2 \boldsymbol{\eta}'^{(3)}\right]:d\boldsymbol{\eta}' \tag{3.3.43}$$

Thus, owing to Eqs. (3.3.38) and (3.3.43), we may obtain

$$\boldsymbol{\sigma} = K\varepsilon_V \mathbf{1} + \frac{2}{3}\frac{\Sigma}{E_{SG}}\boldsymbol{\varepsilon}', \quad \text{that is} \quad \boldsymbol{\sigma}' = \frac{2}{3}\frac{\Sigma}{E_{SG}}\boldsymbol{\varepsilon}' \tag{3.3.44}$$

$$\boldsymbol{\tau}' = \sum_{n=1}^{3}\boldsymbol{\tau}'^{(n)}, \quad \boldsymbol{\tau}'^{(n)} = \frac{\Sigma}{E_{SG}}l_n^2 \boldsymbol{\eta}'^{(n)}, \quad n = 1,2,3 \quad \text{(no sum on } n) \tag{3.3.45}$$

As a result of Eqs. (3.3.44) and (3.3.45), and using Eq. (3.3.27), we can prove that

$$\Sigma^2 = \frac{3}{2}\boldsymbol{\sigma}': \boldsymbol{\sigma}' + l_1^{-2}\boldsymbol{\tau}'^{(1)}:\boldsymbol{\tau}'^{(1)} + l_2^{-2}\boldsymbol{\tau}'^{(2)}:\boldsymbol{\tau}'^{(2)} + l_3^{-2}\boldsymbol{\tau}'^{(3)}:\boldsymbol{\tau}'^{(3)} \tag{3.3.46}$$

The Legendre transformation is made for Eq. (3.3.38), which is similar to deducing Eq. (3.2.36) from Eq. (3.2.34); it is obtained by

$$\boldsymbol{\varepsilon} = \frac{\partial\phi(\boldsymbol{\sigma}, \boldsymbol{\tau}')}{\partial\boldsymbol{\sigma}}, \quad \boldsymbol{\eta}' = \frac{\partial\phi(\boldsymbol{\sigma}, \boldsymbol{\tau}')}{\partial\boldsymbol{\tau}'} \tag{3.3.47}$$

where

$$\phi(\boldsymbol{\sigma}, \boldsymbol{\tau}') = \boldsymbol{\sigma}:\boldsymbol{\varepsilon} + \boldsymbol{\tau}':\boldsymbol{\eta}' - w(\boldsymbol{\varepsilon}, \boldsymbol{\eta}') \tag{3.3.48}$$

3.4.1 Experimental Law for Strain Gradient Plasticity Theory

Nix and Gao (1998) clarified the significance of material characteristic length l, which was introduced by Fleck and Hutchinson (1993), by analyzing indentation experimental data; they proposed an experimental law required by the microscale MSG plasticity theory. They started from a Taylor relation given subsequently, which was used to describe the shear strength and dislocation density of materials:

$$\tau = \alpha\mu b\sqrt{\rho_T} = \alpha\mu b\sqrt{\rho_S + \rho_G} \tag{3.4.1}$$

where ρ_T is the total dislocation density, ρ_S is the density of SSD, ρ_G is the density of GND, μ is the shear modulus, b is the Burgers vector, and α is an empirical constant, which is the quantity 1.0 (Nix and Gibeling, 1985). Because the strain gradient is related only to the GND, the effective strain gradient η can be defined as

$$\eta = \rho_G b \tag{3.4.2}$$

According to this expression, the η can be explained as a curvature for the bending problem and twist angle at a unit length for the torsion problem.

For most tough materials, the relation of uniaxial tension stress-strain can be written as a power form:

$$\sigma = \sigma_{\text{ref}}\varepsilon^N \tag{3.4.3}$$

where N $(0 < N < 1)$ is the plastic hardening exponent and σ_{ref} is the referenced stress. For a polycrystalline material, tension flow stress σ is $M = 3.06$ times the shear flow stress (Taylor, 1938). Under uniaxial tension, the strain gradient is 0. Eq. (3.4.3) is reflects hardening resulting from statistically stored dislocation. Thus, ρ_S can be determined by the stress-strain relation of uniaxial tension. Eq. (3.4.1) provides the hardening law of strain gradient plasticity:

$$\sigma = \sigma_{\text{ref}}\sqrt{\varepsilon^{2N} + l\eta} \tag{3.4.4}$$

and

$$l = M^2\alpha^2 b\left(\frac{\mu}{\sigma_{\text{ref}}}\right)^2 \tag{3.4.5}$$

where l is an intrinsic material length for the strain gradient plasticity theory. For a tough metal, l is at the micrometer scale. This is the same quantity compared with the torsion test of thin copper wire by Fleck et al. (1994) and the bending test of thin nickel beam by Stolken and Evans (1998). Nix and Gao (1998) used $M = 3$ in Eq. (3.4.5) according to the Mises criterion for an isotropic solid.

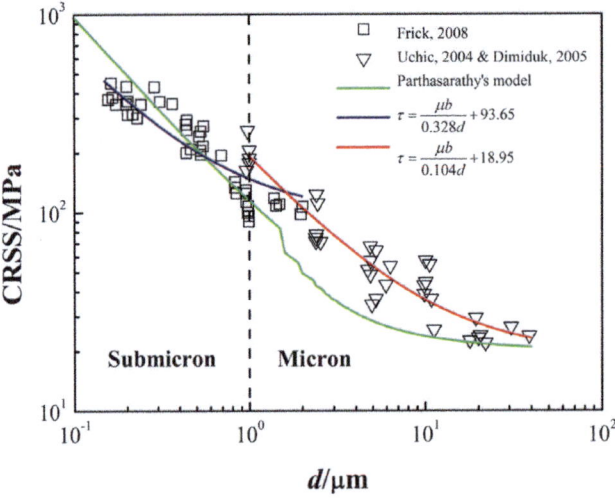

Figure 5.2
Results of fitted experimental data of submicron and micron pillars with Eq. (5.2.7) show different ratios of effective source length to diameter at submicron $\bar{\lambda}/d \approx 0.328$ and micron pillars $\bar{\lambda}/d \approx 0.104$. *CRSS,* critical shear stress. *Reprinted from Lin, P., Liu, Z.L., Cui, Y.N., Zhuang, Z., 2015. A stochastic crystal plasticity model with size-dependent and intermittent strain bursts characteristics at micron scale. International Journal of Solids and Structures 69—70, 267—276. Copyright 2015, with permission from Elsevier.*

Table 5.1: Source Number *n* for Different Pillar Diameter in Eq. (5.2.9)

$d/\mu m$	n
<1.75	1
1.75—2.14	2
2.14—2.47	3

As mentioned earlier, to derive Eq. (5.2.8), it is assumed that each pinning point has one dislocation arm to free surface and the dislocations do not interact with each other in the statistical model. This is true for submicron pillars, in which there are only a few pins, as shown in Table 5.1. However, in micron pillars, because of forest dislocations, there are many pins and the dislocation arms will interact with each other, forming a more complex microstructure and reducing the effective source length (Fig. 5.3).

In Fig. 5.3, the black line represents an SAD source on one slip plane whereas the dots represent forest dislocations on the other slip planes. For submicron pillars, because there are only a few forest dislocations, the SAD source is rarely influenced by other dislocations during the operation. For micron pillars, because the dislocation arm will interact with a large number of forest dislocations, more complex dislocation networks will be formed. Thus, $\bar{\lambda}$ is not actually the effective source length but a characteristic length of microstructures. The average space between dislocations can be considered the

Strain gradient law (3.4.4) was used by Nix and Gao (1998) to calculate microindentation hardness. They demonstrated that the relational expression for indentation hardness H could be deduced from Eq. (3.4.4), which is given by

$$\frac{H}{H_0} = \sqrt{1 + \frac{h^*}{h}} \tag{3.4.6}$$

where H_0 is a hardness without considering the strain gradient effect and h is the indentation depth. Therefore, there is

$$h^* = \frac{81}{2} ba^2 \tan^2 \theta \left(\frac{\mu}{H_0}\right)^2 \tag{3.4.7}$$

where θ is an angle between the cone-shaped surface of the indenter head and surface of body under pressure. The results given by Eq. (3.4.6) agree well with the experimental data obtained by McElhaney et al. (1998), who carried out microindentation experiments using single crystals and machining polycrystalline copper along the (111) direction. The results also agree well with experimental data obtained by Ma and Clarke (1995), who conducted microindentation experiments using single-crystal silver. The calculation results of Eq. (3.4.6) and the experimental data obtained by McElhaney et al. (1998) for the relation between hardness and the indentation depth are plotted in Fig. 3.2. In contrast, the results given by classical plasticity theory are a horizontal line, also shown in Fig. 3.2, which proves that there is no relation between hardness and the indentation depth at a macroscale.

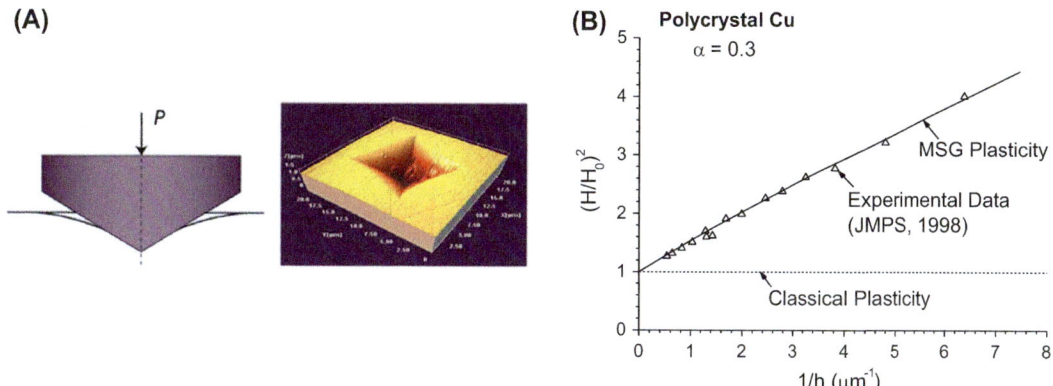

Figure 3.2

(A) Microindentation tests extracting material mechanical properties; (B) Calculation results from Eq. (3.4.6) and experimental data obtained by McElhaney et al. (1998) for the relation between hardness and indentation depth. *MSG,* mechanism-based strain gradient.

3.4.2 Motivation for Microscale Mechanism-Based Strain Gradient Plasticity Theory

Although Fleck and Hutchinson (1993, 1997) proposed the strain gradient plasticity theory motivated from the dislocation theory, they used high-order effective stress and strain to replace effective stress and strain in a conventional plastic constitutive relation, as described in Section 3.2.6. However, high-order strain involves a strain gradient and one or a few material lengths need to be determined by the experiments. In other words, the theory proposed by Fleck-Hutchinson is mainly through the measured uniaxial stress-strain relations at a macroscale whereas some tests, such as the indentation test, microtorsion test, and bending test at a microscale, do not work with developing a theory process, except for determining material lengths. However, good results are given by Eq. (3.4.4), used in microindentation tests for various materials; thus, there is a linear relation between H^2 and $1/h$, which indicates that the materials have essential and intrinsic deformation characteristics. This provides a strong motivation to set up another theory based on a fundamental assumption in Eq. (3.4.4). Gao et al. (1999) and Huang et al. (1999) proposed a multiscale and multilevel theoretical framework to determine the connection between plasticity theory and dislocation theory.

The strain field in each microscale cell varies owing to a linear rule. Each internal point in the cell acts as a microcell element, as shown in Fig. 3.3. The interactions of dislocations in the microcell element approximatively obey the Taylor law. Thus, the strain gradient plastic law in Eq. (3.4.4) can be used. In the microcell element, the term η is a measure of

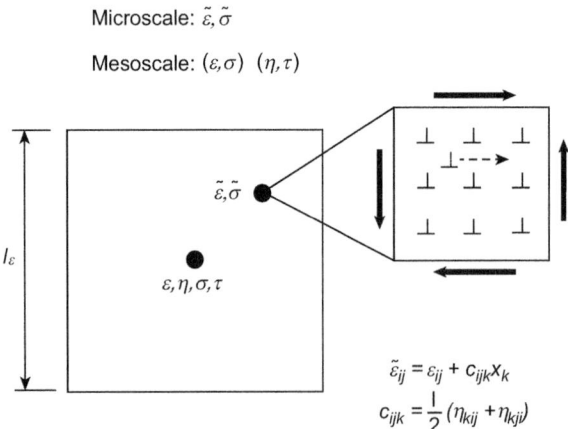

Figure 3.3
Multiscale computation framework in mechanism-based strain gradient theory.

the density of GND, whereas the accumulated GND results in increased flow stress strictly following the Taylor model. It is supposed that plastic flow at the microscale occurs owing to the glide of SSD under a GND background. It is supposed that microscale plastic deformation follows the Taylor work-hardening relation and conventional plastic associated rule. The concept of GND associated with a strain gradient is established on the level of the microscale cell element. For the plasticity theory set up at a microscale level, high-order stress appears as a thermodynamics conjunction quantity, which guarantees this theory satisfies the continuum Clausius-Duhem thermodynamic limitation. This kind of multilevel model provides a method for establishing microscale constitutive theory. Through equalization dealt with by the microscale plastic law in a representative cell element, the microscale constitutive relation is obtained. Although this kind of theory just satisfies the mathematic framework of phenomenological theory established by Fleck and Hutchinson (1997), its start point is based on the microscale mechanism rather than existing phenomenological theories.

3.4.3 Microscale Computation Framework

The multiscale computation framework in MSG theory is illustrated in Fig. 3.3. At the microscale cell element, the size of the element is smaller than the variational area of the strain field. The interaction of dislocations can be described as the glide of SSD under the GND background. According to the strain gradient law in Eq. (3.4.4), the flow stress at a microscale is affected by the GND. At the microscale cell element, stress and strain are defined conventionally, written as $\tilde{\sigma}$ and $\tilde{\varepsilon}$, respectively. In the higher-level analysis, which is called microscale analysis, for the concept related to strain gradient plasticity, for example for high-order stress and the strain gradient, the model is guaranteed to satisfy the requirements of thermodynamic limitations.

Why one requires different levels framework in strain gradient plasticity theory? One reason is to establish the theory related to the situation predicted by the Taylor model at the microscale. According to the Taylor model, flow stress is defined as the critical force needed to make dislocations slip between barriers. Other reasons are to establish the relationship between the GND and strain gradient at the microscale and to construct the constitutive framework to satisfy thermodynamic limitations. In the other words, the multiscale framework can form the GND to become the necessary part of the constitutive model. The variables at a microscale include stress σ and strain ε, as well as a high-order stress τ and strain gradient η.

The high-order framework established by Fleck and Hutchinson (1997) is adopted when the form of MSG theory is set up at a microscale, based on early-stage work by Toupin (1962), Koiter (1964), and Mindlin (1964, 1965). In this framework, the generalized strain tensor includes the symmetry strain tensor ε in Eq. (3.2.16) and displacement second-order

gradient η in Eq. (3.3.1). For simplicity, elastic deformation and compressible material are ignored in the following analyses. According to the theory given by Fleck and Hutchinson (1997), the incompressibility can be expressed by $\varepsilon_{ii} = 0$ and $\eta_{ikk} = 0$. Thus, the arbitrary variation of displacement u results in the variation in strain energy per unit volume, which is given by

$$\delta w = \sigma'_{ij}\delta\varepsilon_{ij} + \tau'_{ijk}\delta\eta_{ijk} \tag{3.4.8}$$

To set up a relation between strain gradient plasticity at the microscale and the Taylor hardening law at the microscale, the following basic assumptions are adopted in the MSG theory framework:

1. It is supposed that flow stress is controlled by dislocation motion at the microscale and obeys the Taylor hardening relation given by the strain gradient law in Eq. (3.4.4):

$$\tilde{\sigma} = \sigma_y\sqrt{f^2(\tilde{\varepsilon}) + l\eta} \tag{3.4.9}$$

2. Strain gradient plasticity describes dislocation motion at the microscale, which can be deduced from the plastic law of dislocation at the microscale. Selection of the microscale cell element is as small as possible to guarantee that the strain has an approximative linear variation and is also large enough to guarantee that the Taylor model can be used. The high-order strain gradient is ignored at the microscale. The connection between the microscale and microscale is equivalent to plastic work:

$$\int_{V_{\text{cell}}} \tilde{\sigma}'_{ij}\delta\tilde{\varepsilon}'_{ij}\mathrm{d}V = \left(\sigma'_{ij}\delta\varepsilon_{ij} + \tau'_{ijk}\delta\eta_{ijk}\right)V_{\text{cell}} \tag{3.4.10}$$

3. It is supposed that the fundamental structure of conventional plasticity is established at the microscale cell element. For this purpose, it is assumed that plastic flow is the glide of the SSD under the GND background. At the microscale, the J_2 deformation theory can be expressed by

$$\tilde{\sigma}'_{ij} = \left(2\tilde{\varepsilon}_{ij}/3\varepsilon_{\text{eq}}\right)\tilde{\sigma}_{\text{eq}} \tag{3.4.11}$$

where ε_{eq} is the equivalent strain and $\tilde{\sigma}_{\text{eq}}$ is the equivalent stress at the microscale. They are given as

$$\tilde{\sigma}_{\text{eq}} = \sqrt{\frac{3}{2}\tilde{\sigma}'_{ij}\tilde{\sigma}'_{ij}} \quad \tilde{\varepsilon}_{\text{eq}} = \sqrt{\frac{2}{3}\tilde{\varepsilon}_{ij}\tilde{\varepsilon}_{ij}} \tag{3.4.12}$$

The yield condition at a microscale is

$$\tilde{\sigma}_{\text{eq}} = \tilde{\sigma} \tag{3.4.13}$$

where $\tilde{\sigma}$ is given in Eq. (3.4.9).

3.4.4 Dislocation Model

Before deducing the constitutive law of MSG theory based on the assumptions given in Section 3.4.3, the relation between η in Eq. (3.4.2) and strain gradient $\boldsymbol{\eta}$ must be established. Gao et al. (1999) used the invariant of $\boldsymbol{\eta}$ to define the equivalent strain gradient and used η measure GND density:

$$\eta = \sqrt{c_1' \eta_{iik}\eta_{jjk} + c_2' \eta_{ijk}\eta_{ijk} + c_3' \eta_{ijk}\eta_{kji}} \tag{3.4.14}$$

Because experimental data on the strain gradient effect are lacking, they used several different dislocation models to determine the constants c_1', c_2' and c_3' instead of using test data to fit them. The three models include plane strain bending, pure torsion, and two-dimensional axial symmetry void growth.

There are a few corresponding images of GND for a given continuum strain field. Thus, these constants may not be uniquely determined. If the orientations of slip systems are different, the corresponding images of GND are also different. Gao et al. (1999) did not consider these complexities. For a given deformation field, they considered a few kinds of slip systems to provide the minimum density of GND as the most effective dislocation image. Three dislocation models of plane strain bending, pure torsion, and two-dimensional axial symmetry void growth provide three equations corresponding to c_1', c_2' and c_3':

$$c_1' = 0, \quad c_2' = \frac{1}{4}, \quad c_3' = 0 \tag{3.4.15}$$

Consequently, the equivalent strain gradient corresponding to the most effective dislocation image is obtained by

$$\eta = \sqrt{\frac{1}{4}\eta_{ijk}\eta_{ijk}} \tag{3.4.16}$$

The form in Eq. (3.4.16) is similar to $\varepsilon = \sqrt{\frac{2}{3}\varepsilon_{ij}\varepsilon_{ij}}$ in conventional plastic theory. According to the relation between c_i' and l_i $(i = 1, 2, 3)$ in Eq. (3.3.28), $l_1^2 = l_2^2 = l_3^2 = 1/4$. Thus, Eq. (3.4.16) can be written as

$$\eta = \sqrt{\frac{1}{4}\left(\eta_{ijk}^{(1)}\eta_{ijk}^{(1)} + \eta_{ijk}^{(2)}\eta_{ijk}^{(2)} + \eta_{ijk}^{(3)}\eta_{ijk}^{(3)}\right)} \tag{3.4.17}$$

3.4.5 Constitutive Equation of Mechanism-Based Strain Gradient Plasticity Theory

The new strain gradient plasticity theory framework can be established in this section according to assumptions adopted from the MSG theory framework in Section 3.4.3. There is only the MSG deformation theory.

Considering a cell element at the microscale, a length at each side is l_ε, as shown in Fig. 3.3. It is assumed that the displacement variation in this cell element is

$$\tilde{u}_k = \varepsilon_{ik}x_i + \frac{1}{2}\eta_{ijk}x_ix_j + O(x^3) \qquad (3.4.18)$$

where x_i indicates the local coordinate as a cell center to coordinate the original point and η_{ijk} is the second-order gradient of the displacement field. When the cell element is small enough, the high-order gradient can be ignored. Thus, variation in the strain field can be linearly expressed by

$$\tilde{\varepsilon}_{ij} = \varepsilon_{ij} + \frac{1}{2}\left(\eta_{kij} + \eta_{kji}\right)x_k \qquad (3.4.19)$$

In this way, strain $\tilde{\varepsilon}_{ij}$ at the microscale has to be linked to strain ε_{ij} and strain gradient η_{ijk} at the microscale. Cell size l_ε at the microscale must be much smaller than the intrinsic material size l of strain gradient plasticity in Eq. (3.4.5).

Substituting the constitutive relation (3.4.11) and kinematic assumption at the microscale,

$$\delta\tilde{\varepsilon}_{ij} = \delta\varepsilon_{ij} + \frac{1}{2}\left(\delta\eta_{kij} + \delta\eta_{kji}\right)x_k \qquad (3.4.20)$$

into plastic work Eq. (3.4.10), and taking the coefficients of $\delta\varepsilon_{ij}$ and $\delta\eta_{ijk}$ to be respectively equivalent to both sides of the equation, the deformation theory constitutive equation of the MSG can be obtained as

$$\sigma'_{ij} = \frac{1}{V_{\text{cell}}}\int_{V_{\text{cell}}} \tilde{\sigma}'_{ij}dV \qquad (3.4.21)$$

From the coefficients of $\delta\eta_{ijk}$, which are respectively equivalent to both sides of equation,

$$\tau'_{ijk} = \frac{1}{V_{\text{cell}}}\text{Dev}\left[\frac{1}{2}\int_{V_{\text{cell}}} \left(\tilde{\sigma}'_{jk}x_i + \tilde{\sigma}'_{ik}x_j\right)dV\right] \qquad (3.4.22)$$

where $\text{Dev}[\cdots]$ indicates the deviating part of $[\cdots]$. This definition is similar to a form of $\boldsymbol{\eta} = \boldsymbol{\eta}^{\text{Hy}} + \boldsymbol{\eta}'$ in Eq. (3.3.5).

Because an original point of coordinate x_i is located on the center of the square cell, the integral in Eq. (3.4.22) can be carried out according to the rule

$$\frac{1}{V_{\text{cell}}}\int_{V_{\text{cell}}} dV = 1, \quad \int_{V_{\text{cell}}} x_k dV = 0, \quad \frac{1}{V_{\text{cell}}}\int_{V_{\text{cell}}} x_k x_m dV = \frac{1}{12}l_\varepsilon^2 \delta_{km} \qquad (3.4.23)$$

The lowest-order term of l_ε is retained and then the constitutive equation at a microscale is obtained as

$$\sigma'_{ij} = \frac{2}{3}\frac{\sigma}{\varepsilon}\varepsilon_{ij} \qquad (3.4.24)$$

$$\tau'_{ijk} = l_\varepsilon^2 \left[\frac{\sigma}{\varepsilon}(\Lambda_{ijk} - \Pi_{ijk}) + \frac{1}{\sigma}N\sigma_{\text{ref}}^2\varepsilon^{2N-1}\Pi_{ijk} \right] \tag{3.4.25}$$

where

$$\sigma = \sigma_{\text{ref}}\sqrt{\varepsilon^{2N} + l\eta} \tag{3.4.26}$$

$$\Lambda_{ijk} = \frac{1}{72}\left[2\eta_{ijk} + \eta_{kji} + \eta_{kij} - \frac{1}{4}(\delta_{ik}\eta_{ppj} + \delta_{jk}\eta_{ppi}) \right] \tag{3.4.27}$$

$$\Pi_{ijk} = \frac{1}{54}\frac{\varepsilon_{mn}}{\varepsilon^2}\left[\varepsilon_{ik}\eta_{jmn} + \varepsilon_{jk}\eta_{imn} - \frac{1}{4}(\delta_{ik}\varepsilon_{jp} + \delta_{jk}\varepsilon_{ip})\eta_{pmn} \right] \tag{3.4.28}$$

$O\left(l_\varepsilon^2\right)$ is omitted in σ'_{ij} and $O\left(l_\varepsilon^4\right)$ is omitted in τ'_{ijk}. In these equations, ε indicates the equivalent strain of conventional plasticity theory and η is referenced in Eq. (3.4.16):

$$\varepsilon = \sqrt{\frac{2}{3}\varepsilon_{ij}\varepsilon_{ij}}, \quad \eta = \sqrt{\frac{1}{4}\eta_{ijk}\eta_{ijk}} \tag{3.4.29}$$

The kind of multiscale method used here can reflect the microscale constitutive law of MSG plasticity theory deduced by the dislocation mechanism-based microscale plastic law in the Taylor model. Therefore, there is no need to suppose that there is a strain energy density function. In fact, the strain energy density function does not exist in MSG plasticity theory, which is discussed next.

One important question is whether strain energy density function $w(\boldsymbol{\varepsilon}, \boldsymbol{\eta})$ exists in MSG plasticity gradient theory. Thus, the constitutive equation may be expressed by $w(\boldsymbol{\varepsilon}, \boldsymbol{\eta})$ (Eq. 3.3.8):

$$\boldsymbol{\sigma}' = \frac{\partial w(\boldsymbol{\varepsilon}, \boldsymbol{\eta})}{\partial \boldsymbol{\varepsilon}}, \quad \boldsymbol{\tau} = \frac{\partial w(\boldsymbol{\varepsilon}, \boldsymbol{\eta})}{\partial \boldsymbol{\eta}} \tag{3.4.30}$$

Obviously, Eq. (3.4.30) is impossible because MSG plastic constitutive Eqs. (3.4.24) and (3.4.25) do not satisfy reciprocal relations:

$$\frac{\partial \sigma'_{ij}}{\partial \eta_{kmn}} \neq \frac{\partial \tau'_{kmn}}{\partial \varepsilon_{ij}} \tag{3.4.31}$$

whereas the existing reciprocal relations of Eq. (3.4.31) are necessary conditions for holding the strain energy density function. This raises the question of whether the fracture mechanics J integral could be used in MSG plasticity theory. However, the nonexistence of the strain energy density function is closely related to the GND at the microscale.

3.4.6 Size of Cell Element at the Microscale

It is hard to understand a size l_ε of a cell element at the microscale in MSG plasticity theory because there is no similar concept in conventional plasticity theory. Gao et al. (1999)

pointed out that it is a fundamental parameter related to plastic deformation behavior but not introduced for convenience. It is an intrinsic material size used to describe strain gradient and flow stress accurately and simultaneously. As discussed in Section 3.4.5, l_ε is a resolution ratio parameter used to control precision when calculating the strain gradient in each cell element. Therefore, it must be small enough to guarantee the precision of the strain gradient, which is given by

$$l_\varepsilon \ll l = M^2 \alpha^2 b \left(\frac{\mu}{\sigma_{ref}}\right)^2 \tag{3.4.32}$$

where l is an intrinsic material length in strain gradient theory, as given in Eq. (3.4.5). On the other hand, it is assumed that the Taylor relation is held in the cell element. The cell element must be large enough so that sufficient numbers of dislocations are included. Thus, it needs

$$l_\varepsilon \gg l_{yield} \sim \frac{\mu}{\sigma_y} b \tag{3.4.33}$$

where l_{yield} is an average interval of SSD when the material yields, which is about $(\mu/\sigma_y)b$. The yielding stress σ_y corresponds to 0.2% material yielding strain. Comparing these two requirements of mutual conditions, Eqs. (3.4.32) and (3.4.33), a suitable choice of l_ε was obtained by Gao et al. (1999):

$$l_\varepsilon = \beta \frac{\mu}{\sigma_y} b \tag{3.4.34}$$

where β is a coefficient determined by the experiment.

The following representative mechanical characteristic of toughness in a material could be used to estimate the order of β:

$$N = 0.2, \quad \varepsilon_y = \sigma_y/E = 0.2\%, \quad \sigma_{ref}\Big/\sigma_y = \varepsilon_y^{-N},$$
$$\eta/E = 3/8, \quad \alpha = 1, \quad M = 3.06 \tag{3.4.35}$$

where ε_y is the yielding stress and E is the Young's modulus. The intrinsic material length l can be estimated by Eqs. (3.4.5) and (3.4.32):

$$l = 27,400b \tag{3.4.36}$$

For copper, $b = 0.255$ nm; the length l is obtained as $l \cong 7$ μm, which is the same order of material length in the strain gradient phenomenological theory given by Fleck and Hutchinson (1997). An average interval of SSD is given when the material yields:

$$l_{yield} = 188b \tag{3.4.37}$$

This corresponds to the order $10^{14}\,\mathrm{m}^{-2}$ of dislocation density. For copper, $l_{\mathrm{yield}} = 50\,\mathrm{nm}$ in Eq. (3.4.37), which is equal to an interval length of double-slip surfaces. Thus, the order of β should be 10, i.e., $\beta \sim 10$.

Huang et al. (1999) studied a few classical examples of MSG plasticity theory, including thin beam bending, thin wire torsion, microvoid growth, void instability, and bimaterial shear. These results indicated that the macroscopic physical quantity is insensitive to coefficient β but the local deformation field is strongly affected by β.

3.4.7 Mechanism-Based Strain Gradient Plasticity Predicts Stress Singularity at Crack Tip

MSG and SG theories were generalized to compressible materials based on a new decomposition of strain gradient tensors, because the original MSG and SG theories were suitable only for incompressible materials. The results for microindentation and fracture showed that the influence of compressibility cannot be ignored. The main influence regarding the fracture of MSG plasticity is that the crack tip stress singularity does not exceed the square root singularity.

Finite deformation theories of strain gradient were developed for both MSG and SG theory. The constitutive relations were established in the initial configuration by two different methods. The kinematics and tensors transfer relations were provided between the initial and current configurations. The equilibrium equation and boundary conditions were developed in both configurations. Microindentation and fracture were studied by FEM. The microindentation results showed that both theories for finite and infinitesimal deformation agreed well with experimental data by suitably chosen material lengths. The results for fracture showed that the difference between finite and infinitesimal deformation existed only very close to the crack tip.

The fracture studies focused on the increase in stress level around the crack tip to explain the observed cleavage fracture in ductile materials. The size effect of the fracture and cleavage fracture mechanism in the toughness of the material were successfully explained by the MSG theory. In 1994, a sapphire-niobium bimaterial fracture test demonstrated an edge level with an atomic order at the crack tip even though it was subjected to a large plastic deformation, as shown in Fig. 3.4A. Because the stress level predicted by classical plasticity theory does not separate atoms at the crack tip, this phenomenon is called imaginary by international mechanics societies. Using MSG theory, the singularity curve was solved by Jiang et al. (2001) at the log-log coordinate. This curve can be used to illustrate equivalent stress along the crack extension line corresponding to the crack tip distance, as shown in Fig. 3.4B. At a distance from the crack tip less than $0.3l$, results calculated by MSG plasticity also predict that the crack tip stress singularity is not only

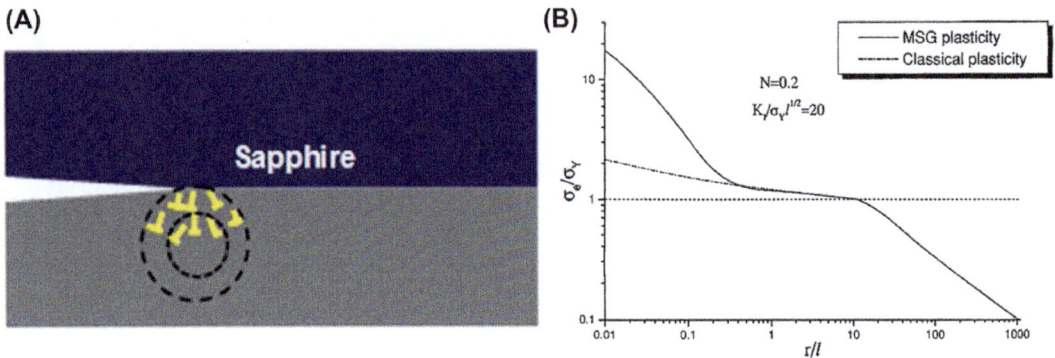

Figure 3.4

(A) Sapphire-niobium bimaterial crack tip field; (B) Mode I crack tip stress singularity curve calculated by strain gradient plasticity theory (Jiang et al., 2001). *MSG, mechanism-based strain gradient.* *(B) Reprinted from Jiang, H., Huang, Y., Zhuang, Z., Hwang, K.C., 2001. Fracture in mechanism-based strain gradient plasticity. Journal of the Mechanics and Physics of Solids 49 (5), 979—993. Copyright 2001, with permission from Elsevier.*

larger than that in the HRR field but also equals or exceeds the square root singularity of the stress intensity factor (K). Here, l is the material characteristic length in the constitutive model, which is about 0.4 μm for copper. The equivalent stress is more than twice it is for conventional theory. The mechanism of cleavage fracture in ductile materials was successfully explained by this solution. It proves that the strain gradient theory has bridged the dislocation model and continuum plasticity theory, which may provide the stress to separate atoms sufficiently during cleavage fracture. Consequently, the imaginary phenomenon has been successfully explained internationally. Moreover, the crack tip stress singularity is independent of the plastic work-hardening exponent. In conjunction with the other model, the results provide a multiscale view of cleavage fracture in ductile materials.

3.5 Summary

In this chapter, three strain gradient plasticity theories were described: CS in Section 3.2, SG theory in Section 3.3, and MSG deformation theory in Section 3.4. Fleck and Hutchinson (1993, 1997) proposed that the CS and SG strain gradient plasticity theories were motivated by the dislocation theory. They used high-order effective stress and strain to replace effective stress and strain in a conventional plastic constitutive relation. CS theory can be regarded as a special case of SG theory with the material constant combination given in Eq. (3.3.32). However, high-order strain involves a strain gradient and one or a few material lengths need to be determined by the experiments. Good results given by Eq. (3.4.4) are used in microindentation tests for various materials; thus, there is a linear relation between H^2 and $1/h$ indicating that the materials have essential and

intrinsic deformation characteristics. This provides a strong motivation to set up another theory based on a fundamental assumption in Eq. (3.4.4). Gao et al. (1999) and Huang et al. (1999) proposed a multiscale and multilevel theoretical framework to realize the connection between plasticity theory and dislocation theory. Nix and Gao (1998) clarified the significance of material characteristic length l introduced by Fleck and Hutchinson (1993) by analyzing the indentation experimental data and proposed the experimental law required by the microscale MSG plasticity theory. According to the Taylor relation, they described the shear strength and dislocation density of materials in Eq. (3.4.1), named the MSG deformation theory.

The limitation of the strain gradient plasticity theory is that it involves only a material characteristic size from 10 μm to a few hundred microns. In the torsion test of thin copper wire, the torsion strength increases three times when the diameter of the copper wire is reduced from d = 170 μm to d = 12 μm, as shown in Fig. 3.1A. In contrast, there is no size effect for stress-strain curves under uniaxial tension without a strain gradient even though the diameters vary from d = 170 μm to d = 12 μm, as shown in Fig. 3.1B. The reason originates from the Taylor assumption for the strain gradient only related to the GND. Material intrinsic size l is a few microns, which cannot describe a problem with a material intrinsic size less than a micrometer at a submicron scale. The question is, what is the sample size if it has a size effect under pure tension or compression? To answer this question, the discrete dislocation mechanism-based crystal plasticity theory and simulation method were developed and will be given in the following chapters.

Dislocation-Based Single-Crystal Plasticity Model

Chapter Outline

Dislocation Mechanism-Based Crystal Plasticity. https://doi.org/10.1016/B978-0-12-814591-3.00004-2

4.1 Introduction

Macroscopic continuum plasticity theories are introduced in Chapter 2, in which crystallographic information such as the anisotropy of crystal lattices and their underlying dislocation microstructure are not considered. The aim of this chapter is to present single-crystal constitutive formulations that characterize the micromechanics of crystals accurately.

In this chapter, conventional single-crystal plasticity theory and recent progress are introduced. First, the conventional constitutive model for single crystals, including the kinematics of single-crystal plasticity, dislocation density evolution, and stress updates, is described in detail in Section 4.2. Then, a new dislocation-based crystal plasticity model is developed via the continuum description of the collective behavior of dislocations in Section 4.3. Compared with conventional crystal plasticity models, a microforce balance equation is involved by combining the equation of dislocations motion and Orowan's law. Then, as an application of the dislocation-based crystal plasticity model, the size effect and Bauschinger effect in thin films with a passivation layer during loading and unloading are numerically studied in Section 4.4. Finally, summaries are given in Section 4.5.

4.2 General Constitutive Model for Single Crystals
4.2.1 Basic Kinematics of Crystal Plasticity

The details of a kinematic description of the deformation of single crystals are provided in this section. As indicated in Fig. 4.1, the structure of a crystalline solid at the micron scale may be viewed as an aggregate of volume elements, or lattice blocks, which are undeformed parts of the lattice, potentially delineated by dislocations. For crystalline solids subjected to a continuous deformation gradient $\mathbf{F}$, two basic mechanisms of deformation are assumed to take place concurrently: slip between lattice blocks, followed by elastic lattice distortions; and lattice rotations that also arise under a general deformation gradient $\mathbf{F}$ and account for observable phenomena such as the geometric softening of crystals (Asaro, 1979). Decomposition of gradient $\mathbf{F}$ into these deformation and rotation mechanisms would thus facilitate the constitutive description of single crystals.

As illustrated in Fig. 4.1, the kinematics of single-crystal plasticity are described following the works of Hill and Rice (1972) and Asaro (1983). They are based on the Kröner-Lee decomposition (Kroner, 1960; Lee, 1969) of deformation gradient tensor $\mathbf{F}$ as

$$\mathbf{F} = \mathbf{F}^e \cdot \mathbf{F}^p$$
$$det\mathbf{F}^p = 1 \tag{4.2.1}$$
$$J = \det\mathbf{F} = det\mathbf{F}^e$$

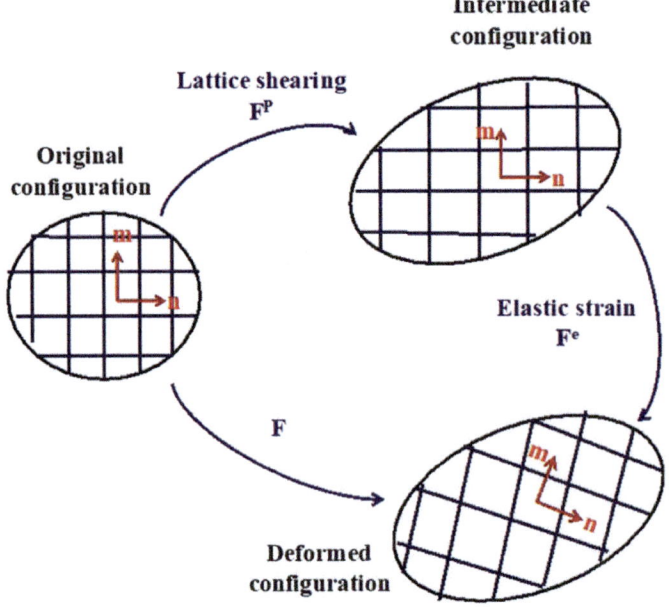

Figure 4.1
Decomposition of deformation in a crystal into elastic and plastic parts.

where $\mathbf{F}^e$ represents small elastic stretches and rigid-body rotations of the lattice, $\mathbf{F}^p$ involves plastic deformation owing to dislocations motion, and J is a ratio of the deformed volume to the undeformed volume under the assumption that plastic deformation does not change the volume. Based on Eq. (4.2.1), the velocity gradient tensor $\mathbf{L}$ is decomposed as

$$
\begin{aligned}
\mathbf{L} &= \dot{\mathbf{F}} \cdot \mathbf{F}^{-1} = \mathbf{L}^e + \mathbf{L}^p \\
\mathbf{L}^e &= \dot{\mathbf{F}}^e \cdot \mathbf{F}^{e-1} \\
\mathbf{L}^p &= \mathbf{F}^e \cdot \dot{\mathbf{F}}^p \cdot \mathbf{F}^{p-1} \cdot \mathbf{F}^{e-1}
\end{aligned}
\tag{4.2.2}
$$

Plastic deformation can be defined as the superposition of all crystallographic slip rates $\dot{\gamma}^{(\alpha)}$ as (Rice, 1971)

$$
\dot{\mathbf{F}}^p \cdot \mathbf{F}^{p-1} = \sum_{\alpha} \dot{\gamma}^{(\alpha)} \mathbf{s}^{(\alpha)} \otimes \mathbf{m}^{(\alpha)}
\tag{4.2.3}
$$

where $\mathbf{s}^{(\alpha)}$ and $\mathbf{m}^{(\alpha)}$ represent the slip direction and slip plane, respectively, normal in slip system α attached to the lattice space in the intermediate configuration. When the single crystal deforms, the lattice stretches and rotates. The slip systems are updated by Nanson's formula as

$$
\begin{aligned}
\widetilde{\mathbf{s}}^{(\alpha)} &= \mathbf{F}^e \cdot \mathbf{s}^{(\alpha)} \\
\widetilde{\mathbf{m}}^{(\alpha)} &= \det \mathbf{F}^e \cdot \mathbf{m}^{(\alpha)} \cdot \mathbf{F}^{e-1}
\end{aligned}
\tag{4.2.4}
$$

where $\widetilde{\mathbf{s}}^{(\alpha)}$ and $\widetilde{\mathbf{m}}^{(\alpha)}$ are in the deformed configuration. Usually, the elastic deformation is small and approximation det $\mathbf{F}^e = 1$ is made in Eq. (4.2.4). Then, the following formulas are used:

$$
\begin{aligned}
\widetilde{\mathbf{s}}^{(\alpha)} &= \mathbf{F}^e \cdot \mathbf{s}^{(\alpha)} \\
\widetilde{\mathbf{m}}^{(\alpha)} &= \mathbf{m}^{(\alpha)} \cdot \mathbf{F}^{e-1}
\end{aligned}
\tag{4.2.5}
$$

and the rate form can be written as

$$
\begin{aligned}
\dot{\widetilde{\mathbf{s}}}^{(\alpha)} &= \mathbf{L}^e \cdot \widetilde{\mathbf{s}}^{(\alpha)} \\
\dot{\widetilde{\mathbf{m}}}^{(\alpha)} &= -\widetilde{\mathbf{m}}^{(\alpha)} \cdot \mathbf{L}^e
\end{aligned}
\tag{4.2.6}
$$

Combining Eqs. (4.2.2), (4.2.3), and (4.2.5), the plastic velocity gradient tensor $\mathbf{L}^p$ is rewritten as

$$
\mathbf{L}^p = \sum_{\alpha} \dot{\gamma}^{(\alpha)} \widetilde{\mathbf{s}}^{(\alpha)} \otimes \widetilde{\mathbf{m}}^{(\alpha)}
\tag{4.2.7}
$$

The rate of deformation tensor $\mathbf{D}$ and spin tensor $\mathbf{W}$ are defined as the symmetric and antisymmetric parts of $\mathbf{L}$, respectively. They can also be decomposed into elastic and plastic parts as

$$
\begin{aligned}
\mathbf{D} &= \frac{1}{2}\left(\mathbf{L} + \mathbf{L}^T\right), \quad \mathbf{W} = \frac{1}{2}\left(\mathbf{L} - \mathbf{L}^T\right) \\
\mathbf{D}^e &= \frac{1}{2}\left(\mathbf{L}^e + \mathbf{L}^{eT}\right), \quad \mathbf{W}^e = \frac{1}{2}\left(\mathbf{L}^e - \mathbf{L}^{eT}\right) \\
\mathbf{D}^p &= \frac{1}{2}\left(\mathbf{L}^p + \mathbf{L}^{pT}\right), \quad \mathbf{W}^p = \frac{1}{2}\left(\mathbf{L}^p - \mathbf{L}^{pT}\right)
\end{aligned}
\tag{4.2.8}
$$

If, as shown in Fig. 4.1, slip occurs in multiple planes, small gaps, overlaps, or displacement discontinuities will arise between lattice blocks, which will result in a noncompact intermediate configuration (Kroner, 1981): that is, one that is not simply connected, so that $\mathbf{F}^p$ cannot be a gradient of a plastic displacement field. Deformation map $\mathbf{F}^p$ is thus generally incompatible with the existence of a unique displacement field (Kroner, 1981). $\mathbf{F}^e$ maps from the noncompact intermediate configuration to the current configuration and consequently is also incompatible. As with any continuum, however, it is necessary for the compact reference configuration to be mapped by a compatible deformation $\mathbf{F}$ to a compact current configuration: that is, one without holes, overlaps, or discontinuities. Hence, $\mathbf{F}^p$ and $\mathbf{F}^e$ must evolve in a consistent manner such that their product $\mathbf{F}^e\mathbf{F}^p$ is compatible. This compatibility condition may be expressed for any closed path taken around the deforming material point as (Kroner, 1981):

$$
\delta = \oint_C \mathbf{F}^e\mathbf{F}^p \mathrm{d}X = 0
\tag{4.2.9}
$$

4.2.2 Slip Rate and Dislocation Density Evolution

Compared with the plasticity theory in previous chapters, the most difficult part of single-crystal plasticity is how to decide the crystallographic slip rates $\dot{\gamma}^{(\alpha)}$ in Eq. (4.2.3). In this section, the dislocation density and dislocation speed are introduced to express the crystallographic slip rates during the deformation of single crystals.

There are approximately $10^6 \sim 10^{12}$ dislocations in every square centimeter of metal crystals, depending on the processing history of the material, such as the degree of deformation that has occurred and the various microstructural features formed during heat treatment (Hull and Bacon, 2006; Polmear, 2006). To predict the engineering performance of materials, it is convenient to smear individual dislocations into densities, so as to study their evolution (causing plastic deformation) in a continuum approximation. A dislocation density, ρ, is defined as the total dislocation line length of all dislocations per unit volume $[\text{m} \cdot \text{m}^{-3}] = [\text{m}^{-2}]$.

It may be assumed that during any increment of strain exceeding the elastic limit, a change ensues in the dislocation structure within multiple slip systems, which defines the overall plastic activity of a crystal. The dislocation structure in each slip system is further taken to be potentially composed of mobile (glissile) and immobile (sessile) components. That is, for a given state of the material, the dislocation structure in slip system a, which is represented by total dislocation density ρ^α, is also decomposed into a mobile density (ρ_m^α) and an immobile density (ρ_{im}^α), as (Estrin et al., 1996):

$$\rho^\alpha = \rho_m^\alpha + \rho_{im}^\alpha \tag{4.2.10}$$

This decomposition follows from the view that dislocation motion accounts for plastic deformation and therefore ductility, whereas immobile dislocations contribute to the strengthening or hardening of crystals. A single dislocation line may have both mobile and immobile segments at any one time. All plastic mechanisms activated by local straining of a material point will then be seen as resulting from the generation or annihilation of mobile and immobile dislocation densities in their respective slip systems, as well as their interaction across slip systems. Furthermore, during deformation, mobile dislocations can be immobilized and immobile dislocations can be released, which suggests that evolutionary laws for mobile and immobile dislocations need to be coupled. For a large class of crystalline materials, dislocation density evolutionary laws could thus take the form (Estrin and Kubin, 1986)

$$\begin{aligned} \dot{\rho}_m^\alpha &= \dot{\rho}_{\text{generatation}}^\alpha - \dot{\rho}_{\text{interaction}}^\alpha \\ \dot{\rho}_{im}^\alpha &= \dot{\rho}_{\text{annihilation}}^\alpha + \dot{\rho}_{\text{interaction}}^\alpha \end{aligned} \tag{4.2.11}$$

To define the terms completely in Eq. (4.2.11), specific assumptions often need be made. However, fairly general evolutionary laws that apply to metals of different crystal

structures were derived in Estrin and Kubin (1986) and enhanced by subsequent research (Kameda and Zikry, 1996; Rezvanian et al., 2007; Shi and Zikry, 2009; Shanthraj and Zikry, 2011).

A material point, which is at the micron scale, is always assumed to represent sufficient crystalline structure such that the generation, annihilation, and interaction mechanisms of dislocation densities may arise. An example of evolutionary laws, which capture the terms of Eq. (4.2.11) at a material point, is the following linear superposition of various mobile and immobile mechanisms (Estrin et al., 1996; Shanthraj and Zikry, 2011):

$$\dot{\rho}_m^{(\alpha)} = \left|\dot{\gamma}^{(\alpha)}\right|\left(g_s^{(\alpha)}\left(\rho_{im}^{(\alpha)}\rho_m^{(\alpha)}\right)^{-1} - g_{m0}^{(\alpha)}\rho_m^{(\alpha)} - g_{im0}^{(\alpha)}\sqrt{\rho_{im}^{(\alpha)}}\right) \tag{4.2.12a}$$

$$\dot{\rho}_{im}^{(\alpha)} = \left|\dot{\gamma}^{(\alpha)}\right|\left\{g_{m1}^{(\alpha)}\rho_m^{(\alpha)} - g_{im1}^{(\alpha)}\sqrt{\rho_{im}^{(\alpha)}} - g_r^{(\alpha)}\rho_{im}^{(\alpha)}\exp\left(\frac{-\Delta H_0}{kT}\left(1 - \sqrt{\frac{\rho_{im}^{(\alpha)}}{\rho_{im}^{sat}}}\right)\right)\right\} \tag{4.2.12b}$$

where ΔH_0 is an activation enthalpy for dislocation mechanisms, k is the Boltzmann constant, T is the current absolute temperature, and ρ_{sat}^{im} is an experimentally determined saturation value for immobile dislocation densities.

A set of coefficients in Eq. (4.2.12) was specified for the material dependence of the degree of activity of specific dislocation mechanisms: g_s, for the generation of mobile dislocation densities; g_{m0}, for the annihilation of mobile dislocations by mobile dislocation self-interactions; g_{m1}, for the dislocation loop and debris formation of the result of the self-interaction of mobile dislocations; g_r, for the recovery of immobile dislocations; and g_{im1} and g_{im0}, for the immobilization of mobile dislocations owing to immobile dislocation interaction with mobile dislocations. Explicit expressions for these coefficients in terms of the generation, annihilation, and interaction mechanisms are outlined in Table 4.1, along with typical values for metals (Estrin and Kubin, 1986).

Table 4.1: Coefficients of Dislocation Density Evolutionary Equations

Coefficient	Equation	Typical Value
g_s^α	$\frac{\phi^{(\alpha)}}{|\boldsymbol{b}^{(\alpha)}|}\cdot\sum_\eta c_{\alpha\eta}\left(\rho_{im}^\eta\right)^{1/2}$	$\frac{1}{|\boldsymbol{b}^{(\alpha)}|}\times 2.76e^{-5}$
g_{m0}^α	$\phi^{(\alpha)}\cdot l_c\cdot\sum_\eta\left(\sqrt{a_{(\alpha)\eta}}\left[\frac{\rho_m^\eta}{\rho_m^{(\alpha)}|\boldsymbol{b}^{(\alpha)}|} + \frac{\dot{\gamma}^\eta}{\dot{\gamma}^{(\alpha)}|\boldsymbol{b}^\eta|}\right]\right)$	$\frac{1}{|\boldsymbol{b}^{(\alpha)}|}\times 5.53$
g_{im0}^α	$\frac{\phi^{(\alpha)}\cdot l_c}{\left(\rho_{im}^{(\alpha)}\right)^{1/2}}\cdot\sum_\eta\left(\sqrt{a_{(\alpha)\eta}}\cdot\rho_{im}^\eta\right)$	$\frac{1}{|\boldsymbol{b}^{(\alpha)}|}\times 0.0127$
g_{m1}^α	$\frac{\phi^{(\alpha)}\cdot l_c}{|\boldsymbol{b}^{(\alpha)}|\dot{\gamma}^{(\alpha)}\rho_m^{(\alpha)}}\cdot\sum_{\eta,\kappa\leq\eta}\left(Z_{(\alpha)}^{\eta\kappa}\sqrt{a_{\eta\kappa}}\left[\frac{\rho_m^\eta\dot{\gamma}^\kappa}{|\boldsymbol{b}^\kappa|} + \frac{\rho_m^\kappa\dot{\gamma}^\eta}{|\boldsymbol{b}^\eta|}\right]\right)$	$\frac{1}{|\boldsymbol{b}^{(\alpha)}|}\times 5.53$
g_{im1}^α	$\frac{\phi^{(\alpha)}\cdot l_c}{|\boldsymbol{b}^{(\alpha)}|\dot{\gamma}^{(\alpha)}\left(\rho_{im}^{(\alpha)}\right)^{1/2}}\cdot\sum_{\eta,\kappa\leq\eta}\left(Z_{(\alpha)}^{\eta\kappa}\sqrt{a_{\eta\kappa}}\cdot\rho_m^\kappa\dot{\gamma}^\eta\right)$	$\frac{1}{|\boldsymbol{b}^{(\alpha)}|}\times 0.0127$
g_r^α	$\frac{\phi^{(\alpha)}\cdot l_c}{\dot{\gamma}^{(\alpha)}}\cdot\sum_\eta\left(\sqrt{a_{(\alpha)\eta}}\cdot\frac{\dot{\gamma}^\eta}{|\boldsymbol{b}^\eta|}\right)$	$6.69e^5$

Eq. (4.2.12) and these coefficients are only an example of possible dislocation density evolutionary laws; various other laws could be found in the literature, depending on the material and plastic phenomenon of interest.

Next, an explanation and derivation are presented for the specific dislocation density evolutionary laws selected from Eq. (4.2.12). These laws are not unique in the literature, and the discussion presented here is merely to illustrate a possible choice of evolutionary laws with a fair level of generality. The dislocation mobility, generation, annihilation, and interaction will be briefly discussed.

4.2.2.1 Dislocation Mobility

Dislocation motion is generally of two types, conservative and/or nonconservative, depending on whether the number of atoms in the neighborhood of a dislocation core changes as a result of the motion (Bulatov and Cai, 2006). As shown in Fig. 4.2, an edge dislocation (A) can move conservatively by gliding in the slip direction (B), as discussed earlier, or nonconservatively by climbing in a direction normal to the slip plane (C). For an edge dislocation to climb upward, a line of atoms must be continually emitted from above the dislocation core; whereas for climb to proceed downward, as shown in the figure, a line of atoms must be continually absorbed under the dislocation core. Such a process may occur in real crystals when atomic motion is assisted by substantial thermal

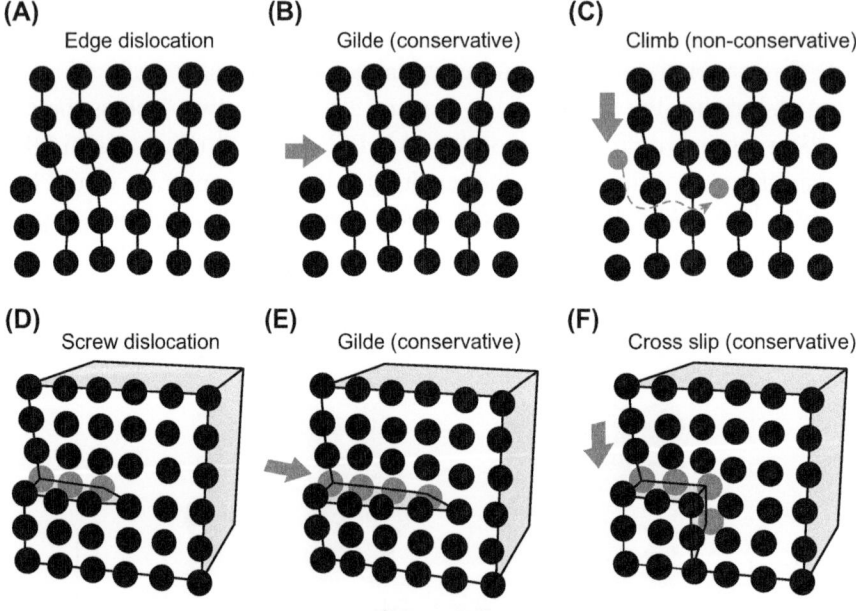

Figure 4.2
Conservative and/or nonconservative dislocation motion.

agitation (i.e., deformation proceeds at a high temperature) and when the crystal possesses a population of point defects (interstitials and vacancies) (Fig. 4.3), which may become attracted to dislocation cores by virtue of their interacting stress fields (Bulatov and Cai, 2006), as shown in Fig. 4.2C.

Conversely, screw dislocations can migrate from one slip system to another that shares the same slip direction, by the mechanism of cross-slip, as shown in Fig. 4.2D–F. This process may be understood in view of the fact that screw dislocations are defined by parallel dislocation lines and Burgers vectors, such that no unique slip plane normal exists. Therefore, for many crystals (e.g., body-centered cubic structures), multiple planes can accommodate screw dislocation motion and cross-slip may be observed to evolve along the different slip systems for the same applied load, as shown in Fig. 4.2E and F. Because no atoms need to be added or removed by this process, cross-slip and screw dislocation motion in general are conservative.

Thus, both the screw and edge components of mixed dislocations can move during deformation. In general, their motion evolves at a rate form depending on the exact mechanism: for example, glide, climb, and cross-slip. Orowan's rule summarizes this result by relating micron-scale plastic shear rate $\dot{\gamma}^{\alpha}$ to mobile dislocations ρ_m^{α}, Burgers vector $b^{(\alpha)}$, and effective dislocation speed $v^{(\alpha)}$ by (Hull and Bacon, 2006):

$$\dot{\gamma}^{\alpha} = \rho_m^{\alpha} \left| b^{(\alpha)} \right| v^{(\alpha)} \tag{4.2.13}$$

This relation may be understood by considering Fig. 4.4. For a crystal of volume $V = Lwh$, when a single dislocation traverses the entire crystal, an average shear strain $\gamma^{\alpha} = |b^{(\alpha)}|h^{-1}$ is created (Fig. 4.4, right). For a dislocation that travels only partway across the crystal, the average shear strain would rescale to $\gamma^{\alpha} = \Delta L \cdot L^{-1} \cdot |b^{(\alpha)}|h^{-1}$ (Fig. 4.4, middle). For N such dislocations in the slip system, the resulting average shear strain would be $\gamma^{(\alpha)} = N^{(\alpha)}\Delta L \cdot L^{-1} \cdot |b^{(\alpha)}|h^{-1}$. We can rehash this equation as $\gamma^{(\alpha)} = N^{(\alpha)}wV^{-1} \cdot |b^{(\alpha)}|\Delta L$, where V is the volume and $wV^{-1} = L^{-1}h^{-1}$. Recognizing that $\rho^{\alpha} = N^{\alpha}wV^{-1}$,

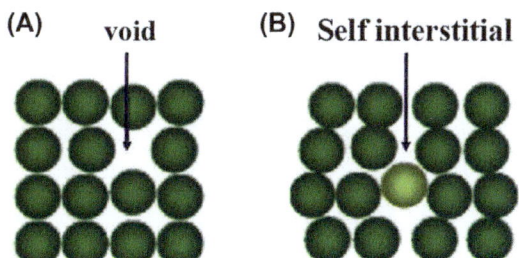

Figure 4.3

(A) Vacancy and (B) Interstitial. *Adapted from Bulatov V., Cai W., 2006. Computer Simulations of Dislocations. Oxford University Press, New York.*

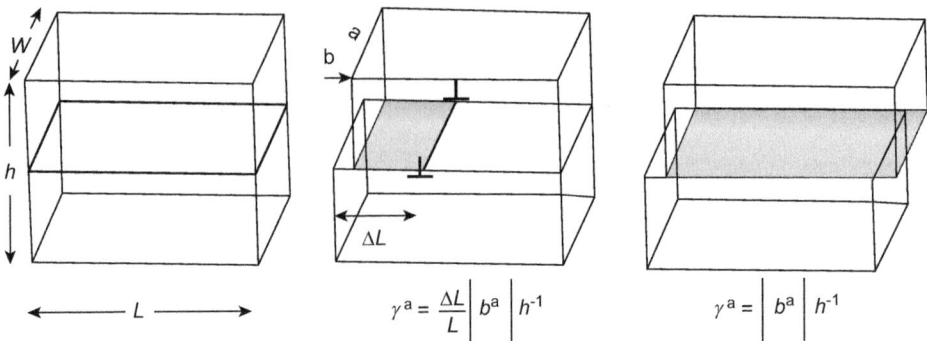

Figure 4.4
Mechanism behind Orowan's rule.

i.e., dislocation density is the total length of dislocations per unit volume, we find that $\gamma^{(\alpha)} = \rho^{(\alpha)}|b^{(\alpha)}|\Delta L$. Differentiating in time yields Eq. (4.2.13).

Various equations characterize dislocation velocity v^{α}, which must be determined in Eq. (4.2.13), depending on the details of the evolving dislocation mechanism (Caillard and Martin, 2003). Generally, speed v^{α} can be described by a relation of the form (Caillard and Martin, 2003)

$$v^{\alpha} = v_0 \left(\frac{\tau^{\alpha}}{\tau_{\text{ref}}}\right)^n \exp\left(-\frac{\Delta E}{kT}\right) \tag{4.2.14}$$

where τ^{α} is the effective resolved shear stress, and $\tau^{\alpha} = \widetilde{\mathbf{s}}^{(\alpha)} \cdot \boldsymbol{\sigma} \cdot \widetilde{\mathbf{m}}^{(\alpha)}$ is the projection of the Cauchy stress tensor $\boldsymbol{\sigma}$ in slip system α. Superscript α runs through all active slip systems of a crystal. n is the stress exponent and ΔE is the activation energy for dislocation motion. k is the Boltzmann constant and T is the current absolute temperature. For conservation and nonconservation motion, different values for ΔE and v_0 would be needed.

4.2.2.2 Dislocation Generation

Dislocation densities tend to increase during any deformation beyond the elastic limit, because they are the bearers of plasticity in a crystal. The generation of dislocations proceeds from various sources in a crystal and takes the form (Estrin and Kubin, 1986; Estrin et al., 1996; Shanthraj and Zikry, 2011)

$$\dot{\rho}^{(\alpha)}_{\text{generation}} = \rho^{(\alpha)}_{\text{source}} \frac{v^{(\alpha)}}{l_c} \tag{4.2.15}$$

where $\rho^{(\alpha)}_{\text{source}}$ is the local density of dislocation generation sites in slip system α, $v^{(\alpha)}$ is the average speed at which dislocations are emitted from the sources, and l_c is the characteristic spacing between sources.

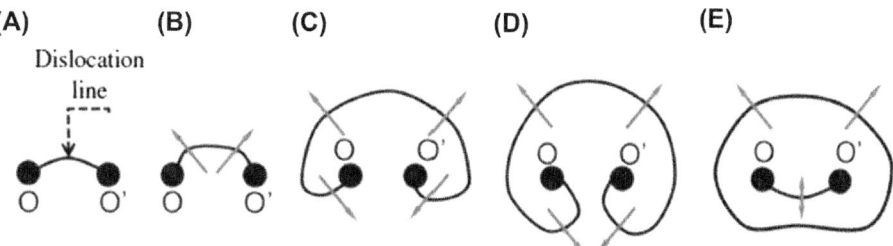

Figure 4.5
Frank-Read mechanism.

It is difficult to determine $\rho_{source}^{(\alpha)}$ because multiple sites for generating dislocation may arise under different loading conditions and for different microstructures. Commonly, immobile dislocations form a connected network called a Frank net, which extends throughout the crystal (Gurtin, 2006). In slip system α, segments of mobile dislocation lines will inevitably become pinned and immobilized by this net. By means of the Frank-Read mechanism, these immobilized segments often act as sources of new dislocations (Estrin and Kubin, 1986; Gurtin, 2006).

As shown in Fig. 4.5, a dislocation segment immobilized and pinned at both ends by obstacles (e.g., a Frank net) can bow from under an applied shear stress to form a dislocation loop that is emitted away from the obstacles. In this process, the immobile segment reforms and remains pinned so that the dislocation loop formation and emission mechanism may be sustained. In this case, it may be assumed that $\rho_{source}^{(\alpha)} = \phi^{(\alpha)} \cdot \rho_{im}^{(\alpha)}$, where $\phi^{(\alpha)}$ is the probability immobile dislocations in slip system α, are conductive to the activation of a Frank-Read mechanism. Emission speed $v^{(\alpha)}$ used in Eq. (4.2.15) may be computed from Orowan's relation, whereas distance l_c for a Frank net can be determined

from $l_c = \left(\sum_{\eta} \left(\rho_{im}^{\eta} \right)^{1/2} \right)^{-1}$.

4.2.2.3 Dislocation Annihilation

Dislocation annihilation proceeds as the result of a process known as dynamic recovery (Estrin and Kubin, 1986; Hull and Bacon, 2006), which is a thermally activated mechanism that evolves with a large deformation, in which immobile dislocations are released to cancel out immediately with neighbors of opposite signs. Together with dislocation generation, annihilation accounts for the experimental observation that dislocation densities tend to saturate at large strain. It can be shown that recovery obeys an Arrhenius-type equation of the form (Shanthraj and Zikry, 2011)

$$\dot{\rho}_{annihilation}^{(\alpha)} = \dot{A}^{(\alpha)} \rho_{im}^{(\alpha)} \exp\left(-\frac{\Delta H^{(\alpha)}}{kT} \right) \tag{4.2.16}$$

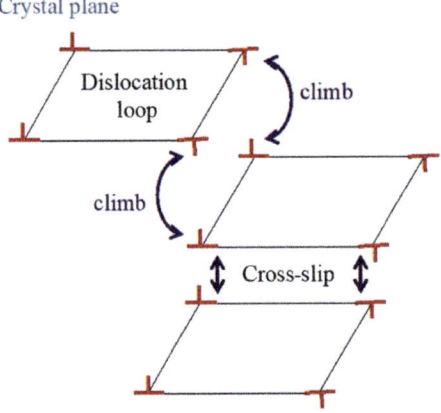

Figure 4.6

Climb and cross-slip. *Adapted from Poirier, J.P., 1976. On the symmetrical role of cross-slip of screw dislocations and climb of edge dislocations as recovery processes controlling high temperature creep, Revue De Physique Applique'e, 11.*

where $\dot{A}^{(\alpha)}$ is the attempt rate of dislocation annihilation by its fundamental mechanisms of climb and cross-slip, as indicated in Fig. 4.6 for idealized rectangular dislocation loops composed of two edge components (thin lines) and two split-screw components of mobile edge dislocations (thick lines). Thus, $\dot{A}^{(\alpha)}$ is measured in s^{-1} and reflects: (1) the rate by which segments of mobile edge dislocations of opposite signs climb to annihilate one another, and (2) the rate by which segments of mobile screw dislocation of opposite signs cross-slip to annihilate one another. The attempt rate is therefore related to Orowan's velocities for both conservative and nonconservative motions and may be heuristically given by (Shanthraj and Zikry, 2011)

$$\dot{A}^{(\alpha)} = \phi^{(\alpha)} \sum_{\eta} \left(\sqrt{a_{(\alpha)\eta} l_c} \; \dot{\gamma}^{\eta} |b^{\eta}|^{-1} \right) \tag{4.2.17}$$

where $\phi^{(\alpha)}$ is the probability of interacting dislocations being annihilated at a given temperature. ΔH in Eq. (4.2.16) is the enthalpy of activation of the climb and cross-slip mechanisms responsible for annihilation. It may be taken to decrease from reference value ΔH_0 for a crystal to 0, in proportion to the ratio of immobile dislocation densities to their observed saturation value, such that the process of annihilation reaches its maximum rate as dislocation densities saturate, which is consistent with experimental observation. That is (Shanthraj and Zikry, 2011),

$$\Delta H^{\alpha} = \Delta H_0 \left(1 - \sqrt{\rho_{im}^{\alpha} \left(\rho_{im}^{sat} \right)^{-1}} \right) \tag{4.2.18}$$

4.2.2.4 General Dislocation Interaction

As mobile dislocations move, they interact among themselves as well as with immobile dislocations, so that the resulting dislocation structure changes in ways other than by means of the generation and the thermally activated annihilation mechanisms described previously. The net effect of these general interactions is usually immobilization of a fraction of the mobile densities, which on the one hand decreases mobile dislocation densities and on the other hand adds to immobile dislocation junctions. The reduction of mobile dislocation densities may be expressed as a result of mobile-mobile and mobile-immobile interactions as (Shanthraj and Zikry, 2011)

$$\dot{\rho}^{(\alpha)-}_{\text{interaction}} = \phi^{(\alpha)} \cdot l_c \sum_{\eta} \rho_m^{(\alpha)} \left(\rho_m^\eta \bar{v}_{\alpha\eta} + \rho_{im}^\eta v_\alpha \right) \tag{4.2.19}$$

where velocity $\bar{v}_{\alpha\eta} = \frac{1}{2}(v_\alpha + v_\eta)$ measures the average mobility of dislocations in slip systems α and η and may be determined from Orowan's relations described earlier.

To determine the rate at which stable dislocation junctions are formed, Frank's energetic criterion needs be applied to all interaction combinations. The criterion specifies whether it is energetically favorable for intersecting dislocations from slip systems β and γ to form a junction in system α. Favorability is measured by a set of binary coefficients $z_\alpha^{\beta\gamma} \in [0, 1]$ that take a value of 0 if the formation of a junction is unfavorable, and unity when it is favorable. The test for favorability may be stated as (Shanthraj and Zikry, 2011)

$$G|\mathbf{b}^\alpha|^2 < G\left(|\mathbf{b}^\beta|^2 + |\mathbf{b}^\gamma|^2\right) \quad \text{and} \quad \mathbf{b}^\alpha = \mathbf{b}^\beta + \mathbf{b}^\gamma \tag{4.2.20}$$

where it is recognized that strain energy scales with $G|\mathbf{b}^\eta|^2$. Once the set $z_\alpha^{\beta\gamma}$ has been determined, the dislocation junction formation rate in slip system α may be given by (Shanthraj and Zikry, 2011)

$$\dot{\rho}^{(\alpha)+}_{\text{interaction}} = \phi^{(\alpha)} \cdot l_c \sum_{\beta,\gamma \leq \beta} z_\alpha^{\beta\gamma} \sqrt{\alpha_{\beta\gamma}} \left(\rho_m^\beta \rho_m^\gamma \bar{v}_{\beta\gamma} + \rho_m^\beta \rho_{im}^\gamma v_\beta + \rho_m^\gamma \rho_{im}^\beta v_\gamma \right) \tag{4.2.21}$$

Finally, combining and manipulating the equations derived here for dislocation density generation, annihilation, and interaction would yield the terms in Eq. (4.2.12).

4.2.3 Plastic Stress Required for Dislocation Motion

This section uses the dislocation densities evolved in Section 4.2.2 to determine crystal slip resistance, which is an essential component in the stress update algorithm of single crystals.

As mentioned earlier, around each dislocation core there must exist a stress field associated with the distortion it brings to crystalline order. Thus, if a mobile dislocation is to move in a crystal that already contains a number of immobile dislocations, it must interact with and overcome the additional stress fields surrounding these dislocations.

Immobile dislocations in this sense act as obstacles to mobile dislocations, and their density in large part determines the threshold plastic stress required for dislocation motion within a slip system. The threshold value is known as the critical resolved shear stress (τ_{ref}^{α}) and defines the strength of a crystal, whereas its rate of change characterizes the hardening of the crystal and determines the shape of the stress strain curve (Fig. 1.1B). If τ_{ref}^{α} increases, the crystal is said to harden; if it decreases, the crystal is said to soften. The resolved shear stress τ^{α} must always be compared with the τ_{ref}^{α}; the condition for slip to occur could be stated as

$$|\tau^{\alpha}| > \tau_{ref}^{\alpha} \tag{4.2.22}$$

In addition to the density of immobile dislocations, the value of τ_{ref}^{α} generally depends on the crystal structure, the point defects that interact with slip systems to hinder dislocation motion, as well as the temperature and other internal variables (Hull and Bacon, 2006). Taylor (1934) showed that the necessary τ_{ref}^{α} for single slip to occur may be given by

$$\tau_{ref}^{\alpha} = \frac{G|\boldsymbol{b}^{(\alpha)}|}{d^{(\alpha)}} = G|\boldsymbol{b}^{(\alpha)}|\sqrt{\rho_{im}^{\alpha}} \tag{4.2.23}$$

where ρ_{im}^{α} is the density of immobile dislocations in slip system α and $d^{(\alpha)}$ is the average dislocation spacing in the slip system. This relation elucidates the need for Eq. (4.2.22) to define the evolving strength of a crystal and the corresponding shape of the stress-strain curve, and determines the connection between crystal plasticity theory and dislocation theory. Various modifications to Eq. (4.2.23) have been proposed to better account for dislocation interaction in the case of poly-slip, as well as thermal effects (Elkhodary et al., 2011):

$$\tau_{ref}^{\alpha} = \left(\tau_0^{\alpha} + G\sum_{\beta=1}^{nss}|\mathbf{b}^{\beta}|\sqrt{a_{\alpha\beta}\rho_{im}^{\beta}}\right)\left(\frac{T}{T_0}\right)^{-\xi} \tag{4.2.24}$$

where τ_0^{α} is the initial slip resistance on slip system (α), *nss* is the number of slip systems, $|\mathbf{b}^{\beta}|$ is the magnitude of the Burgers vector, and Taylor coefficients $a_{\alpha\beta}$ reflect the strength of interaction between slip systems, which vary between 0 and 1 and could be approximated from the relation $a_{\alpha\beta} = 2\sqrt{P_{ij}^{\alpha}P_{ij}^{\beta}}$ for simplicity (Elkhodary et al., 2011). *T* is the current temperature, T_0 is a reference temperature, and *x* is the thermal softening exponent.

4.2.4 Update of Cauchy Stress in Single-Crystal Plasticity

Formulations from Asaro and Needleman (1985a) are adopted to the constitutive law of Cauchy stress. Defining the Kirchhoff stress $\tau = J\sigma$, the Jaumann rate of Kirchhoff stress formed on axes that spin with the lattice is related to the elastic deformation rate as

$$\nabla_{\tau}^{e} = \dot{\tau} - \mathbf{W}^{e}\cdot\tau + \tau\cdot\mathbf{W}^{e} = \mathbf{C} : \mathbf{D}^{e} \tag{4.2.25}$$

where $\mathbf{C}$ is a fourth-order elasticity tensor. For an isotropic material, $\mathbf{C}$ is in the form $C_{ijkl} = \mu(\delta_{ik}\delta_{jl} + \delta_{il}\delta_{jk}) + \lambda\delta_{ij}\delta_{kl}$, where μ and λ are Lamé parameters. Then, the rate of Cauchy stress can be obtained by

$$\dot{\sigma} = J^{-1}\mathbf{C}:\mathbf{D}^e - (\operatorname{tr}\mathbf{D}^e)\sigma + \mathbf{W}^e \cdot \sigma - \sigma \cdot \mathbf{W}^e \qquad (4.2.26)$$

4.3 Higher-Order Dislocation Dynamics-Based Crystal Plasticity Model

Lots of crystal plasticity models have been developed by expanding the basic constitutive model of single crystals given in Section 4.2 to explain the unconventional mechanical behaviors of crystalline materials at a microscale. These theoretical models can be roughly divided into two types: one involves the evolution equation of crystal slip and the other involves the evolution equation of dislocation density.

A new dislocation-based crystal plasticity model is developed in this section, which is motivated by the continuum description of the collective behavior of dislocations from Groma et al. (2003). Compared with conventional crystal plasticity models, a microforce balance equation is involved by combining the equation of dislocations motion and Orowan's law. The microforce balance equation contains a geometrically necessary dislocations (GND)-related back stress to consider the short-range interactions of dislocations. The resulting equation is for crystal slip instead of dislocation density, so it is more convenient when coupled with the macroforce balance equation in conventional crystal plasticity theories. Furthermore, a new model has a formulation similar to the familiar coupled thermomechanical problems. Thus, it has many advantages in dealing with additional boundary conditions, just like convection and diffusion boundary conditions in standard thermal conduction problems.

4.3.1 Governing Equations of Macroforces

Similar to conventional continuum mechanics, Cauchy stress σ satisfies the equilibrium equation

$$\nabla \cdot \sigma + \mathbf{f} = 0 \qquad (4.3.1)$$

where ∇ denotes divergence with respect to the material point in the deformed configuration and $\mathbf{f}$ is body force. The boundary conditions are given by

$$\begin{aligned} \sigma \cdot \mathbf{n} &= \bar{\mathbf{t}} \quad \text{on } S_t \\ \mathbf{u} &= \bar{\mathbf{u}} \quad \text{on } S_u \end{aligned} \qquad (4.3.2)$$

where $\mathbf{n}$ is a unit normal to the surface of the deformed body, $\bar{\mathbf{t}}$ is prescribed traction on the surface S_t, and $\bar{\mathbf{u}}$ is prescribed displacement on surface S_u, whereas S $(=S_t + S_u)$ is a surface of the deformed body.

4.3.2 Governing Equations of Microforces

4.3.2.1 Equilibrium Equation of Microforces

Because plastic deformation is caused by the motion of dislocations, it is important to derive a continuum description of the behavior of dislocations. Starting from the equation of dislocations motion, a continuum description of collective dislocations of slip system α can be derived as (Groma et al., 2003)

$$Bv^{(\alpha)} = b\left[\tau^{(\alpha)} - \tau_b^{(\alpha)} - g^{(\alpha)}\text{sgn}\left(\tau^{(\alpha)} - \tau_b^{(\alpha)}\right)\right] \tag{4.3.3}$$

where B is the drag coefficient, $v^{(\alpha)}$ is the average dislocation velocity along the slip direction, b is the magnitude of the Burgers vector, $\tau^{(\alpha)}$ is the external driving stress, $\tau_b^{(\alpha)}$ is the back stress resulting from dislocation interactions, and $g^{(\alpha)}$ is slip resistance including lattice friction and other dissipative factors. sgn is the sign function. For simplicity, $g_s^{(\alpha)}$ is used to denote $g^{(\alpha)}\text{sgn}\left(\tau^{(\alpha)} - \tau_b^{(\alpha)}\right)$ in the following sections. Eq. (4.3.3) is valid only when

$$\left|\tau^{(\alpha)} - \tau_b^{(\alpha)}\right| > g^{(\alpha)} \tag{4.3.4}$$

Otherwise, $v^{(\alpha)} = 0$. The average dislocation velocity $v^{(\alpha)}$ can be evaluated by dislocation density $\rho^{(\alpha)}$ and slip rate $\dot{\gamma}^{(\alpha)}$ via Orowan's equation:

$$v^{(\alpha)} = \dot{\gamma}^{(\alpha)}\Big/\left(\rho^{(\alpha)}b\right) \tag{4.3.5}$$

As discussed in Kuroda and Tvergaard (2008), back stress can be expressed as the divergence of a vector quantity $\xi^{(\alpha)}$, which is the so-called microstress (Gurtin, 2000):

$$\tau_b^{(\alpha)} = -\nabla\cdot\xi^{(\alpha)} \tag{4.3.6}$$

By substituting Eqs. (4.3.5) and (4.3.6) into Eq. (4.3.3), the final equilibrium equation yields

$$D\dot{\gamma}^{(\alpha)} = \nabla\cdot\xi^{(\alpha)} + \tau^{(\alpha)} - g_s^{(\alpha)} \tag{4.3.7}$$

where D is a drag coefficient. Eq. (4.3.7) shows a linear viscosity, whereas a power function is used to define rate dependency in conventional plasticity theories (Peirce et al., 1983). To model rate-independent crystals, a small value of D should be used. The linear viscosity comes from the dislocation dynamics, as described in Eq. (4.3.3). Linear viscosity was also used in the plasticity model based on a statistical description of dislocation systems (Yefimov et al., 2004). Eq. (4.3.7) preserves some nature of dislocation dynamics and describes the motion of dislocations in a continuum manner; it will be applied to characterize the plastic behavior at micron or submicron scales in the following studies. Details of the evolution of $\xi^{(\alpha)}$, $\tau^{(\alpha)}$, and $g^{(\alpha)}$ are discussed in the following sections.

To solve Eq. (4.3.7), boundary conditions are needed:

$$\xi^{(\alpha)} \cdot \mathbf{n} = \overline{q}^{(\alpha)} \text{ on } S_q$$
$$\gamma^{(\alpha)} = \overline{\gamma}^{(\alpha)} \text{ on } S_\gamma \tag{4.3.8}$$

where $\mathbf{n}$ is a unit normal to the surface of the deformed body, $\overline{q}^{(\alpha)}$ is prescribed microtraction on the surface S_q, and $\overline{\gamma}^{(\alpha)}$ is prescribed slip on surface S_γ, whereas $S \ (=S_q + S_\gamma)$ is the surface of the deformed body. The boundary conditions in Eq. (4.3.8) control the different behaviors of dislocations at the surface. Two specific cases are worth noticing. When $\overline{q}^{(\alpha)} = 0$, there is no microtraction, so dislocations can escape from the surface freely. When $\overline{\gamma}^{(\alpha)} = 0$, according to Eq. (4.3.5), the dislocations cannot pass through the surface; in other words, the surface blocks dislocations from escaping.

4.3.2.2 Geometrically Necessary Dislocations

In plastically nonhomogeneous single crystals, the dislocations are stored in them to accommodate the deformation gradients. These dislocations are called GNDs (Cottrell, 1964). They contribute to work hardening in two ways: by acting as individual obstacles to slip and by creating a back stress (Ashby, 1970). In dislocation-based crystal plasticity theories, GND density is related to the gradient of slip (Gurtin, 2002). For simplicity, a two-dimensional (2D) model is established and only edge dislocations need to be considered (the equations are similar for 3D models and screw dislocations). In this case, the rate of GND density is governed by the following differential equation (Kuroda and Tvergaard, 2008):

$$\dot{\rho}_{\text{GND}}^{(\alpha)} = -\frac{1}{b} \nabla \dot{\gamma}^{(\alpha)} \cdot \widetilde{\mathbf{s}}^{(\alpha)} \tag{4.3.9}$$

Thus, the GND density $\rho_{\text{GND}}^{(\alpha)}$ is the integral of $\dot{\rho}_{\text{GND}}^{(\alpha)}$ with respect to time t:

$$\rho_{\text{GND}}^{(\alpha)} = \rho_{\text{GND}}^{(\alpha)} \Big|_{t=0} + \int_0^t \dot{\rho}_{\text{GND}}^{(\alpha)} dt \tag{4.3.10}$$

4.3.2.3 Constitutive Laws of Microforces

There are different ways to consider the effect of GNDs in crystal plasticity. Some models add GNDs to statistically stored dislocations (SSDs) and GNDs contribute to hardening by the Taylor hardening relation (Fleck et al., 1994; Gao et al., 1999). It has been pointed out that the size-dependent flow stress in micropillar compression cannot be attributed to GNDs (Akarapu et al., 2010; Tang et al., 2007; Volkert and Lilleodden, 2006). However, back stress associated with GNDs can influence the arrangement of dislocations (Groma et al., 2003).

This back stress comes from interactions between dislocations. It can change the distribution of stress in the pillar and lead to a size-dependent deformation morphology. It is formulated as follows. The accumulated GNDs will add defect energy (Gurtin, 2002) to the conventional free energy given by local elastic strain energy in the crystal. This energy can be evaluated by the GND density in the intermediate configuration via the equation (Gurtin, 2008; Kuroda and Tvergaard, 2008)

$$\psi_G = \frac{1}{2} S_0 l^2 b^2 \sum_\alpha \left(\rho_{GND}^{(\alpha)} \right)^2 \tag{4.3.11}$$

where S_0 is usually taken to be the same as the initial slip resistance and l is a length parameter characterizing the scale of microstructures. The defect energy may take different forms (Gurtin, 2008, 2010). Hurtado and Ortiz (2012) and Wulfinghoff et al. (2015) took the energy to be proportional to the dislocation density and the quadratic defect energy as assumed in our model.

The rate of ψ_G is equal to the power created by microstress $\xi^{(\alpha)}$ and its work-conjugate term $\nabla_x \dot{\gamma}^{(\alpha)}$

$$\sum_\alpha \xi^{(\alpha)} \cdot \nabla \dot{\gamma}^{(\alpha)} = J^{-1} \dot{\psi}_G \tag{4.3.12}$$

Based on Eqs. (4.2.19), (4.2.21), and (4.2.22), the constitutive relation of microstress $\xi^{(\alpha)}$ can be derived as

$$\xi^{(\alpha)} = -J^{-1} S_0 l^2 b \rho_{GND}^{(\alpha)} \widetilde{\mathbf{s}}^{(\alpha)} \tag{4.3.13}$$

with its rate form

$$\dot{\xi}^{(\alpha)} = J^{-1} \mathbf{C}_\gamma^{(\alpha)} \cdot \nabla \dot{\gamma}^{(\alpha)} - (\mathrm{tr}\, \mathbf{D}^e) \xi^{(\alpha)} + \mathbf{L}^e \cdot \xi^{(\alpha)} \tag{4.3.14}$$

where $\mathbf{C}_\gamma^{(\alpha)} = S_0 l^2 \widetilde{\mathbf{s}}^{(\alpha)} \otimes \widetilde{\mathbf{s}}^{(\alpha)}$ is a two-order tensor. If $\mathbf{D}^e$ is small compared with $\mathbf{W}^e$, Eq. (4.3.14) can be approximately replaced by

$$\dot{\xi}^{(\alpha)} = J^{-1} \mathbf{C}_\gamma^{(\alpha)} \cdot \nabla \dot{\gamma}^{(\alpha)} - (\mathrm{tr}\, \mathbf{D}^e) \xi^{(\alpha)} + \mathbf{W}^e \cdot \xi^{(\alpha)} \tag{4.3.15}$$

Eq. (4.3.15) is the constitutive relation of microstress $\xi^{(\alpha)}$ used in our model for finite deformation.

The external driving stress $\tau^{(\alpha)}$ is the resolved shear stress of Cauchy stress σ as in conventional crystal plasticity theories

$$\tau^{(\alpha)} = \widetilde{\mathbf{m}}^{(\alpha)} \cdot \sigma \cdot \widetilde{\mathbf{s}}^{(\alpha)} \tag{4.3.16}$$

with its rate form

$$\dot{\tau}^{(\alpha)} = \tilde{\mathbf{m}}^{(\alpha)} \cdot (\dot{\boldsymbol{\sigma}} - \mathbf{L}^{\mathrm{e}} \cdot \boldsymbol{\sigma} + \boldsymbol{\sigma} \cdot \mathbf{L}^{\mathrm{e}}) \cdot \tilde{\mathbf{s}}^{(\alpha)} \qquad (4.3.17)$$

The form of slip resistance $g^{(\alpha)}$ is taken from Forest (1998). The material exhibits a short strain-hardening period followed by strain softening. The following nonlinear hardening law is used:

$$g^{(\alpha)} = g_0^{(\alpha)} + p_1 \sum_{\beta} h^{(\alpha)(\beta)} \left[1 - \exp\left(-k_1 \gamma_{\mathrm{c}}^{(\beta)} \right) \right] + p_2 \left[1 - \exp\left(-k_2 \gamma_{\mathrm{c}}^{(\alpha)} \right) \right] \qquad (4.3.18)$$

where $g_0^{(\alpha)}$ is the initial slip resistance; p_1, k_1, p_2, and k_2 are material parameters; $h^{(\alpha)(\beta)}$ is interaction matrix and is taken to be $\delta^{(\alpha)(\beta)}$ in the simulations here; and $\gamma_{\mathrm{c}}^{(\alpha)}$ is cumulative shear strain on slip system α:

$$\gamma_{\mathrm{c}}^{(\alpha)} = \int_0^t \left| \dot{\gamma}^{(\alpha)} \right| \mathrm{d}t \qquad (4.3.19)$$

The material-softening part in Eq. (4.3.18) is helpful to trigger localization from the material defect region and was used in some earlier work (Brechet et al., 1996; Forest, 1998) to analyze shear band formation. The physical meaning of the softening can be found in molecular dynamics simulations (Horstemeyer et al., 2001; Xu et al., 2013). Because the sample starts with a perfect lattice in these simulations, it needs a large stress to activate a certain number of dislocation sources. Dislocations are emitted from the activated dislocation sources and take most of the deformation. The plastic deformation created by movements of dislocations release the large stress and lead to the softening part. Therefore, in our continuum approach, a softening part is introduced in the constitutive law to represent a large increase in dislocation after the dislocation sources are created.

4.3.3 Coupling of Macroscopic and Microscopic Equations

Macroscopic equations (4.3.1) and (4.3.2) and microscopic equations (4.3.7), (4.3.8), (4.3.15), (4.3.17), and (4.3.18) are capable of solving size-dependent problems at small scales. The externally applied stress and displacement are controlled by Eq. (4.3.2), whereas the effect of the surface on dislocations is controlled by Eq. (4.3.8). First, the stress field is solved by Eq. (4.3.1). Then, stress is applied to the microscopic equations by driving stress $\tau^{(\alpha)}$. The dislocations motion is characterized by Eq. (4.3.7) via the evolution of slip $\gamma^{(\alpha)}$. In return, the plastic deformation caused by the motion of dislocations is included in the constitutive relation of Cauchy stress. Except for replacing discrete dislocations by the continuum description, the procedure is similar to the 3D *finite element* analysis model (Liu et al., 2009) and is illustrated in Fig. 4.7.

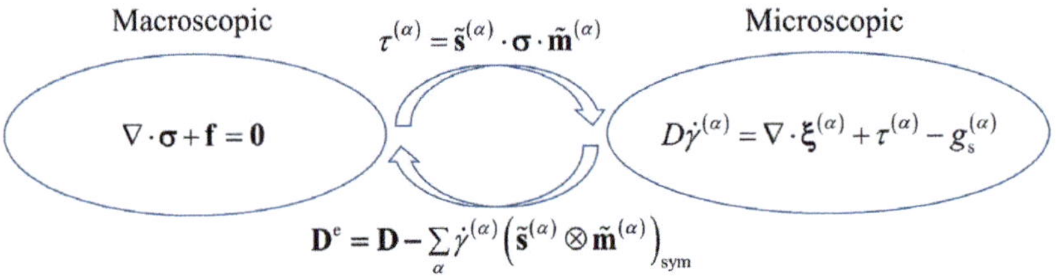

Macroscopic $\qquad\quad \tau^{(\alpha)} = \tilde{\mathbf{s}}^{(\alpha)} \cdot \boldsymbol{\sigma} \cdot \tilde{\mathbf{m}}^{(\alpha)}$ $\qquad$ Microscopic

$$\nabla \cdot \boldsymbol{\sigma} + \mathbf{f} = \mathbf{0}$$

$$D\dot{\gamma}^{(\alpha)} = \nabla \cdot \boldsymbol{\xi}^{(\alpha)} + \tau^{(\alpha)} - g_{s}^{(\alpha)}$$

$$\mathbf{D}^{e} = \mathbf{D} - \sum_{\alpha} \dot{\gamma}^{(\alpha)} \left(\tilde{\mathbf{s}}^{(\alpha)} \otimes \tilde{\mathbf{m}}^{(\alpha)} \right)_{\text{sym}}$$

Figure 4.7

Coupling between macroscopic and microscopic equations.

4.4 Size and Bauschinger Effect in Passivated Thin Films

At the microscale, thin films with a passivation layer exhibit a strong size effect during loading and a prominent Bauschinger effect during reverse loading. It is commonly recognized that hardening, which is caused by GNDs near the passivation layer, is responsible for these phenomena. However, GNDs may contribute to hardening either by slip resistance via a generalized Taylor equation or by back stress originating from elastic interaction between dislocations. Which one dominates thin film plasticity is not well-understood. In this section, both slip resistance and back stress hardening are included in the dislocation-based crystal plasticity model. The model is applied to simulate passivated thin films with different thicknesses and the results are compared with experimental data.

4.4.1 Two Hardening Mechanisms Caused by Geometrically Necessary Dislocations

The passivation layer prevents dislocations from escaping, leading to the pileups of GNDs in its vicinity (Xiang and Vlassak, 2006). GNDs may contribute to hardening in different ways by different theoretical models. In our model, both slip resistance hardening and back stress hardening are included. Each is considered singly to distinguish its effect on passivated thin films.

Slip resistance hardening is introduced via a generalized Taylor equation by adding the density of GNDs to that of SSDs (Han et al., 2005; Shu et al., 2001):

$$g^{(\alpha)} = g_{0}^{(\alpha)} + \alpha\mu b \sqrt{\rho_{\text{SSD}}^{(\alpha)} + k\rho_{\text{GND}}^{(\alpha)}} \tag{4.4.1}$$

where $g_{0}^{(\alpha)}$ is the initial slip resistance including lattice friction stress, α is a parameter in the range of 0.3–0.5, μ is the shear modulus, and k is a parameter indicating whether to involve GND hardening. The evolution of SSD density follows the model proposed by Devincre et al. (2008) as

$$\dot{\rho}_{\text{SSD}}^{(\alpha)} = \frac{1}{b} \left(\frac{1}{L^{(\alpha)}} - 2y_{c}\rho_{\text{SSD}}^{(\alpha)} \right) \left| \dot{\gamma}^{(\alpha)} \right| \tag{4.4.2}$$

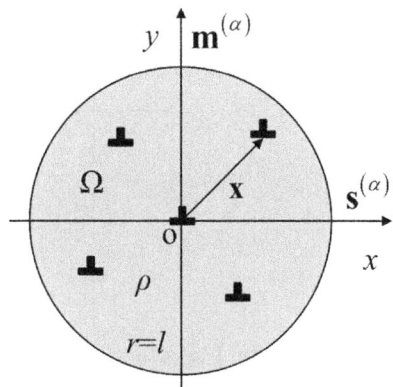

Figure 4.8
Back stress at material point O contains stress caused by dislocations within area Ω.

where $L^{(\alpha)}$ is the dislocation mean free path, generally $L^{(\alpha)} = K \Big/ \sqrt{\rho^{(\alpha)}_{\mathrm{SSD}}}$, and y_c represents the critical annihilation distance for dislocations.

Back stress hardening is formulated based on the elastic interaction between dislocations (Bayley et al., 2006; Liu et al., 2011). As shown in Fig. 4.8, a 2D planar model is used and only edge dislocation is considered. Back stress at material point O can be evaluated by summing the stress caused by dislocations within area Ω:

$$\tau^{(\alpha)}_b = \int_\Omega \rho^{(\alpha)}_{\mathrm{GND}}(\mathbf{x}) \frac{\mu b}{2\pi(1-\nu)} \frac{x(x^2-y^2)}{(x^2+y^2)^2} \, dxdy \tag{4.4.3}$$

where $\rho^{(\alpha)}_{\mathrm{GND}}(\mathbf{x})$ presents the GND density at coordinate $\mathbf{x}$. $\rho^{(\alpha)}_{\mathrm{GND}}(\mathbf{x})$ can be approximated by its Taylor expansion around the point O as

$$\rho^{(\alpha)}_{\mathrm{GND}}(\mathbf{x}) = \rho^{(\alpha)}_{\mathrm{GND}} + \nabla\rho^{(\alpha)}_{\mathrm{GND}} \cdot \mathbf{x} + \frac{1}{2!}\nabla^2\rho^{(\alpha)}_{\mathrm{GND}} : (\mathbf{x} \otimes \mathbf{x}) + \cdots \tag{4.4.4}$$

where $\rho^{(\alpha)}_{\mathrm{GND}}$ is the GND density at point O. Substituting Eq. (4.4.4) into Eq. (4.4.3) and keeping only the first nonvanishing terms, we can obtain the relation between back stress and GND density as

$$\tau^{(\alpha)}_b = \frac{\mu}{8(1-\nu)} l^2 b \nabla\rho^{(\alpha)}_{\mathrm{GND}} \cdot \mathbf{s}^{(\alpha)} \tag{4.4.5}$$

where μ is a lame constant and l is a characteristic length parameter of the film.

4.4.2 Model Description

The simulation model is sketched in Fig. 4.9. A 2D plain strain model is used with height h and width w. It has three slip systems oriented at 0, 60, and 120 degrees from the x-axis.

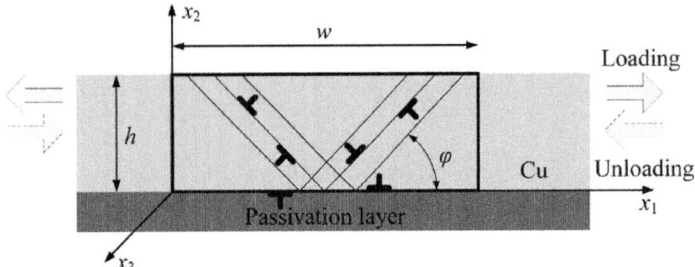

Figure 4.9

Thin film with one passivation layer and three slip systems is considered under loading and unloading conditions.

The lower surface is passivated, which is simulated as the dislocation impenetrable boundary. The boundary conditions are chosen as follows. Because it is periodic along the x-axis, the corresponding conditions are chosen for the left and right sides, respectively:

$$u_1(w, x_2) = u_1(0, x_2) + U$$
$$u_2(w, x_2) = u_2(0, x_2) \qquad\qquad (4.4.6)$$
$$\gamma^{(\alpha)}(w, x_2) = \gamma^{(\alpha)}(0, x_2) \quad \alpha = 1, 2, 3$$

where U is the applied displacement. The upper surface is a free surface, given by

$$\sigma_{22}(x_1, h) = \sigma_{12}(x_1, h) = 0$$
$$\xi_1^{(\alpha)}(x_1, h) = \xi_2^{(\alpha)}(x_1, h) = 0 \quad \alpha = 1, 2, 3 \qquad (4.4.7)$$

The lower surface is a dislocation impenetrable boundary, given by

$$\sigma_{22}(x_1, 0) = \sigma_{12}(x_1, 0) = 0$$
$$\gamma^{(\alpha)}(x_1, 0) = 0 \quad \alpha = 1, 2, 3 \qquad (4.4.8)$$

To compare the results with experimental data from Xiang and Vlassak (2006), the material parameters of copper are used, which can be found in the literature (Bayley et al., 2006), as listed in Table 4.2. These parameters are the same for all simulations with different film thicknesses.

The choice of the initial slip resistance g_0 is made under the following considerations. In the experiment (Xiang and Vlassak, 2006), films without a passivation layer also have a

Table 4.2: Material (Cu) Parameters in Simulations

Young's modulus E	144 GPa	Mean free path coefficient K	26
Poisson's ratio ν	0.33	Annihilation distance y_c	1.6 nm
Burgers vector b	0.256 nm	Taylor hardening coefficient α	0.3

Table 4.3: Initial Slip Resistance With Film Thickness

$h/\mu m$	0.89	0.67	0.61	0.44	0.34
g_0/MPa	108.38	115.23	129.77	150.08	158.31

slight size effect. This part of the size effect comes from the decreasing grain size with a decrease in the film thickness and follows the Hall-Petch relation. It is much weaker than the size effect caused by the passivation layer. In our simulations, the size effect caused by grain size is directly considered in the initial slip resistance. Therefore, the initial slip resistance g_0 is first calibrated according to the yield strength at a 0.2% offset strain of unpassivated films. Because the unpassivated films deform uniformly, there are no GNDs. This process excludes the size effect caused by non-GND factors. The resulting initial slip resistance is listed in Table 4.3.

4.4.3 Size Effect of Passivated Thin Films Under Tension

4.4.3.1 Slip Resistance Hardening Model

In this case, only the contribution of GNDs to slip resistance is considered by taking $l = 0$ in Eq. (4.4.5). The parameter k in Eq. (4.4.1) is calibrated by the result of passivated film with thickness $h = 0.89$ μm. Then the obtained value $k = 13$ is used in films with the other thickness to find whether the predicted size effect matches the experiment (Xiang and Vlassak, 2006). Simulation results of passivated thin films with a thickness ranging from 0.31 to 0.89 μm are plotted in Fig. 4.10.

Fig. 4.10A shows can be found that the hardening is indeed increased by considering GNDs in slip resistance. In Fig. 4.10B, the predicted yield stresses show a certain amount of size effect and compare well with experimental results when the thicknesses are large

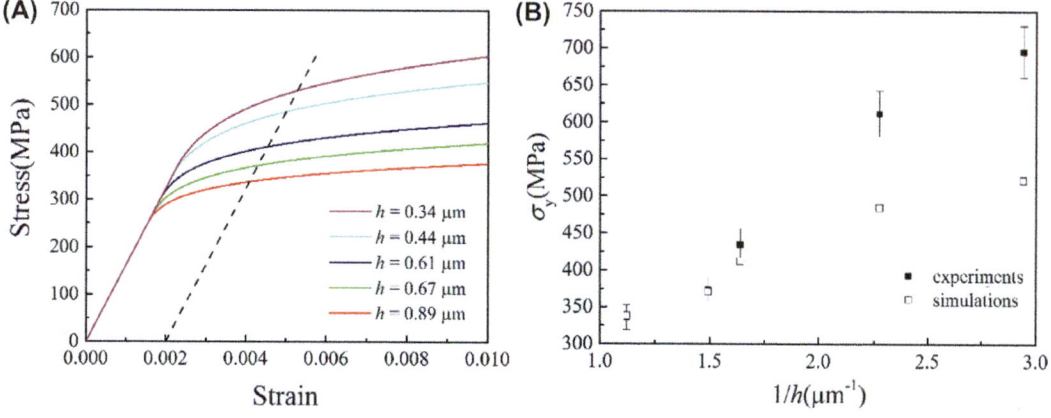

Figure 4.10
Results of passivated films in tension by slip resistance hardening model: (A) Stress-strain curves, and (B) Yield stress at 0.2% offset strain.

($h = 0.89$, 0.67, and 0.61 μm). However, as the thickness further decreases, the predicted yield stresses are much lower than the experimental data, which indicates that involving GNDs in slip resistance via Taylor hardening does not effectively reflect the size effect in passivated thin films.

4.4.3.2 Back Stress Hardening Model

In this case, only the contribution of GNDs to back stress is considered by taking $k = 0$ in Eq. (4.4.1). Characteristic length l in Eq. (4.4.5) is calibrated as the result of passivated film with a thickness of $h = 0.89$ μm and the corresponding $l = 0.498$ μm. The same value is used in films with other thicknesses. The simulation results, together with experimental data (Xiang and Vlassak, 2006), are provided in Fig. 4.11.

Fig. 4.11A shows that obvious hardening appears when back stress includes the contribution of GNDs. In Fig. 4.11B, the predicted yield stresses match well with experimental data. This indicates that back stress hardening caused by GNDs is dominated by the size effect for passivated thin films at a submicron scale. This proposition is consistent with conclusions from the work of Mayeur and McDowell (2013), who assessed several hardening models for incorporating GNDs by comparing them with dislocation dynamics simulations.

4.4.4 Bauschinger Effect of Passivated Thin Films During Unloading

This section discusses the plastic behavior of passivated thin films during unloading. Experimental observations (Xiang and Vlassak, 2006) revealed a prominent Bauschinger effect during unloading, with significant reverse plastic flow even when overall stress in the film was still tension. Both slip resistance hardening and back stress hardening are

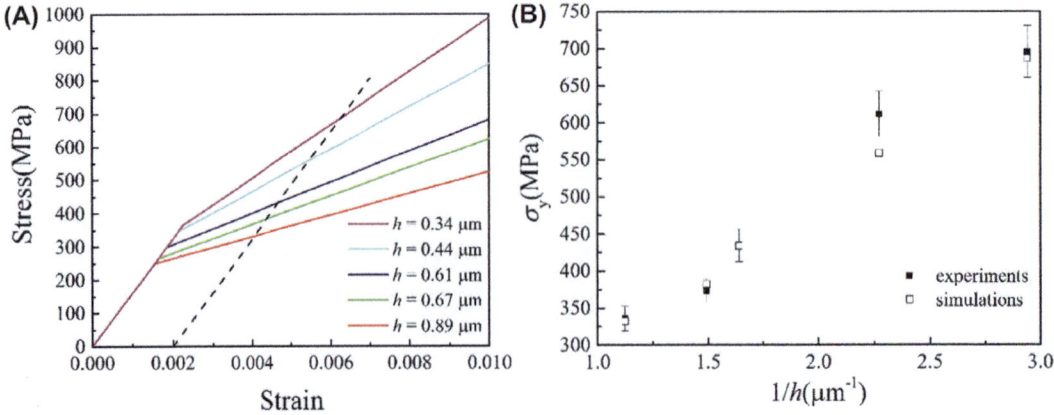

Figure 4.11

Results of passivated films in tension by back stress hardening model: (A) Stress-strain curves, (B) Yield stress at 0.2% offset strain.

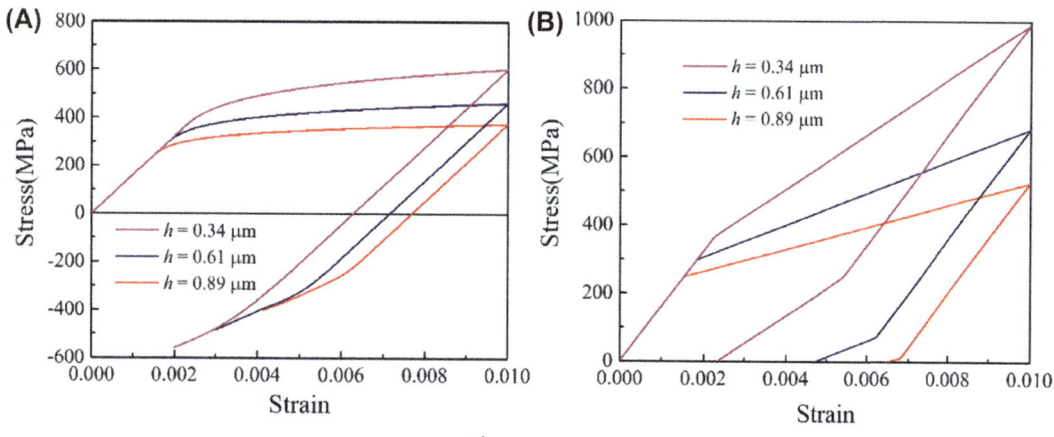

Figure 4.12
Unloading stress-strain curves of passivated films with different thickness, (A) By slip resistance hardening model, (B) By back stress hardening model.

adopted to study the contributions of GNDs to the Bauschinger effect, respectively. Changes in the Bauschinger effect on film thickness and prestrain (strain at the beginning of unloading) are investigated.

4.4.4.1 Bauschinger Effect With Film Thickness

First, the plastic behavior of passivated films during unloading or reverse loading is simulated with thicknesses $h = 0.89$, 0.61, and 0.34 μm, respectively. The prestrain is $\varepsilon_u = 0.01$. The results obtained using the slip resistance hardening model are shown in Fig. 4.12A, whereas results using the back stress hardening model are shown in Fig. 4.12B.

For the slip resistance hardening model, as shown in Fig. 4.12A, there is only a slight Bauschinger effect for all thicknesses. Reverse plastic flow happens only when the films are under large compression stress. The thinner the film, the larger the compression stress. This is not consistent with experimental data (Xiang and Vlassak, 2006), in which thinner films showed a larger Bauschinger effect.

For the back stress hardening model, as shown in Fig. 4.12B, the Bauschinger effect observed in the experiments is revealed. For example, for the film with a thickness of $h = 0.34$ μm, reverse plastic flow begins when the overall large tension stress is about 250 MPa. Fig. 4.12B also shows that the Bauschinger effect increases as the film thickness decreases, which matches well with experimental observations.

4.4.4.2 Bauschinger Effect With Pre-strain

Next, the plastic behavior of passivated films is simulated during unloading from different prestrains, $\varepsilon_u = 0.010$, 0.012, 0.014, and 0.016, respectively. The simulation results of films with a thickness of $h = 0.89$ μm are illustrated in Fig. 4.13.

As an application, the strong size effect and Bauschinger effect in thin films with a passivation layer during loading and unloading were numerically studied using the proposed crystal plasticity model. Simulation results showed that the passivation layer impeded the motion of dislocations, leading to dislocation pileups, which resulted in strong hardening in the flow stress. Further studies showed that the hardening came from dislocations interaction-induced back stress rather than slip resistance. The reason for the Bauschinger effect was the nonuniform distribution of stress. During unloading, high compression stress may have existed in some areas, leading to reverse plastic deformation.

Revealing the Size Effect in Micropillars by Dislocation-Based Crystal Plasticity Theory

Chapter Outline

5.1 Introduction

Experimental studies showed that crystalline materials exhibit many features at the micro- to nanoscales. The most widely known one is size effect, which means that the properties of materials are significantly different as the size changes. Size-dependent phenomena appear when the sample dimension is on a small scale. At these scales, dislocation-surface and dislocation-dislocation interactions have an important role in the plastic deformation of crystals. These phenomena can be considered in two ways. First, for the overall response, size-dependent flow stress and serrated stress-strain curves of crystal micropillars are observed in uniaxial compression tests (Greer et al., 2005; Kiener and Minor, 2011a;

Dislocation Mechanism-Based Crystal Plasticity. https://doi.org/10.1016/B978-0-12-814591-3.00005-4
Copyright © 2019 Tsinghua University Press. Published by Elsevier Inc.

Uchic et al., 2004; Volkert and Lilleodden, 2006). Second, the local deformation of the pillar, such as the deformation morphology and microstructure evolution, is size-dependent.

In this chapter, the two categories of size-dependent deformations in micropillars are studied by applying the dislocation-based crystal plasticity model described in Chapter 4. Size effect and strain burst in compressed single-crystal micropillars and the size-dependent deformation morphology of compressed single-crystal micropillars are numerically studied in Sections 5.2 and 5.3, respectively.

5.2 Strain Burst and Size Effect in Compression Micropillars

With the aid of scanning electron microscopy (SEM) and transmission electron microscopy techniques, many significant phenomena have been revealed by experiments indicating that the mechanical properties of crystal materials at small scales are different from their counterparts at macroscales. Uniaxial compression tests of single crystals at small scales enable a new understanding of plastic behavior: (1) A strong size effect exists. The yield strength increases roughly with the inverse square root of the pillar diameter up to 15 times the bulk material. A size-dependent hardening rate is also observed (Dimiduk et al., 2005; Norfleet et al., 2008; Shan et al., 2008; Uchic et al., 2004; Volkert and Lilleodden, 2006). (2) The stress-strain curves are jerky with interspersing strain bursts and elastic segments. The occurrence of each burst under a given stress is not a deterministic process; and neither is the burst size (Dimiduk et al., 2006; Friedman et al., 2012a,b; Ng and Ngan, 2008b; Wang et al., 2012a,b). The size-dependent strength and stochastic signature in the stress-strain data during plastic flow will certainly influence the design of microelectromechanical systems. The size-dependent strength, or the "smaller is stronger" phenomenon may be beneficial in designing small-scale structures capable of supporting high stress before permanent deformation or failure. However, the stochastic discrete strain bursts would cause difficulties in controlling homogeneous plastic forming at small scales. Thus, it is important to reveal the internal mechanism causing these distinct phenomena at small scales.

5.2.1 Stochastic Crystal Plasticity Model

According to conventional crystal plasticity theory, plastic deformation is modeled as a smooth and continuous process, which is not able to describe intermittent flow in crystal plasticity at small scales. As mentioned earlier, dislocation avalanches lead to strain bursts and exhibit a shock-aftershock, earthquake-like behavior over a period. In the continuum model formulated by Zhang and Shang (Zhang and Shang, 2014; Zhang et al., 2014), the hybrid loading mode (including a loading stage, burst slip, and holding stage) is considered and strain bursts are well-captured by using different boundary conditions. Different from their model, strain burst is introduced into our

model by the constitutive equation. The main idea is that plastic strain does not change continuously but consists of a series of strain bursts. The size of each strain burst can be determined by a power-law distribution function and a constitutive equation controls the numbers of strain bursts during a time increment. Each strain burst can be considered a characteristic strain of an internal dislocation structure evolution event. This method is similar to the model proposed by Michael and Paolo (2005); it includes randomness in the local stress-strain relationships to consider intrinsic fluctuations. However, in our model, strain burst is adopted instead of a local stress fluctuation. On the other hand, in conventional crystal plasticity theory, slip resistance is usually determined by a phenomenological way that is size-independent. In our model, however, the single-arm dislocation (SAD) model with a modified parameter is incorporated into slip resistance to characterize the size effect. Except for these modifications, most of the equations are similar to conventional crystal plasticity theory.

We confine attention to the small strain condition, in which geometry changes are neglected. The kinematics of face-centered cubic structure (FCC) crystals are well-established in the literature (Asaro and Lubarda, 2006). The Cauchy stress rate $\dot{\sigma}$ is associated with the elastic strain rate $\dot{\varepsilon}^e$:

$$\dot{\sigma} = \mathbf{C} : \dot{\varepsilon}^e = \mathbf{C} : (\dot{\varepsilon} - \dot{\varepsilon}^p) \tag{5.2.1}$$

where $\mathbf{C}$ is the stiffness matrix, $\dot{\varepsilon}$ is the total strain rate, and $\dot{\varepsilon}^p$ is the plastic strain rate. $\dot{\varepsilon}^p$ is a combination of slip rate $\dot{\gamma}^{(\alpha)}$ in all slip systems:

$$\dot{\varepsilon}^p = \frac{1}{2} \sum_\alpha \dot{\gamma}^{(\alpha)} \left(s^{(\alpha)} \otimes m^{(\alpha)} + m^{(\alpha)} \otimes s^{(\alpha)} \right) \tag{5.2.2}$$

where $s^{(\alpha)}$ and $m^{(\alpha)}$ are the slip direction and slip plane normal, defining the αth slip system.

Different from conventional plasticity theory, slip increment $\Delta\gamma^{(\alpha)}$ during each time increment Δt is determined by a series of strain bursts s_i in our model, given by

$$\Delta\gamma^{(\alpha)} = \mathrm{sgn}\left(\tau^{(\alpha)}\right) \sum_i^n s_i \tag{5.2.3}$$

where $\tau^{(\alpha)} = s^{(\alpha)} \cdot \sigma \cdot m^{(\alpha)}$ is the resolved shear stress. Two variables should be determined to complete Eq. (5.2.3): the number of strain bursts n during time increment Δt and the size of each strain burst s_i. Here, n is related to the burst rate $r = n/\Delta t$, which is given by a constitutive equation

$$r = \frac{\dot{\gamma}_0}{\langle s \rangle} \left(\frac{\tau^{(\alpha)}}{g^{(\alpha)}} \right)^m \tag{5.2.4}$$

where $\dot{\gamma}_0$ is a reference slip rate; and m is a rate-sensitivity exponent, which usually has a large value (≥ 20). $g^{(\alpha)}$ is a size-dependent slip resistance including dislocation source operation stress, lattice friction stress, and Taylor hardening stress. $\langle s \rangle$ is the average size of strain bursts. The size of each burst s_i and the average size of strain bursts $\langle s \rangle$ are discussed in detail in the following sections. From Eqs. (5.2.3) and (5.2.4), it can be seen that although the burst rate is determined in one time increment, the slip increment $\Delta\gamma^{(\alpha)}$ is undetermined because s_i is different from one to another following a power-law distribution. This stochastic feature makes it possible to capture intermittent plastic behavior.

5.2.2 Determination of Size-Dependent Slip Resistance

Slip resistance $g^{(\alpha)}$ is determined by slip rate $\dot{\gamma}^{(\beta)}$ in conventional crystal plasticity theory (Asaro and Needleman, 1985) as

$$g^{(\alpha)} = g_0 + \int_t h_{\alpha\beta} \left| \dot{\gamma}^{(\beta)} \right| dt \tag{5.2.5}$$

where g_0 is the initial slip resistance and $h_{\alpha\beta}$ is the hardening modulus. g_0 is usually considered to be a constant. However, as mentioned earlier, at small scales the yield stress is strongly affected by internal microstructures and shows a so-called size effect. In our model, to consider the size effect, initial slip resistance g_0 is derived from the SAD model (Parthasarathy et al., 2007) with a modified parameter.

5.2.2.1 Dislocation Source-Controlled Initial Slip Resistance

The origins of an experimentally observed size effect in uniaxial compression of submicron and micron pillars have been explained by the dislocation starvation mechanism (Greer et al., 2005). However, the exact relation between strength and the pillar diameter has not been well-studied. There is widespread agreement that the size effect can be expressed as power-law d^{-x} with diameter d (Derlet and Maaß, 2014; Uchic et al., 2009a,b):

$$g_0 = Ad^{-x} \tag{5.2.6}$$

By fitting this to experimental data, it has been claimed that scaling exponent x is approximately 0.6 for FCC metals (Korte and Clegg, 2011; Uchic et al., 2009a,b). Derlet and Maaß (2014) pointed out that the size effect embodied by the power law originates from a combination of the size effect in stress derived from the extreme value statistics of an assumed distribution of critical stress and size effect in strain derived from finite scaling associated with scale-free dislocation activity. On the other hand, Dunstan and Bushby (2013) suggested using an exponent $x = 1$ to consider the operation stress of dislocation sources, which is associated with the effective dislocation source length.

According to the model proposed by Parthasarathy (Parthasarathy et al., 2007), critical shear stress consists of three parts: the stress required to operate a single-arm source, the stress from the dislocation forest given by the Taylor equation, and friction stress τ_0, expressed as (Parthasarathy et al., 2007)

$$g_0 = \frac{k\mu b}{\bar{\lambda}} + \alpha\mu b\sqrt{\rho_0} + \tau_0 \tag{5.2.7}$$

where k is a geometrical constant usually taken as 1.0, α is a parameter in the range 0.3–0.5, μ is the shear modulus, b is the magnitude of the Burgers vector, $\bar{\lambda}$ is the effective source length, and ρ_0 is the initial dislocation density. In Dunstan's analysis (Dunstan and Bushby, 2013), effective source length $\bar{\lambda}$ was taken as the radius of pillar $d/2$, whereas in Parthasarathy's work it was derived by a statistical model.

In our model, initial slip resistance takes the form of Eq. (5.2.7). Because Eq. (5.2.7) is derived from the SAD model, it has physical meaning only in small samples (<1 μm), in which there are only a few dislocation sources. For large samples (5–40 μm), which are modeled in this section, Eq. (5.2.7) is only a simplification and $\bar{\lambda}$ is not actually the effective source length but rather a characteristic length of microstructures. Thus, the relation between effective source length $\bar{\lambda}$ and pillar diameter d will be different in the submicron and micron ranges. To obtain a fine ratio of $\bar{\lambda}$ to d, experimental data from the literature are used to find the best-fit parameters.

5.2.2.2 Relation Between Effective Source Length and Pillar Diameter

At first, Parthasarathy's statistical model for small samples is briefly introduced to enable a better understanding of Eq. (5.2.7) in the submicron and micron ranges. For small samples, the double-pinned Frank-Read sources interact with the free boundary and result in truncated SAD sources. Image stresses from the free surface tend to rotate the dislocation line until it is normal to the surface; thus, the effective source length can be considered the shortest distance from a pinning point to a free surface. Based on this, Parthasarathy et al. proposed a statistical model to obtain $\bar{\lambda}$ (Parthasarathy et al., 2007):

$$f = \frac{\bar{\lambda}}{d} = \int_0^{1/2} 2nx\left[1 - (1-2x)\left(\frac{1}{\cos\beta} - 2x\right)\cos\beta\right]^{n-1} \times \left(1 + \frac{1}{\cos\beta} - 4x\right)\cos\beta \, dx \tag{5.2.8}$$

where n is a number of pins in the pillar and β is the angle between the primary slip plane and the compression axis. Thus, for a given loading orientation, the ratio f of $\bar{\lambda}$ to d is related only to n, as demonstrated in Fig. 5.1. This ratio increases to a stable value of approximately 0.5 as the number of pins increases. It changes apparently only when there are a few pins. Assuming the average length of each dislocation segment is equal to the

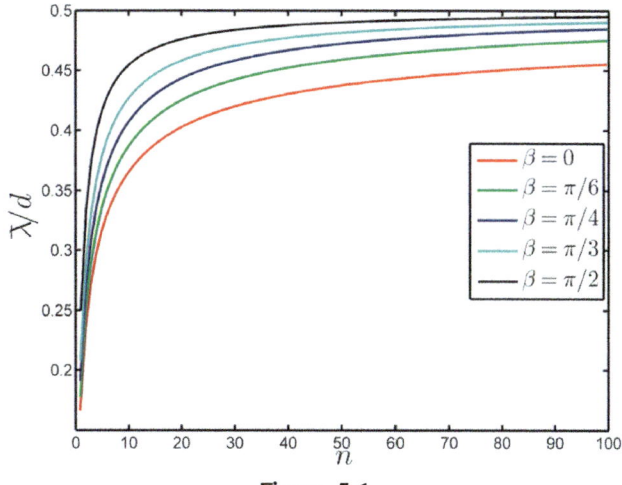

Figure 5.1

Relation of effective source length to number of pins from Parthasarathy's model. *Reprinted from Lin, P., Liu, Z.L., Cui, Y.N., Zhuang, Z., 2015. A stochastic crystal plasticity model with size-dependent and intermittent strain bursts characteristics at micron scale. International Journal of Solids and Structures 69—70, 267—276. Copyright 2015, with permission from Elsevier.*

pillar radius, the number of pins n is related to sample size d and dislocation density ρ (Parthasarathy et al., 2007):

$$n = \text{Integer} \left[\frac{\rho \pi d h}{2 n_{\text{slip}}} \right], \tag{5.2.9}$$

where h is the height of micron pillars and n_{slip} is the number of slip systems, and is equal to 12 for FCC crystals. With a constant dislocation density about $10^{12}\ \text{m}^{-2}$, it can be seen that ratio $\bar{\lambda}/d$ increases from 0.2 to 0.5 as diameter d (number of pins n) increases.

The critical shear stress predicted by this model with Eq. (5.2.7) was compared with experimental data in micron and submicron compression tests (Dimiduk et al., 2005; Frick et al., 2008; Uchic et al., 2004), as shown in Fig. 5.2. The result showed that the predicted values fit the experimental data well at a submicron scale, but for micron pillars the predicted values were much lower than the experimental data. Meanwhile, the statistical model showed discrete features in the low microregime in Fig. 5.2. This can be understood as follows. In Eq. (5.2.9), source number n is an integer and is a function of the pillar diameter. Using the parameter from Parthasarathy's work (Parthasarathy et al., 2007), in which $\rho = 2 \times 10^{12}/\text{m}^2$, $h = 2.5d$, and $n_{\text{slip}} = 12$, n can be obtained for different pillar diameter d as in Table 5.1. Together with Fig. 5.1, we can see that when n changes from 1 to 2, the ratio of the effective source length to the pillar diameter will increase dramatically in Fig. 5.2. Table 5.1 also shows that in Parthasarathy's statistical model, only a few dislocation sources exist in small samples.

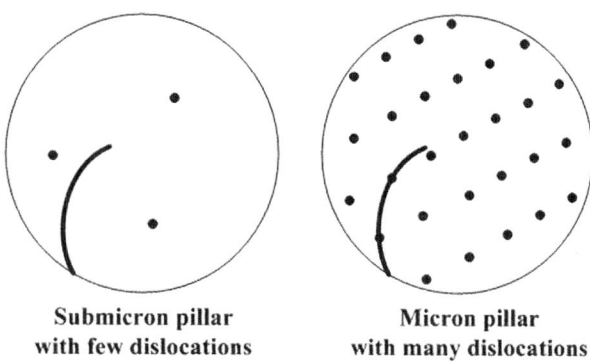

**Submicron pillar
with few dislocations**   **Micron pillar
with many dislocations**

Figure 5.3
Single-arm dislocation source on the slip plane interacts with dislocations on other slip planes.
For a submicron pillar, the single arm is rarely affected by forest dislocations. For a micron pillar,
the single arm is truncated by forest dislocations. *Reprinted from Lin, P., Liu, Z.L., Cui, Y.N.,
Zhuang, Z., 2015. A stochastic crystal plasticity model with size-dependent and intermittent strain bursts
characteristics at micron scale. International Journal of Solids and Structures 69–70, 267–276. Copyright
2015, with permission from Elsevier.*

divide between the two mechanisms and can be evaluated by $1/\rho$. Experiments (Maaß and
Uchic, 2012; Norfleet et al., 2008) showed that the dislocation density increases with the
decreasing pillar diameter in the range $10^{11} \sim 1^3$ m^{-2}. Therefore, the space between
dislocations can be considered about $0.3 \sim 3$ μm. It is difficult to decide the exact size
regarding whether using an SAD source is reasonable because the microstructures change
gradually from small samples to large ones. Thus, here an approximate value of 1 μm is used
to distinguish between the submicron and micron scales. The theoretical ratio of the effective
source length to the pillar diameter is reasonable in a sample much smaller than 1 μm but not
in a sample much larger than 1 μm. When the sample size is around 1 μm, it is not clear.

Because the statistical model is not suitable for large pillars, $\bar{\lambda}$ in Eq. (5.2.7) is obtained by
fitting the experimental data. To compare the different effective source lengths at
submicron and micron scales, the experimental data of submicron pillars (Frick et al.,
2008) and micron pillars (Dimiduk et al., 2005) are adopted. The size effect can be
modified by the form according to Eq. (5.2.7):

$$g_0 = \frac{\mu b}{fd} + \tau_a \qquad (5.2.10)$$

where f is a ratio of the effective source length $\bar{\lambda}$ to diameter d, μ is the shear modulus, b
is the magnitude of the Burgers vector, and τ_a is a source-independent resolve shear stress
consisting of Taylor hardening and lattice friction stress. Assuming f and τ_a are constants in
both submicron and micron pillars individually, and using the least square method, both τ_a
and f are obtained by fitting. The results are plotted in Fig. 5.2, together with the predicted
results of the statistical model.

Fig. 5.2 shows that for submicron pillars, the ratio of the effective source length to the diameter is $f = 0.328$, and for micron pillars, $f = 0.104$. Using discrete dislocation dynamics (DD) simulations and a theoretical model, Cui et al. (2014) pointed out that a stable value of $n \approx 3$ (the number is larger than that in Table 5.1 owing to the use of a higher dislocation density) exists in compression tests of submicron pillars 200–800 nm in diameter. Considering dislocation multiplication by a single-arm source and escape from a free surface, a constant value of $\bar{\lambda}/d = 0.317$ is derived, which is close to the fitted data here. This confirms that the statistical model based on SAD (Parthasarathy et al., 2007) can calculate the effective source length in Eq. (5.2.7) effectively for small samples. However, for large samples, $\bar{\lambda}/d \approx 0.5$ is obtained using the statistical model in Fig. 5.1, whereas the fitted result is 0.104 here. The overestimate of $\bar{\lambda}/d$ is explained in Fig. 5.2. The exact theoretical expression of $\bar{\lambda}/d$ has not yet been established and needs further study. In the following discussions, $\bar{\lambda} = 0.104d$ is taken to consider the size effect of micron pillars.

On the other hand, the source-independent resolved shear stress τ_a of submicron pillars ($\tau_a = 93.65$ MPa) is much larger than that of micron pillars ($\tau_a = 18.95$ MPa). For a bulk material with an initial dislocation density of $2 \times 10^{12} \text{m}^2$ and a lattice friction stress of $\tau_0 = 11$ MPa, $\tau_a = 18.73$ MPa is obtained and is almost the same with the fitted results of micron pillars. The high shear stress of $\tau_a = 93.65$ MPa for submicron pillars indicates that the threshold stress to activate the dislocation source may be also size-dependent and sensitive when the diameter is below 1 μm.

This is only a simplification to distinguish between submicron and micron artificially and to fit experimental data with different parameters. As mentioned earlier, there is no exact value to separate them. Thus, the fitted results cannot be covered along the full range of d, especially in the range near 1 μm, where the SAD mechanism gradually changes to a collective dislocation network-dominated mechanism. Because our model is applied to very large samples (5–40 μm), it is valid to use $\bar{\lambda} = 0.104d$ in Eq. (5.2.7) to consider the size effect phenomenologically.

5.2.2.3 Strain Hardening for Micron Pillars

In our model, strain hardening is introduced by dislocation accumulation via the Taylor hardening relation as (Taylor, 1934)

$$g^{(\alpha)} = g_0 + \alpha\mu b \sqrt{\rho^{(\alpha)}} \tag{5.2.11}$$

where dislocation density $\rho^{(\alpha)}$ is a function of $\gamma^{(\alpha)}$.

For submicron pillars, as dislocations escape from free surface, the dislocation density decreases to a stable value, so strain hardening is absent (Cui et al., 2014). For micron pillars, the dislocation density increases as a result of dislocation interactions and causes

Taylor hardening during deformation. The rate at which dislocations accumulate under strain is given by the following relation (Devincre et al., 2008):

$$\frac{d\rho}{d\gamma} = \frac{1}{b}\left(\frac{1}{L} - 2y_c\rho\right) \tag{5.2.12}$$

where L is a mean free path, b is the Burgers vector, ρ is the total dislocation density, and y_c is proportional to the critical annihilation distance for screw dislocations. The first term on the right side of Eq. (5.2.12) represents the dislocation multiplication owing to forest dislocations and the second term describes the effect of a mechanism called dynamic recovery. Generally, $L = K/\rho$, where K is a dimensionless constant and dependent on the material.

y_c is the capture radius for annihilation; it has a significant effect only when dislocations accumulate to a high level. In our simulation, y_c is taken as 0 for simplification as many other researchers (Mayeur and McDowell, 2013). Therefore, only forest hardening is considered in our simulations.

5.2.3 Strain Bursts at Small Scales

The statistical features of strain bursts have been well-understood through both experiments and theoretical models. Two statistical features of strain bursts are incorporated into our stochastic model: the power-law distribution of burst size s_i and the size-dependent cutoff strain s_0.

5.2.3.1 Determination of Burst Size and Average Size

Strain bursts in microcrystals arise from the collective avalanche-like motion of dislocations. The dislocations may trap each other into jammed configurations during motion. When local stress exceeds a critical value, the jammed configurations at that site are destroyed. Then, the long-range interactions among dislocations may trigger the motion of other dislocations and make the destruction of jammed configurations a collective avalanche-like process. This process can be described using a self-organized branching model (Zapperi et al., 1995). Through theoretical analysis of this model, dislocation avalanches can be characterized by power-law size distribution, which is confirmed by a large number of experiments (Dimiduk et al., 2006; Friedman et al., 2012a,b; Ng and Ngan, 2008b) and three-dimensional (3D) DD simulations (Csikor et al., 2007) from multiple samples with different but statistically equivalent initial configurations. The distribution function has the general form (Csikor et al., 2007)

$$P(s) = Cs^{-\omega} \exp\left[-(s/s_0)^2\right] \tag{5.2.13}$$

where C is a normalization constant, $\omega \cong 1.5$ is a scaling exponent, and s_0 is the characteristic strain of the largest avalanche, called the cutoff strain. The distribution function is plotted in Fig. 5.4. It can be seen that small bursts occur more frequently than do large ones.

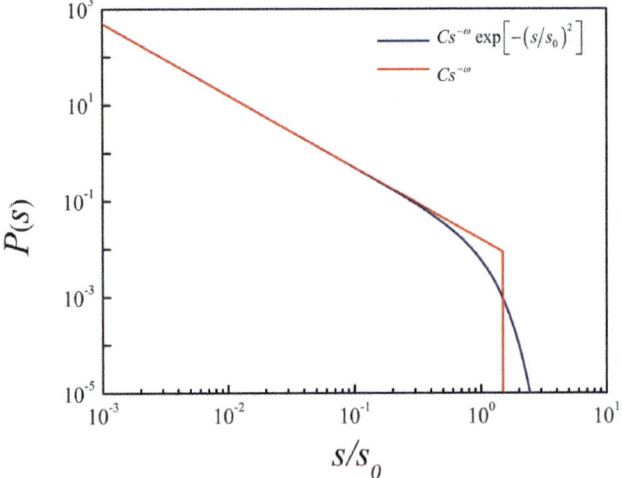

Figure 5.4
Strain burst distribution function in original and simplified forms. The simplified form differs from the original form only when the burst size is close to the cutoff strain. *Reprinted from Lin, P., Liu, Z.L., Cui, Y.N., Zhuang, Z., 2015. A stochastic crystal plasticity model with size-dependent and intermittent strain bursts characteristics at micron scale. International Journal of Solids and Structures 69–70, 267–276. Copyright 2015, with permission from Elsevier.*

Thus, strain bursts s_i in Eq. (5.2.3) can be derived from the distribution function Eq. (5.2.13). For simplicity, we modify Eq. (5.2.13) in the following ways. First, as we see in Fig. 5.4, the possibility $P(s)$ of extreme large burst s (e.g., $10s_0$) can be neglected, so there is no need to consider such a strain burst. We choose an interval $[s_{min}, s_{max}]$ as the boundary of strain bursts s. The choice of s_{min} and s_{max} will be discussed later. Second, with the exponent term $\exp\left[-(s/s_0)^2\right]$ in Eq. (5.2.13), the distribution function decreases significantly as the burst size increases close to the cutoff strain. However, when the strain burst is small, it has little effect on the distribution function. Therefore, to reduce the complexity of generating strain bursts in mathematics, Eq. (5.2.13) is simplified to the form

$$P(s) = Cs^{-\omega} \quad s_{min} \leq s \leq s_{max} \tag{5.2.14}$$

Both Eq. (5.2.13) and Eq. (5.2.14) are plotted in Fig. 5.4 for comparison.

Considering Eq. (5.2.13), as the burst size asymptotically approaches 0, the value of $P(s)$ diverges. That means that only amplitudes above the lower limit of s_{min} follow the power-law behavior. In the current work, we take into account only strain bursts that satisfy the statistical distribution described by Eq. (5.2.13). Accordingly, s_{min} was expressed as $s_{min} = 4\,b/L$ by Zhang and Shang (Zhang and Shang, 2014). The cutoff strain s_0 is discussed in detail in Section 5.2.3.2. s_{max} in Eq. (5.2.14) is slightly larger than s_0 and taken as $1.5s_0$ to obtain a better simplification.

If we use the distribution function as the form in Eq. (5.2.14) and take ω as 1.5, the average size of strain bursts $\langle s \rangle$ used in Eq. (5.2.4) is determined by

$$\langle s \rangle = \int_{s_{\min}}^{s_{\max}} sP(s)ds = \sqrt{s_{\min}s_{\max}} \propto \sqrt{s_0} \qquad (5.2.15)$$

It can be seen that the average size $\langle s \rangle$ is proportional to the square root of cutoff strain s_0.

It is easy to generate a random number, *rand*, with a uniform distribution in [0, 1] by a random number generator. To generate strain burst s following distribution function $P(s)$ in $[s_{\min}, s_{\max}]$, we have the following equation from statistical theory:

$$\frac{\int_{s_{\min}}^{s} P(s)ds}{\int_{s_{\min}}^{s_{\max}} P(s)ds} = rand \qquad (5.2.16)$$

Using Eqs. (5.2.14) and (5.2.16) together, we obtain the relation between burst size s and a uniform distributed number, *rand*:

$$s = s_{\min} \left\{ 1 + \left[\left(\frac{s_{\max}}{s_{\min}} \right)^{1-\omega} - 1 \right] rand \right\}^{\frac{1}{1-\omega}} \qquad (5.2.17)$$

So far, we have established the stochastic model to generate strain burst following the distribution function $P(s)$. Next, we can obtain slip increment $\Delta\gamma^{(\alpha)}$ in Eq. (5.2.3) by generating a series of strain bursts s_i via Eq. (5.2.17).

5.2.3.2 CutOff Strain s_0

By tracing the spatiotemporal dynamics of intermittent plasticity directly at small scales, Maaβ et al. (2013, 2015) showed that the slip size magnitude and slip velocities are independent of applied stress over a range of up to $\sim$400 MPa and are independent of the plastic strain and sample size. If the slip size remains within the same bounds for differently sized crystals, the strain burst increases proportionally to the inverse of the sample size. Thus, the cutoff strain s_0 in Eq. (5.2.13) can be obtained by the following equation according to Csikor's work (Csikor et al., 2007):

$$s_0 \propto \frac{bE}{L_c(\Theta + \Gamma)} \qquad (5.2.18)$$

where b is the magnitude of the Burgers vector, E is the Young's modulus, Θ is the strain hardening coefficient, and Γ is the effective stiffness of the specimen-machine system.

Eq. (5.2.18) shows that the smaller the specimen size L_c, the larger the cutoff strain s_0, and there are larger fluctuations of the stress-strain curves. For a bulk material, the cutoff strain is small, so the strain bursts size also tends to be small according to Eq. (5.2.13). This is the plasticity exhibits a uniform and smooth process at a macroscale.

On the other hand, other researchers (Dahmen et al., 2009; Derlet and Maaß, 2013; Friedman et al., 2012a,b; Ng and Ngan, 2008c) pointed out that s_0 is related not only to sample size but also to current stress τ:

$$s_0 \propto (g - \tau)^{-2} \tag{5.2.19}$$

where g is the slip resistance. Eq. (5.2.19) can also be derived from our model under the following consideration. In the stochastic model established in Section 5.2.1, if a small time increment, Δt, is taken, the resolution of strain is high enough to distinguish each strain burst. In such a time increment Δt, the following equation can be obtained from Eq. (5.2.4):

$$\frac{1}{\langle s \rangle} = \frac{1}{\dot{\gamma}_0 \Delta t} \left(\frac{\tau}{g} \right)^{-m} = \frac{1}{\dot{\gamma}_0 \Delta t} \left(1 - \frac{g - \tau}{g} \right)^{-m} \tag{5.2.20}$$

In a quasistatic loading, resolved shear stress τ is close to slip resistance g when the slip plane begins to glide, so we have

$$\frac{1}{\langle s \rangle} \approx \frac{1}{\dot{\gamma}_0 \Delta t} \left(1 + m \frac{g - \tau}{g} \right) \propto g - \tau. \tag{5.2.21}$$

In the *zero hardening* case (Dahmen et al., 2009), slip resistance g is a constant and equals τ_c. Meanwhile, from Eq. (5.2.15), we have $\langle s \rangle \propto \sqrt{s_0}$. Therefore, Eq. (5.2.21) derived from our model is the same as Eq. (5.2.19).

It can be concluded that the disagreements among different models result from different treatments of strain bursts. Eq. (5.2.18) is the statistical feature of the whole loading history whereas Eq. (5.2.19) describes the characteristic of strain bursts at a particular stress. In our model, the effects of both specimen size and stress can be considered.

5.2.4 Application to the Compression of Single-Crystal Ni Micron Pillars

As shown in Fig. 5.5, a single-crystal nickel micron pillar with diameter D is subjected to uniaxial compression. Height H is taken to be $2.5D$ in all simulations. The crystal is considered for symmetric double-slip systems oriented with angle $\phi^{(\alpha)}$ between the slip plane and compression axis, which are $\phi^{(1)} = \phi$ and $\phi^{(2)} = \pi - \phi$.

The slip direction and slip normal can be expressed as

$$\begin{aligned} \mathbf{s}^{(1)} &= \cos \phi \mathbf{e}_1 + \sin \phi \mathbf{e}_2, \quad \mathbf{m}^{(1)} = -\sin \phi \mathbf{e}_1 + \cos \phi \mathbf{e}_2, \\ \mathbf{s}^{(2)} &= -\cos \phi \mathbf{e}_1 + \sin \phi \mathbf{e}_2, \quad \mathbf{m}^{(2)} = -\sin \phi \mathbf{e}_1 - \cos \phi \mathbf{e}_2 \end{aligned} \tag{5.2.22}$$

Then, the resolved shear stress $\tau^{(\alpha)} = \mathbf{s}^{(\alpha)} \cdot \sigma \cdot \mathbf{m}^{(\alpha)}$ can be obtained by

$$\begin{aligned} \tau^{(1)} &= \frac{1}{2} (\sigma_{22} - \sigma_{11}) \sin 2\phi + \sigma_{12} \cos 2\phi \\ \tau^{(2)} &= -\frac{1}{2} (\sigma_{22} - \sigma_{11}) \sin 2\phi + \sigma_{12} \cos 2\phi \end{aligned} \tag{5.2.23}$$

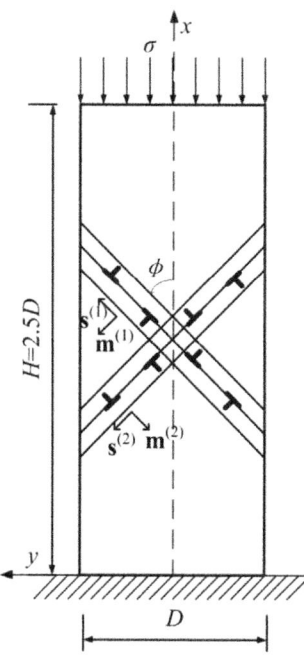

Figure 5.5
Uniaxial compression of single-crystal Ni micron pillar with symmetric double-slip systems.
Reprinted from Lin, P., Liu, Z.L., Cui, Y.N., Zhuang, Z., 2015. A stochastic crystal plasticity model with size-dependent and intermittent strain bursts characteristics at micron scale. International Journal of Solids and Structures 69—70, 267—276. Copyright 2015, with permission from Elsevier.

Considering $\sigma_{22} = \sigma_{12} = 0$, $\sigma_{11} = -\sigma$ in uniaxial compression, we obtain

$$\tau^{(1)} = -\tau^{(2)} = \frac{\sigma}{2} \sin 2\phi = \tau \tag{5.2.24}$$

and we have $\gamma^{(1)} = -\gamma^{(2)} = \gamma$.

During load-controlled compression, after each load increment $\Delta\sigma$, the elastic and plastic strain increment can be calculated by Eq. (5.2.2) as

$$\Delta\varepsilon^e = \frac{\Delta\sigma}{E}, \quad \Delta\varepsilon^p = \Delta\gamma \sin 2\phi \tag{5.2.25}$$

First, the effective source length can be estimated by $\overline{\lambda} = fd$, where $f = 0.104$ is taken for micron pillars. Then, the slip resistance can be obtained by Eqs. (5.2.10) and (5.2.11). The number of strain bursts, n, is calculated during time increment Δt by Eq. (5.2.4). Then, n strain bursts, s_i, are generated by Eq. (5.2.17). The slip increment and plastic strain can be calculated by Eqs. (5.2.3) and (5.2.25), respectively. Dislocation density is updated by Eq. (5.2.12) at the end of the increment.

Table 5.2: Material Parameters in Simulations

Pillar Diameter D	5—40 μm	Characteristic Length L_c	D
Pillar height H	2.5D	Strain hardening coefficient Θ	197.6 MPa
Shear modulus μ	76 GPa	Effective stiffness Γ	0 MPa
Poisson's ratio v	0.3	Mean free path coefficient K	26
Burgers b	0.24 nm	Annihilation distance y_c	0
Initial dislocation density ρ_0	3×10^{12} m^{-2}	Slip plane angle ϕ	45°
Lattice friction stress τ_0	11 MPa	Taylor hardening coefficient α	0.3

Parameters chosen in the simulations are listed in Table 5.2. Most of the parameters are chosen to be the same as in the computation models of Ni in the literature (Mayeur and McDowell, 2013). According to Csikor's work (Csikor et al., 2007), strain hardening coefficient Θ is taken as E/1000 and effective stiffness Γ is taken as 0 in the case of load-controlled compression. Mean free path coefficient K is according to Devincre's work (Devincre et al., 2008).

The simulation results of engineering stress-strain for Ni pillars with a diameter of 5—40 μm are plotted in Fig. 5.6A. The stress-strain curves successfully capture the characteristics of plastic flow at different sizes and are comparable to those of experimental data (Uchic et al., 2004). It can be seen that the size of the pillar affects plastic flow in two ways: (1) The strength of yield stress increases dramatically as the diameter decreases. The smaller the diameter, the bigger the yield stress. For diameter larger than 40 μm, the strengthening effect still exists but is no longer evident. (2) The postyield deformation is also strongly influenced by the size. For large pillars, the stress-strain curves are continuous and smooth, as shown in Fig. 5.6B. This means that the evolution of the

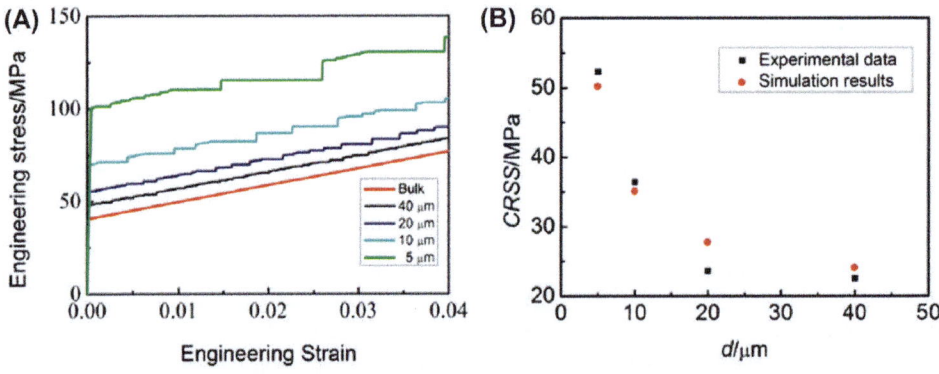

Figure 5.6

(A) Stress-strain curves of an Ni pillar with a diameter of 5—40 μm. (B) Yield stress of simulation results and experimental data. *CRSS*, critical shear stress. *Reprinted from Lin, P., Liu, Z.L., Cui, Y.N., Zhuang, Z., 2015. A stochastic crystal plasticity model with size-dependent and intermittent strain bursts characteristics at micron scale. International Journal of Solids and Structures 69—70, 267—276. Copyright 2015, with permission from Elsevier.*

underlying microstructures proceeds in a stable way. For small pillars, plastic flow is characterized by intermittent strain bursts and the stress-strain curves are staircase-shaped under conditions of stress control. As the size decreases, this phenomenon is much more evident. The large strain bursts lead to nonhomogeneous deformation and certainly influence the stability of the structures.

A further quantitative comparison with SEM experimental data (Uchic et al., 2004) is made in Fig. 5.6B. Flow stress is converted into resolved shear stress τ to facilitate the comparison. In the experiment, the resolved shear stress values are taken from [134] orientated compression Ni pillars and the Schmid factor is 0.471. For small pillars, we take the average value for each pillar because the flow stress fluctuates strongly at a small length scale. A good quantitative match is achieved.

Next, three random tests were carried out for each diameter to study stochastic mechanical behavior at the micron and submicron scales. The stress-strain curves for each pillar diameter are plotted in Fig. 5.7. Even with the same diameter, the stress-strain curves

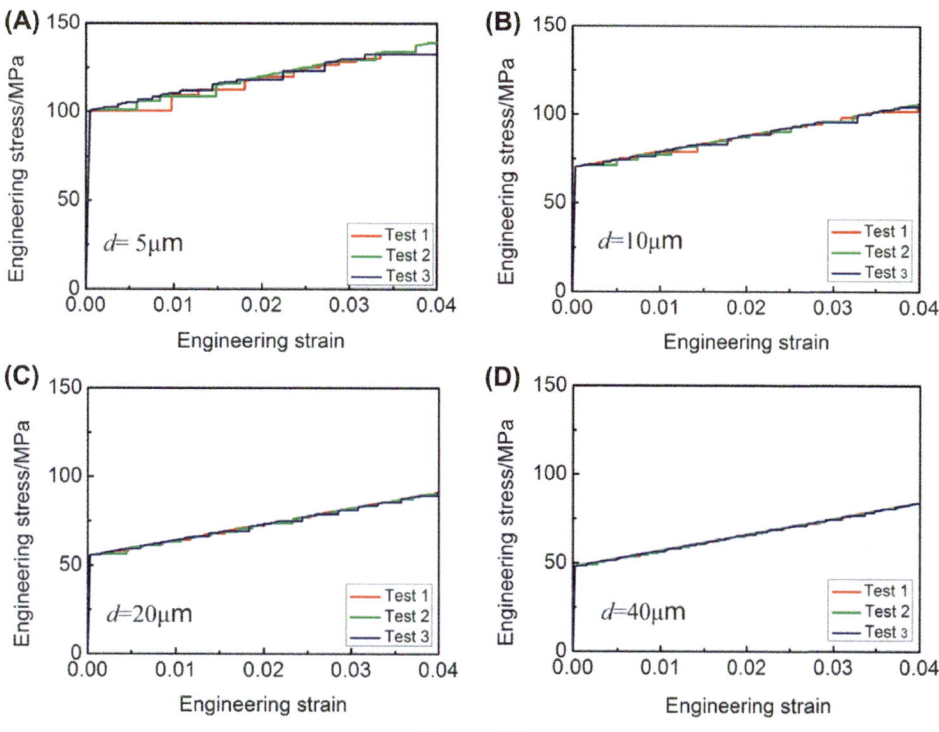

Figure 5.7

Results of three random simulations for each diameter. Stress-strain curves in small pillars show fluctuations and deviation, but in large pillars they are smooth and almost the same. *Reprinted from Lin, P., Liu, Z.L., Cui, Y.N., Zhuang, Z., 2015. A stochastic crystal plasticity model with size-dependent and intermittent strain bursts characteristics at micron scale. International Journal of Solids and Structures 69–70, 267–276. Copyright 2015, with permission from Elsevier.*

were different for each random test, as obtained in the experiments. For large pillars (e.g., $D = 40$ μm) the strain bursts were not evident and the stress-strain curves were almost the same for the three tests. However, as the size decreased (e.g., $D = 5$ μm), the strain bursts were much more evident and stress under a certain strain was different in the three tests, which means that deformation was indeterminate and unpredictable. This can be understood as explained subsequently. Because of the randomness of dislocations, small pillars exhibit stochastic and unpredictable plastic behavior. For large pillars, dislocations will interact with each other and form a stable microstructure. The mechanical behavior is the average response of many dislocations rather than some particular dislocations. Thus, large pillars exhibit almost the same plastic behavior. From Fig. 5.7, we can infer that below 20 μm, plastic behavior is uncontrollable. All of the results show that the stochastic model proposed in this section can predict both size effect and randomness at a micron scale.

5.3 Size-Dependent Deformation Morphology of Micropillars

Greer (2006) pointed out that small pillars under compression lacked stage II work-hardening: that is, small samples tended to deform on an individual slip system instead of activating multiple slip systems. After deformation, large pillars had barrel shapes with an almost homogeneous deformation whereas small ones had severe shear bands (Dimiduk et al., 2005) (Fig. 5.8A). Even loaded in a high-symmetry orientation <100> (Kiener and

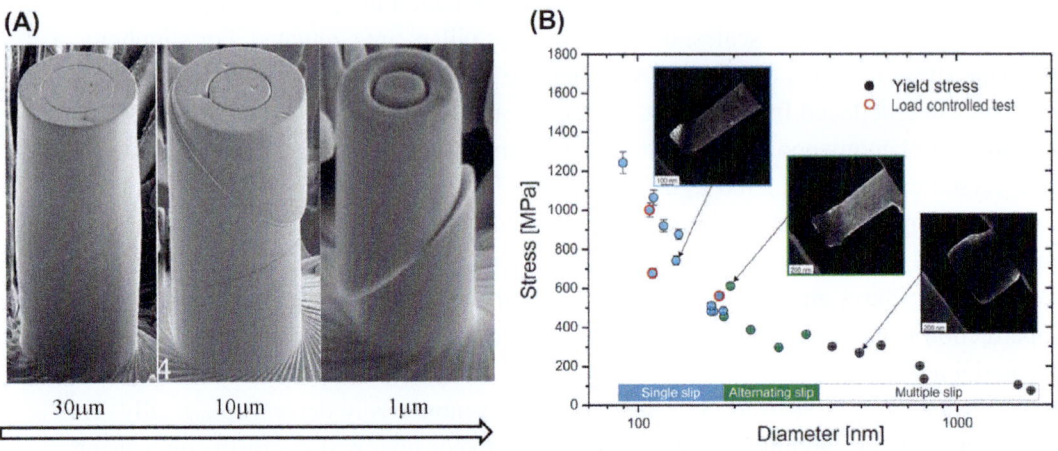

Figure 5.8

(A) Transition from barrel shape to shear band with decreasing pillar size, from experiments of Dimiduk et al. (2005); (B) Besides a size effect on strength, a change in the deformation morphology is shown, as indicated by differently *colored circles* and *color bars* indicating the regimes, from the experiments of Kiener and Minor (2011a). *(A) Reprinted from Dimiduk, D.M., Uchic, M.D., Parthasarathy, T.A., 2005. Size-affected single-slip behavior of pure nickel microcrystals. Acta Materialia 53, 4065–4077. Copyright 2005, with permission from Elsevier. (B) Reprinted from Kiener, D., Minor, A.M., 2011a. Source-controlled yield and hardening of Cu (100) studied by in situ transmission electron microscopy. Acta Materialia 59 (4), 1328–1337. Copyright 2011, with permission from Elsevier.*

Minor, 2011a), the pillars revealed a transition from multiple to single slip with decreasing diameter (Fig. 5.8B).

Many researchers attempted to study the underlying mechanism of size-dependent flow stress and serrated stress-strain curves. Lots of theoretical and numerical models were proposed (Akarapu et al., 2010; Akasheh et al., 2007; Csikor et al., 2007; Cui et al., 2014; Greer et al., 2005; Lin et al., 2015; Ohashi et al., 2007; Parthasarathy et al., 2007; Tang et al., 2007) and size-dependent flow stress and serrated stress-strain curves have been well-understood by these models. On the other hand, the size-dependent deformation morphology has not gained enough attention. So far, most studies focused on external factors. For example, the lateral constraint of the test system had a positive effect on the activation of multiple slip systems (Greer, 2006; Shade et al., 2009) and the deformation mode was observed to depend on the aspect ratio of the pillar (Milne et al., 2011; Ouyang et al., 2009). However, the influence of these external factors is not size-dependent and cannot reveal the underlying internal deformation mechanism.

In the next section, the dislocation-based crystal plasticity model proposed in Chapter 4 is used to study the role of back stress in size-dependent deformation morphology in micropillar compressions.

5.3.1 Simulation Setups

To study the mechanism of the size-dependent deformation morphology in pillar compression tests at small scales, different sizes of pillars are simulated. For simplicity, a 2D plane strain model is employed. Although the simplification may lose some information, the conclusion deduced from plain strain conditions can shed some light on understanding the physical phenomenon. Furthermore, under some circumstances, a rectangular strip in an FCC crystal can be considered to deform under plain strain conditions (Sluys and Estrin, 2000). To trigger localization, imperfection should be introduced. There are two kinds of imperfections in the literature: geometry (Niordson and Tvergaard, 2005) and material (Forest, 1998; Sluys and Estrin, 2000). To study the intrinsic mechanism, the latter imperfection is employed in our simulations. In general, the imperfection can be considered a material defect in the sample. For example, a dislocation source is located in the middle and as dislocations emit from the source, the center more easily deforms plastically. This is only a rough understanding, because it is difficult to consider real microstructures using continuum modeling. Details of the model are sketched in Fig. 5.9.

Deformation takes place in the x-y plane and the direction of compression corresponds to the x-axis. Different sizes of pillars have the same aspect ratio of $h/d = 2.5$ and the mesh is 80×200. Each pillar has two symmetric slip systems oriented at $\pm 60°$ from the x-axis. There is a material defect region with a radius of $r_0 = 0.1d$ in slip system $\alpha = 1$ in the middle of the pillar. We put the imperfection in the center because we want to exclude the

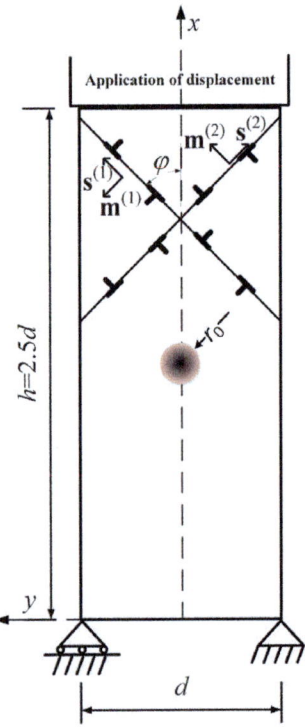

Figure 5.9

Schematic of plane strain model for a pillar of width *d* and height *h*, oriented for symmetric double-slip systems at ±φ from the *x* axis. A material defect region (*dark circle* in the middle) exists on slip system α = 1 with a radius of r_0. *Reprinted from Lin, P., Liu, Z., Zhuang, Z., 2016. Numerical study of the size-dependent deformation morphology in micropillar compressions by a dislocation-based crystal plasticity model. International Journal of Plasticity 87, 32—47. Copyright 2016, with permission from Elsevier.*

external differences of the two slip systems, e.g., the boundary and the geometric asymmetry, and focus on the influence of internal interactions between dislocations by back stress only. The initial slip resistance is slightly smaller in the material defect region and is given by

$$\widehat{g}_0(x,y) = ag_0 + \frac{\sqrt{(x - h/2)^2 + y^2}}{r_0}(1 - a)g_0, \tag{5.3.1}$$

where $\widehat{g}_0$ is the modified initial slip resistance in the material defect region and *a* is the ratio of residual initial slip resistance. $a = 0.9$ is taken in the simulations. According to Eq. (5.3.1), the modified initial slip resistance changes linearly from the defect center ($\widehat{g}_0 = ag_0$) to the defect boundary ($\widehat{g}_0 = g_0$). Instead of introducing material defect to one element (Forest, 1998; Sluys and Estrin, 2000), using Eq. (5.3.1) can eliminate the

Table 5.3: Material Parameters in Simulations

Young's Modulus E	200 GPa	Slip Resistance Parameter p_2	22 MPa
Poisson's ratio ν	0.3	Slip resistance parameter k_2	1800
Initial slip resistance $g_0^{(\alpha)}$	50 MPa	Back stress parameter S_0	50 MPa
Slip resistance parameter p_1	−45 MPa	Magnitude of Burgers vector b	0.25 nm
Slip resistance parameter k_1	210		

dependence on mesh size. The two vertical edges are traction-free whereas the bottom and top are controlled by displacement as

$$
\begin{aligned}
u &= 0, \quad \text{when } x = 0, \\
u &= -U, \quad \text{when } x = h,
\end{aligned}
\tag{5.3.2}
$$

where U is the magnitude of applied displacement. The displacement boundary condition applied allows rotation of the crystal axis. All microboundary conditions are chosen to let dislocations escape freely from boundaries as

$$
\xi^{(\alpha)} \cdot \bar{\mathbf{n}} = 0.
\tag{5.3.3}
$$

The other material parameters are listed in Table 5.3. The elastic properties and initial slip resistance are chosen to be those of copper. The hardening and softening of slip resistance are taken from Forest (1998).

5.3.2 Size-Dependent Deformation Morphology

The cumulative shear strain $\gamma_c \left(= \gamma_c^{(1)} + \gamma_c^{(2)} \right)$ is plotted in Fig. 5.10 to reveal the plastic deformation of pillars with different diameters. The three pillars exhibit different deformations. For the largest pillar ($l/d = 0.1$), after the plastic deformation starts from the material defect region, it propagates to the neighborhood in a homogeneous way, indicating that both slip systems operate during deformation. After deformation, the pillar is barrel-shaped with no obvious lateral displacement. The plastic zone does not proceed along a crystallographic direction. As the size of pillar decreases, plastic deformation tends to propagate along slip system $\alpha = 1$, which has a material defect region, and the operation of the other slip system, $\alpha = 2$, is suppressed. For the smallest pillar ($l/d = 1.0$), plastic deformation is carried by almost all of slip system $\alpha = 1$, even though the two slip systems are symmetrically oriented from the compression axis. After deformation, the pillar shows a clear lateral displacement and a shear band along slip system $\alpha = 1$ exists. These results are qualitatively consistent with experiments (Dimiduk et al., 2005; Kiener and Minor, 2011a) and DD simulations (Kiener et al., 2011). Experimental evidence (Maass et al., 2013, 2015) shows that slip offsets across different sample sizes are similar in size and the apparent homogeneous slip results only because more slip planes are available. Therefore, length parameter l used in our model has a physical meaning and is

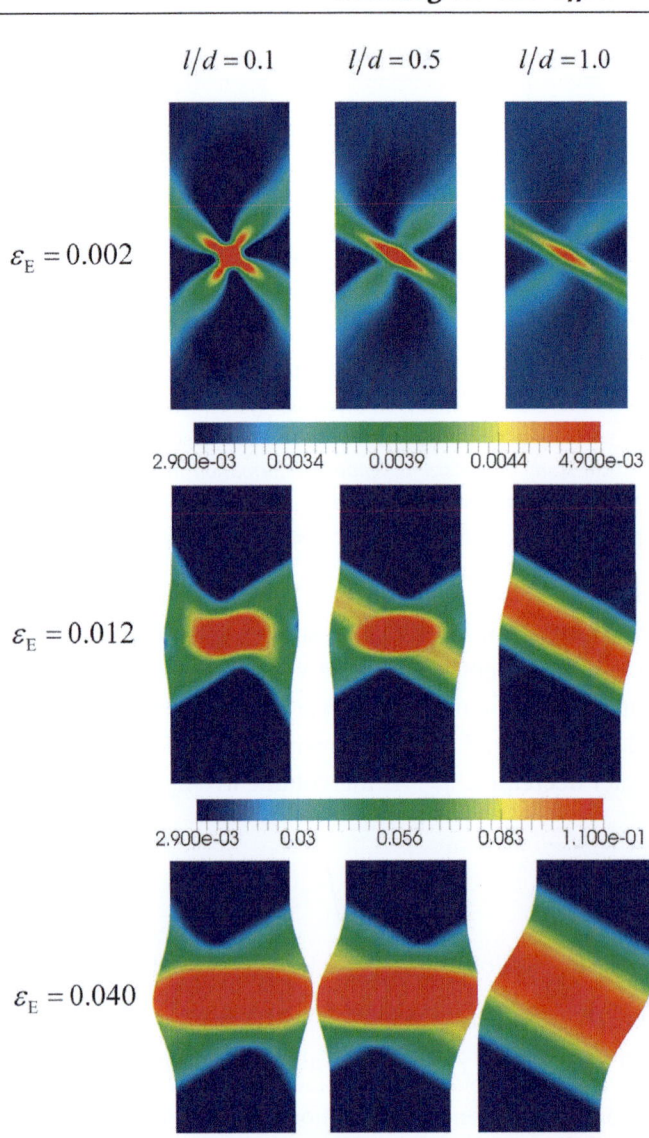

Figure 5.10

Cumulative shear strain of all slip systems $\gamma_c = \gamma_c^{(1)} + \gamma_c^{(2)}$. The same pillar during loading is plotted in the same column, whereas different sizes of pillars under the same engineering strain ε_E are plotted in the same row. Deformations are multiplied by a factor of three. *Reprinted from Lin, P., Liu, Z., Zhuang, Z., 2016. Numerical study of the size-dependent deformation morphology in micropillar compressions by a dislocation-based crystal plasticity model. International Journal of Plasticity 87, 32—47. Copyright 2016, with permission from Elsevier.*

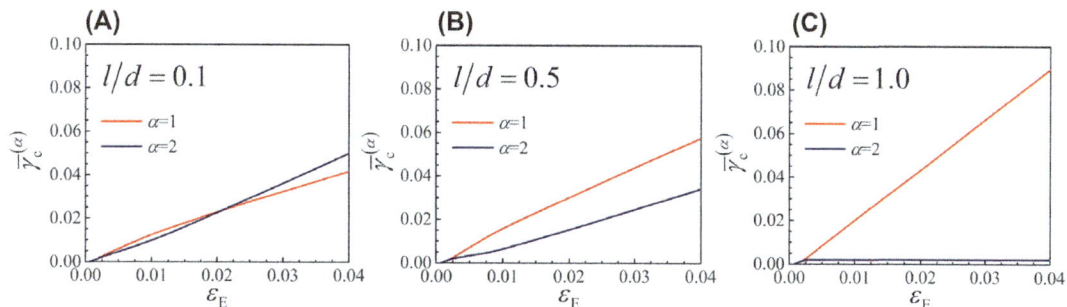

Figure 5.11

Volume averaged cumulative shear strain of two slip systems during loading. Transition from double slip to single slip is observed as the diameter decreases. *Reprinted from Lin, P., Liu, Z., Zhuang, Z., 2016. Numerical study of the size-dependent deformation morphology in micropillar compressions by a dislocation-based crystal plasticity model. International Journal of Plasticity 87, 32—47. Copyright 2016, with permission from Elsevier.*

associated with the distance between slip planes. The phenomenon from homogeneous deformation to severe shear is similar to that of strain bursts. Although the absolute value of the slip event is size-independent, the relative value to the sample size is size-dependent. The evolutions of the cumulative shear strain on each slip system are demonstrated more clearly in Fig. 5.11. The transition from double slip to single slip is observed as the diameter decreases. $\overline{\gamma}_c^{(\alpha)}$ is the volume averaged cumulative shear strain of the pillar, as

$$\overline{\gamma}_c^{(\alpha)} = \frac{1}{V} \int_V \gamma_c^{(\alpha)} dV. \tag{5.3.4}$$

The values of $\overline{\gamma}_c^{(\alpha)}$ at the end of loading ($\varepsilon_E = 0.040$) are listed in Table 5.4. When $l/d = 0.1$, $\overline{\gamma}_c^{(1)}$ is nearly equal to $\overline{\gamma}_c^{(2)}$. As the size decreases, $\overline{\gamma}_c^{(1)}$ increases while $\overline{\gamma}_c^{(2)}$ decreases, but the sum of $\overline{\gamma}_c^{(1)}$ and $\overline{\gamma}_c^{(2)}$ remains almost the same.

The activation of only one slip system (and not more) in some experiments (Ng and Ngan, 2008a; Parthasarathy et al., 2007) is based only on the appearance of the slip trace on the surface. By in situ Laue and electron backscatter diffraction, deformation details are revealed (Maass and Uchic, 2012; Maass et al., 2008a,b; Maass et al., 2009). These experiments show that lattice rotation and strain gradient exist at the early stage of deformation and the slip

Table 5.4: Volume Averaged Cumulative Shear Strain $\overline{\gamma}_c^{(\alpha)}$ at End of Loading

	$\overline{\gamma}_c^{(1)}$	$\overline{\gamma}_c^{(2)}$	$\overline{\gamma}_c^{(1)} + \overline{\gamma}_c^{(2)}$
$l/d = 0.1$	0.04172	0.04996	0.09168
$l/d = 0.5$	0.05749	0.03428	0.09177
$l/d = 1.0$	0.08953	0.00221	0.09174

systems that are geometrically unexpected are activated at the beginning of yield. This is consistent with our simulation results. As shown in Fig. 5.11 and Table 5.4, slip system $\alpha = 2$ is activated at the early stage even in the smallest pillar (Fig. 5.11C). Although $\overline{\gamma}_c^{(2)}$ is small compared with $\overline{\gamma}_c^{(1)}$, it is not exactly 0. Therefore, the single slip case in this chapter means that most of the slip strain is created by one slip system and the double slip case means the slip strains of the two slip systems are comparable.

5.3.3 Role of Short-Range Back Stress

Experimental evidence show that the geometrically necessary dislocation (GND) density (Maass and Uchic, 2012) or total density (Norfleet et al., 2008) is size-dependent. The results in our simulations are in accordance with these works. The results in Section 5.3.2 can be understood by studying the short-range interactions of dislocations. First, normalized GND densities $\left(\widehat{\rho}_G = \left(\left|\rho_G^{(1)}\right| + \left|\rho_G^{(2)}\right|\right)l^2\right)$ of different pillars are shown in Fig. 5.2A for comparison. High-GND densities are found at the boundary of the intersection of the two slip systems, and the distributions are size-dependent. The high-GND density regions in the largest pillar ($l/d = 0.1$) are disperse, indicating that localization takes place on a number of slip planes. These results are similar to the DD simulations from Kiener et al. (2011). The resulting back stress from Eqs. (4.3.6) and (4.3.13) of Chapter 4 is illustrated in Fig. 5.12B and C, showing that back stress is also size-dependent. The larger pillar has smaller back stress and high back stress happens at the boundary of the intersection of the two slip systems. Thus, the size-dependent morphology can be explained as follows. When localization starts at the material defect region and develops into a slip band along slip system $\alpha = 1$, stress inside the slip band decreases. To satisfy the macroforce equilibrium equation, the stress of Point P near but outside the slip band should also decrease (Fig. 5.13A). There are two deformation paths through which to accomplish this. One is that slip system $\alpha = 2$ at Point P operates and develops into another slip band; the other is that Point P unloads elastically (Fig. 5.13B). Which one happens can be decided by Eq. (4.3.4) in Chapter 4. Because back stress is small in the large pillar, $\left|\tau^{(2)} - \tau_b^{(2)}\right| > g^{(2)}$ is easily satisfied. As a result, the large pillar prefers to deform along Path 1, and then double slip appears. On the contrary, high back stress exists in the small pillar and prevents slip system $\alpha = 2$ from operating. Thus, Path 2 is more appropriate, and therefore single slip appears. These deductions can be confirmed by activating slip systems plotted in Fig. 5.14.

5.3.4 Critical Transition Size

A series of pillars with l/d in the range 0.1–1.0 were simulated to find the critical size of the transition from double to single slip. The ratio of the volume averaged cumulative

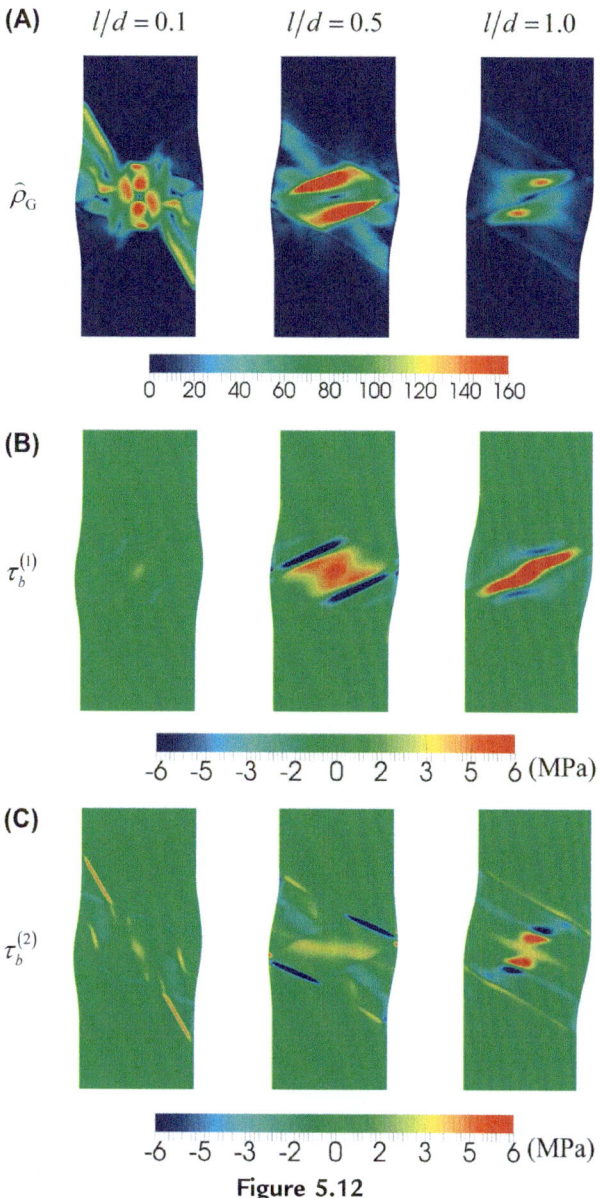

Figure 5.12

Normalized geometrically necessary dislocation density $\widehat{\rho}_G = \left(\left|\rho_G^{(1)}\right| + \left|\rho_G^{(2)}\right|\right)l^2$ and back stress $\tau_b^{(\alpha)}$ at $\varepsilon_E = 0.012$. $\widehat{\rho}_G$ shows different distributions at various sizes. Back stress in smaller pillars is larger than in larger pillars. Deformations are multiplied by a factor of three. *Reprinted from Lin, P., Liu, Z., Zhuang, Z., 2016. Numerical study of the size-dependent deformation morphology in micropillar compressions by a dislocation-based crystal plasticity model. International Journal of Plasticity 87, 32—47. Copyright 2016, with permission from Elsevier.*

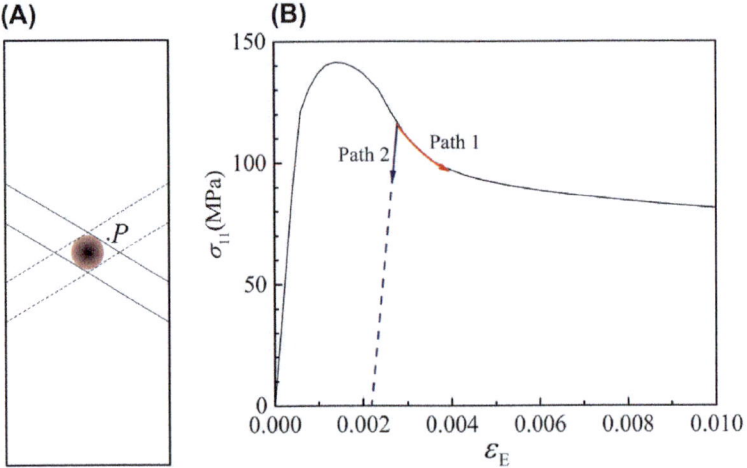

(A)

(B)

Figure 5.13

Illustration of whether the second slip band forms. (A) *P* is a material point near but outside the first slip band after localization appearing from the material defect region. (B) Two deformation paths exist at Point *P*: plastic localization by the second slip band (Path 1) and elastic unloading (Path 2). *Reprinted from Lin, P., Liu, Z., Zhuang, Z., 2016. Numerical study of the size-dependent deformation morphology in micropillar compressions by a dislocation-based crystal plasticity model. International Journal of Plasticity 87, 32—47. Copyright 2016, with permission from Elsevier.*

shear strain of two slip systems $\overline{\gamma}_c^{(2)} / \overline{\gamma}_c^{(1)}$ is plotted in Fig. 5.15. When the ratio is about 1, it means the two slip systems operate equally. When the ratio tends towards 0, only slip system $\alpha = 1$ is activated during plastic deformation. Consistent with the conclusions in Section 5.3.2, the ratio decreases as the diameter d decreases. Critical size d_c is easily obtained from Fig. 5.15. $D_N = (D\dot{U})/(g_0h)$ is a dimensionless representation of drag coefficient D in Eq. (4.3.7) of Chapter 4, which is used to discuss the influence of viscosity. $\dot{U}$ is the rate of applied displacement at the top of the pillar. If normalized drag coefficient D_N is chosen as 0.028 (the red curve in Fig. 5.15), l/d_c is supposed to be approximately 0.85. Once length parameter l is properly chosen, critical size d_c is determined. A length scale of $\mu b/g_0$ was proposed by Gurtin (2010) as the natural material length scale in back stress. Using the material parameters in our simulations, $l = \mu b/g_0 = 385$ nm is obtained. It is just a small slip plane spacing. Thus, critical size d_c is 453 nm. In experiments from Kiener and Minor (2011a), single slip, double slip ,and multiple slip appear in pillars with a diameter of 136, 208, and 535 nm, respectively. Critical size d_c obtained from our simulation is between double slip and multiple slip compared with experiments. The results from our simulations so far are under a 2D plane strain case and only two slip systems are considered. d_c is considered more precisely as the critical size from multiple slip to single slip in 3D models with real slip systems. In this sense, critical size d_c obtained from our simulations is consistent with the experiments.

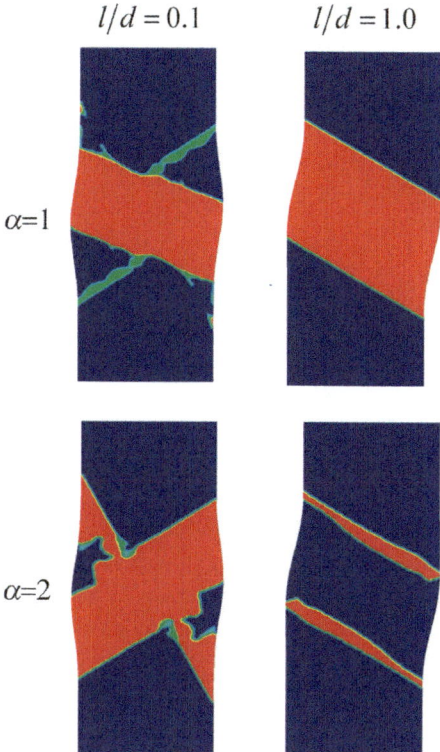

Figure 5.14
Activation of slip systems at $\varepsilon_E = 0.012$. *Red* indicates that the slip system is active; *blue* indicates that the slip system is inactive. Slip system $\alpha = 2$ develops into a slip band in a large pillar whereas it is inactive in most regions of the small pillar. Deformations are multiplied by a factor of three. *Reprinted from Lin, P., Liu, Z., Zhuang, Z., 2016. Numerical study of the size-dependent deformation morphology in micropillar compressions by a dislocation-based crystal plasticity model. International Journal of Plasticity 87, 32–47. Copyright 2016, with permission from Elsevier.*

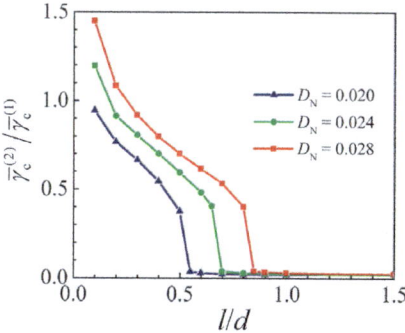

Figure 5.15
Ratio of volume averaged cumulative shear strain of two slip systems with respect to pillar size d and normalized drag coefficient D_N. *Reprinted from Lin, P., Liu, Z., Zhuang, Z., 2016. Numerical study of the size-dependent deformation morphology in micropillar compressions by a dislocation-based crystal plasticity model. International Journal of Plasticity 87, 32–47. Copyright 2016, with permission from Elsevier.*

Another finding from Fig. 5.15 is that viscosity has an obvious effect on the critical size. The larger the drag coefficient, the smaller the critical size. In other words, viscosity makes the pillars more likely to be in a multiple than a single slip state. Plastic slip tends to be homogeneous rather than taking place in a particular slip band owing to large viscosity. This is consistent with the early results of Watanabe et al. (1998), who suggested that strong viscosity tends to suppress shear band formation. The drag coefficient is size-independent and is different for different materials. Thus, different materials may have different critical sizes.

The critical size in our simulations is only a rough result under some simplifications. It is specific to the investigated system and can hardly be used to make general conclusions. It can only offer a better understanding of factors that influence size-dependent morphology qualitatively.

5.3.5 Discussions of Material Softening

To understand the formation of slip bands, simulations were performed with or without strain softening, determined by material parameter p_1 in Eq. (4.3.18) of Chapter 4. In the strain-softening case, p_1 is a negative value as in Table 5.3. In the no-strain softening case, p_1 is 0. The profiles of cumulative shear strain along $y = 0$ at different loading stages are plotted in Figs. 5.16 and 5.17. In Fig. 5.16, the formation and propagation of slip

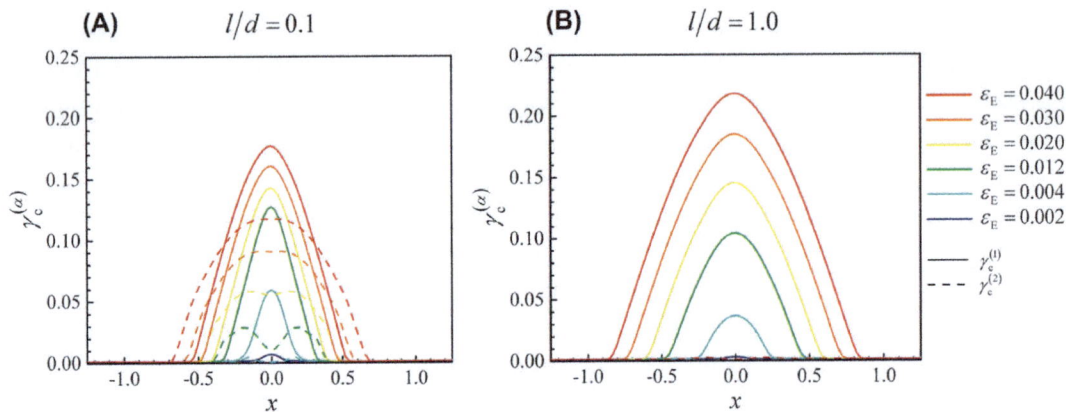

Figure 5.16

Profiles of cumulative shear strain along $y = 0$ with strain softening. *Different colors* represent different loading stages. *Solid lines* and *dashed lines* represent slip systems $\alpha = 1$ and $\alpha = 2$, respectively. ε_E is the engineering strain of the pillar. $\gamma_c^{(\alpha)}$ is the cumulative shear strain of the pillar on slip system α. l is the length parameter, and d is pillar diameter. *Reprinted from Lin, P., Liu, Z., Zhuang, Z., 2016. Numerical study of the size-dependent deformation morphology in micropillar compressions by a dislocation-based crystal plasticity model. International Journal of Plasticity 87, 32–47. Copyright 2016, with permission from Elsevier.*

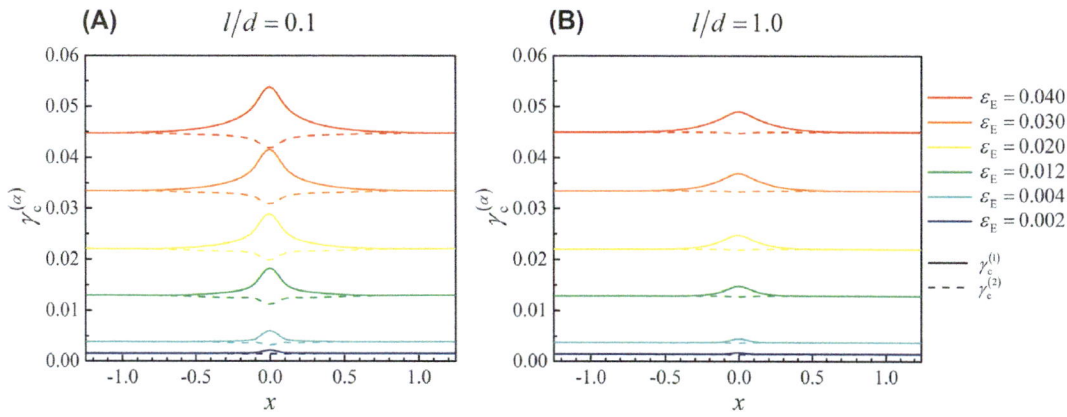

Figure 5.17

Profiles of cumulative shear strain along $y = 0$ without strain softening. *Different colors* represent different loading stages. *Solid lines* and *dashed lines* represent slip systems $\alpha = 1$ and $\alpha = 2$, respectively. ε_E is the engineering strain of the pillar. $\gamma_c^{(\alpha)}$ is the cumulative shear strain of the pillar on slip system α. l is the length parameter, and d is pillar diameter. *Reprinted from Lin, P., Liu, Z., Zhuang, Z., 2016. Numerical study of the size-dependent deformation morphology in micropillar compressions by a dislocation-based crystal plasticity model. International Journal of Plasticity 87, 32–47. Copyright 2016, with permission from Elsevier.*

bands are clearly revealed. A slip band of slip system $\alpha = 1$ originates from the material defect region and the width of the slip band increases during loading. No slip occurs outside the slip band. As discussed in Section 5.3.2, the activation of slip system $\alpha = 2$ is size-dependent. The second slip band appears in the large pillar and originates from the boundary of the first slip band, as shown by a line representing $\varepsilon_E = 0.004$ in Fig. 5.16A. It is the place where Point P in Fig. 5.13A is located and confirms the mechanism revealed in Section 5.3.3. In the small pillar, only the first slip band forms and propagates, and the second slip band does not exist (Fig. 5.16B). If there is no strain softening (Fig. 5.17), the introduced material defect region cannot trigger a slip band and slip take place in the whole region but is much smaller than in the strain-softening case. Therefore, both slip systems operate as in a homogeneous material and the introduced material defect region does not have much of an effect. Therefore, it is necessary to include material softening in our model to study size-dependent localization behavior, and it has a physical basis, as described earlier.

The slip band phenomenon has been addressed by many researchers over the years (Forest, 1998; Niordson and Redanz, 2004; Watanabe et al., 1998). Most models were proposed to study the necking of thin sheets in tension. The strain gradient plasticity theory predicts the delayed onset of localization compared with the conventional theory, and the width of the shear band is influenced by the strain gradient. In our model, the shear bands are simulated in compressed micropillars. The width of the shear bands shows

the same trend as in models mentioned previously. However, the suppression of a second slip system by back stress observed in our model has not been mentioned before (Fig. 5.16). The results are consistent with phenomena observed in experiments and may enable a better understanding of deformation in micropillar compressions.

5.4 Summary

In this chapter, the size-dependent deformation of micropillars at micro- to nanoscales were numerically studied via a dislocation-based crystal plasticity model. Conclusions regarding the simulation results include:

1. Size effect and strain burst in compressed single-crystal micropillars. The study shows that the reason for the increase in yield stress is that the dislocation sources are truncated by the surface, forming a short SAD source that has higher operating stress. Moreover, the strain burst phenomenon comes from either the intermittent operation of the dislocation source or dislocation avalanches. By introducing an SAD mechanism and strain burst mechanism, a stochastic plasticity model is proposed that well-captures the phenomena found by experiments. In addition, the model is capable of predicting uncontrollable deformation.

2. Size-dependent deformation morphology of compressed single-crystal micropillars. The study shows that as the size decreases, the deformation changes from a barrel shape to severe shear and crystal slip changes from multiple to single slip. Further analysis indicates that the transition of deformation morphology results from competition between applied external stress and back stress induced by short-range interactions of dislocations. When the size is large, there is little interaction of dislocation-induced back stress, so the second slip system is easily operated. However, when the size decreases, back stress increases and prevents the second slip system from operating. Furthermore, a critical transition size is obtained by a large amount of simulations and matches well with experimental data.

Microscale Crystal Plasticity Model Based on Phase Field Theory

Chapter Outline

6.1 Introduction

In Chapter 4, the dislocation mechanism-based crystal plasticity theory was developed to describe the size effect and Bauschinger effect by generally describing the collective behavior of dislocations. However, it is still hard to capture single dislocation behavior within the framework of continuum crystal plasticity theory, which is important for studying the mechanical properties of crystals at a microscale.

Several dislocation modeling methods completely based on continuum mechanics framework have been developed. The most important advantage of those methods is that the equilibrium field is solved directly. Based on field dislocation mechanics, a general crystal plasticity model was developed by Acharya (2001). Dislocation density serves as the primary internal variable and a finite element implementation of the theory is represented (Roy and Acharya, 2005). It has been used to study the mechanical response

Dislocation Mechanism-Based Crystal Plasticity. https://doi.org/10.1016/B978-0-12-814591-3.00006-6

of multilayer thin films (Puri et al., 2011). Inspired by crack modeling using the extended finite element method (FEM), the dislocation is modeled by a Heaviside function away from the core but enrichment is employed around the core (Belytschko and Gracie, 2007; Gracie et al., 2007). However, such core enrichment is not available for anisotropic materials (Gracie et al., 2008; Oswald et al., 2009).

Among all of the dislocation modeling methods, the phase field method (PFM), which represents discontinuities associated with dislocations by regularization, has received increasing attention (Khachaturyan, 1983; Hu and Chen, 2001; Wang et al., 2001; Rodney et al., 2003; Koslowski et al., 2004). The advantages of PFM are that there is no need to track dislocation segments explicitly and that it can automatically take into account elastic interaction between individual dislocations as well as the interaction between dislocations and interfaces within a unified framework (Wang et al., 2003; Lei et al., 2013). Currently, most phase field models of dislocations are based on the microelasticity theory of Khachaturyan and Shatalov (Khachaturyan, 1967, 1983; Khachaturyan and Shatalov, 1969). In the model, it takes stress-free inelastic strain to represent dislocations. The elastic strain energy caused by dislocations is expressed as a function of inelastic strain in a closed form through the exact Green's function. However, the analytical Green's function solution is not available for complex structures or complex boundary conditions. For elastically inhomogeneous structures, the virtual misfit strain that is considered to be additional phase fields has to be introduced and increases the number of equations to be solved (Wang et al., 2002). Dislocation dynamics in heteroepitaxial thin films have been simulated using the phase field microelasticity model (Wang et al., 2003). The same elastic modulus between the film and substrate is considered only to avoid dealing with the additional misfit strain between the film and substrate (Wang et al., 2003). On the other hand, the equations are solved by the fast Fourier transform method, which restricts the application to complex structures such as core-shell nanopillars. In addition, based on a Ginzburg-Landau-type evolution equation, the phase field microelasticity model is fully viscous in nature and does not differentiate between stored and dissipated energy (Wang et al., 2001, 2003; Lei et al., 2013).

In this chapter, a microscale crystal plasticity model based on phase field theory was developed to overcome difficulties in the phase field microelasticity model. In contrast, the stress-free inelastic strain is directly considered to be the plastic strain based on crystal plasticity theory in the microscale crystal plasticity model. The plastic slip associated with each slip system is described by a phase field to model a single dislocation. Elastic strain energy is expressed as a function of elastic strain through the crystal plasticity constitutive model. Based on a thermodynamically consistent framework that differentiates between energetic and dissipative mechanisms during plastic deformation, the coupled balances of quasistatic stress equilibrium and plastic slip evolution are derived from the principle of

virtual power. Then, the boundary value problem is solved directly by FEM. It can be used for complex structures or complex boundary conditions in which the analytical Green's function solution is unavailable, which is an advantage of the proposed model. Another advantage is that the elastic modulus mismatch in heteroepitaxial structures is easily treated without additional complications. Moreover, with numerical implementation by FEM, it is flexible for dealing with finite plastic deformation in heteroepitaxial structures at a microscale.

This chapter is structured as follows. The theoretical model is developed in Section 6.2. Three computational demonstrations are presented in Section 6.3, which include a screw dislocation near a free surface, a screw dislocation in an anisotropic material, and an edge dislocation interacting with a bimaterial interface. Through these examples, the accuracy of new model is studied by comparison with analytical solutions. Dislocations in heteroepitaxial structures are simulated in Section 6.4. The summary is provided in Section 6.5.

6.2 Theoretical Model

6.2.1 Basic Equations of Crystal Plasticity Theory

In this study, small strain conditions are considered for simplicity, in which changes in crystal geometry are neglected, although the approach can easily be extended to large deformation. Thus, according to additive decomposition, total strain $\boldsymbol{\varepsilon}$ can be divided into elastic and plastic parts, $\boldsymbol{\varepsilon}^{\mathrm{e}}$ and $\boldsymbol{\varepsilon}^{\mathrm{p}}$, as (Gurtin, 2002; Kuroda and Tvergaard, 2008a,b)

$$\boldsymbol{\varepsilon} = \nabla_s \mathbf{u} = \boldsymbol{\varepsilon}^{\mathrm{e}} + \boldsymbol{\varepsilon}^{\mathrm{p}} \tag{6.2.1}$$

where $\mathbf{u}$ is the displacement field and ∇_s is the symmetric gradient operator. The plastic strain is given by (Gurtin, 2002; Kuroda and Tvergaard, 2008a,b)

$$\boldsymbol{\varepsilon}^{\mathrm{p}} = \sum_\alpha \mathbf{P}_\alpha \gamma_\alpha, \quad \mathbf{P}_\alpha = \frac{1}{2}(\mathbf{s}_\alpha \otimes \mathbf{m}_\alpha + \mathbf{m}_\alpha \otimes \mathbf{s}_\alpha) \tag{6.2.2}$$

where γ_α is a scalar field that represents the plastic slip on slip system α, and $\mathbf{s}_\alpha$ and $\mathbf{m}_\alpha$ are the slip direction and slip plane normal unit vectors of slip system α, respectively. The expression of plastic slip γ_α associated with each slip system α is derived in Section 4.2 of Chapter 4.

The constitutive equation that describes the elastic stress-strain relation is

$$\boldsymbol{\sigma} = \mathbf{C} : \boldsymbol{\varepsilon}^{\mathrm{e}} = \mathbf{C} : (\boldsymbol{\varepsilon} - \boldsymbol{\varepsilon}^{\mathrm{p}}) \tag{6.2.3}$$

where $\boldsymbol{\sigma}$ is Cauchy stress and $\mathbf{C}$ is the elastic modulus tensor, which may be either isotropic or anisotropic.

6.2.2 Phase Field Description of Plastic Slip

To extend the capability of conventional crystal plasticity theory to model a single dislocation, the same description for dislocations as in the phase field is adopted here. In phase field microelasticity theory, elastic strain $\boldsymbol{\varepsilon}^e$ is expressed as the difference between total strain $\boldsymbol{\varepsilon}$ and inelastic strain $\boldsymbol{\varepsilon}^{\text{inelast}}$ (Khachaturyan, 1967, 1983; Khachaturyan and Shatalov, 1969):

$$\boldsymbol{\varepsilon}^e = \boldsymbol{\varepsilon} - \boldsymbol{\varepsilon}^{\text{inelast}} \tag{6.2.4}$$

Compared with Eq. (6.2.1), inelastic strain is directly considered to be plastic strain in our microscale crystal plasticity model.

A dislocation is regarded as the boundary between a slipped and an unslipped region or between two differently slipped regions. By taking ϕ_α as the phase field of slip system α that interpolates between a slipped and an unslipped region, inelastic strain associated with dislocations from different slip systems can be expressed (Nabarro, 1951; Khachaturyan, 1983; Shen and Wang, 2003), i.e.,

$$\boldsymbol{\varepsilon}^{\text{inelast}} = \sum_\alpha \boldsymbol{\varepsilon}_\alpha^{\text{dis}} \phi_\alpha, \quad \boldsymbol{\varepsilon}_\alpha^{\text{dis}} = \frac{b_\alpha}{2d}(\mathbf{s}_\alpha \otimes \mathbf{m}_\alpha + \mathbf{m}_\alpha \otimes \mathbf{s}_\alpha) \tag{6.2.5}$$

where $\boldsymbol{\varepsilon}_\alpha^{\text{dis}}$ is the eigenstrain of a dislocation loop enclosing the slipped region on slip system α, d is the interplanar distance of the slip planes, and b_α is the magnitude of the Burgers vector.

Compared with Eq. (6.2.2), the plastic slip γ_α associated with each slip system α can be expressed by the phase field ϕ_α:

$$\gamma_\alpha = \frac{b_\alpha}{d}\phi_\alpha \tag{6.2.6}$$

Here, the value of phase field ϕ_α represents the amount of plastic slip with respect to a perfect crystal in unit b_α/d.

Compared with the conventional crystal plasticity theory, the plastic slip γ_α associated with each slip system α is described by a continuous smooth phase field ϕ_α independently. A dislocation with the location is identified in which the phase field value changes smoothly in the crystal plane and thus has a finite dislocation core width, as shown in Fig. 6.1A. It has a regularized dislocation topology that is different from the Volterra dislocation model, as shown in Fig. 6.1B. Regularization is governed by the interplanar distance of the slip planes d and gives $d \to 0$ for the Volterra dislocation.

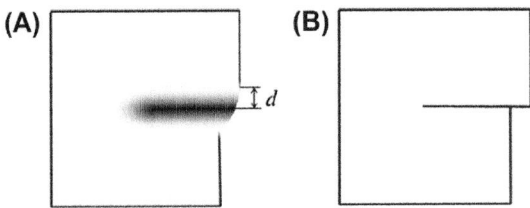

Figure 6.1

(A) Phase field dislocation model; (B) Volterra dislocation model. *Reprinted from Wang, L.Y., Liu, Z.L., Zhuang, Z., 2016. Developing micro-scale crystal plasticity model based on phase field theory for modeling dislocations in heteroepitaxial structures, 17, 267–283. Copyright 2016, with permission from Elsevier.*

6.2.3 Stored Energy and Dissipated Energy

We now focus on time-dependent dislocation evolution by exploiting this phase field description of plastic slip. The evolution of dislocation is driven by energetic driving forces, as outlined next. The stored energy of the dislocation system consists of four parts: elastic energy E_{el}, crystalline energy E_{cryst}, gradient energy E_{grad}, and surface energy E_{surf}:

$$E = E_{el} + E_{cryst} + E_{grad} + E_{surf} \qquad (6.2.7)$$

Elastic energy is associated with elastic strain of the crystal lattice caused by dislocation and external loading. It is expressed as a function of elastic strain through constitutive Eq. (6.2.3):

$$E_{el} = \int_V \frac{1}{2} \boldsymbol{\varepsilon}^e : \mathbf{C} : \boldsymbol{\varepsilon}^e dV \qquad (6.2.8)$$

The formulation of elastic energy here is different from the one in the microelasticity phase field model, which is based on the theory of Khachaturyan and Shatalov and is a closed-form function of inelastic strain using the exact Green's function (Khachaturyan, 1967, 1983; Khachaturyan and Shatalov, 1969).

Different from the conventional crystal plasticity theory, two additional energy terms, including crystalline energy E_{cryst} and gradient energy E_{grad}, are introduced here to describe recoverable strong nonlinear deformation behavior in the vicinity of the dislocation core. Crystalline energy represents potential energy in a crystal subjected to a general slip produced by arbitrary linear combinations of simple slips associated with all slip systems. It is a periodic function reflecting the crystal's natural preference to have the slip vector constrained to integer multiples of the lattice vectors. For a particular slip plane, crystalline energy can be directly related to generalized stacking fault energy. The general expression of the multiperiodical function can always be presented by a Fourier series. For full dislocation, crystalline energy can be approximated by a sinusoidal function.

The simplest approximation by decoupling different slip systems was given by Wang et al. (2001):

$$E_{\text{cryst}} = A \sum_{\alpha} \int_V \sin^2(\pi\phi_\alpha) dV \tag{6.2.9}$$

where $A = \mu(\varepsilon^0)^2 / 2\pi^2$ is a constant (the unstable stacking fault energy) and is obtained by fitting at a small strain limit to shear modulus μ for the corresponding slip system and $\varepsilon^0 = b/d$ for typical shear strain.

The gradient energy is proportional to the dislocation length and vanishes over the slip plane. This can be achieved by using the term $\mathbf{m}_\alpha \times \nabla\phi_\alpha$, which penalizes only the in-plane components of the field gradients. The simplest approximation by decoupling different slip systems was given by Wang et al. (2005):

$$E_{\text{grad}} = B \sum_{\alpha} \int_V (\mathbf{m}_\alpha \times \nabla\phi_\alpha)^2 dV \tag{6.2.10}$$

where B is a positive constant. The gradient energy helps to smooth and widen the dislocation core for numerical stability. Because the phase field represents the amount of strain that is proportional to the displacement gradient, the gradient energy actually accounts for contributions from high-order derivatives of the displacement. Thus, the gradient energy is physical rather than phenomenological.

The surface effect becomes accentuated as heteroepitaxial structures become increasingly small. A form of stored energy generated by the surface steps as dislocations exit a crystal was described by Hurtado and Ortiz (2012). A diffusional surface model was also developed to simulate surface roughening in heteroepitaxial films (Wang et al., 2004). Currently, surface energy is neglected for simplicity. The specific form of surface energy, E_{surf}, will be considered in future work.

Then, substituting Eqs. (6.2.8)–(6.2.10) into Eq. (6.2.7) and taking the derivative of Eq. (6.2.7) with respect to time, the energy storage rate can be achieved as:

$$\dot{E} = \int_V \left[\boldsymbol{\sigma} : \dot{\boldsymbol{\varepsilon}}^e + \sum_{\alpha} \left(A\pi \sin(2\pi\phi_\alpha)\dot{\phi}_\alpha + 2B\mathbf{m}_\alpha \times \nabla\phi_\alpha \times \mathbf{m}_\alpha \cdot \nabla\dot{\phi}_\alpha \right) \right] dV \tag{6.2.11}$$

On the other side, dislocation evolution is also an energy dissipation process. In this work, it is achieved by introducing a dissipation function, which depends only on the rate of phase fields for simplicity:

$$D = \frac{1}{2} k \sum_{\alpha} \int_V (\dot{\phi}_\alpha)^2 dV \tag{6.2.12}$$

where k is the dissipation coefficient. More complex dissipation functions (also depending on the gradient of phase fields) can be used.

6.2.4 Principle of Virtual Power

With regard to the displacement field, the boundary surface of a body is decomposed via $S = S_u \cup S_t$, where displacement $\mathbf{u}_0$ and surface traction $\mathbf{t}$ are prescribed on S_u and S_t, respectively. Regarding the phase fields, the boundary surface of a body is decomposed via $S = S_{\phi_\alpha} \cup S_{q_\alpha}$, where phase fields $\overline{\phi}_\alpha$ and higher-order surface force q_α are prescribed on S_{ϕ_α} and S_{q_α}, respectively. Clearly, there are $S_u \cap S_t = 0$ and $S_{\phi_\alpha} \cap S_{q_\alpha} = 0$.

Next, external power is expressed as:

$$P = \int_V \mathbf{f} \cdot \dot{\mathbf{u}} \, dV + \int_{S_t} \mathbf{t} \cdot \dot{\mathbf{u}} \, dS + \sum_\alpha \int_{S_{q_\alpha}} q_\alpha \dot{\phi}_\alpha \, dS \tag{6.2.13}$$

where $\mathbf{f}$ is a given body force per unit volume.

Based on a thermodynamically consistent framework, macroscopic and microscopic balance equations are derived from the principle of virtual power. With energy storage rate Eq. (6.2.11), dissipation function Eq. (6.2.12), and external power Eq. (6.2.13) in hand, we demand

$$\delta \dot{E} + \delta D = \delta P \tag{6.2.14}$$

for all admissible rates $\delta \dot{\mathbf{u}}$ and $\delta \dot{\phi}_\alpha$ of the displacement field and phase fields. Balance Eq. (6.2.14) equilibrates the internal and external virtual powers, in which the former includes stored and dissipated parts. With these formulations, we have

$$\int_V \left[\boldsymbol{\sigma} : \delta \dot{\boldsymbol{\varepsilon}}^e + \sum_\alpha \left(k \dot{\phi}_\alpha \delta \dot{\phi}_\alpha + 2B \mathbf{m}_\alpha \times \nabla \phi_\alpha \times \mathbf{m}_\alpha \cdot \nabla \delta \dot{\phi}_\alpha + A\pi \sin(2\pi\phi_\alpha) \delta \dot{\phi}_\alpha \right) \right] dV$$
$$= \int_S \mathbf{t} \cdot \delta \dot{\mathbf{u}} \, dS + \int_V \mathbf{f} \cdot \delta \dot{\mathbf{u}} \, dV + \sum_\alpha \int_S q_\alpha \delta \dot{\phi}_\alpha \, dS \tag{6.2.15}$$

Interestingly, Eq. (6.2.15) has some similarities with the higher-order strain gradient crystal plasticity model. Energy balance in the higher-order crystal plasticity can be expressed as (Gurtin, 2002; Kuroda and Tvergaard, 2008a,b)

$$\int_V \left[\boldsymbol{\sigma} : \delta \dot{\boldsymbol{\varepsilon}}^e + \sum_\alpha \left(\tau_\alpha^p \delta \dot{\gamma}_\alpha + \boldsymbol{\xi}_\alpha \cdot \nabla \delta \dot{\gamma}_\alpha \right) \right] dV$$
$$= \int_S \mathbf{t} \cdot \delta \dot{\mathbf{u}} \, dS + \int_V \mathbf{f} \cdot \delta \dot{\mathbf{u}} \, dV + \sum_\alpha \int_S q_\alpha \delta \dot{\gamma}_\alpha \, dS \tag{6.2.16}$$

where $\dot{\gamma}_\alpha$ is the plastic slip rate on slip system α; and τ_α^p, $\boldsymbol{\xi}_\alpha$, and q_α are the slip resistance, higher-order force, and higher-order surface force on slip system α, respectively. Comparing Eq. (6.2.15) with Eq. (6.2.16), we can find that item $k\dot{\phi}_\alpha$ corresponds to slip resistance τ_α^p, and item $2B\mathbf{m}_\alpha \times \nabla \phi_\alpha \times \mathbf{m}_\alpha$ corresponds to higher-order force $\boldsymbol{\xi}_\alpha$, respectively. Eq. (6.2.15) has one more item $A\pi \sin(2\pi\phi_\alpha) \delta \dot{\phi}_\alpha$ than Eq. (6.2.16). With this item,

the proposed microscale crystal plasticity model can be used at a smaller size scale in which a single dislocation behavior is important compared with the conventional higher-order strain gradient crystal plasticity model.

6.2.5 Coupled Balance Equations

By applying the Gauss theorem, Eq. (6.2.15) is further expressed as

$$\int_V \left[(\nabla \cdot \boldsymbol{\sigma} + \mathbf{f}) \cdot \delta \dot{\mathbf{u}} + \sum_\alpha \left(\boldsymbol{\varepsilon}_\alpha^{\text{dis}} : \boldsymbol{\sigma} - k\dot{\phi}_\alpha + 2B(\mathbf{m}_\alpha \times \nabla)^2 \phi_\alpha - A\pi \sin(2\pi\phi_\alpha) \right) \delta\dot{\phi}_\alpha \right] dV$$
$$- \int_S (\mathbf{n} \cdot \boldsymbol{\sigma} - \mathbf{t}) \cdot \delta\dot{\mathbf{u}} dS - \sum_\alpha \int_S [\mathbf{n} \cdot (2B\mathbf{m}_\alpha \times \nabla\phi_\alpha \times \mathbf{m}_\alpha) - q_\alpha] \delta\dot{\phi}_\alpha dS = 0$$

$$(6.2.17)$$

Because the choices of all admissible rates $\delta\dot{\mathbf{u}}$ and $\delta\dot{\phi}_\alpha$ of the displacement field and phase fields are arbitrary, the following coupled balance equations are obtained:

$$\nabla \cdot \boldsymbol{\sigma} + \mathbf{f} = 0 \tag{6.2.18}$$

$$\boldsymbol{\varepsilon}_\alpha^{\text{dis}} : \boldsymbol{\sigma} - k\dot{\phi}_\alpha + 2B(\mathbf{m}_\alpha \times \nabla)^2 \phi_\alpha - A\pi \sin(2\pi\phi_\alpha) = 0 \tag{6.2.19}$$

and subjected to boundary conditions

$$\mathbf{u} = \mathbf{u}_0 \quad \text{on } S_u \tag{6.2.20}$$

$$\mathbf{n} \cdot \boldsymbol{\sigma} = \mathbf{t} \quad \text{on } S_t \tag{6.2.21}$$

$$\phi_\alpha = \overline{\phi}_\alpha \quad \text{on } S_{\phi_\alpha} \tag{6.2.22}$$

$$\mathbf{n} \cdot (2B\mathbf{m}_\alpha \times \nabla\phi_\alpha \times \mathbf{m}_\alpha) = q_\alpha \quad \text{on } S_{q_\alpha} \tag{6.2.23}$$

where $\mathbf{n}$ is the outward unit normal vector of the boundary surface.

In addition, Eq. (6.2.19) is supplemented with initial conditions for phase fields:

$$\phi_\alpha(\mathbf{r}, 0) = \phi_{\alpha 0}(\mathbf{r}) \quad \mathbf{r} \in V \tag{6.2.24}$$

The initial phase fields can be used to model preexisting dislocations, as discussed in Section 6.3.

Compared with the Khachaturyan-type phase field models (Khachaturyan, 1983; Chen and Khachaturyan, 1991; Wang et al., 1993) in which the stress field is expressed as a function of inelastic strain through the exact Green's function, in the proposed model the stress field is directly calculated by solving the boundary value problem expressed by Eqs. (6.2.18)−(6.2.24). It is convenient to deal with complex structures or complex boundary conditions in which the analytical Green's function solution is unavailable. Moreover, the elastic modulus mismatch in heteroepitaxial structures is easily treated without additional complications.

6.2.6 Finite Element Discretization

The variational structure of coupled problem Eq. (6.2.15) is exploited in a spatial discretization by FEM. Using the Galerkin method, the explicit representations of $\mathbf{u}(\mathbf{r}, t)$ and $\phi_\alpha(\mathbf{r}, t)$ in terms of the base functions and nodal variables are

$$u_i(\mathbf{r}, t) = \sum_{I=1}^{n_u} N^I(\mathbf{r}) u_i^I(t) \tag{6.2.25}$$

$$\phi_\alpha(\mathbf{r}, t) = \sum_{J=1}^{n_\phi} \psi_\alpha^J(\mathbf{r}) \phi_\alpha^J(t) \tag{6.2.26}$$

where n_u and n_ϕ are the dimensions of discrete spaces, $N^I(\mathbf{r})$ and $\psi_\alpha^J(\mathbf{r})$ are the standard shape functions, i is the spatial degree-of-freedom number, and $u_i^I(t)$ and $\phi_\alpha^J(t)$ are the nodal values of $u_i(\mathbf{r}, t)$ and $\phi_\alpha(\mathbf{r}, t)$, respectively. When Eqs. (6.2.25) and (6.2.26) are substituted into weak formulation Eq. (6.2.15) and integration is performed, the discrete FEM formulation is achieved.

The following conventions

$$\{\mathbf{u}\} = \left\{ (\mathbf{u}^1)^T, \cdots, (\mathbf{u}^{n_\alpha})^T \right\}^T, \{\phi_\alpha\} = \left\{ \phi_\alpha^1, \cdots, \phi_\alpha^{n_\phi} \right\}^T, \mathbf{N}^I = \mathbf{I} N^I \tag{6.2.27}$$

and Voigt notation are used. Thus, matrices $\mathbf{B}^I$ and $\mathbf{B}_\alpha^J$ can be defined as

$$\mathbf{B}^I = \begin{bmatrix} N_{,x}^I & 0 & 0 \\ 0 & N_{,y}^I & 0 \\ 0 & 0 & N_{,z}^I \\ N_{,y}^I & N_{,x}^I & 0 \\ 0 & N_{,z}^I & N_{,y}^I \\ N_{,z}^I & 0 & N_{,x}^I \end{bmatrix}, \quad \mathbf{B}_\alpha^J = \begin{bmatrix} \psi_{\alpha,x}^J \\ \psi_{\alpha,y}^J \\ \psi_{\alpha,z}^J \end{bmatrix} \tag{6.2.28}$$

Thus, the matrix form of discrete FEM formulation can be achieved:

$$[\mathbf{K}_{uu}]\{\mathbf{u}\} - \sum_\alpha \left[\mathbf{K}_{u\phi}^\alpha\right]\{\phi_\alpha\} = \{\mathbf{f}_{\text{ext}}\} \tag{6.2.29}$$

$$-\left[\mathbf{K}_{u\phi}^\beta\right]^T \{\mathbf{u}\} + \sum_\alpha \left[\mathbf{K}_{\phi\phi}^{\beta\alpha}\right]\{\phi_\alpha\} + \left[\mathbf{K}_{\phi\phi}^\beta\right]\{\phi_\beta\} + \left[\mathbf{\Xi}_\beta\right]\{\dot{\phi}_\beta\} + \left\{\mathbf{Q}_{\text{cryst}}^\beta\right\} = \left\{\mathbf{Q}_{\text{ext}}^\beta\right\} \tag{6.2.30}$$

where

$$(\mathbf{K}_{uu})^{IJ} = \int_{\Omega_0} (\mathbf{B}^I)^T [\mathbf{C}] \mathbf{B}^J d\Omega_0 \tag{6.2.31}$$

$$\left(\mathbf{K}_{u\phi}^{\alpha}\right)^{IJ} = \int_{\Omega_0} \left(\mathbf{B}^I\right)^T [\mathbf{C}] \left[\boldsymbol{\varepsilon}_{\alpha}^{\text{dis}}\right] \psi_{\alpha}^J \mathrm{d}\Omega_0 \tag{6.2.32}$$

$$\left(\mathbf{K}_{\phi\phi}^{\beta\alpha}\right)^{IJ} = \int_{\Omega_0} \psi_{\beta}^I \left[\boldsymbol{\varepsilon}_{\beta}^{\text{dis}}\right]^T [\mathbf{C}] \left[\boldsymbol{\varepsilon}_{\alpha}^{\text{dis}}\right] \psi_{\alpha}^J \mathrm{d}\Omega_0 \tag{6.2.33}$$

$$\left(\mathbf{K}_{\phi\phi}^{\beta}\right)^{IJ} = 2B \int_{\Omega_0} \left(\mathbf{B}_{\beta}^I\right)^T \left[\mathbf{I} - \mathbf{m}_{\beta}\mathbf{m}_{\beta}^T\right] \mathbf{B}_{\beta}^J \mathrm{d}\Omega_0 \tag{6.2.34}$$

$$\left(\boldsymbol{\Xi}_{\beta}\right)^{IJ} = k \int_{\Omega_0} \psi_{\beta}^I \psi_{\beta}^J \mathrm{d}\Omega_0 \tag{6.2.35}$$

$$\left(\mathbf{Q}_{\text{cryst}}^{\beta}\right)^I = \int_{\Omega_0} A\pi\psi_{\beta}^I \sin\left(2\pi \sum_{J=1}^{n_\phi} \psi_{\beta}^J \phi_{\beta}^J\right) \mathrm{d}\Omega_0 \tag{6.2.36}$$

$$\left(\mathbf{f}_{\text{ext}}\right)^I = \int_{\Gamma_0^t} \mathbf{N}^I \mathbf{t} \mathrm{d}\Gamma_0 + \int_{\Omega_0} \mathbf{N}^I \mathbf{f} \mathrm{d}\Omega_0 \tag{6.2.37}$$

$$\left(\mathbf{Q}_{\text{ext}}^{\beta}\right)^I = \int_{\Gamma_0^{q_\beta}} \psi_{\beta}^I q_{\beta} \mathrm{d}\Gamma_0 \tag{6.2.38}$$

The time integration scheme consists of implicit time-stepping and equilibrium iterations in each time step. At time t, the integral evaluations in Eqs. (6.2.31)−(6.2.38) are performed once. We first estimate $\{\phi_\beta\}$ at $t + \Delta t$: $\{\phi_\beta\}_{t+\Delta t}^0$ by applying the forward Euler time integration scheme to Eq. (6.2.30). Then, we iterate to satisfy equilibrium Eq. (6.2.29) at $t + \Delta t$ while simultaneously correcting $\{\phi_\beta\}$ to $\{\phi_\beta\}_{t+\Delta t}^i$ to satisfy Eq. (6.2.30). Iteration $i + 1$ is assumed to be converged when $\max_J \left| \{\phi_\beta^J\}_{t+\Delta t}^{i+1} - \{\phi_\beta^J\}_{t+\Delta t}^i \right| \leq 10^{-6}$.

6.3 Computational Demonstrations

In this section, three examples are given to demonstrate the accuracy of the proposed microscale crystal plasticity model. To model a preexisting dislocation, the phase field of the relevant slip system is initially defined as 1 on the slip plane where dislocation has slipped and transformed to 0 gradually across the slip plane. The transition width is at least of several grids to keep numerical stability. It is chosen as one grid on both sides of the slip plane in the following examples. The quadratic Lagrange element is used for the displacement field and the linear Lagrange element is used for the phase field for compatibility. The subsequent examples show that they are sufficient for the model.

In the simulation, mesh size l is always greater than the real interplanar distance of slip planes d. To maintain an accurate strain field, typical shear strain should be chosen as $\varepsilon^0 = b/l$ rather than $\varepsilon^0 = b/d$. Dimensionless parameters $A^* = A/\mu(\varepsilon^0)^2$, which controls

the crystalline energy, $B^* = B/\mu(\varepsilon^0)^2 l^2$, which influences the dislocation core width, and $\Delta t^* = \mu(\varepsilon^0)^2 \Delta t/k$, which decides the stable time step, are defined in the simulations. $A^* = 1/2\pi^2$, $B^* = 0.07$, and $\Delta t^* = 0.1$ provide reasonable values for the subsequent dislocation systems.

6.3.1 Dislocation Near a Free Surface

To examine the accuracy of this method, a screw dislocation near a free surface of a semiinfinite domain is first considered. The free surface is located at $y = 0$ and the semiinfinite domain occupies domain $y > 0$, as shown in Fig. 6.2A. The infinitely long screw dislocation of Burgers vector **b** is along the positive z-axis and located at a distance of $L = 100b$ from the free surface. The slip plane is parallel to the free surface. The stress field has an analytical solution in which the stress-free condition of the surface is satisfied by placing one image screw dislocation of opposite sign at a distance L from the surface (Hirth and Lothe, 1982):

$$\sigma_{xz} = -\frac{\mu b}{2\pi}\left[\frac{(y-L)}{x^2+(y-L)^2} - \frac{(y+L)}{x^2+(y+L)^2}\right]$$

$$\sigma_{yz} = \frac{\mu b}{2\pi}\left[\frac{x}{x^2+(y-L)^2} - \frac{x}{x^2+(y+L)^2}\right]$$

$$\sigma_{xy} = \sigma_{xx} = \sigma_{yy} = \sigma_{zz} = 0$$

(6.3.1)

where μ is the shear modulus.

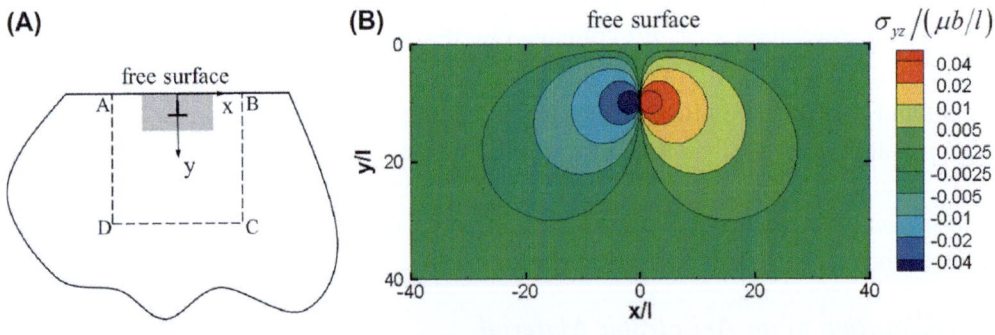

Figure 6.2

(A) Screw dislocation near a free surface of a semiinfinite domain, including the numerical simulation subdomain ABCD. (B) Scaled stress $\sigma_{yz}/(\mu b/l)$ in the *gray* region labeled in (A). *Reprinted from Wang, L.Y., Liu, Z.L., Zhuang, Z., 2016. Developing micro-scale crystal plasticity model based on phase field theory for modeling dislocations in heteroepitaxial structures, 17, 267–283. Copyright 2016, with permission from Elsevier.*

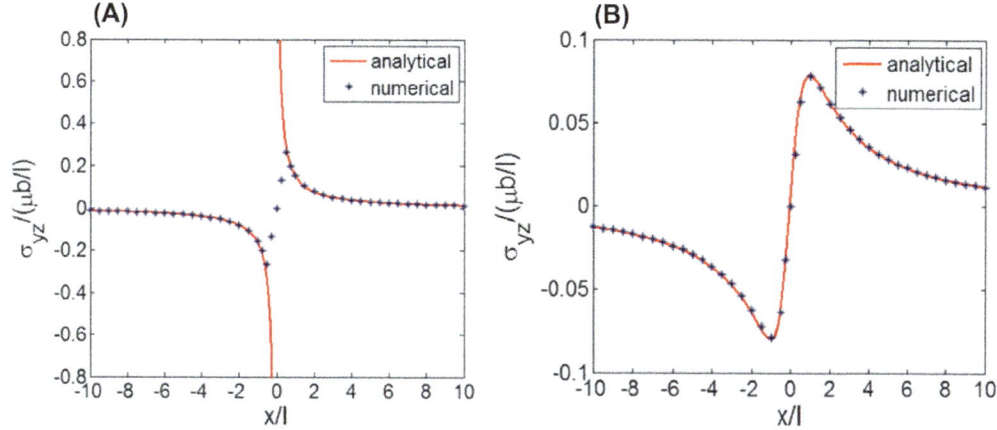

Figure 6.3

Quantitative comparisons of $\sigma_{yz}/(\mu b/l)$ between numerical and analytical solutions: (A) along the line in slip plane $y = 10l$; (B) along the line of a one-grid distance from slip plane $y = 9l$. *Reprinted from Wang, L.Y., Liu, Z.L., Zhuang, Z., 2016. Developing micro-scale crystal plasticity model based on phase field theory for modeling dislocations in heteroepitaxial structures, 17, 267–283. Copyright 2016, with permission from Elsevier.*

The problem is solved on subdomain ABCD with dimensions $5000b \times 5000b \times 10b$, as plotted in Fig. 6.2A, which can be approximately regarded as a semiinfinite domain. The periodic boundary condition is applied at surfaces $z = 0$ and $z = 10b$ to simulate the infinite length along the z-direction. Displacements in all three directions are constrained at surface $y = 5000b$ to restrict rigid body motion. The free boundary condition is applied on the other three boundary surfaces. A hexahedral mesh with an element size of $l = 10b$ is used. The elastic modulus is $E = 26$ GPa and the Poisson's ratio is $\nu = 0.3$.

In this case, the only nonvanishing stress component that contributes to surface traction is σ_{yz}. The scaled stress field $\sigma_{yz}/(\mu b/L)$ obtained by this method in an $80l \times 40l$ portion that is labeled by the gray region in Fig. 6.2A is plotted in Fig. 6.2B. To compare the numerical and analytical solutions quantitatively, the results along the line in slip plane $y = 10l$ and along the line that is of a one-grid distance from slip plane $y = 9l$ are plotted in Fig. 6.3A and B, respectively. The numerical solutions agree well with the analytical solutions outside the dislocation core. Within the core, the linear elastic analytical solution is singular whereas the numerical solution is not.

6.3.2 Dislocation in an Anisotropic Material

To verify the accuracy of the method for anisotropic materials, screw dislocation in an infinite anisotropic material is considered, as shown in Fig. 6.4A. The infinitely long screw dislocation of Burgers vector **b** is along the positive z-axis and is located in the middle of the domain. The slip plane is parallel to the x-axis. The stress field has an analytical

(A) **(B)** $\sigma_{yz}/(\mu b/l)$

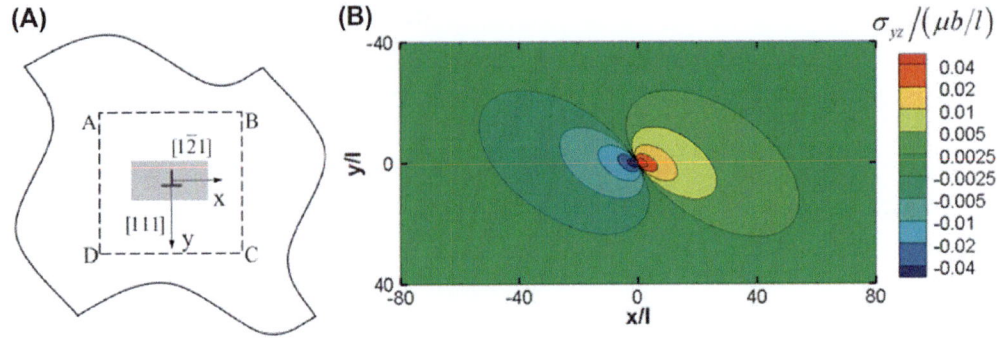

Figure 6.4

(A) Screw dislocation in an infinite domain including numerical simulation subdomain ABCD. (B) Scaled stress $\sigma_{yz}/(\mu b/l)$ in the *gray region labeled in (A). Reprinted from Wang, L.Y., Liu, Z.L., Zhuang, Z., 2016. Developing micro-scale crystal plasticity model based on phase field theory for modeling dislocations in heteroepitaxial structures, 17, 267–283. Copyright 2016, with permission from Elsevier.*

solution (Hirth and Lothe, 1982). The same geometry, mesh, and boundary conditions are used as those in Section 6.0. The material selected is face-centered cubic anisotropic copper and the elastic constants are $c_{11} = 168.4$ GPa, $c_{12} = 121.4$ GPa, and $c_{44} = 75.4$ GPa when the coordinate axes coincide with the cubic axes, which are taken from Hirth and Lothe (1982). Because the x, y, and z axes here are along the [1$\bar{2}$1], [111], and [$\bar{1}$01] directions, respectively, the elastic constant matrix needs to be transformed. The stresses are normalized by $\mu b/l$, where $\mu = 54.6$ GPa is the Voigt average elastic constant (Hirth and Lothe, 1982).

Scaled shear stress $\sigma_{yz}/(\mu b/l)$ obtained by this method in a $160l \times 80l$ portion, which is labeled by the gray region in Fig. 6.4A, is plotted in Fig. 6.4B. Compared with Fig. 6.2B, the contour plot is not symmetric owing to elastic anisotropy. To compare the numerical and analytical solutions quantitatively, the results along a line in slip plane $y = 0$ and along a line that is of a one-grid distance from slip plane $y = l$ are plotted in Fig. 6.5A and B, respectively. The numerical solutions agree well with the analytical solutions outside the dislocation core, as in the isotropic case.

6.3.3 Dislocation Near a Bimaterial Interface

Next, an edge dislocation near a bimaterial interface is considered, as shown in Fig. 6.6A. The bimaterial interface between two semiinfinite domains is located at $x = 0$. The infinitely long edge dislocation of Burgers vector **b** is along the positive z-axis and located at a distance of $L = 95b$ from the interface. The glide plane is perpendicular to the interface. The solution to this problem was given by Head (1953a,b); further clarification of the solution was provided by Lubarda (1997) in the context of dislocation arrays near the bimaterial interface.

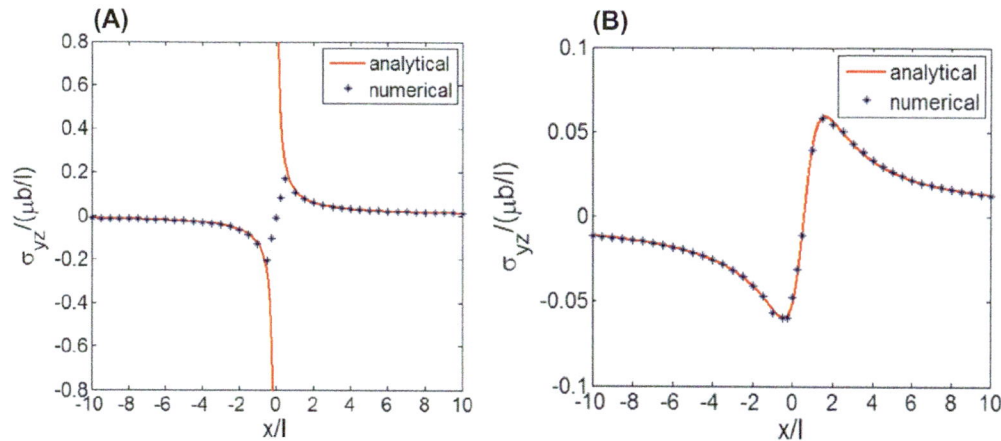

Figure 6.5
Quantitative comparisons of $\sigma_{yz}/(\mu b/l)$ between the numerical and analytical solutions:
(A) along the line in the slip plane $y = 0$; (B) along the line that is of a one-grid distance from
slip plane $y = l$. *Reprinted from Wang, L.Y., Liu, Z.L., Zhuang, Z., 2016. Developing micro-scale crystal
plasticity model based on phase field theory for modeling dislocations in heteroepitaxial structures, 17,
267−283. Copyright 2016, with permission from Elsevier.*

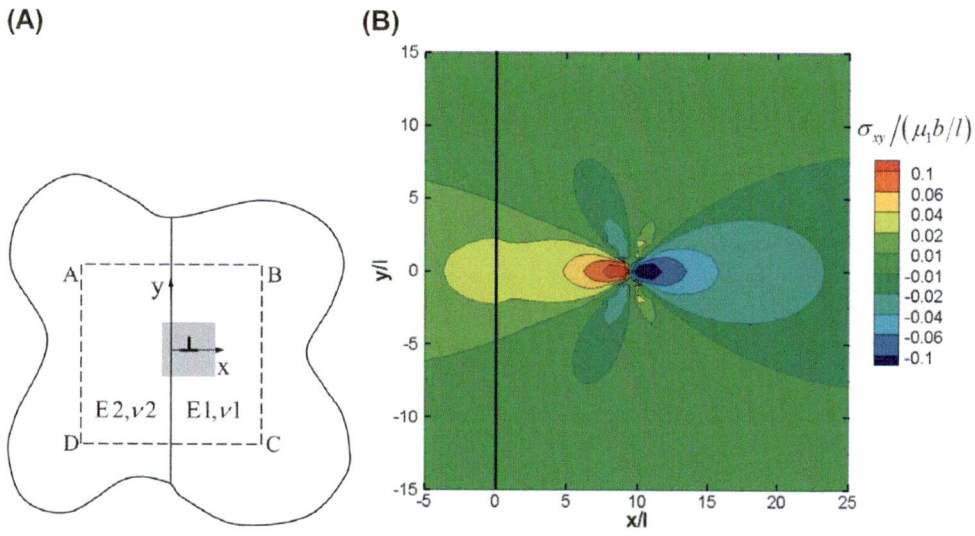

Figure 6.6
(A) Edge dislocation near a bimaterial interface between two semiinfinite domains including
numerical simulation subdomain ABCD. (B) Scaled stress $\sigma_{xy}/(\mu_1 b/l)$ in the *gray* region labeled
in (A). *Reprinted from Wang, L.Y., Liu, Z.L., Zhuang, Z., 2016. Developing micro-scale crystal plasticity
model based on phase field theory for modeling dislocations in heteroepitaxial structures, 17, 267−283.
Copyright 2016, with permission from Elsevier.*

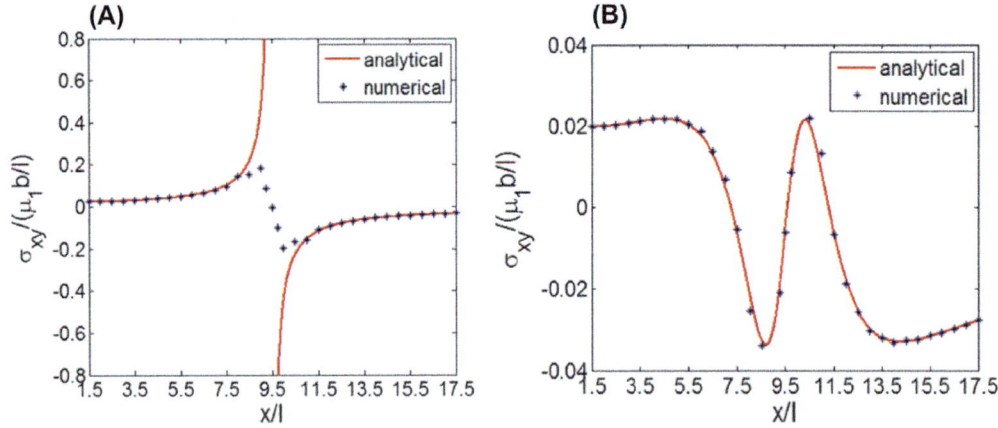

Figure 6.7

Quantitative comparisons of $\sigma_{xy}/(\mu_1 b/l)$ between numerical and analytical solutions: (A) along the line in slip plane $y = 0$; (B) along the line that is of a two-grid distance from slip plane $y = 2l$. *Reprinted from Wang, L.Y., Liu, Z.L., Zhuang, Z., 2016. Developing micro-scale crystal plasticity model based on phase field theory for modeling dislocations in heteroepitaxial structures, 17, 267−283. Copyright 2016, with permission from Elsevier.*

It can be regarded as a plane strain problem and solved on subdomain ABCD with dimensions of $10{,}000b \times 10{,}000b$, as shown in Fig. 6.6A. The displacements in all three directions are constrained at the bottom line, $y = -5000b$, to restrict rigid body motion. The free boundary condition is applied to the other boundary lines. A structured quadrilateral mesh with an element size of $l = 10b$ is used. In subdomain $x > 0$, elastic modulus $E_1 = 31$ GPa and Poisson's ratio $\nu_1 = 0.276$; in subdomain $x < 0$, elastic modulus $E_2 = 94.7$ GPa and Poisson's ratio $\nu_2 = 0.276$.

Scaled shear stress $\sigma_{xy}/(\mu_1 b/l)$ obtained by this method in a $30l \times 30l$ portion, which is labeled by the gray region in Fig. 6.6A, is plotted in Fig. 6.6B. μ_1 is the shear modulus of subdomain $x > 0$. $\sigma_{xy}/(\mu_1 b/l)$ is continuous across the bimaterial interface, as expected. To compare the numerical and analytical solutions quantitatively, the results along the line in slip plane $y = 0$ and along the line that is of a two-grid distance from slip plane $y = 2l$ are plotted in Fig. 6.7A and B, respectively. The numerical solutions agree well with the analytical solutions outside the dislocation core. Within the core, the linear elastic analytical solution is singular whereas the bounded solution is predicted in the calculation.

6.4 Applications to Heteroepitaxial Structures

Heteroepitaxial structures (e.g., epitaxial films or core-shell nanopillars) are of considerable interest owing to their important applications in engineering, such as

electronics, optoelectronics, and solar cells (Freund, 2000; Fu et al., 2004; Panda and Tseng, 2013). However, dislocation-free heteroepitaxial structures cannot be grown with an arbitrary thickness and misfit dislocations will form above a critical thickness. The dislocations and their strong interactions with the material interface not only influence mechanical properties such as strength and fracture toughness but also change the electrical or optical properties of heteroepitaxial crystalline materials (Bennett, 2010; Zbib et al., 2011; Liu et al., 2013b; Abdolrahim et al., 2014). Understanding the underlying dynamics of dislocations in heteroepitaxial structures is crucial to providing better insight into material design and property prediction. However, there is still no effective tool for studying dislocation behaviors in complex heteroepitaxial structures.

In this section, two examples are given to demonstrate the applicability of a microscale crystal plasticity model for dislocations in typical heteroepitaxial structures.

6.4.1 Critical Shell Thickness of Core-Shell Nanopillars

Heteroepitaxial core-shell nanopillars are considerably interesting because of their important applications in electronics, optoelectronics, and solar cells. When the shell is first deposited, the interface between the core and the shell is coherent and free of dislocations. During thickening of the shell layer, stored strain energy caused by lattice misfit strain increases. A stable misfit dislocation forms when the shell reaches a critical thickness beyond which strain energy induced by the lattice misfit strain exceeds the formation energy of the misfit dislocation. The critical shell thickness of Ge/Si core-shell nanopillars is predicted by employing the microscale crystal plasticity model developed in Section 6.2.

In addition to energetic considerations, it is important to know kinetic processes that cause misfit dislocation, to predict critical shell thickness. The glide-relaxation mechanism is pursued here for Ge/Si core-shell nanopillars, which is consistent with experimental studies (Goldthorpe et al., 2008; Dayeh et al., 2013). For Ge/Si core-shell nanopillars that have diamond crystal structures, $a_{Si}/2\langle 110 \rangle$ perfect dislocations nucleate from the surface of the Si shell and glide on the inclined $\{111\}$ slip planes to deposit elliptical dislocation loops at the core-shell interface. For a given growth direction, slip systems with no resolved shear stress (RSS) are not considered. Of the remaining slip systems, the one that provides the smallest critical shell thickness is that most likely activated in actuality.

In the calculation, [111] growth direction is considered. For convenience, a coordinate system is defined with x, y, and z axes along the $[\bar{1}01]$, $[1\bar{2}1]$, and $[111]$ directions, respectively, as shown in Fig. 6.8A. In a Ge/Si core-shell nanopillar, the lattice misfit strain of Si with respect to Ge is $\varepsilon_m = (a_{Ge} - a_{Si})/a_{Si} = 0.0418$, where $a_{Si} = 0.5433$ nm

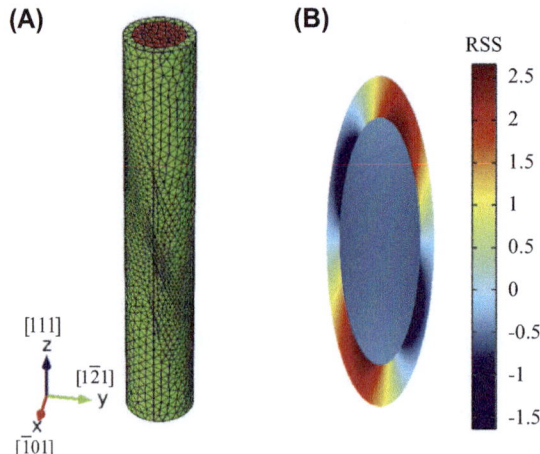

Figure 6.8

(A) Geometry and mesh of core-shell nanopillar with [111] growth direction. The core radius is 15 nm and the shell thickness is 5 nm. (B) Resolved shear stress (RSS) in a $(1\bar{1}1)$ plane crossing the nanopillar in (A) for the [110] $(1\bar{1}1)$ slip system. *Reprinted from Wang, L.Y., Liu, Z.L., Zhuang, Z., 2016. Developing micro-scale crystal plasticity model based on phase field theory for modeling dislocations in heteroepitaxial structures, 17, 267–283. Copyright 2016, with permission from Elsevier.*

and $a_{Ge} = 0.5660$ nm (Schwarz, 1999b) are the lattice constants of Si and Ge crystals, respectively. To calculate the elastic field caused by the lattice misfit strain, initial strain field $\boldsymbol{\varepsilon}_{initial} = \varepsilon_m \mathbf{I}$ is introduced in the shell to match the lattice of shell to the core. Thus, the total strain field is given by $\boldsymbol{\varepsilon} = \nabla_s \mathbf{u} + \boldsymbol{\varepsilon}_{initial}$. Then, the elastic field can be obtained by solving balance Eq. (6.2.18) and boundary conditions (6.2.20) and (6.2.21). In this section, the anisotropic elastic constants are used. $c_{11} = 128.4$ GPa, $c_{12} = 48.2$ GPa, and $c_{44} = 66.7$ GPa correspond to core material Ge and $c_{11} = 166.2$ GPa, $c_{12} = 64.4$ GPa, $c_{44} = 79.8$ GPa correspond to shell material Si when the coordinate axes coincide with the cubic axes (Freund and Suresh, 2003).

The core radius and shell thickness are set to 15 and 5 nm, respectively. The height of the core-shell nanopillar is set to 240 nm. An FEM mesh generated by the tetrahedral elements is used with an average element size of $l = 1.5$ nm, as shown in Fig. 6.8A. To identify the preferred nucleation sites, the RSS calculated by the proposed method is first plotted in Fig. 6.8B, for a dislocation with a $(1\bar{1}1)$ slip plane and $\mathbf{b} = a_{Si}/2[110]$ Burgers vector. A large positive value of RSS indicates a preferred site for dislocation nucleation and propagation. On the other hand, a 0 or negative RSS will suppress the nucleation and propagation of that dislocation. By inspecting the RSS map in Fig. 6.8B, we expect dislocations to nucleate at the position where the RSS is maximum and to propagate in the Si shell.

To determine the critical shell thickness, we consider the reduction in stored energy of core-shell nanopillar ΔE from a dislocation-free state to a misfit dislocation state. The misfit dislocation state is energetically favorable when the following criterion is satisfied:

$$\Delta E = E_m - E_0 \leq 0 \tag{6.4.1}$$

where E_m is the stored energy in the core-shell nanopillar in the presence of a misfit dislocation loop at the core-shell interface and E_0 is the stored energy owing only to lattice misfit strain. The critical shell thickness is then evaluated when the reduction in stored energy ΔE equals 0. For a Ge core radius of 15 nm, the reduction in stored energy ΔE with a variation of Si shell thickness from 2 to 5 nm is plotted in Fig. 6.9A. Thus, the critical shell thickness is 3.5 nm at a core radius of 15 nm. The result is consistent with the experimental result for Ge/Si core-shell nanopillars, which was reported to be ~3 nm (Dayeh et al., 2013). Moreover, the result predicted by the proposed method is closer to the experimental result than the analytical prediction made by Chu et al. (2011), as shown in Fig. 6.9B.

The predicted critical shell thicknesses for Ge/Si core-shell nanopillars with respect to different core radiuses are shown in Fig. 6.9B. The results suggest that variations in the critical shell thickness depend on the core radius, and reductions in the core radius can increase the critical shell thickness. When the core radius is sufficiently small, the misfit dislocation will no longer nucleate whereas the critical shell thickness tends to a constant as the core radius increases. These trends are in good agreement with the theoretical results (Chu et al., 2013).

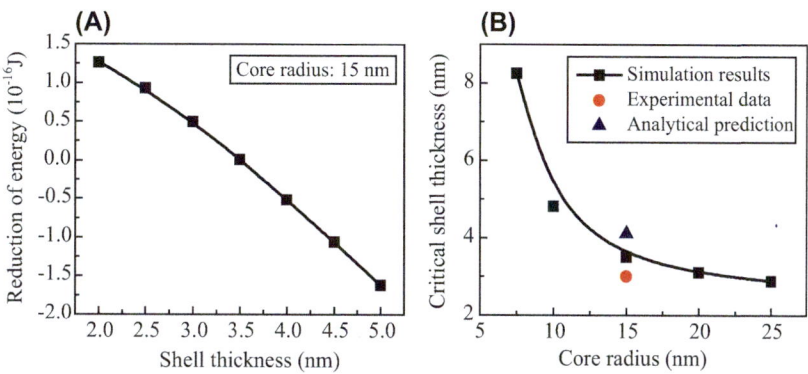

Figure 6.9

(A) Reduction in stored energy ΔE from a dislocation-free state to a misfit dislocation state for a Ge core radius of 15 nm and Si shell thickness from 2 to 5 nm. (B) Dependence of critical shell thickness on the core radius predicted by the proposed method. The experimental and analytical critical shell thicknesses at a core radius of 15 nm are also plotted for comparison. *Reprinted from Wang, L.Y., Liu, Z.L., Zhuang, Z., 2016. Developing micro-scale crystal plasticity model based on phase field theory for modeling dislocations in heteroepitaxial structures, 17, 267–283. Copyright 2016, with permission from Elsevier.*

6.4.2 Dislocations in Heteroepitaxial Thin Films

Like core-shell nanopillars, a dislocation-free film cannot be grown with an arbitrary thickness on a substrate and misfit dislocation will form above a critical thickness. Here, the proposed microscale crystal plasticity model is employed to simulate dislocation behaviors in an $Si_{1-x}Ge_x$ heteroepitaxial thin film grown on an Si substrate. In particular, the operation of a dislocation source that can generate dislocation loops is considered.

The stress-free lattice constant of $Si_{1-x}Ge_x$ is well-approximated by Vegard's law, $a_{SiGe} = a_{Si}(1 - x) + a_{Ge}x$, where x is the atomic percentage of Ge in the $Si_{1-x}Ge_x$ film, and $a_{Si} = 0.5433$ nm and $a_{Ge} = 0.5660$ nm (Schwarz, 1999b) are the lattice constants of Si and Ge crystals, respectively. We assume $x = 0.24$ in the simulation. The lattice misfit strain, defined as $\varepsilon_m = (a_{SiGe} - a_{Si})/a_{Si} = 0.0418x$, generates internal stresses in the film and substrate. If the film thickness is larger than the critical thickness, this internal stress can be relieved by the nucleation and motion of dislocations. The way in which this occurs and the configuration of dislocations that results are greatly important to determining whether the relaxed film is technologically useful.

Here, we use the thermal expansion process to achieve internal stresses caused by the misfit strain (Cui et al., 2015a). Both the $Si_{1-x}Ge_x$ film and Si substrate are assumed to be isotropic, with a Young's modulus of $E_{Si} = 130$ GPa (Shen, 2008) and $E_{Ge} = 102.5$ GPa (Neuberger, 1971), respectively, as well as the same Poisson ratio $\nu_{Si} = \nu_{Ge} = 0.28$. The Young's modulus and Poisson ratio of $Si_{1-x}Ge_x$ are approximated by Vegard's law, $E_{SiGe} = E_{Si}(1 - x) + E_{Ge}x$ and $\nu_{SiGe} = \nu_{Si}(1 - x) + \nu_{Ge}x$, respectively. Then, the equivalent thermal expansion coefficients of the film and substrate are given as (Freund and Suresh, 2003):

$$\alpha_f = \varepsilon_m \frac{h_s M_s}{h_f M_f + h_s M_s}$$

$$\alpha_s = -\varepsilon_m \frac{h_f M_f}{h_f M_f + h_s M_s}$$

(6.4.2)

where h_f and h_s are the thicknesses of the film and substrate, and $M_f = E_{SiGe}/(1 - \nu_{SiGe})$ and $M_s = E_{Si}/(1 - \nu_{Si})$ are the biaxial modulus of the film and substrate, respectively. Then, internal stresses caused by misfit strain are equal to those caused by thermal expansion in one unit.

In the simulation, a side length of 3 μm is used for both the film and substrate. The thicknesses of the film and substrate are set to be $h_f = 0.3$ μm and $h_s = 0.6$ μm, respectively. Out-of-plane displacement on the bottom surface of substrate is constrained and a lateral periodic boundary condition is used to exclude lateral bend. A diamond cubic film-substrate system with a (100) interface is considered. Supposing there is a source in

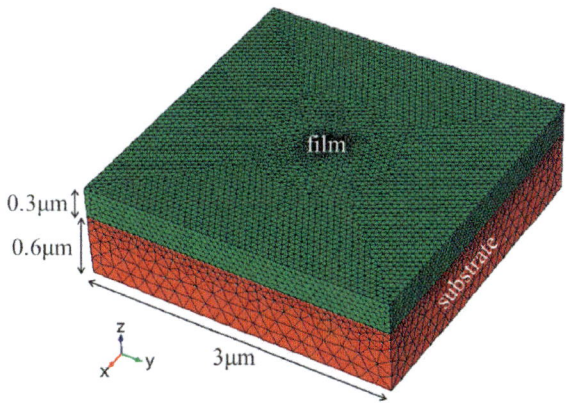

Figure 6.10
Geometry and mesh of film-substrate system with (100) interface. *Reprinted from Wang, L.Y., Liu, Z.L., Zhuang, Z., 2016. Developing micro-scale crystal plasticity model based on phase field theory for modeling dislocations in heteroepitaxial structures, 17, 267—283. Copyright 2016, with permission from Elsevier.*

the (111) slip plane that can generate dislocation loops of 0.15 μm with a Burgers vector **b** of magnitude 0.236 nm along $\left[\bar{1}01\right]$ in the middle of the film, it naturally multiplies under the effect of internal stress caused by the misfit strain and image force from the free surface with no a priori constraint. An FEM mesh generated by the tetrahedral elements is used with an average element size of $l = 60b$, as shown in Fig. 6.10.

The evolution of the dislocation source in the (111) cross-section of the three-dimensional (3D) epitaxial system is illustrated in Fig. 6.11A—D at various moments. As the dislocation loop touches the free surface and the film-substrate interface, it forms two threading arms. The threading arms move away from the source and deposit a misfit dislocation segment at the film-substrate interface. Because of the attraction by image force from the free surface, which is accurately captured through this model, the threading arms near the free surface try to remain perpendicular to the free surface during motion. The internal stress of the opposite sign in the substrate prevents the misfit segments from entering it. The slip-steps on the free surface of the film produced by the passage and exit of the dislocations are described in Fig. 6.12. The deformation is enlarged 300 times and the scaled displacement nephogram of z direction w/b is also plotted in Fig. 6.12.

For comparison, the operation of the same dislocation source in a misfit periodic multilayer is considered. The same model geometry is used as that of epitaxial film, except that the periodic boundary condition is used in the vertical direction. The simulated motion of the dislocation source in the (111) cross-section of the 3D epitaxial system is illustrated in Fig. 6.11E—H at various moments. In this system, there is no free surface and dislocation motion is driven only by internal stresses caused by misfit strain. As the

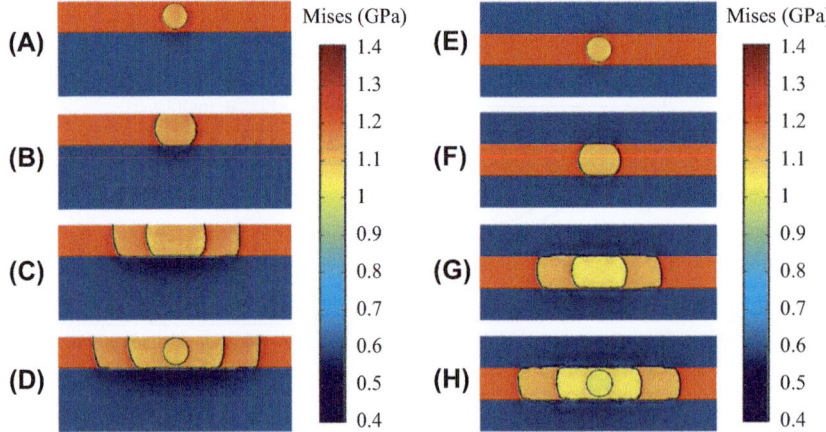

Figure 6.11

Simulated operation of a dislocation source in a (111) cross-section of the epitaxial film and the multilayer at various time moments, as well as Mises stress: (A and E) Time step = 0; (B and F) Time step = 200; (C and G) Time step = 1130; (D and H) Time step = 1550. *Reprinted from Wang, L.Y., Liu, Z.L., Zhuang, Z., 2016. Developing micro-scale crystal plasticity model based on phase field theory for modeling dislocations in heteroepitaxial structures, 17, 267–283. Copyright 2016, with permission from Elsevier.*

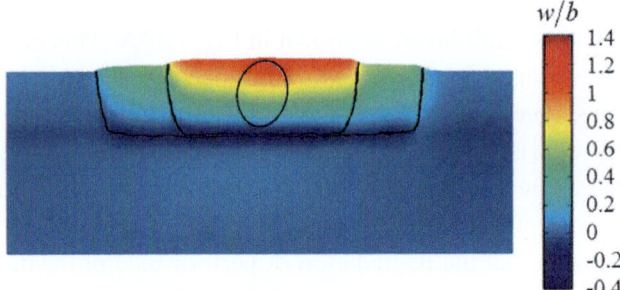

Figure 6.12

Scaled displacement nephogram of z direction w/b of epitaxial film at time step = 1550; deformation is enlarged 300 times. *Reprinted from Wang, L.Y., Liu, Z.L., Zhuang, Z., 2016. Developing micro-scale crystal plasticity model based on phase field theory for modeling dislocations in heteroepitaxial structures, 17, 267–283. Copyright 2016, with permission from Elsevier.*

dislocation loop touches the film-substrate interfaces, it forms two threading arms. The threading arms move away from the source and deposit two misfit dislocation segments at the interfaces. The internal stress of the opposite sign in substrates also prevents misfit segments from entering them. The threading arms bow out symmetrically, which is different from the situation with the free surface. The results are consistent with the other studies (Wang et al., 2003; Cui et al., 2015a).

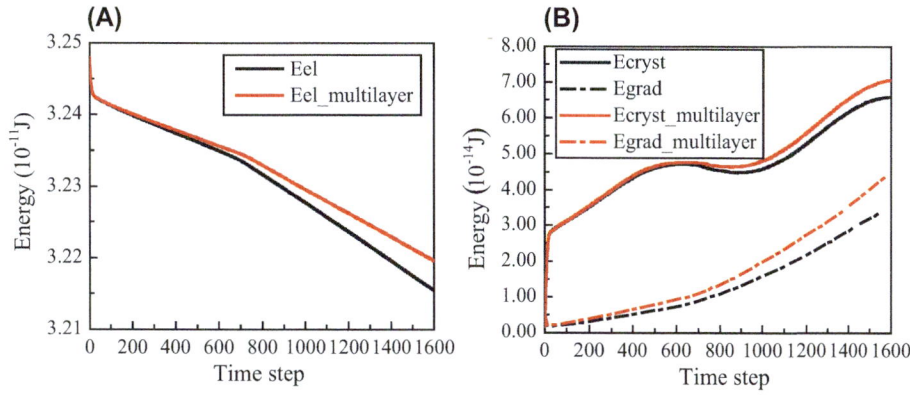

Figure 6.13
(A) Elastic energy (Eel) of two systems. (B) Crystalline energy (Ecryst) and gradient energy (Egrad) of two systems. *Reprinted from Wang, L.Y., Liu, Z.L., Zhuang, Z., 2016. Developing micro-scale crystal plasticity model based on phase field theory for modeling dislocations in heteroepitaxial structures, 17, 267–283. Copyright 2016, with permission from Elsevier.*

By comparing the two cases, it can be found that the threading arms move faster in the epitaxial film than those in the multilayer, as illustrated in Fig. 6.11. This can be explained from the energy point of view. The stored elastic energy is initially the same in the two systems Whereas the decrease in elastic energy caused by dislocation slip in the epitaxial film is larger than that in the multilayer, as shown in Fig. 6.13A. This means that the dislocation slip velocity is larger in the epitaxial film than in the multilayer. There is an extra misfit dislocation segment created in the multilayer system, resulting in more dislocation core energy consisting of crystalline energy and gradient energy, as shown in Fig. 6.13B. In both cases, the stress in the film can be relieved by dislocation slip. The decreased amount of Mises stress in the epitaxial film is smaller than that in the multilayer, which illustrates that the degradation of performance of the device is smaller when epitaxial film is used compared with the multilayer. However, the Mises stress at the film-substrate interface increases when misfit dislocations pile up at the interface as the source operates. Such locations may serve as preferable sites for crack nucleation.

6.5 Summary

Under a thermodynamically consistent framework, a microscale crystal plasticity model was developed based on phase field theory. Compared with widely used Khachaturyan-type phase field dislocation models, the new model has some major advantages for the analysis of dislocations in complex heteroepitaxial structures. It can be used for complex structures or boundary conditions for which the analytical Green's function solution is unavailable. Moreover, the elastic modulus mismatch in heteroepitaxial

structures is easily treated without additional complications. Besides, with numerical implementation by FEM, it is flexible to deal with finite deformation in heteroepitaxial structures at the microscale. Compared with the micron-scale crystal plasticity model, the proposed model can be used at a smaller-sized scale where a single dislocation behavior is important.

The accuracy of the model was studied in three problems for which analytical solutions are available: a screw dislocation interacting with a free surface, a screw dislocation in an infinite anisotropic material, and an edge dislocation interacting with a bimaterial interface. Because the method presented here uses FEM to determine the total stress field, the mesh must be sufficiently refined around the dislocation core. The solutions match well with analytical solutions outside the dislocation core. Within the core, the linear elastic analytical solution is singular, whereas bounded solutions are predicted in all cases.

The critical shell thickness of Ge/Si core-shell nanopillars is predicted by employing the proposed model. The result is consistent with the experimental result and more accurate than the analytical prediction. The results suggest that variations in critical shell thickness depend on the core radius and a reduction in core radius can increase critical shell thickness. The model also captures the fundamental physical trends that misfit dislocation will not nucleate when the core radius is sufficiently small and the critical shell thickness tends toward a constant as the core radius increases.

The proposed model is also employed to simulate the motion of a dislocation source in heteroepitaxial thin films. It shows that the image forces of a free surface and film-substrate interface can be properly estimated. At the same time, structure deformation is calculated and slip-steps on the free surface of film produced by the passage and exit of dislocations are well-captured. Finally, the reason why threading arms move faster in an epitaxial film than those in a multilayer is explained by comparing the energy of two systems.

Discrete Dislocation Mechanism-Based Crystal Plasticity

Discrete-Continuous Model of Crystal Plasticity at the Submicron Scale

Chapter Outline

From this chapter, the book comes into Part II, which focuses on discrete dislocation-based mechanism of crystal plasticity.

7.1 Discrete Dislocation Dynamics

Trends involving integrated computational materials engineering accentuate the need for developing effective bottom-up and top-down modeling methods (Council, 2008; Matouš et al., 2017; McDowell et al., 2011). At an atomistic scale, molecular dynamics simulation methods are well-developed. Substantial molecular dynamics simulations were conducted that gave many valuable insights into material deformation mechanisms (Prakash et al., 2017; Xu et al., 2013; Zhou et al., 1998; Zuo and Ngan, 2006). Previously, spatial and

Dislocation Mechanism-Based Crystal Plasticity. https://doi.org/10.1016/B978-0-12-814591-3.00007-8

temporal scales that could be accessed by molecular dynamics simulations were about 10^{-9} to 10^{-6} m and 10^{-12} to 10^{-6} s (Matouš et al., 2017). Meanwhile, the high strain rate, which is used to speed computation, is still controversial owing to its influences on microstructural behavior (Needleman, 2000). At the macroscale and microscale, continuum models are typically used to assess material properties, as described in Chapters 2 and 3, respectively. They are typically used at a coarse-grained microstructure scale ($>10^{-5}$ m) or macroscopic engineering scales ($10^{-2}-10^2$ m) (Matouš et al., 2017), with time scales larger than 10^{-6} s. The spatial and temporal gap between atomistic simulations and continuum models requires the development of a microscale simulation method. The discrete dislocation dynamics (DDD) simulation method deals directly with the evolution of collective dislocation lines. It raises the possibility of linking atomistic simulations and continuum models.

DDD simulation was developed based on the elastic theory of dislocations. Plastic deformation is described according to the movement of dislocation lines. Early in 1968, Forman et al. studied the interaction between discrete dislocations and rigid obstacles based on a two-dimensional (2D) model. With improvements in computational conditions, lots of 2D-DDD simulations have been carried out since the 1980s (Duesbery et al., 1992; Ghoniem and Amodeo, 1988; Lepinoux and Kubin, 1987). Generally, 2D-DDD considers only infinitely long straight edge dislocations, represented by discrete points in the plane perpendicular to the dislocation lines. The finite deformation framework of 2D-DDD was developed by Deshpande et al. (2003, 2005). In the past few decades, 2D-DDD has been widely used to reveal the size effect, strain burst behavior, and strain hardening mechanism during the plastic deformation of thin films (Nicola et al., 2003, 2005b) and micropillars (Ouyang et al., 2009; Papanikolaou et al., 2017b). Crack growth and failure problems (O'day and Curtin, 2005) and fatigue features have also been studied within the 2D-DDD plasticity framework. Although the 2D treatment of dislocations leads to higher computational efficiency, 2D-DDD generally cannot consider cross-slip and the 3D formation and destruction of dislocation junctions (Benzerga et al., 2004). To solve this problem, Benzerga (2009) developed 2.5D-DDD by introducing some 3D dislocation reaction laws and to some extent captured dislocation starvation hardening.

To consider different kinds of dislocation reactions with minimal ad hoc assumptions and fundamentally resolve the problems in 2D techniques, 3D-DDD was developed. Dislocation network is discretized into segments (Fig. 7.1) based on two main methods. One depends on the underlying lattice and the dislocation lines are discrete into edge-screw or edge-mixed-screw dislocation segments. It is convenient to ensure that the movement of the dislocation segment is always along the crystal orientation, and it exhibits a high computation efficiency (Kubin et al., 1992). The other is continuous description. The dislocation lines are discrete into multiple straight (Bulatov and Cai, 2006; Liu et al., 2009b) or curved dislocation segments (El-Awady et al., 2008;

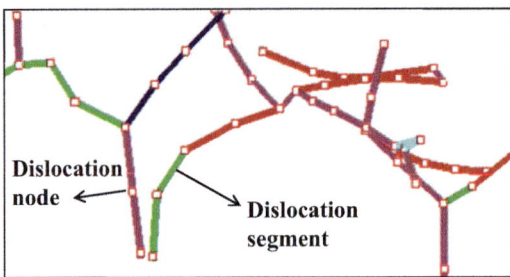

Figure 7.1

Dislocation network discretized in segments (different *colors* correspond to different slip planes) that are connected by nodes.

Ghoniem et al., 2000b; Po et al., 2014). This method is better at dealing with complex dislocation configurations. 3D-DDD is described in detail in Bulatov and Cai (2006), Kubin (2013) and Po and Ghoniem (2015). In the following discussion, we provide a brief description of only the most salient features.

7.1.1 Dislocation Kinetic Equation

For an arbitrary dislocation segment *i*, such as that shown in Fig. 7.1, its kinetic equation obeys the relationship

$$\mathbf{v}_i = F(\mathbf{f}_i) \qquad (7.1.1)$$

where $\mathbf{v}_i$ and $\mathbf{f}_i$ are the nodal velocity vector and nodal force vector, respectively. F represents a function of argument $\mathbf{f}_i$. $\mathbf{f}_i = \int_l \mathbf{N}^T \mathbf{f} dl$, where $\mathbf{N}$ is the shape function and $\mathbf{f}$ is the total force vector acting per unit length of the dislocation line, which is calculated as

$$\mathbf{f} = (\mathbf{\sigma}_i \cdot \mathbf{b}_i) \times \mathbf{\xi}_i + \mathbf{f}^{self} + \mathbf{f}^{image} \qquad (7.1.2)$$

The first term on the right side of Eq. (7.1.2) represents the Peach-Koehler force, where $\mathbf{\sigma}_i$ is the stress field caused by other dislocations and external boundary conditions, $\mathbf{b}_i$ is the Burgers vector of segment *i*, and $\mathbf{\xi}_i$ is the unit vector describing its direction. The second term, $\mathbf{f}^{self}$, is a line tension, which is computed by the negative derivative of the segment total energy with respect to its position. The third term, $\mathbf{f}^{image}$, is the image force induced by the free surface or interface. Regarding traditional DDD simulations, a periodic boundary condition is usually applied without considering image force.

For submicron crystals, image force induced by the free surface has an important role in influencing the behavior and evolution of dislocations. Solutions by Hirth and Lothe (1982) and Yoffe (1961) are widely used to estimate the image force induced by the free surface. Further discussion about the image force calculation is given in Section 7.3.2.

Function $F(\mathbf{f}_i)$ in Eq. (7.1.1) generally has different expressions for different crystal structures. For most face-centered cubic (FCC) crystals, lattice resistance to dislocation motion is typically negligible compared with the applied force. Eq. (7.1.1) can be expressed as

$$\mathbf{M}_d \dot{\mathbf{v}}_i + B\mathbf{v}_i = \mathbf{f}_i \tag{7.1.3}$$

where $\mathbf{M}_d$ and $\mathbf{B}$ are the corresponding effective mass matrix and drag coefficient matrix, respectively. At room temperature and under static loading condition, dislocation motion is assumed to be in the overdamped regime (Nadgornyi, 1998). That is, dislocation velocity can rapidly reach the stable value and the first inertia term on the left side of Eq. (7.1.3) can be ignored. Assuming that dislocation mobility is isotropic, $\mathbf{B}$ is expressed as $\mathbf{B} = \int_l B_0 \mathbf{N}^T \mathbf{N} dl$. Here, B_0 is the static drag coefficient matrix.

With respect to body-centered cubic (BCC) crystals, the kinetic equation is much more complex owing to their particular core structure of screw dislocations. Generally, the motion of screw dislocations needs to overcome the high Peierls barrier. Thermally activated kink-pair formation must be carefully taken into account so as to capture the twinning-antitwinning and tension-compression asymmetries of the yield and flow stresses, as well as the strong temperature and strain rate dependence. Several phenomenological mobility laws have been proposed along this line (Chaussidon et al., 2008; Naamane et al., 2010; Po et al., 2016; Tang et al., 1998; Tang and Marian, 2014; Wang and Beyerlein, 2011) based on analytical theory, molecular dynamics simulations, and experimental data.

The plastic behavior of hexagonal close-packed (HCP) crystals involves a complex interplay between dislocation slip and twinning. Considering solely dislocation plasticity, dislocations preferentially glide in basal planes (e.g., Cd, Zn, Mg, and Be), or prismatic planes (e.g., Ti, Zr, and Hf), because these slip planes correspond to the shortest Burgers vector. For prismatic slip, lattice friction on screw dislocations has an important role in dislocation mobility, as discussed in detail in Section 7.3.3 in Kubin (2013). Nonbasal slip on pyramidal planes, as a secondary or additional slip system, sometimes is important as well. More detail can be found in the review article by Yoo et al. (2002). Only a few large-scale dislocation dynamics studies were performed that investigated the plastic response of HCP crystals (Aubry et al., 2016; Bertin et al., 2014; Fan et al., 2016; Monnet et al., 2004).

7.1.2 Dislocation Interactions and Topology Update

During each step of DDD, apart from calculating the dislocation velocity, the topology also needs to be updated to deal with short-range interactions between dislocations. Generally, some criteria (Rhee et al., 1998) and operators (Bulatov and Cai, 2006) need to be given. As shown in Fig. 7.2, these dislocation reactions are achieved through the merge and split operators among the dislocation segments and nodes. According to the relation of the Burgers vectors and slip planes of interacting dislocations, the dislocation reactions that are captured by DDD are mainly classified as (Groh and Zbib, 2009):

1. Mutual annihilation: Two dislocations of opposite Burgers vector directions exist in the same slip plane.
2. Collinear annihilation: The Burgers vector directions of two dislocations are collinear. Each dislocation is in the cross-slip plane of the other dislocation.
3. Hirth lock: Two dislocations of perpendicular Burgers vector directions exist in the intersection slip plane.
4. Glissile junction: The sum of the Burgers vectors of two dislocations is parallel to one slip plane. Their slip planes intersect.
5. Lomer lock: The sum of the Burgers vectors of two dislocations is not parallel to either slip plane. Their slip planes intersect.

In addition, when dislocations glide out of the crystal, surface annihilation should be considered, as shown in Fig. 7.3. Alternatively, the annihilated dislocations can be canceled by introducing virtual dislocation loops with the same Burgers vector and opposite line direction. This treatment needs to be coupled with boundary value problem corrections, but it is good at capturing surface roughness induced by dislocation motion. Readers interested in more detailed information are referred to Hussein and El-Awady (2016) and Po et al. (2014).

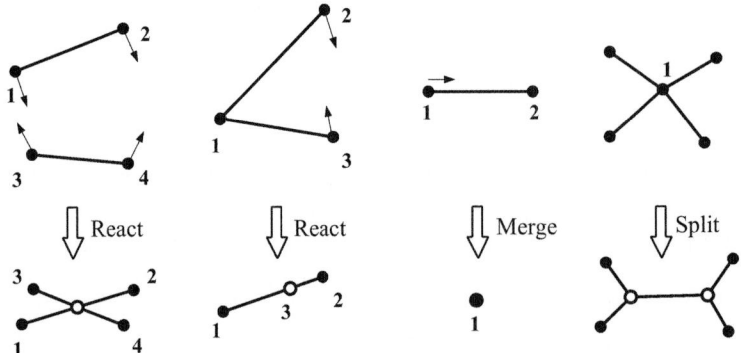

Figure 7.2
Merge and split operators for dislocation segments and nodes.

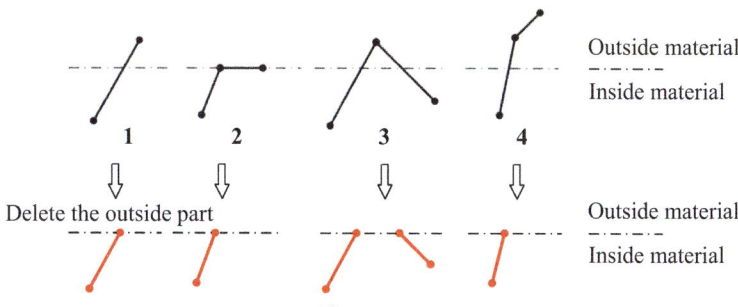

Figure 7.3

Four possible geometries and corresponding topology update schemes when dislocations are annihilated at the free surface.

7.1.3 Dislocation Cross-Slip

Cross-slip is another important dislocation mechanism, especially for crystals with medium to high stacking fault energy (Madec et al., 2002b; Motz et al., 2009; Rao et al., 2011; Wang et al., 2007; Wei and Wei, 2012). It frequently can be induced by a local heterogeneous stress state (Motz et al., 2009), and significantly influences spatiotemporal developments in the dislocation microstructure under both quasistatic and shocking loads (Wang et al., 2007). High cross-slip activity is inclined to make the substructure morphology appear cellular instead of planar (Wang et al., 2007) and promotes dislocation self-organization instead of uniform dislocation distributions (Madec et al., 2002b). Therefore, it is important to introduce a reasonable cross-slip model into DDD simulations.

Numerous models have been built to describe the physical process of cross-slip (Püschl, 2002; Ramírez et al., 2012). Generally, the screw dislocation dissociates into a pair of partial dislocations separated by a lattice stacking fault. For most densely packed planes, such as (111) for FCC, the dislocation core prefers to extend. The occurrence of cross-slip requires the stacking fault ribbon to be compressed to a critical length, either by applied stress or by thermal fluctuation. Then, the dislocation may bow out in the cross-slip plane or redissociate if the cross-slip plane is a close-packed one (Cui et al., 2013).

From a simulation point of view, the process of cross-slip usually can be modeled only phenomenologically because the properties of the dislocation core are involved (Devincre, 1996). Three conditions need to be met for cross-slip to occur (Cui et al., 2013):

1. The resolved shear stress in the cross-slip plane should be larger than that in initial glide plane, because cross-slip will take energy and barely occur unless screw dislocations have low mobility in the usual slip plane.

2. The screw segments of concern should be larger than a 0.1-μm restricted length set. This corresponds to the condition for stacking fault ribbon constriction (Devincre, 1996).

3. As mentioned, the probability of a cross-slip event in each discrete time step is determined by the Monte-Carlo method (Kubin et al., 1992):

$$P = \beta \frac{L}{L_0} \frac{\delta t}{\delta t_0} \exp\left(V \cdot \frac{\tau - \tau_{III}}{kT}\right) \tag{7.1.4}$$

where V is the activation volume, τ_{III} is the resolved shear stress at the onset of stage III during a tension test, k is the Boltzmann constant, and L_0 and δt_0 represent the length and time, respectively. For aluminum (Al) (Groh et al., 2009), $V = 300b^3$, b is the Burgers vector magnitude, $\tau_{III} = 5$ MPa, $L_0 = 1$ μm, and $\delta t_0 = 1$ s. Probability P is set to 1 at room temperature by adjusting normalizing coefficient β when the screw dislocation with length $L = L_0$ is subjected to resolved shear stress $\tau = \tau_{III}$. Cross-slip occurs only when the calculated P is larger than a randomly generated number N between 0 and 1. Actually, if τ is much larger than τ_{III}, the probability function will become inoperative, which implies that cross-slip is thermally activated.

7.1.4 Current Three-Dimensional Discrete Dislocation Dynamics Simulations

There are about a dozen 3D-DDD codes in use (Arsenlis et al., 2007; Devincre et al., 2011; Po and Ghoniem, 2014; Schwarz, 1999; Verdier et al., 1998; Zbib et al., 2003). Information about most widespread 3D-DDD simulation codes is summarized in work by Kubin (2013) and Sills et al. (2016). 3D-DDD simulations provide a great deal of physical insight into the fundamental mechanisms of dislocation plasticity. For example, they reveal the dislocation starvation mechanism and source truncation mechanism resulting from the limited sample size (Fertig and Baker, 2009; Liu et al., 2009a,b; Rao et al., 2008) and demonstrate the relation between strain burst and dislocation dynamical behavior. In addition, DDD results are scaled up to understand macroscopic bulk deformation behavior. For example, Groma (1997) (Yefimov et al., 2004) developed a high-order crystal plasticity model based on a 2D statistical-mechanics description of the collective behavior of dislocations, in which total dislocation density and net-Burgers vector density are used as two critical constitutive variables. Devincre et al. (2006, 2008) (Madec et al., 2002a) used the outcome of large-scale 3D-DDD simulations to estimate the strengthening constant in the Taylor relation to describe the strain hardening effect induced by an increase in dislocation density.

To simulate the experiments as faithfully as possible, several problems are worth mention. The first is the reasonable generation of an initial dislocation configuration. Numerous 3D-DDDs initially provide a Frank-Read source with a specific source length distribution (El-Awady et al., 2011; Rao et al., 2008). The non-destructible pinning points inevitably lead to an overestimation of the source lifetime and artificially increase the dislocation

density (Lee et al., 2013). Therefore, to consider the generation and destruction of dislocation sources physically, the initial equilibrium dislocation configurations should be generated via a relaxation procedure to approximate a real thermal annealing process (Cui et al., 2016c; Lee et al., 2013; Liu et al., 2009b; Motz et al., 2009; Tang et al., 2008). For example, consider the initial randomly created dislocation loops and straight dislocation lines spread on all available slip systems for FCC or BCC crystals. For straight dislocation lines, both ends terminate at the free surfaces. Then, the starting dislocation structure evolves dynamically without external loading until the dislocation density remains stable and further dislocation activity is not noticeable. Fig. 7.4 gives an example of dislocation configurations before and after stress relaxation in an Ni micropillar with $d = 200$ nm (Cui et al., 2014). The dislocation junctions emerge as a natural outcome of the dislocation interaction. Compared with the initial fixed pinning points, this kind of simulation method can consider more physical formation and destruction of dislocation junctions.

Another point is that the image force effect induced by the free surface is important, especially when the surface-to-volume ratio is high. The image force can influence source operation stress at the near-surface region (Rao et al., 2007), promote dislocation surface annihilation (Weinberger and Cai, 2007), and trigger cross-slip for surface dislocation (Zhou et al., 2011). However, conventional DDD simulations used periodic boundary conditions, and the dislocation interaction is described based on the elastic stress field in infinite media. To date, how to capture the image force induced by a complex surface or interface accurately remains difficult.

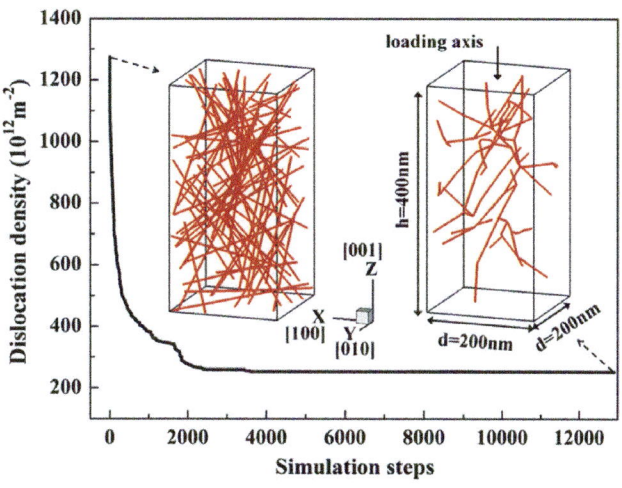

Figure 7.4

Dislocation density evolution during stress relaxation for $d = 200$ nm. Insets show the corresponding dislocation snapshots before and after stress relaxation. *Reprinted from Cui, Y., Lin, P., Liu, Z., Zhuang, Z., 2014. Theoretical and numerical investigations of single arm dislocation source controlled plastic flow in FCC micropillars. International Journal of Plasticity 55, 279–292, Copyright 2014, with permission from Elsevier.*

In addition, 3D-DDD modeling alone generally has some limitations: (1) It cannot consider the finite deformation of the computational cell and is suitable only for small deformation; (2) it is difficult to describe the strong nonlinear effect around a dislocation core accurately; (3) the mechanism of dislocation climb (nucleation) is still not fully understood; and (4) computation costs increase rapidly with an increase in the dislocation segment number, so the length scale and strain range that can be considered remain limited. To solve these problems, 3D-DDD is coupled with other simulation methods (such as molecular dynamics and continuum mechanics) to realize multiscale simulations.

7.2 Coupling Discrete Dislocation Dynamics With Finite Element Method

To consider the finite deformation of computational cells and deal with complex boundary conditions and the surface effect, DDD is usually coupled with the finite element method (FEM) (Liu et al., 2009b; Po et al., 2014; Van der Giessen and Needleman, 1995; Zbib et al., 2002) or boundary element method (El-Awady et al., 2008; Zhou et al., 2010a). These coupling procedures are mainly divided into two categories: one is the superposition method (SPM) and the other is so-called discrete-continuous model (DCM). In the following discussion, only streamlined presentations of the framework for these two methods are described. Details of the methods are described elsewhere (Cui et al., 2015b; El-Awady et al., 2008; Gao et al., 2010; Lemarchand et al., 2001; Liu et al., 2009b; Po et al., 2014; Van der Giessen and Needleman, 1995; Vattré et al., 2013; Zbib and Diaz de la Rubia, 2002; Zbib et al., 2002).

7.2.1 Superposition Method

SPM was first proposed by Van der Giessen and Needleman (1995). As schematically shown in Fig. 7.5A, total stress field σ in a finite crystal medium is the sum of the analytical stress field of dislocations in an infinite media σ^{∞} and a complementary elastic solution $\widehat{\sigma}$:

$$\sigma = \sigma^{\infty} + \widehat{\sigma} \tag{7.2.1}$$

where σ^{∞} is obtained by an analytical solution generally calculated based on the idea of eigenstrain using the Green function method (de Wit, 1960):

$$\sigma_{ij}^{\infty} = \frac{\mu b_n}{4\pi} \int_c \left[\frac{1}{2} R_{,mpp} \left(\xi_{jmn} dl_i + \xi_{imn} dl_j \right) + \frac{1}{1-\nu} \xi_{kmn} \left(R_{,ijm} - \delta_{ij} R_{,ppm} \right) dl_k \right] \tag{7.2.2}$$

where μ is the shear modulus, ν is Poisson's ratio, b is the Burgers vector, and R is a distance from the point on the dislocation segment to the field point. ξ_{jmn} represents the permutation operator. $()_i$ denotes the ith component and $()_{,i}$ denotes differentiation with respect to x_i. Repeated indices are summed. dl represents infinite small dislocation segments. Eq. (7.2.2) is derived with respect to the closed dislocation loop.

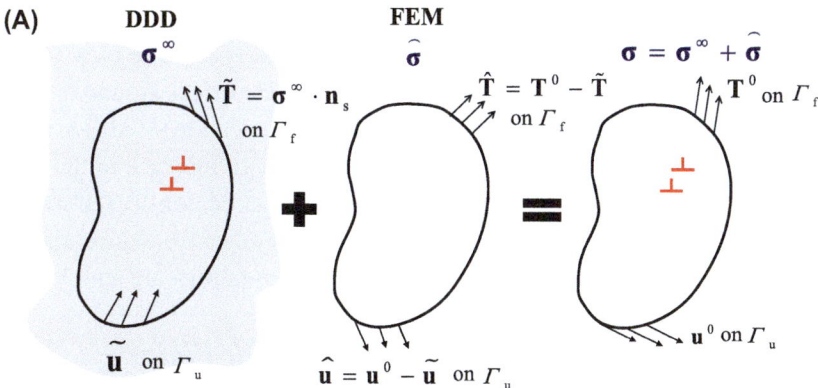

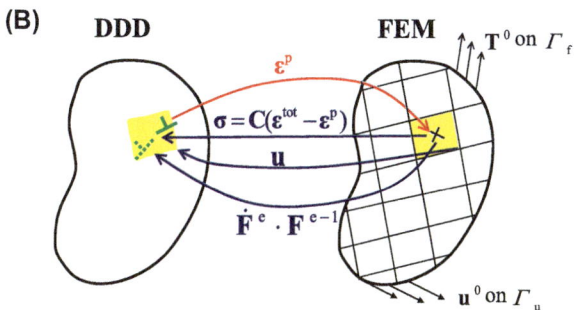

Figure 7.5

(A) Schematic diagram of superposition method (Van der Giessen and Needleman, 1995); (B) Schematic of variable-transferring procedures in improved discrete-continuous model. *DDD*, discrete dislocation dynamics; *FEM*, finite element method. *Reprinted from Cui, Y., Liu, Z., Zhuang, Z., 2015b, Quantitative investigations on dislocation based discrete-continuous model of crystal plasticity at submicron scale. International Journal of Plasticity 69, 54—72, Copyright 2015, with permission from Elsevier.*

σ^∞ will induce surface traction on the sample:

$$\widetilde{\mathbf{T}} = \sigma^\infty \cdot \mathbf{n}_s \qquad (7.2.3)$$

Here, $\mathbf{n}_s$ is the normal direction of the surface. A complementary stress field is used to cancel this surface traction and consider the real boundary condition.

The short-range interaction can be relatively well-captured by SPM (Vattré et al., 2013). However, the analytical stress fields of all dislocations must be recalculated at each time step, which requires extensive computing time. Besides, it is relatively complicated to calculate the displacement field induced by dislocation (de Wit, 1960; Ghoniem and Sun, 1999):

$$u_i = -\frac{b_i \Omega}{4\pi} + \frac{1}{8\pi} \int_c \left[\xi_{ikl} b_l R_{,pp} + \frac{1}{1-\nu} \xi_{kmn} b_n R_{,mi} \right] \mathrm{d}l_k \qquad (7.2.4)$$

where Ω represents the solid angle (Ghoniem and Sun, 1999):

$$\Omega = \int d\Omega = \int_A \frac{\mathbf{e} \cdot d\mathbf{A}}{R^2} = -\frac{1}{2} \int_A R_{,ppi} dA_i \tag{7.2.5}$$

where $d\Omega$ is the solid angle differential, which is a ratio of the projected slip area element $d\mathbf{A}$ to the square of R. The dislocation stress field can be expressed nicely by a line integral (Eq. 7.2.2). Therefore, as long as information about the dislocation position is recorded, the stress field can be calculated. However, in addition to the line integration part, the solid angle part of the displacement field is not uniquely defined for a line segment (Barnett, 1985), so information about the dislocation slip area history has to be recorded to calculate the displacement field accurately in the sample, which is not trivial for 3D dislocation motion, especially when cross-slip and dislocation climb are considered. More important, the concept of plastic strain is not explicitly introduced.

7.2.2 Discrete-Continuous Model

The DCM is based directly on the concept of eigenstrain in micromechanics, which can directly calculate the plastic strain and solve the boundary value problem under a unified framework (Lemarchand et al., 2001; Zbib and Diaz de la Rubia, 2002). In previous work (Gao et al., 2010; Liu et al., 2009b), it mainly contained three information-transfer procedures, as shown in Fig. 7.5B: (1) Calculate the plastic strain $\boldsymbol{\varepsilon}^p$ induced by the glide of dislocations using a DDD simulation. Then, plastic strain is localized to the continuum material point, which is crucial in the whole calculation procedure. This replaces the conventional phenomenological constitutive law to calculate the total stress:

$$\overset{\triangledown}{\boldsymbol{\sigma}}{}^e = \mathbf{C}^e : (\dot{\boldsymbol{\varepsilon}} - \dot{\boldsymbol{\varepsilon}}^p) \tag{7.2.6}$$

where $\overset{\triangledown}{\boldsymbol{\sigma}}{}^e$ is the Jaumann rate of Cauchy stress $\boldsymbol{\sigma}$, $\mathbf{C}^e$ is the tensor of elastic modulus, and $\dot{\boldsymbol{\varepsilon}}$ is the total strain rate tensor. (2) The equilibrium stress field associated with the homogenized plastic strain is calculated by FEM under a specific boundary condition in a unified continuum mechanics framework. It is expressed as (Gao et al., 2010; Liu et al., 2009b)

$$\mathbf{M\ddot{u}} + \mathbf{f}^{\text{int}} = \mathbf{f}^{\text{ext}} \tag{7.2.7}$$

$$\mathbf{M} = \int_\Omega \rho \mathbf{N}^T \mathbf{N} d\Omega$$

$$\mathbf{f}^{\text{int}} = \int_\Omega \mathbf{B}_e^T \boldsymbol{\sigma} d\Omega \tag{7.2.8}$$

$$\mathbf{f}^{\text{ext}} = \int_\Gamma \mathbf{B}_e^T \mathbf{T}^0 d\Gamma + \int_\Omega \mathbf{B}_e^T \boldsymbol{\sigma}^0 d\Omega$$

$$\mathbf{u} = \mathbf{u}^0 \quad \Gamma \in \Gamma_u$$
$$\mathbf{T} = \mathbf{T}^0 \quad \Gamma \in \Gamma_f, \quad \mathbf{T} = 0 \quad \Gamma \notin (\Gamma_u \cup \Gamma_f) \tag{7.2.9}$$

where $\mathbf{M}$ is the mass matrix, $\mathbf{N}$ is the shape function, $\mathbf{B}_e = \text{grad}[\mathbf{N}]$, and $\mathbf{f}^{\text{int}}$ is an internal force and $\mathbf{f}^{\text{ext}}$ is an external force resulting from the applied traction $\mathbf{T}^0$ and initial stress field $\boldsymbol{\sigma}^0$, which is introduced to represent preexisting stationary dislocations. Then, stress field $\boldsymbol{\sigma}$ calculated by FEM is transferred to DDD and serves as the applied stress to drive dislocation line motion. (3) Displacement field $\mathbf{u}$ of the FEM cell is transferred into the DDD cell to update the geometry configuration. In DCM, time increment Δt_{DDD} in the DDD model is set to be a small value (10^{-10}–10^{-12} s), which can be equal to or smaller than that in the FEM model.

Although both DCM and SPM have been largely investigated (Lemarchand et al., 2001; O'day and Curtin, 2005; Vattré et al., 2013; Zbib et al., 2002), some important problems remain that are not yet well-clarified, especially for DCM. The next section mainly focuses on DCM with respect to the following critical issues. The first for DCM is the regularization method used to localize discrete plastic strains to continuum material points. Different researchers proposed various regularization methods (Lemarchand et al., 2001; Liu et al., 2009b; Vattré et al., 2013; Zbib and Diaz de la Rubia, 2002). However, a quantitative comparison among them and how to select slip system-dependent adjustment parameters are still not clear. The second issue is calculation of the so-called image force. At small scales, the image force caused by the free surface attracts the dislocations toward the surface and thus promotes dislocation starvation (Weinberger and Cai, 2007), triggers cross-slip of surface dislocation (Zhou et al., 2011), etc. To investigate submicron plasticity, special attention must be paid to calculating the dislocation image force. Generally, SPM is supposed to capture the short-range interaction and image force effect effectively. Thus, SPM and DCM are sometimes used together in the multiscale model with the aim of taking full advantage of both methods (Gao et al., 2010; Zbib and Diaz de la Rubia, 2002). However, does this kind of treatment double-count the contribution of the image force? How can the accuracy of the DCM alone capture the effect of the free surface? These analyses will provide useful guidelines for effectively correcting image force calculation in DCM. The third issue is the reproduction of deformed configurations in DCM, especially for analyzing the failure process. In micropillar compression experiments, deformation is usually observed to be localized in a few slip bands (Dimiduk et al., 2005). This leads to significant variations in the surface configuration and further influences their stress distribution and failure process. For example, small variations in a surface configuration can lead to stress concentration sufficient to promote the initiation of cracks. El-Awady et al. (2008) coupled DDD and the boundary element method to investigate the deformed shape of micropillars induced by the operation of Frank-Read sources. The simulation work of Zbib et al. also reproduced the character of deformation bands and the formation of ledges on the surface (Akarapu et al., 2010; Zbib and Diaz de la Rubia, 2002).

Gao et al. (2010) captured buckling configuration when micropillars were subjected to uniaxial compression without friction between the pillar and the indenter. The next section presents algorithmic details to capture the deformed configuration of the material and lattice rotation in a 3D-DDD model, which enables the DCM model to tackle true finite-strain problems.

7.3 Improved Discrete-Continuous Model

7.3.1 Efficient Regularization Method

In this section, different regularization methods are briefly reviewed. The calculated stress field of one prismatic dislocation loop is compared with its analytical solution to validate all of these methods.

As shown in Fig. 7.6, supposing that dislocation segment AB slips to $A'B'$ without rotating during a time increment, the swept area is S_{AB}. According to Orowan's law, total plastic shear increment $\Delta\gamma$ can be explicitly expressed as a function of the area swept by dislocation motion:

$$\Delta\gamma = \frac{bS_{AB}}{V} \tag{7.3.1}$$

where b is a magnitude of the Burgers vector and V is the representative volume. This plastic strain increment will be localized to the material points (or integration points) of the FEM element around the swept area by some regularization methods.

At the center of the swept surface, a local coordinate system can be established, as shown in Fig. 7.6, where **n** is the normal direction of the slip plane, **g** is the glide direction, and ξ is the dislocation line direction. From a view along the ξ direction, the different regularization methods are schematically presented in Figs. 7.7 and 7.8. The main differences between the regularization methods reside in the choice of representative

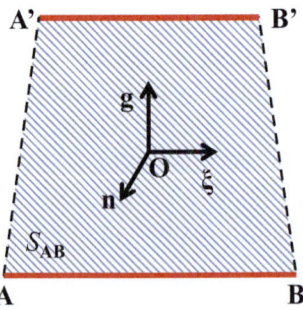

Figure 7.6

Dislocation segment AB glides to $A'B'$, generating swept surface S_{AB}; a local coordinate system is built at the center of S_{AB} (Cui, 2016).

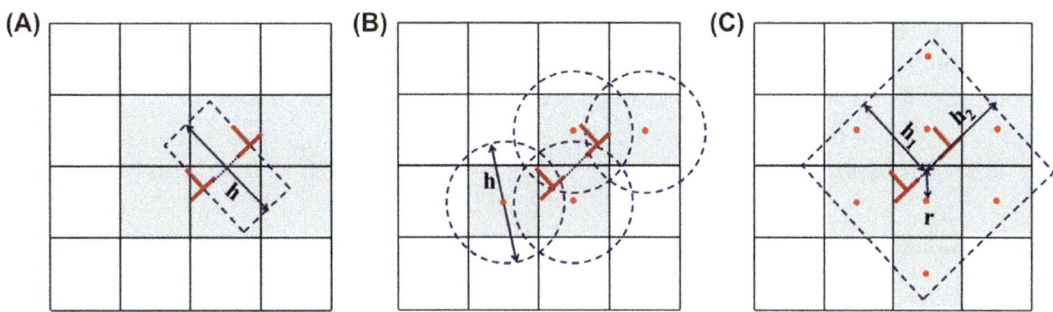

Figure 7.7

Plastic strain induced by dislocation glide is localized in the *shaded elements* with different methods: (A) used by Lemarchand et al. (2001), (B) used by Vattré et al. (2013), and (C) used by Liu et al. (2009b). The *red points* in (B) and (C) represent integration points; and the linear element is used for clarity of presentation. *Reprinted from Cui, Y., Liu, Z., Zhuang, Z., 2015b, Quantitative investigations on dislocation based discrete-continuous model of crystal plasticity at submicron scale. International Journal of Plasticity 69, 54—72, Copyright 2015, with permission from Elsevier.*

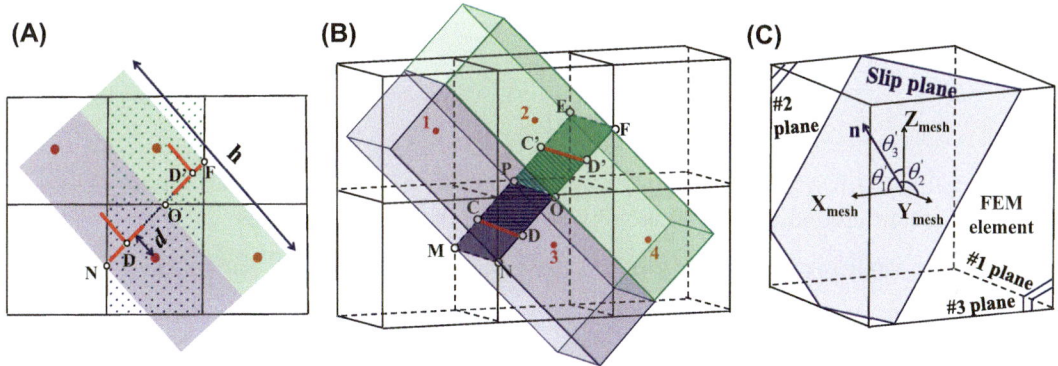

Figure 7.8

(A and B) Schematic showing the new regularization method from a view along dislocation line direction and three-dimensional view, respectively. The *solid dot* represents the centroid of the concerned integration point volume. (C) Schematic definition of X_{mesh}, Y_{mesh}, and Z_{mesh}, which are normal directions of the #1, 2 and 3 finite element method (FEM) element planes, respectively. θ'_1, θ'_2, and θ'_3 are their angles with the normal direction of slip plane **n**, respectively. *Reprinted from Cui, Y., Liu, Z., Zhuang, Z., 2015b, Quantitative investigations on dislocation based discrete-continuous model of crystal plasticity at submicron scale. International Journal of Plasticity 69, 54—72, Copyright 2015, with permission from Elsevier.*

volume *V* in Eq. (7.3.1) and the way of distributing total plastic shear increment $\Delta\gamma$ to the multiple integration points.

In Eq. (7.3.1), *V* is usually set to be the elementary volume V^{int} associated with each integration point in the FEM element (Lemarchand et al., 2001; Liu et al., 2009b). In this case, the total plastic shear increment, $\Delta\gamma$, is expressed as $\Delta\gamma^{int} = bS_{AB}/V^{int}$.

The first typical regularization method was proposed by Lemarchand et al. (2001). Based on the Volterra-like procedure, as discussed by Mura (1987), the elementary slip events in the glide plane are extended over a slab of finite thickness h (Fig. 7.7A). Physically speaking, in this method, dislocations AB and A′B′ are considered to be plate-like inclusions with a cross-section area equal to swept area SAB and thickness equal to h. Total plastic shear increment $\Delta\gamma^{\mathrm{int}}$ is then localized to each integration point according to the intersection volume $\Delta V_s^{(i)}$ between the sheared slab and elementary volume V^{int}:

$$\Delta\gamma^{(i)} = \frac{\Delta V_s^{(i)}/h}{S_{\mathrm{AB}}} \cdot \Delta\gamma^{\mathrm{int}} \tag{7.3.2}$$

Hereafter, superscript "(i)" means the variables associated with the ith integration point. $\Delta V_s^{(i)}/h$ can be considered an effective area corresponding to the ith integration point.

Vattré et al. (2013) presented the algorithmic details of the regularization procedure and made some improvements with respect to the work of Lemarchand et al. In their work, the elementary slip event is also considered to be a plate-like inclusion surrounding its swept surface, but the inclusion is further thought of as the union of overlapping elementary spheres of diameter h. They take V in Eq. (7.3.1) to be a volume of the representative sphere V^{sphere}, and total plastic shear increment $\Delta\gamma$ is expressed as $\Delta\gamma^{\mathrm{sphere}} = bS_{\mathrm{AB}}/V^{\mathrm{sphere}}$. Each sphere centers within the swept surface and corresponds to a homogeneously distributed plastic shear increment. Once an integration point is located within one or more elementary spheres, its plastic shear increment is nonzero. This procedure can be equivalent to the following process. Considering a sphere with its center at the ith integration point and a diameter equal to h, as shown in Fig. 7.7B; the plastic shear increment at each integration point is calculated according to the intersection area between this sphere and the swept surface $S_{\mathrm{sphere}}^{(i)}$:

$$\Delta\gamma^{(i)} = \frac{S_{\mathrm{sphere}}^{(i)}}{S_{\mathrm{AB}}}\Delta\gamma^{\mathrm{sphere}} \tag{7.3.3}$$

The form of Eq. (7.3.3) seems similar to that of Eq. (7.3.2). However, in Eq. (7.3.2), parameter h is introduced during the localization process, whereas in Eq. (7.3.3), h is used to calculate the representative volume. The use of the spherical shape makes it convenient to treat problems with internal interfaces (Vattré et al., 2013). However, the sum of localized plastic strain increment $\sum_i \Delta\gamma^{(i)}$ for all elements participating in regularization is not as straightforward as in other methods because some swept area may be included in two or more elementary spheres.

Different from these regularization methods, Liu et al. (2009b) localized total plastic shear increment $\Delta\gamma^{\mathrm{int}}$ according to a weight function $w^{(i)}$:

$$\Delta\gamma^{(i)} = \frac{w^{(i)}}{\sum_{i=1}^{n} w^{(i)}}\Delta\gamma^{\mathrm{int}} \tag{7.3.4}$$

where n is the total number of integration points whose weight function is nonzero. The weight function is expressed as a function of isotropic Burgers vector density function $\omega(r)$, based on the nonsingular continuum theory of dislocations developed by Cai et al. (2006):

$$w^{(i)} = \int_{V^i} \omega(\mathbf{x})dV, \quad \omega(r) = \frac{1}{\pi} \frac{a}{r^2 + a^2} \tag{7.3.5}$$

where V^i is the volume occupied by the ith integration point, a is a spreading radius, and r is the distance between an integration point and the center of the swept surface, as shown in Fig. 7.7C. Three cutoff distances, h_1, h_2, and h_3, are introduced along three axis directions at the local coordinate system. Only if the ith integration point is within these cutoff distances is its weight function calculated by Eq. (7.3.5). Obviously, it is efficient to calculate the weight function numerically. However, this method is valid based on the assumption that dislocations AB and A'B' can be regarded as a dislocation pole. Thus, at each time increment, the slip distance of each dislocation segment d_{slip} should be much smaller compared with its length.

We proposed a novel regularization method, as shown schematically in Fig. 7.8 (Cui et al., 2015b), in the hope of combining the advantages of previous regularization methods. Inspired by the method of Lemarchand et al. (Lemarchand et al., 2001), the regularization region is also considered to be a slab with a thickness equal to h, but plastic strain is not localized according to the intersection volume. Similar to the method of Liu et al., if the centroid of the ith elementary volume is within the slab, or if the elementary volume is passed through by the swept area, its localized plastic strain $\Delta\gamma^{(i)}$ is given by a weight function $m^{(i)}$

$$\Delta\gamma^{(i)} = \frac{m^{(i)}}{\sum\limits_{i}^{n} m^{(i)}} \cdot \Delta\gamma^{int}, \quad m^{(i)} = \frac{S_{swept}^{(k)}}{S_{all}^{(k)}} \int_{V^i} \frac{1}{\pi} \frac{a}{d^{(i)2} + a^2} dV, \quad \text{if} \quad d^{(i)} < \frac{h}{2} \tag{7.3.6}$$

where $d^{(i)}$ is the distance from the centroid of ith elementary volume to the slip plane. Compared with the method of Liu et al., a coefficient $S_{swept}^{(k)} / S_{all}^{(k)}$ is introduced to characterize the slip extent, so dislocations AB and A'B' do not have to meet the assumption of the dislocation pole and there is no constraint on the slip distance for each time increment. Specifically, the slab region is further divided into several subregions by element mesh, as shown in Fig. 7.8. For the subregion corresponding to the kth swept elementary volume, $S_{swept}^{(k)}$ represents the swept area and $S_{all}^{(k)}$ is the total possible swept area. For example, for blue subregion in Fig. 7.8B, the swept elementary volume is labeled "3" and $S_{swept}^{(k)}$ and $S_{all}^{(k)}$ are the area of CDOP and MNOP, respectively. MNFE is

the cross-section of the slab. If one dislocation line sweeps multiple elementary volumes in a time increment, this method still exhibits high efficiency because $S_{\text{swept}}^{(k)}\big/S_{\text{all}}^{(k)} = 1$ for most subregions and it only requires calculation of $S_{\text{swept}}^{(k)}\big/S_{\text{all}}^{(k)}$ for subregions near the boundary of the swept area.

In addition, all of these regularization methods involve parameters (such as h), which are depend on the type of FEM element and information about the slip system (Lemarchand et al., 2001). For example, for dislocation with a slip plane parallel to one of the FEM mesh planes, good results can be obtained when the plastic strain is localized only to the swept elementary volumes. Thus, regularization parameter h^{parallel} is set to $L/2$ for a quadratic hexahedron element with 20 nodes and 8 Gauss points, where L is the element size. However, for tilted dislocation, the regularization region must be enlarged to ensure the continuity of the eigenstrain from element to element. For arbitrary tilted dislocation, good results can be obtained if slip plane-dependent parameter h is determined (Cui, 2016):

$$h = k_{\text{tilt}}h^{\text{parallel}}, k_{\text{tilt}} = 1 + \sin 2\theta_1 + \sin 2\theta_2 + \sin 2\theta_3,$$

$$\theta_1' = \arccos\left(\frac{\mathbf{n}\cdot\mathbf{X}_{\text{mesh}}}{\|\mathbf{n}\|\cdot\|\mathbf{X}_{\text{mesh}}\|}\right), \theta_2' = \arccos\left(\frac{\mathbf{n}\cdot\mathbf{Y}_{\text{mesh}}}{\|\mathbf{n}\|\cdot\|\mathbf{Y}_{\text{mesh}}\|}\right), \theta_3' = \arccos\left(\frac{\mathbf{n}\cdot\mathbf{Z}_{\text{mesh}}}{\|\mathbf{n}\|\cdot\|\mathbf{Z}_{\text{mesh}}\|}\right)$$

$$\theta_1 = \min\left(\theta_1', \pi - \theta_1'\right), \theta_2 = \min\left(\theta_2', \pi - \theta_2'\right), \theta_3 = \min\left(\theta_3', \pi - \theta_3'\right)$$

$$(7.3.7)$$

where $\mathbf{n}$ is the normal vectors of the slip plane; $\mathbf{X}_{\text{mesh}}$, $\mathbf{Y}_{\text{mesh}}$, and $\mathbf{Z}_{\text{mesh}}$ represent the normal directions of the FEM element planes, as shown in Fig. 7.8C; and angles with $\mathbf{n}$ are θ_1', θ_2', and θ_3', respectively. k_{tilt} is the correction coefficient for regularization parameter h.

A simple numerical test is carried out to validate the effectiveness of the new regularization method for tilted dislocation lines. Considering a cubic crystal with a side length of 40 μm, an edge dislocation along the [100] direction sweeps to the middle of the crystal along the (010) slip plane and its Burgers vector is along the [001] direction. An isotropic elastic solid is considered with a shear modulus of $\mu = 51$ GPa and Poisson's ratio of $v = 0.37$ in most of the following simulations, unless otherwise specified. As suggested by Lemarchand et al. (2001), a quadratic hexahedron element with 20 nodes and 8 Gauss points (C3D20R) is used in all of the following calculations. The element size is set to 850 nm. The results with $h = h^{\text{parallel}}$ and $h = k_{\text{tilt}}h^{\text{parallel}}$ are presented in Fig. 7.9A−F, respectively. When the regularization region is small, there will be numerical noise for the tilted dislocation. However, the stress field can be reasonably captured by enlarging the regularization region controlled by the slip plane-dependent parameter.

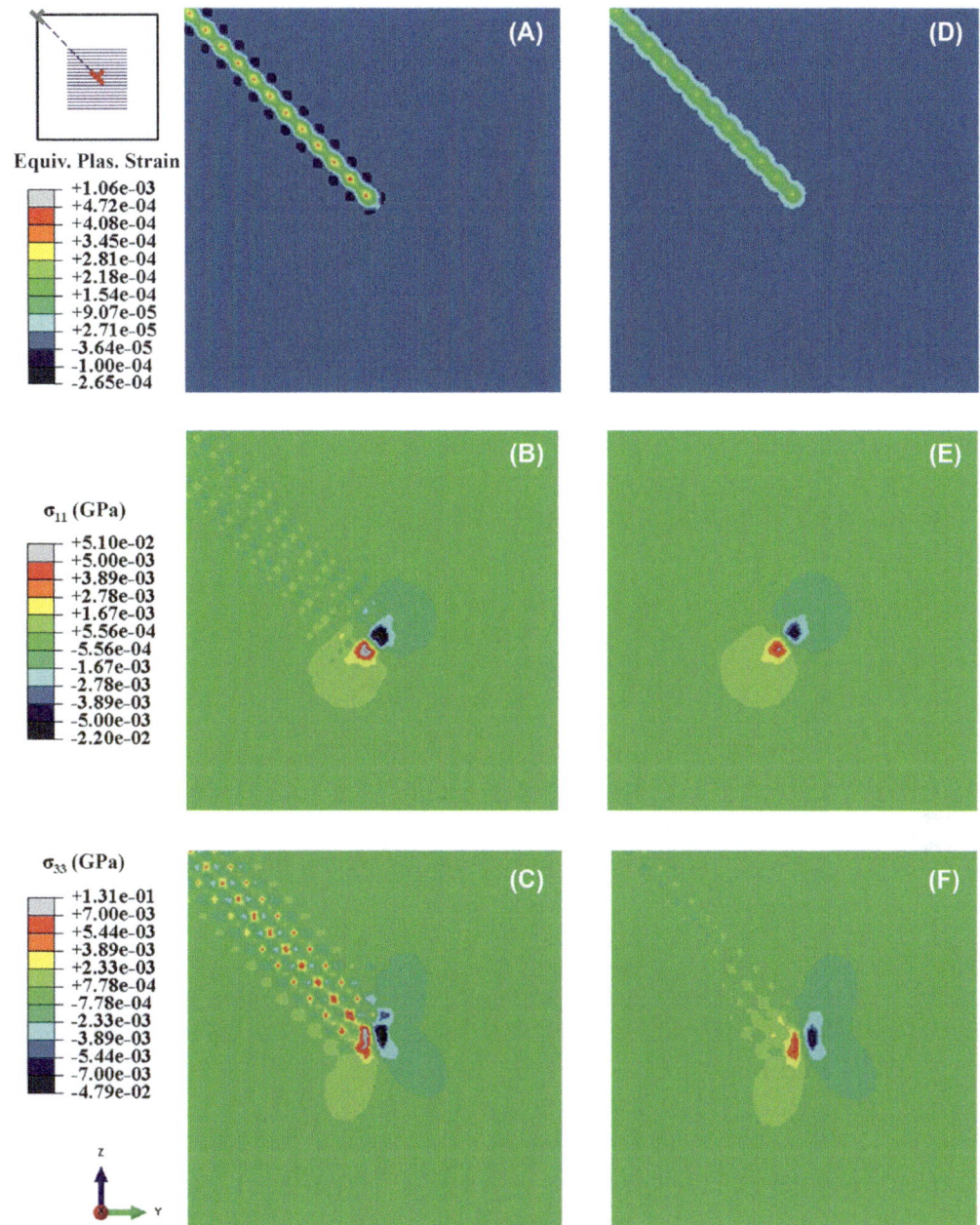

Figure 7.9

For a tilted dislocation, distribution of the equivalent plastic (Equiv. Plas.) strain and stress field in the shadow region in (A). These are calculated using the new regularization method with (A–C) $h = h^{parallel}$ and (D–F) $h = k_{tilt}h^{parallel}$. *Reprinted from Cui, Y., Liu, Z., Zhuang, Z., 2015b, Quantitative investigations on dislocation based discrete-continuous model of crystal plasticity at submicron scale. International Journal of Plasticity 69, 54–72, Copyright 2015, with permission from Elsevier.*

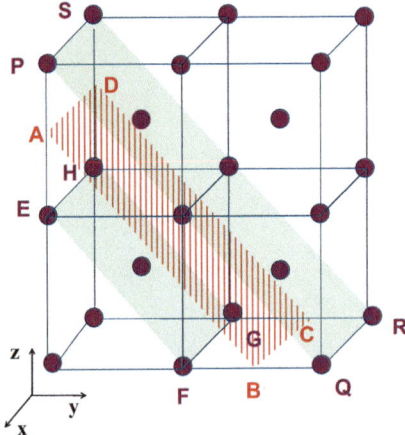

Figure 7.10

Schematic showing crystal lattice-based continuum material point arrangement. *Circle points* represent continuum material points for a BCC crystal.

Alternatively, one does not need to deal with tilted dislocations if the positions of the continuum material point are designed so that no matter which slip plane is considered, multiple points can be found on planes parallel to the slip plane. One way is to design the mesh for the continuum material points according to the relative positions of a lattice corresponding to the crystal structure itself. For example, with respect to a BCC crystal, the continuum material points can be placed as shown in Fig. 7.10. Because the slip plane always corresponds to the closely packed lattice plane, whatever the slip plane is, multiple continuum material points are exactly the same distance from the slip plane. Taking slip plane ABCD in Fig. 7.10 as an example, multiple points can be found on planes parallel to the slip plane: that is, the points on plane EFGH and PQRS. This means that all slip planes can be treated similar to the case of nontilted dislocation. More details about this crystal lattice-based continuum material point arrangement method are described in Cui et al. (2018).

The accuracy of different regularization methods is further compared using a numerical test. Because dislocations in the DCM are represented by the plastic strains caused by their slip, a reasonable regularization method should be able to introduce the self-stress field of dislocation lines well according to the localized strains.

Considering a cubic crystal with the dimension $2250 \times 2000 \times 2250$ nm^3, a square prismatic dislocation loop lies in the (010) slip plane and has four $\langle 100 \rangle$ edge segments of length 250 nm, as illustrated in Fig. 7.11A. The Burgers vector is along the [010] direction and has a magnitude of 0.25 nm. Supposing that this dislocation loop is obtained by the growth of a small loop, the plastic strain inside and around the dislocation loop is nonzero and

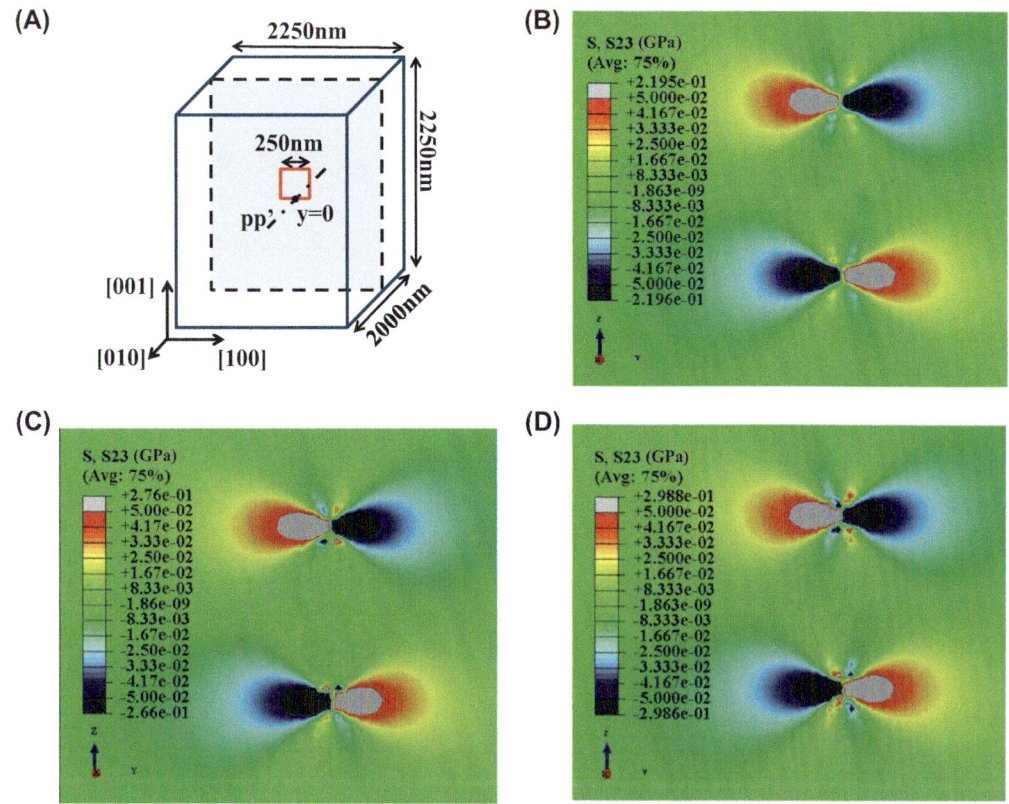

Figure 7.11

(A) Geometrical description of the validation test: a square prismatic dislocation loop in the dashed slip plane; (B—D) For an element size of $L = 16$ nm, distribution of shear stress σ_{23} caused by the prismatic loop using the methods of Lemarchand et al. and Liu et al., and the regularization method of Cui et al., respectively. *Reprinted from Cui, Y., Liu, Z., Zhuang, Z., 2015b, Quantitative investigations on dislocation based discrete-continuous model of crystal plasticity at submicron scale. International Journal of Plasticity 69, 54—72, Copyright 2015, with permission from Elsevier.*

can be localized to the material point by four kinds of regularization methods described previously. Regularization parameter h in the methods of Lemarchand et al. and Vattré et al. is taken to be $3L/2$ (Lemarchand et al., 2001). The cutoff parameter in the method of Liu et al. is $h_1 = h_2 = h_3 = 2L/3$.

Different element sizes L are used in the calculation. Shear stress fields σ_{23} for $L = 16$ nm obtained by different regularization methods are given in Fig. 7.11B—D. The results for the method of Vattré et al. can be obtained from Vattré et al. (2013) and are not shown here. The upper and lower bounds of the stress value in Fig. 7.11B—D are set to the same as in Fig. 10B in Vattré et al. (2013). A similar stress distribution is obtained for different methods.

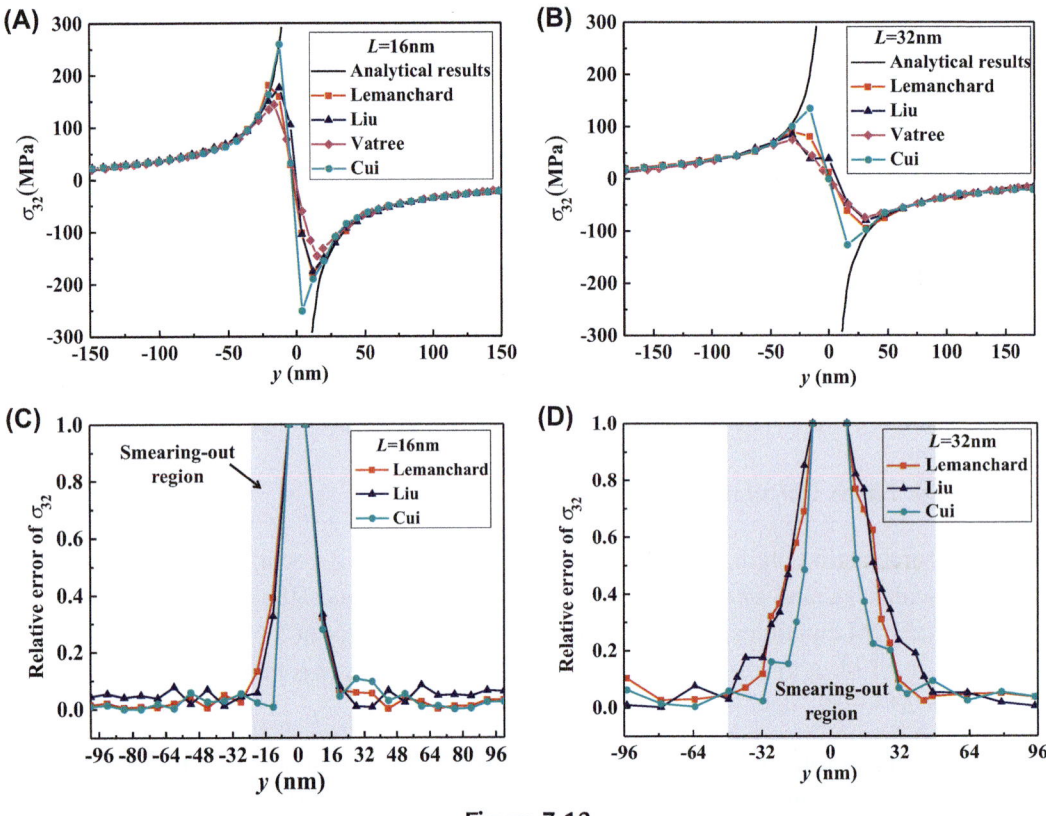

Figure 7.12

(A and B) Stress component σ_{32} caused by the prismatic loop by different regularization methods when element size L is 16 and 32 nm, respectively. Results for the method proposed by Vattré et al. (2013) and element sizes in (A) and (B) are 15 nm × 15.625 nm × 16.25 nm and 30 nm × 31.25 nm × 32.5 nm, respectively. (C and D) Relative errors of σ_{32} when element sizes are 16 and 32 nm, respectively. *Reprinted from Cui, Y., Liu, Z., Zhuang, Z., 2015b, Quantitative investigations on dislocation based discrete-continuous model of crystal plasticity at submicron scale. International Journal of Plasticity 69, 54—72, Copyright 2015, with permission from Elsevier.*

Stress variation along the *dashed line* in Fig. 7.11A is further compared with the analytical solution when element size L is set to 16 and 32 nm, respectively, as given in Fig. 7.12. Fig. 7.12A and B show that different regularization methods display comparable accuracy. The singularity of stress near the dislocation core region is smeared out by all of these regularization methods. This is similar to results derived by the nonsingular continuum theory of dislocations proposed by Cai et al. (2006). To facilitate discussion, this region is denoted as the smearing-out region, as labeled in Fig. 7.12C and D; the other region is denoted as the long-range region. Short-range interaction in the smearing-out region needs to be modified, such as using analytical solutions for dislocation interactions. By comparing Fig. 7.12C with Fig. 7.12D, it can be found that increasing the element size

does not strongly influence the precision in the long-range region, but it enlarges the smearing-out region. Moreover, the results suggest that the smearing-out region corresponds to the region in which the distance to dislocation core is smaller than a critical value $r_{\text{smearing-out}}$. Quantitatively speaking, for the methods of Lemarchand et al. and Vattré et al., $r_{\text{smearing-out}}$ is 1.5 times the FEM element size L, which is exactly equal to the regularization parameter h (Vattré et al., 2013). For the method of Liu et al., $r_{\text{smearing-out}}$ is about twice the cutoff distance. For the method of Cui et al., $r_{\text{smearing-out}}$ is equal to FEM element size L. In the long-range region, all of the calculation results are in good agreement with the analytical results; the relative error is less than 10% (Fig. 7.12) (Vattré et al., 2013). Moreover, Fig. 7.12C and D illustrate that the precision for the method of Cui et al. is even better than that of Lemarchand et al. and Liu et al. Therefore, DCM with the regularization method of Cui et al. is mostly used in the following sections.

7.3.2 Image Force Calculation

The image force acting on dislocations near the free surface is especially important at submicron scales because of the large specific surface area. Generally, it is always believed that DCM cannot capture the image force accurately, or at least, it is not comparable to SPM in this aspect. This section describes two corrections that show the ability of DCM to calculate image force accurately.

In DCM, the image force is lumped into the total stress field calculated by FEM and is passed to DDD according to the stress-transfer procedure (Fig. 7.5B). Specifically, the image force is calculated according to the difference in stress fields for two DCM models. One directly considers the free surface and the other calculates the total stress field in an infinite body by applying surface traction $\widetilde{\mathbf{T}} = \boldsymbol{\sigma}^{\infty} \cdot \mathbf{n}_s$, where $\boldsymbol{\sigma}^{\infty}$ is the analytical stress field of dislocations in infinite media and $\mathbf{n}_s$ is the normal direction of the surface. The stress acting on each dislocation segment is usually taken to be equal to the stress at the integration point of the FEM element where the midpoint of the dislocation segment is located (Gao et al., 2010). Because the image force is strongly sensitive to the distance between the dislocation line and free surface, this kind of stress-transfer rarely gives good results when the dislocation segment does not pass exactly through the integration point. This can be demonstrated by a simple example.

Consider a cubic crystal with a side length of 40 μm, as shown in Fig. 7.13A. An edge dislocation along [100] is located at distance Z below the [001] free surface and its Burgers vector is along the [001] direction. The image force induced by the top surface is calculated for different values of Z using DCM with the new regularization method. FEM element size L is set to 0.85 μm. The lines with triangle points in Fig. 7.13B result when an image force on the dislocation segment is calculated using the stress at the nearest integration point. The results show a large deviation from the analytical solution when the dislocation lines do not pass exactly through the integration points.

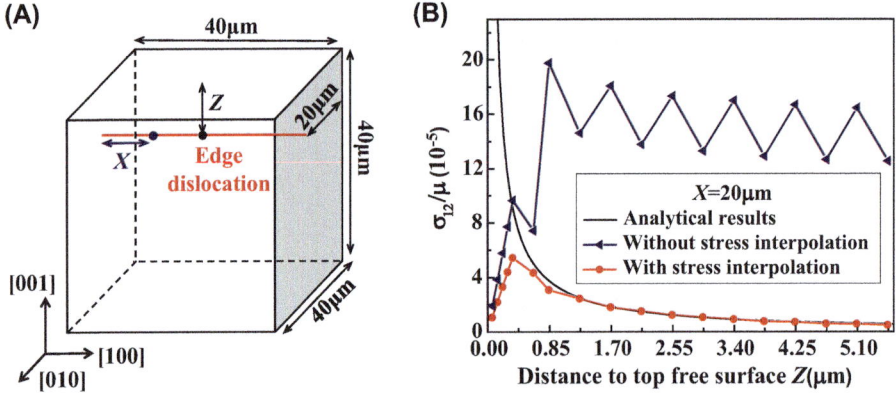

Figure 7.13

(A) Description of the validation test; (B) Comparison of results calculated by the discrete-continuous model with and without stress interpolation. *Reprinted from Cui, Y., Liu, Z., Zhuang, Z., 2015b, Quantitative investigations on dislocation based discrete-continuous model of crystal plasticity at submicron scale. International Journal of Plasticity 69, 54—72, Copyright 2015, with permission from Elsevier.*

It is clear that a reasonable interpolation treatment of the stress field must be introduced to capture the image force well. Good results can be obtained if stress σ at the considered dislocation segment is interpolated from stress at the neighboring integration points (Cui et al., 2015b):

$$
\sigma = \begin{cases} \sigma^{(i)}, & \text{if} \quad d^{(i)}_{seg-int} = \min\left(d^{(1)}_{seg-int}, d^{(2)}_{seg-int}, \cdots\right) < r_{\text{infsmall}} \\[2ex] \dfrac{\sum_i \sigma^{(i)} \big/ d^{(i)}_{seg-int}}{\sum_i 1 \big/ d^{(i)}_{seg-int}}, & \text{if} \quad r_{\text{infsmall}} < d^{(i)}_{seg-int} < r_{\text{cut}} \end{cases} \tag{7.3.8}
$$

where $\sigma^{(i)}$ is the stress at the ith integration point and $d^{(i)}_{seg-int}$ is the distance between the midpoint of dislocation segment and the ith integration point, as shown in Fig. 7.14. If the minimum distance between the midpoint of the dislocation segment and the neighboring integration points is smaller than critical value r_{infsmall}, the dislocation segment almost passes through an integration point and no stress interpolation is required. Otherwise, stress interpolation is carried out. Good results can be obtained when the cutoff radius for interpolation region r_{cut} is set to $0.45L$, in which L is the element size.

The image force calculation with stress interpolation is further carried out for the validation test. The results in Fig. 7.13B highlight the great improvement found after stress interpolation modification is used. Next, the simulations using other regularization methods but with stress interpolation are carried out for the validation test in Fig. 7.13A. The calculated image forces are given in Fig. 7.15A. The results using regularization methods proposed by Liu et al. and Lemarchand et al. are acceptable when distance Z is

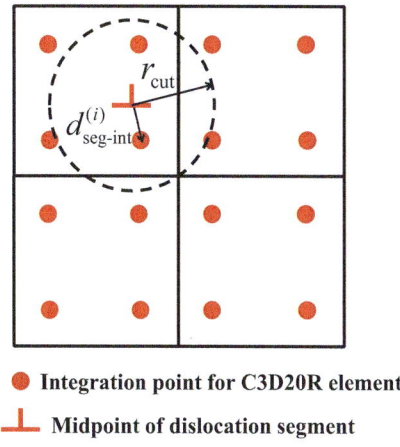

● Integration point for C3D20R element

⊥ Midpoint of dislocation segment

Figure 7.14

Schematic definition of variables used for stress interpolation. *Reprinted from Cui, Y., Liu, Z., Zhuang, Z., 2015b, Quantitative investigations on dislocation based discrete-continuous model of crystal plasticity at submicron scale. International Journal of Plasticity 69, 54–72, Copyright 2015, with permission from Elsevier.*

larger than L and $5L/4$, respectively, whereas the methods of Cui et al. and Vattré et al. exhibit good accuracy until Z is smaller than $3L/4$.

It is also interesting to compare the image force captured by DCM and SPM. Three typical cases are considered here: the one is the same as the subsurface dislocation shown in Fig. 7.13A; the second calculates the image force along the dislocation line that perpendicularly intersects the free surface, as shown in the inset of Fig. 7.15B; and the third calculates the image force along a dislocation loop, as shown in Fig. 7.15C. For these cases, the element size is taken to be 0.85 μm in both the DCM and SPM.

The simulation results for image force for the subsurface dislocation and intersection dislocation are presented in Fig. 7.15A and B, respectively. For these two cases, comparison with analytical results illustrates that using the new regularization method, DCM can calculate the image force well for a distance to the free surface larger than $3L/4$, which is similar to that of SPM. For the third case, because there are no analytical results for comparison, the calculated image force by DCM and SPM along the dislocation loop is compared with results obtained by SPM with a finer element sizes ($L = 600$ nm, which is limited by the large number of elements a single processor can handle). The results are in a good agreement, as shown in Fig. 7.15D. Obviously, as stated in the introduction, the widely used method of introducing SPM to DCM with the aim of correcting the image force is inappropriate and will double-count the image force effect.

Moreover, Fig. 7.15A and B shows that the image force close to the free surface cannot be captured by either DCM or SPM because it is difficult for the linear or quadratic shape

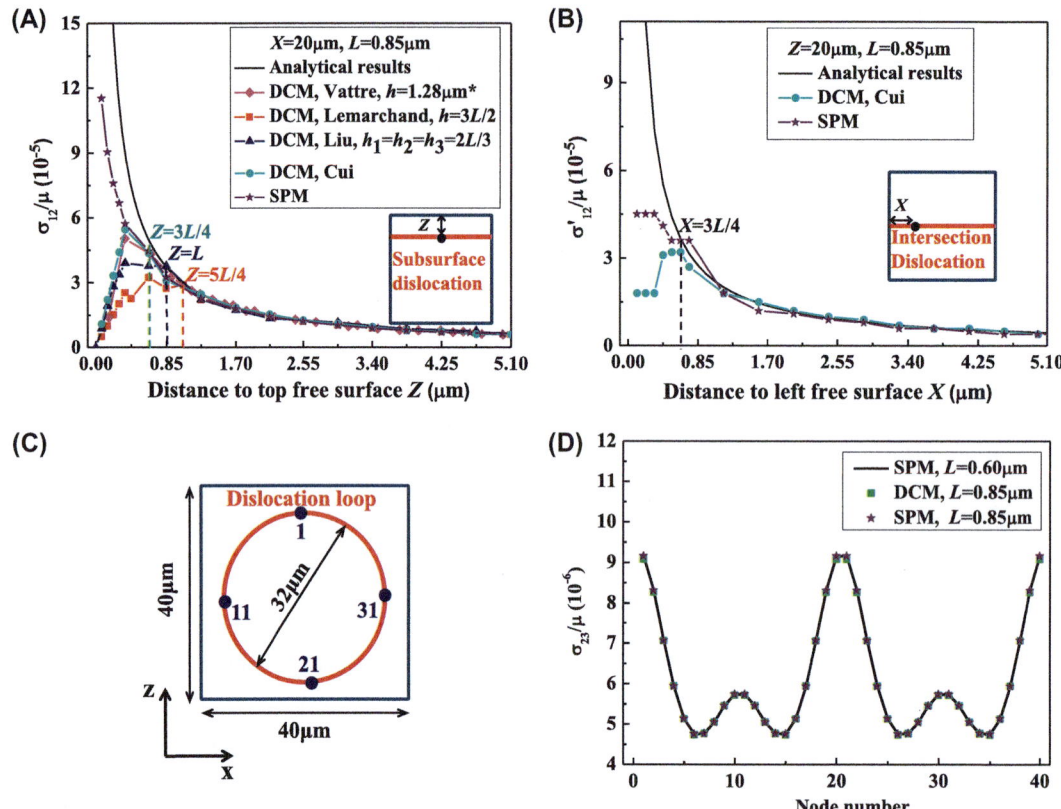

Figure 7.15
(A) Image force induced by the top free surface calculated by the discrete-continuous model (DCM) using different regularization methods and superposition methods (SPM). Results marked with an *asterisk* were obtained from Vattré et al. (2013). (B) Image force σ'_{12} induced by the left free surface when the dislocation segments intersect the left free surface. The definitions of distances Z and X are given in the corresponding inset and Fig. 7.13A. (C) Description of the validation test for a dislocation loop. Four dislocation node numbers are labeled; (D) Comparison of results of image force calculated by DCM and SPM. *Reprinted from Cui, Y., Liu, Z., Zhuang, Z., 2015b, Quantitative investigations on dislocation based discrete-continuous model of crystal plasticity at submicron scale. International Journal of Plasticity 69, 54–72, Copyright 2015, with permission from Elsevier.*

functions of standard FEM to describe the strong nonlinear variation in field variables close to the surface. Therefore, when the dislocation line enters the near-surface region, a reasonable correction in image force must be introduced.

The most straightforward method for correcting image force in DCM is to identify the outermost two-layer elements as a subdomain and then refine the elements in this subdomain. However, this undoubtedly increases the computation time, especially for a 3D problem. Tang et al. (2006) proposed a hybrid SPM method in which the singular part of

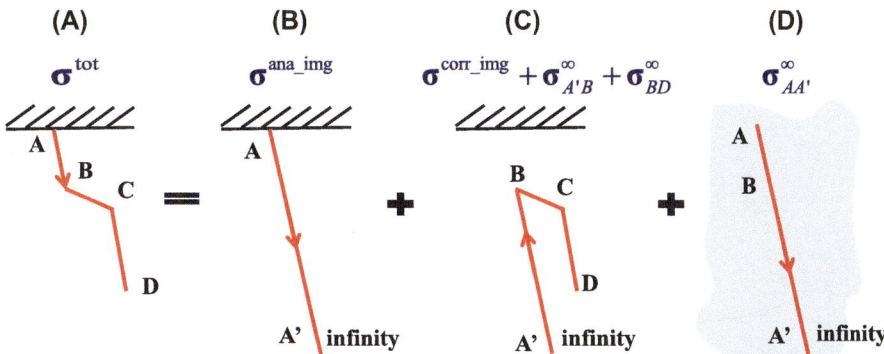

Figure 7.16

Schematic illustration of hybrid discrete-continuous model method when calculating an image force on an arbitrary curved dislocation line in a half-space with only one free surface. *Reprinted from Cui, Y., Liu, Z., Zhuang, Z., 2015b, Quantitative investigations on dislocation based discrete-continuous model of crystal plasticity at submicron scale. International Journal of Plasticity 69, 54—72, Copyright 2015, with permission from Elsevier.*

the image stress is obtained by an analytical solution and the nonsingular part is calculated by SPM. This hybrid scheme can also be used to improve the ability of DCM to capture the image force. Special attention must be paid because the image force is lumped into the total stress field in DCM. Under complex loading conditions, it is difficult to isolate the image force calculated by DCM. Therefore, the stress field in an infinite body must be taken into account in the decomposition and superposition framework, as schematically shown in Fig. 7.16. The singular image stress on AA′ $\sigma^{\text{ana\_imag}}$ in the elastic half-space is calculated according to the Yoffe solution (Yoffe, 1961). The difference in dislocation configurations between Fig. 7.16A and B is shown in Fig. 7.16C, whose total stress field $\sigma^{corr\_img} + \sigma^{\infty}_{A'B} + \sigma^{\infty}_{BD}$ can be calculated by DCM. Then, the stress field caused by semiinfinite dislocation AA′ $\sigma^{\infty}_{AA'}$ should be added, as shown in Fig. 7.16D, according to the Li solution (1964). On the one hand, this part removes the stress field caused by dislocation A′B; on the other hand, it adds the stress field caused by dislocation AB, $\sigma^{\infty}_{AA'} = \sigma^{\infty}_{AB} - \sigma^{\infty}_{A'B}$. Accordingly, the total stress field in the half-space can be expressed as $\sigma^{\text{tot}} = \sigma^{\text{ana\_img}} + \left(\sigma^{corr\_img} + \sigma^{\infty}_{A'B} + \sigma^{\infty}_{BD}\right) + \left(\sigma^{\infty}_{AB} - \sigma^{\infty}_{A'B}\right)$. According to the work of Tang et al. (2006), this hybrid method can also be used to solve a problem with multiple free surfaces.

To show its effectiveness, the image force on a curve dislocation line in the cubic crystal is calculated (see the inset in Fig. 7.17). It is located at the (010) slip plane and its Burgers vector is along the [001] direction. The image force induced by the left free surface is calculated for different values of X using DCM and hybrid DCM. The results in Fig. 7.17 show that the hybrid DCM method can capture the singular image force well near the free surface.

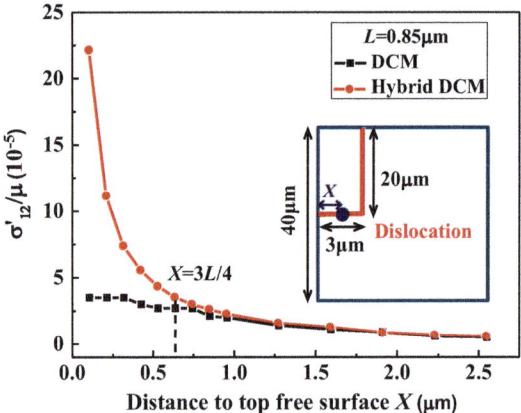

Figure 7.17

Image force σ'_{12} induced by the left free surface. The inset shows the dislocation configuration. *DCM, discrete-continuous model. Reprinted from Cui, Y., Liu, Z., Zhuang, Z., 2015b, Quantitative investigations on dislocation based discrete-continuous model of crystal plasticity at submicron scale. International Journal of Plasticity 69, 54—72, Copyright 2015, with permission from Elsevier.*

7.3.3 Finite Deformation

Capturing deformed configuration by DCM is another important issue for understanding the material deformation and failure process at submicron scales (Deshpande et al., 2003). This section describes the algorithm of calculating deformed configuration in DCM. Two validation tests are carried out to show quantitatively how DCM can be used to capture the strong localized deformation.

7.3.3.1 Deformation Field Transfer and Surface Dislocation Treatment

To reproduce the deformed configuration accurately, the DDD computational cells must deform in accordance with the FEM cells (Gao et al., 2010). Thus, the position of dislocation segments is updated according to its own slip, $\mathbf{u}^I_{\text{slip}}$, and the deformation-induced displacement, $\mathbf{u}^{(I)}_{\text{DDD}}$:

$$\mathbf{u}^{(I)} = \mathbf{u}^{(I)}_{\text{slip}} + \mathbf{u}^{(I)}_{\text{DDD}} \qquad (7.3.9)$$

where subscript "(I)" denotes the node number of the dislocation segment in DDD. $\mathbf{u}^{(I)}_{\text{slip}}$ is directly calculated by the DDD model:

$$\mathbf{u}^{(I)}_{\text{slip}} = \mathbf{v}^{(I)}_{\text{slip}} \Delta t \qquad (7.3.10)$$

where $\mathbf{v}^{(I)}_{\text{slip}}$ is the velocity of the dislocation node and can be calculated according to its kinetic equation:

$$\mathbf{M}\dot{\mathbf{v}}^{(I)}_{\text{slip}} + \mathbf{B}\mathbf{v}^{(I)}_{\text{slip}} = \mathbf{f}^{(I)} \qquad (7.3.11)$$

where $\mathbf{M}$ and $\mathbf{B}$ are the corresponding effective mass matrix and drag coefficient matrix, respectively. The first inertia term on the left side can be ignored when the dislocation

motion is in the overdamped regime. Force vector $\mathbf{f}^{(I)}$ of segment I includes Peach-Koehler force by applied stress and other defects, line tension, and image force.

Deformation-induced displacement $\mathbf{u}_{\mathrm{DDD}}^{I}$ is obtained by FEM calculation and passed to the DDD model. Considering the simplest case, if the dislocation does not move, displacement at dislocation segment node I is just equal to $\mathbf{u}_{\mathrm{DDD}}^{I}$. Specifically, $\mathbf{u}_{\mathrm{DDD}}^{I}$ is obtained by interpolating the displacement of the FEM nodes around dislocation segment node "I". Here, an interpolation scheme similar to Eq. (7.3.8) is used:

$$
\mathbf{u}_{\mathrm{DDD}}^{(I)} = \begin{cases} \mathbf{u}_{\mathrm{DDD}}^{(k)}, & \text{if } d_{\mathrm{seg-node}}^{(k)} = \min\left(d_{\mathrm{seg-node}}^{(1)}, d_{\mathrm{seg-node}}^{(2)}, \cdots\right) < r_{\mathrm{infsmall}} \\ \dfrac{\sum_i \mathbf{u}_{\mathrm{DDD}}^{(k)}\big/ d_{\mathrm{seg-node}}^{(k)}}{\sum_i 1\big/ d_{\mathrm{seg-node}}^{(k)}}, & \text{if } r_{\mathrm{infsmall}} < d_{\mathrm{seg-node}}^{(k)} < r_{\mathrm{cut}} \end{cases} \tag{7.3.12}
$$

where superscript "(k)" denotes the variables associated with the kth FEM node. $d_{\mathrm{seg-node}}^{(k)}$ is the distance between the midpoint of the dislocation segment and the kth FEM node. If the minimum distance between the midpoint of the dislocation segment and the FEM nodes is smaller than an infinitely small value, r_{infsmall}, no displacement interpolation is required. As shown in Fig. 7.14, r_{cut} is also the cutoff radius for the interpolation region.

At the same time, special treatment must be used for surface annihilation and to guarantee that the surface-piercing dislocation segments remain surface-piercing instead of terminating in the interior of the sample. To achieve this, in the FEM model we define the surface element as a separate part used to transfer information conveniently with the boundary of the DDD cell. This makes it easy to update the external geometry of the DDD cell and deal with surface annihilation.

Specifically, during each step, if dislocations slip out of the crystal, the outside part is deleted to deal with surface annihilation. In addition, the intersection nodes between the surface segments and free surface are labeled as surface nodes. If the dislocation line intersects a convex surface, the velocity of the surface node, $\mathbf{v}_{\mathrm{surfnode}}$, is projected to match the local curvature of the free surface and fulfill the constriction of the slip plane (Crone et al., 2014):

$$
\mathbf{n}_{\mathrm{surfnode}} = \frac{\mathbf{n} \times \mathbf{n}_s}{\|\mathbf{n} \times \mathbf{n}_s\|} \tag{7.3.13}
$$

$$
\mathbf{v}_{\mathrm{surfnode}} = (\mathbf{v}_{\mathrm{slip}} \cdot \mathbf{n}_{\mathrm{surfnode}})\mathbf{n}_{\mathrm{surfnode}}
$$

where $\mathbf{n}_{\mathrm{surfnode}}$ is a unit vector indicating the intersection between the surface and slip plane and $\mathbf{n}$ and $\mathbf{n}_s$ are the normal vectors of the slip plane and local free surface, respectively. The introduction of the surface element part in the FEM model makes it easy to calculate $\mathbf{n}_s$.

On the other hand, if the dislocation line slips across a concave surface, such as that induced by an evident slip step, the velocity projection and the displacement correction by the FEM model are insufficient. It is necessary to check whether the surface node remains on the surface after deformation. If the surface dislocations are found to terminate in the bulk, the surface dislocation lines should be extended to intersect with the free surface.

7.3.3.2 Slip System Rotation

The FEM model can conveniently consider the lattice rotation effect. To incorporate the lattice rotation effect in 3D-DDD code, the dislocation slip systems are directly updated according to the large-strain kinematics in crystal plasticity theory (Asaro and Rice, 1977):

$$
\begin{aligned}
\mathbf{b}_{(t)} &= \mathbf{F}^e \cdot \mathbf{b}_{(t_0)} \\
\mathbf{n}_{(t)} &= \mathbf{n}_{(t_0)} \cdot \mathbf{F}^{e-1}
\end{aligned}
\tag{7.3.14}
$$

where $\mathbf{b}$ is the Burgers vector, $\mathbf{n}$ is the normal vector of the slip plane, and subscripts (t) and (t_0) refer to the values of the variables at times t and $t_0 = 0$, respectively. $\mathbf{F}^e$ represents the elastic stretching and rotation of the crystal lattice, which can be determined by the multiplicative decomposition of deformation gradient $\mathbf{F}$:

$$
\begin{aligned}
\mathbf{F} &= \mathbf{F}^e \cdot \mathbf{F}^p \\
\mathbf{L}^p &= \dot{\mathbf{F}}^p \cdot \mathbf{F}^{p-1} = \dot{\gamma} \left(\frac{\mathbf{b}}{\|\mathbf{b}\|} \otimes \frac{\mathbf{n}}{\|\mathbf{n}\|} \right)
\end{aligned}
\tag{7.3.15}
$$

where $\mathbf{F}^p$ is the plastic part of $\mathbf{F}$ induced by the plastic slip, $\mathbf{L}^p$ is the plastic part of the velocity gradient, and $\dot{\gamma}$ is the plastic shear strain rate calculated by the regularization methods. By taking the derivative of Eq. (7.3.14), the following relation can be obtained:

$$
\begin{aligned}
\dot{\mathbf{b}}_{(t)} &= \dot{\mathbf{F}}^e \cdot \mathbf{F}^{e-1} \cdot \mathbf{b}_{(t)} \\
\dot{\mathbf{n}}_{(t)} &= -\mathbf{n}_{(t)} \cdot \dot{\mathbf{F}}^e \cdot \mathbf{F}^{e-1}
\end{aligned}
\tag{7.3.16}
$$

where a superposed dot means the time derivative. Then, combining Eqs. (7.3.15) and (7.3.16), the slip system can be updated by the implicit time-integration

$$
\begin{aligned}
\mathbf{b}_{(t+\Delta t)} &= \left(\mathbf{I} + \dot{\mathbf{F}}^e \cdot \mathbf{F}^{e-1} \Delta t \right) \cdot \mathbf{b}_{(t)} \\
\mathbf{n}_{(t+\Delta t)} &= \mathbf{n}_{(t)} \cdot \left(\mathbf{I} - \dot{\mathbf{F}}^e \cdot \mathbf{F}^{e-1} \Delta t \right) \\
\dot{\mathbf{F}}^e \cdot \mathbf{F}^{e-1} &= \dot{\mathbf{F}} \cdot \mathbf{F}^{-1} - \mathbf{F} \cdot \mathbf{F}^{p-1} \cdot \dot{\mathbf{F}}^p \cdot \mathbf{F}^{-1}, \quad \dot{\mathbf{F}}^p = \mathbf{L}^p \cdot \mathbf{F}^p, \\
\mathbf{F}^p_{(t+\Delta t)} &= \mathbf{F}^p_{(t)} + \dot{\mathbf{F}}^p_{(t)} \Delta t = \left(\mathbf{I} + \mathbf{L}^p \Delta t \right) \cdot \mathbf{F}^p_{(t)}
\end{aligned}
\tag{7.3.17}
$$

where Δt is the time increment and $\mathbf{I}$ is the unit tensor. The result of $\dot{\mathbf{F}}^e \cdot \mathbf{F}^{e-1}$ is transferred from the FEM to the DDD model, as shown in Fig. 7.5B. The corresponding interpolation scheme is similar to Eq. (7.3.8). At the same time, the deformation field transfer from FEM to DDD can naturally consider the update of position and orientation for dislocation lines induced by lattice rotation.

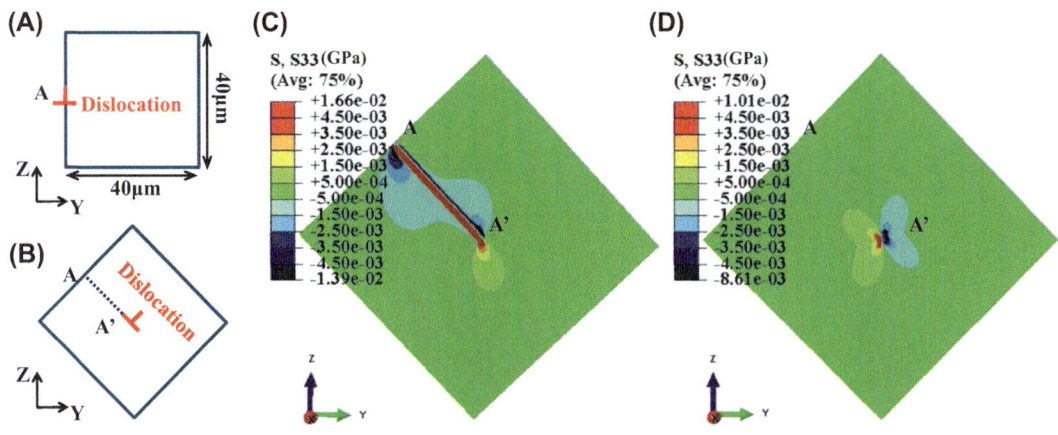

Figure 7.18

(A and B) Sample geometry and dislocation position before and after deformation, respectively. (C and D) Stress field σ_{33} without and with slip system rotation treatment, respectively. *Avg*, average. *Reprinted from Cui, Y., Liu, Z., Zhuang, Z., 2015b, Quantitative investigations on dislocation based discrete-continuous model of crystal plasticity at submicron scale. International Journal of Plasticity 69, 54–72, Copyright 2015, with permission from Elsevier.*

Two validation tests are given next to show its effectiveness. The first is schematically shown in Fig. 7.18A and B. An edge dislocation nucleates from the surface (labeled A) and glides to the middle of the crystal (labeled A′). At the same time, the crystal rotates 45 degrees about the [100] axis. The FEM mesh size is 0.85 μm. If the rotation is not considered, the stress field cannot be captured accurately, as shown in Fig. 7.18C. However, if it is considered according to Eq. (7.3.17), the calculated stress field is reasonable, as shown in Fig. 7.18D.

The second validation test is to investigate the evolution of a Frank-Read source in a bent beam with a length of 50 μm and cross-section area of 10 μm × 10 μm. This Frank-Read source is originally situated along the [100] direction in the (010) slip plane with Burger vector [001]. The FEM element size is 2 μm. To clearly show the evolution of dislocation configuration and save computational time, the strain rate for bending is set to be so large that the dislocation configuration does not have enough time to reach a fully relaxed configuration during deformation. During each time increment of FEM model $\Delta t_{\text{FEM}} = 1.6 \times 10^{-8}\ \text{s}^{-1}$, the displacement increment ΔU_2 is 0.1 μm and the DDD model runs 400 times with time increment $\Delta t_{\text{DDD}} = 4 \times 10^{-11}\ \text{s}^{-1}$. The simulation results are given in Fig. 7.19 when U_2 is 3.5 μm. Fig. 7.19A shows that if the lattice rotation effect is ignored, the position of the dislocation can be updated, but the slip plane is kept at (010). However, if the lattice rotation effect is considered, the slip plane rotates precisely with the rotation of the neutral axis.

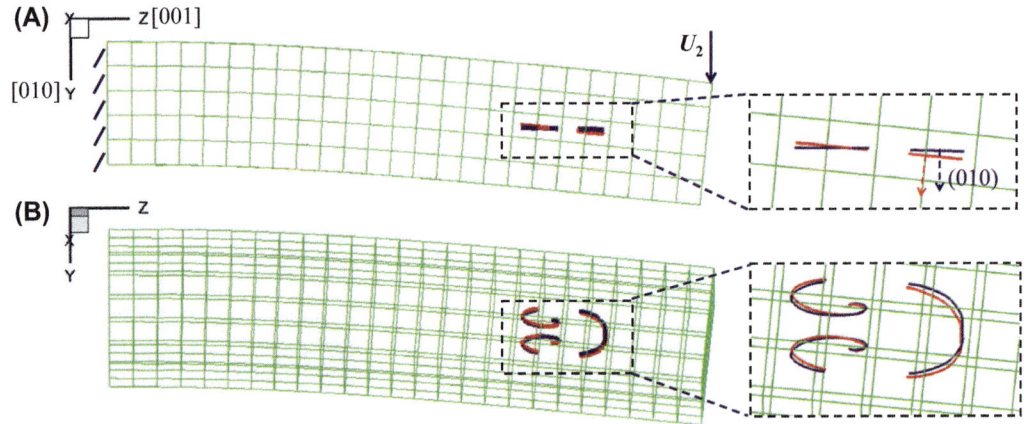

Figure 7.19

Overlay of images of dislocation configurations for bent beam without and with lattice rotation when $U_2 = 3.5$ μm. The *blue and red dislocation lines* correspond to results without and with rotation, respectively. (A and B) Results from different views. *Reprinted from Cui, Y., Liu, Z., Zhuang, Z., 2015b, Quantitative investigations on dislocation based discrete-continuous model of crystal plasticity at submicron scale. International Journal of Plasticity 69, 54—72, Copyright 2015, with permission from Elsevier.*

7.3.3.3 Reproduction of Slip Step

In the following discussion, two validation tests are carried out to investigate the ability of improved DCM to capture deformed configurations.

One validation case is an edge dislocation line sweeping a finite cubic sample, as described in Fig. 7.13A. The element size is 850 nm. The simulation results are given in Fig. 7.20. Here, the displacements are magnified by a factor of 5000 for better visualization. When this dislocation line sweeps half of the slip plane, only the left slip step can be observed (Fig. 7.20A); when this dislocation line sweeps the whole slip plane, one slip step with the magnitude of the Burger vector (0.25 nm) is generated, as expected (Fig. 7.20B). This implies that the displacement field can be well-reproduced by introducing the localized plastic strain from DDD to FEM.

In submicron crystals, it is widely accepted that the operation of single-arm source is the dominated dislocation mechanism (Cui et al., 2014; Rao et al., 2007). Thus, the other validation case is chosen to be a single-arm source sweeping a micropillar with a diameter of 1000 nm and height of 2000 nm. This single-arm source is placed on a (111) slip plane with a Burgers vector along $[10\bar{1}]$. It has one non-destructible pinning point at the central point of the micropillar. The element size of the FEM model is 80 nm, which permits fine resolutions with accurate results and convergence. Uniaxial compression loading is applied on the top of pillar with a constant pressure of 200 MPa. Because the resolved shear stress

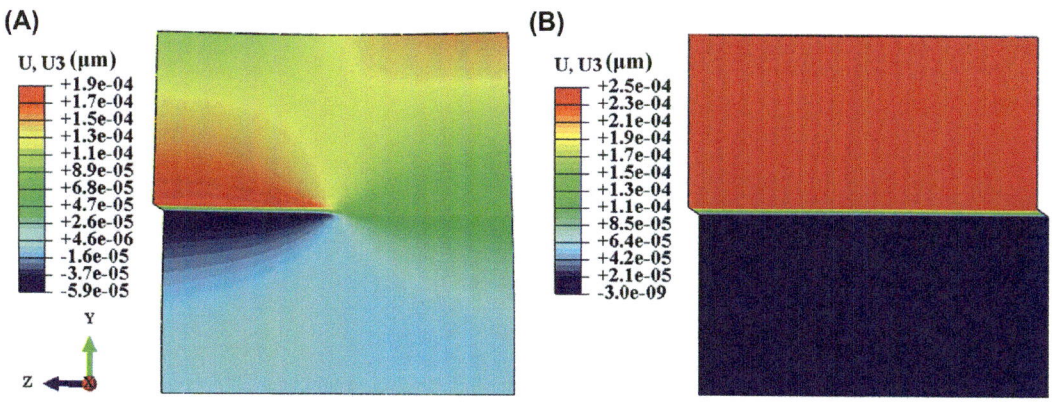

Figure 7.20
Deformed configuration with displacement magnified by a factor of 5000 (A) when an edge dislocation line sweeps half of the sample; (B) when an edge dislocation line sweeps the whole sample. *Reprinted from Cui, Y., Liu, Z., Zhuang, Z., 2015b, Quantitative investigations on dislocation based discrete-continuous model of crystal plasticity at submicron scale. International Journal of Plasticity 69, 54–72, Copyright 2015, with permission from Elsevier.*

needed to active this single-arm source is much less than that generated by the external pressure, the single-arm source can operate continuously, which leads to significant localized deformation (Fig. 7.21A). Compared with the first validation case, this numerical test can be used to check whether the deformed shape can be well-reproduced by DCM when the FEM elements swept by the dislocation segments are irregular, the dislocation line is curved, and localized deformation occurs.

According to Eq. (7.3.1), plastic strain is determined by the sweep area of the dislocation segments. Thus, the distribution of plastic strain is sensitive to the dislocation configuration. Once the surface segments shrink, the value of the plastic strain in the near-surface region is small compared with the other regions. The simulation results in Fig. 7.21B show that the equivalent plastic strain is uniform even in the case of a large deformation and an irregular mesh.

Furthermore, the simulation results were quantitatively evaluated. When the slip plane is swept by a single-arm source n times, the slip distance along a slip step direction (arrow in Fig. 7.21A) is $n|\mathbf{b}|$. For the case under consideration, displacement along x direction U_1 should be equal to $n|\mathbf{b}|/\sqrt{2}$. Taking $n = 1000$ as an example, the value of U_1 should be 0.18 μm. Fig. 7.21C shows that simulated displacement U_1 for the upper part of the sample is extremely close to this value. In addition, the deformed configuration is similar to the experimental observations shown in Fig. 7.21D (Dimiduk et al., 2005). If dislocations exist in the upper part of the pillar shown in Fig. 7.21A, they will shift a displacement equal to U_1 along the x direction according to Eq. (7.3.9).

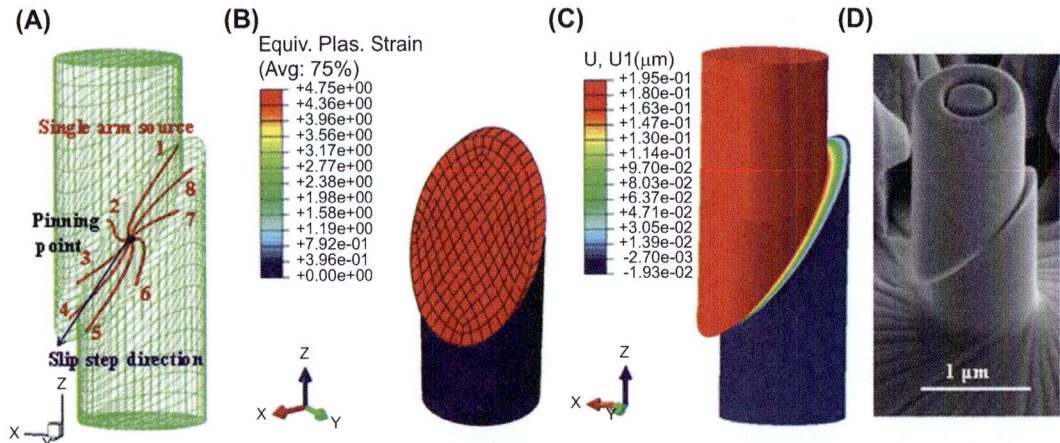

Figure 7.21

For the single-arm source sweeping a slip plane 1000 times: (A) An overlay of images showing the dislocation configuration and deformed shape; (B) distribution of equivalent plastic strain (Equiv. Plas. Strain), which is observed as a cross-section along the (111) plane; and (C) displacement field U_1; (d) Scanning electron microscopy image showing the deformed configuration of a 1-μm-diameter micropillar (Dimiduk et al., 2005). *(C) Reprinted from Cui, Y., Liu, Z., Zhuang, Z., 2015b, Quantitative investigations on dislocation based discrete-continuous model of crystal plasticity at submicron scale. International Journal of Plasticity 69, 54—72, Copyright 2015, with permission from Elsevier.*

7.4 Application to Heteroepitaxial Films

As an example, the modified DCM is applied to study dislocation behavior in heteroepitaxial films, which has received a great deal of attention because of their wide applications in the semiconductor and electron device industries. Thermoelastic analogical calculation is conducted to calculate the internal stress field induced by the lattice misfit between the film and substrate. Dislocation behaviors and their corresponding stress fields are analyzed for thin and thick substrates.

7.4.1 Thermoelastic Calculation to Determine Internal Stress Field

Considering the $Si_{1-x}Ge_x/Si$ film-substrate structure, lattice misfit strain is set to $\varepsilon_m = (a_{SiGe} - a_{si})/a_{si} = 0.0418x = 0.0055$ for $x = 0.13$ (Wang et al., 2003), where a_{SiGe} and a_{si} are the lattice constants of the film and substrate, respectively. The initial stress field caused by misfit strain is usually imposed by analytical solutions, assuming the film and substrate have the same material parameters (Wang et al., 2003). Here, it is calculated by analogizing the lattice misfit as a thermal expansion process for two materials with

different thermal expansion coefficients α. The thermal expansion coefficients of film α_f and substrate α_s, respectively, are set to

$$
\alpha_f = \varepsilon_m \frac{d_s E_{Si}}{d_f E_{SiGe} + d_s E_{Si}}
$$

$$
\alpha_s = -\varepsilon_m \frac{d_f E_{SiGe}}{d_f E_{SiGe} + d_s E_{Si}}
$$

(7.4.1)

Here, both the $Si_{1-x}Ge_x$ film and Si substrate are assumed to be isotropic. $E_{si} = 130$ GPa is the elastic modulus of Si (Shen, 2008). E_{Ge} is the elastic modulus of Ge and is set to 102.5 GPa (Newberger, 1971). The elastic modulus of $Si_{1-x}Ge_x$ $E_{SiGe} = E_{si}(1 - x) + E_{Ge}x$ is approximated by a linear rule of mixtures. d_f is the thickness of the film and is set to 0.3 μm. d_s is the thickness of the substrate and is set to 0.6 and 3 μm to simulate very thin and very thick substrates, respectively. Moreover, the side length of the film and substrate is set to 20 μm. The out-of-plane displacement at the bottom surface of the substrate is fixed to constrain lateral bending. All other surfaces are traction-free. The stress field corresponds to the results when the temperature increases each unit value.

First, the accuracy of this thermoelastic analogical method is verified. When the film is assumed to have the same elastic modulus as substrate E_{si}, the analytical stress field is obtained:

$$
\sigma_f = -\frac{E_{si}}{1 - \nu_{si}} \frac{d_s}{d_f + d_s} \varepsilon_m
$$

$$
\sigma_s = \frac{E_{si}}{1 - \nu_{si}} \frac{d_f}{d_f + d_s} \varepsilon_m
$$

(7.4.2)

where ν_{si} is the Poisson's ratio of Si and is set to 0.28 (Shen, 2008). When d_s is taken to be 0.6 μm, $\sigma_f = -0.662$ GPa and $\sigma_s = 0.331$ GPa. The FEM mesh size is 0.06 μm in the thickness direction and 0.4 μm in the other two directions. The results of the simulation are presented in Fig. 7.22. Figure 7.22A shows that except for the region near the boundary, the stress field is consistent with the analytical results. Fig. 7.22B further compares the stress value along the thickness (dotted line in Fig 7.22A) obtained by the simulation and Eq. (7.4.2). Good quantitative agreement is observed.

7.4.2 Influence of Substrate Thickness on Dislocation Behavior

According to the thermoelastic analogical calculations, the internal stress field can be obtained when the film and substrate have different mechanical properties. Next, the evolution of a Frank-Read source in this internal stress field is simulated by the improved DCM to depict dislocation behavior clearly in a heteroepitaxial film with different thicknesses of substrate.

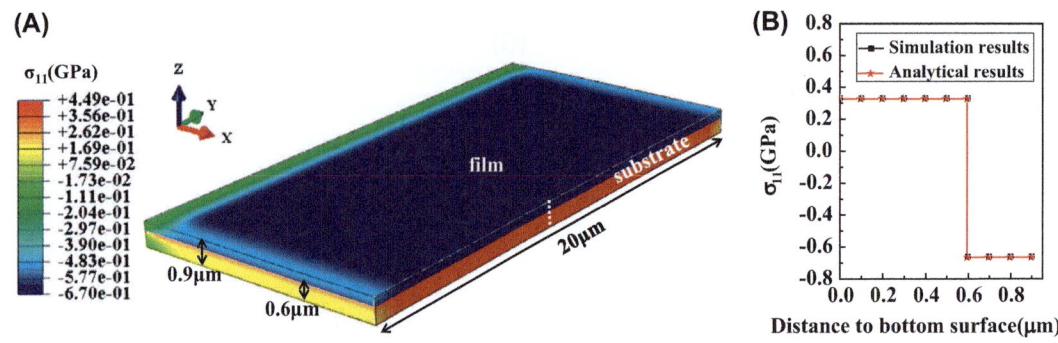

Figure 7.22

(A) [100] Cross-sectional view showing stress field σ_{11} for a heteroepitaxial film and substrate when the film and substrate are assumed to have the same elastic modulus. (B) Comparison of σ_{11} along the thickness (*dotted line* in [A]) for the simulation and analytical results. *Reprinted from Cui, Y., Liu, Z., Zhuang, Z., 2015b, Quantitative investigations on dislocation based discrete-continuous model of crystal plasticity at submicron scale. International Journal of Plasticity 69, 54–72, Copyright 2015, with permission from Elsevier.*

Supposing that initially there is a Frank-Read source in the (111) slip plane with a Burgers vector along $[\bar{1}01]$ in the middle of the film, it naturally multiplies under the action of the free surface and internal stress field with no prior constraint on dislocation motion. The pinning points are indestructible and 200 nm apart. The image force caused by the free surface is modified by the hybrid method described in Section 7.3.2.

The evolution of the calculated dislocation microstructure for the cases of the thin and thick substrate is shown by the black solid lines in Fig. 7.23A–J, respectively. In both cases, threading dislocations form when the bowing-out dislocation segments encounter the free surface, whereas misfit dislocations form when they intersect with the film-substrate interface. With the image force correction, the near-surface threading dislocation segments are almost perpendicular to the free surface, which is consistent with previous phase field method studies (Wang et al., 2003). Comparing Fig. 7.23D with Fig. 7.23I, the misfit dislocations behave differently for different substrate thicknesses. For a very thick substrate, the misfit dislocations can penetrate into the substrate when two misfit dislocations form (Fig. 7.23I and J), whereas for a thin substrate, penetration of the misfit dislocation is not observed in Fig. 7.23E owing to higher compression stress in the substrate. Therefore, for the case of the thin substrate, back stress induced by the misfit dislocations is higher and will inhibit the operation of the Frank-Read source. Although the length of the longest misfit dislocation lines in Fig. 7.23E and J are almost equal, the cycles of Frank-Read source operation are different.

Stress fields without and with dislocations are compared in Fig. 7.24B and C for a thin substrate and Fig. 7.25B and C for a thick substrate, respectively, to illustrate the role of

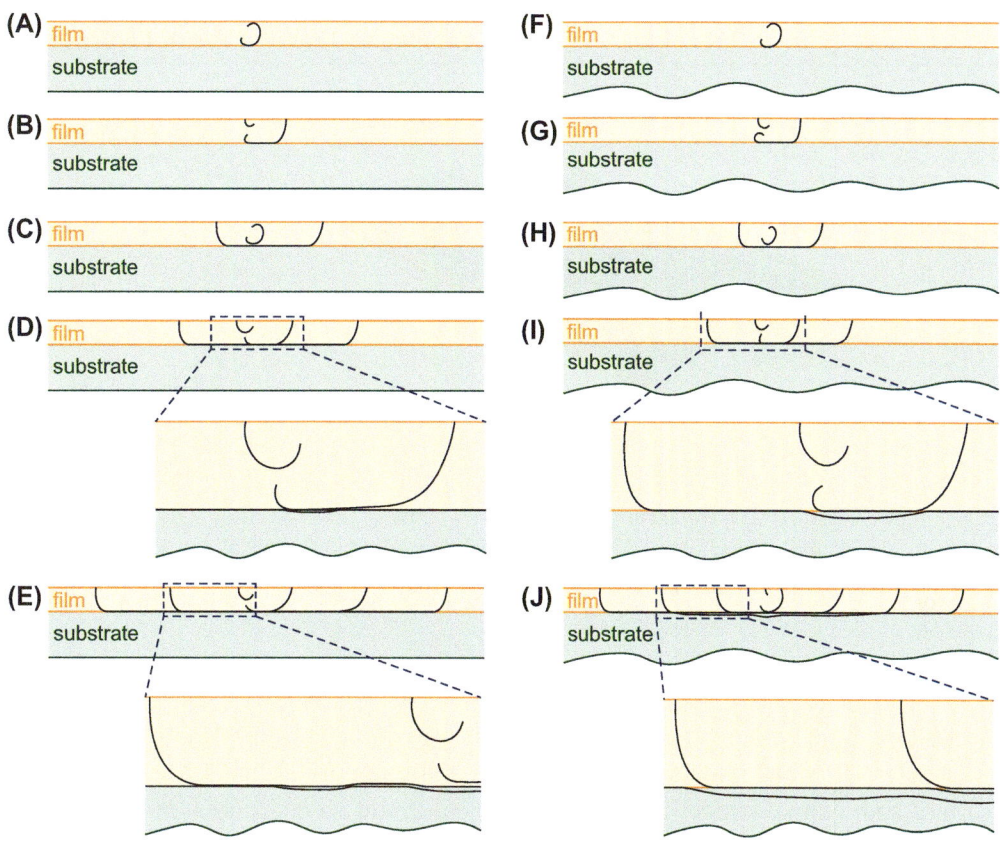

Figure 7.23

(A–E) Dislocation evolution in film with a thin substrate; (F–J) Dislocation evolution in film with a thick substrate. *Black solid lines* represent dislocation lines. *Reprinted from Cui, Y., Liu, Z., Zhuang, Z., 2015b, Quantitative investigations on dislocation based discrete-continuous model of crystal plasticity at submicron scale. International Journal of Plasticity 69, 54–72, Copyright 2015, with permission from Elsevier.*

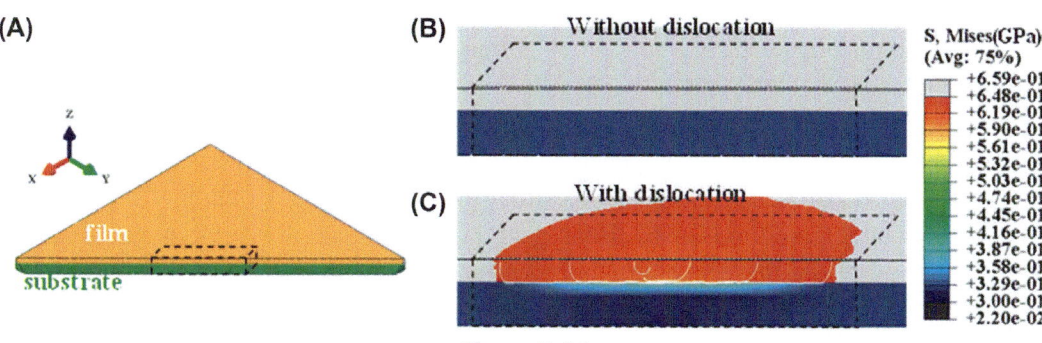

Figure 7.24

(A) (111) plane cross-sectional view showing the concerned region enclosed by *dotted lines* for a film with a thin substrate; (B and C) Mises stress field in the concerned region labeled in (A) before and after the dislocation source operation, respectively. *White solid lines* in (C) represent dislocation lines, corresponding to Fig. 7.23E. *Reprinted from Cui, Y., Liu, Z., Zhuang, Z., 2015b, Quantitative investigations on dislocation based discrete-continuous model of crystal plasticity at submicron scale. International Journal of Plasticity 69, 54–72, Copyright 2015, with permission from Elsevier.*

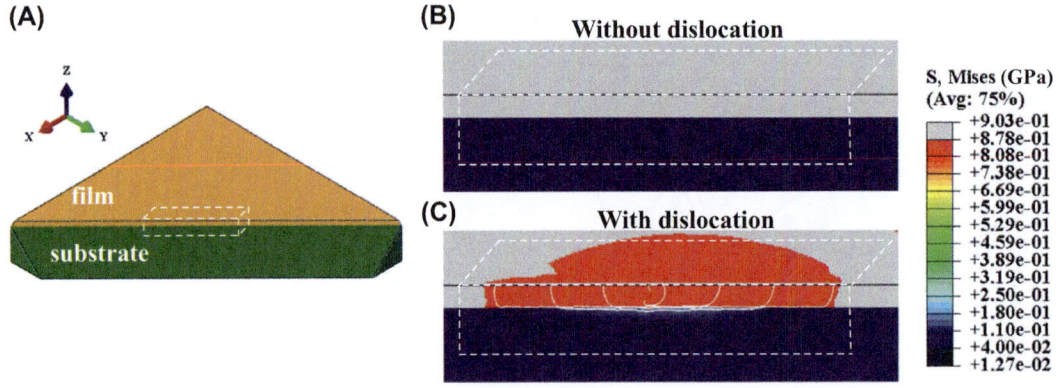

Figure 7.25

(A) (111) plane cross-sectional view showing the concerned region enclosed by *dotted lines* for a film with a thick substrate; (B and C) Mises stress field in the concerned region labeled in (C) before and after the dislocation source operation, respectively. *White solid lines* in (C) represent the dislocation lines corresponding to Fig. 7.23J. *Reprinted from Cui, Y., Liu, Z., Zhuang, Z., 2015b, Quantitative investigations on dislocation based discrete-continuous model of crystal plasticity at submicron scale. International Journal of Plasticity 69, 54—72, Copyright 2015, with permission from Elsevier.*

dislocation evolution in strain relaxation. For both thin and thick substrates, the Mises stress in the film decreases because the misfit strain is relaxed from the glide of dislocations. Comparing Fig. 7.24B and C, it can be found that the piling-up misfit dislocations at the interface lead to an increase in stress in the thin substrate. In contrast, as shown in Fig. 7.25B and C, there is no evident change in stress in the substrate after multiple misfit dislocations form because they can penetrate toward the thick substrate. The simulation implies that crack nucleation and interface delamination are more likely to happen for the thin substrate.

7.5 Application to Irradiated Materials

The computation framework presented earlier can be further extended to investigate the plastic flow localization of materials with dispersed barriers, which is of considerable interest owing to their important applications in designing stronger yet ductile materials. Sometimes, dispersed barriers are intentionally introduced to enhance strength, such as in metal alloys hardened by distributing small precipitates (Papanikolaou et al., 2017a). In other instances, they are brought about as the result of specific service environments, such as defect clusters produced by particle irradiation (Cui et al., 2018) or hydrogen atom clusters forming after metal charging in a hydrogen environment (Gu and El-Awady, 2018; Sills and Cai, 2016). Suppressing deformation localization is a critical goal of failure prevention in these applications. To reveal the possibility of avoiding the loss of ductility in a barrier-strengthening material, the collective interactions between high-density barriers and dislocations have to be fully understood.

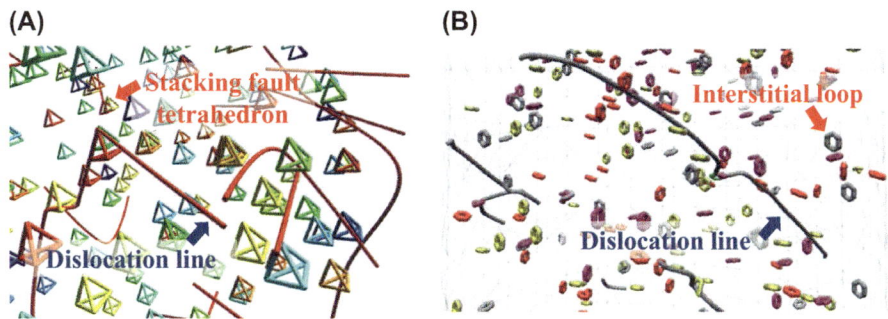

Figure 7.26

Examples showing dislocations interacting with irradiation defects. (A) Dislocations are trapped by stacking fault tetrahedrons in irradiated face-centered cubic crystals; (B) Dislocations form junctions as a result of interactions with interstitial loops in irradiated body-centered cubic crystals. *Different colors* represent different Burgers vectors.

In the following discussion, the irradiated material is taken as an example to illustrate the basic idea of considering a barrier effect in 3D-DDD. When materials are subjected to ion or neutron irradiation, defect clusters are produced and result in strong interactions with dislocations. In FCC metals, stacking fault tetrahedron (SFT) and self-interstitial atom clusters impede dislocation motion. Likewise, dislocations are immobilized by nanovoids and interstitial clusters in BCC metals. It is straightforward to introduce these discrete irradiation defects directly in 3D-DDD (Crosby et al., 2014; Cui et al., 2017a; Diaz de la Rubia et al., 2000), such as Fig. 7.26 shows. Generally, when one SFT is swept by a gliding dislocation line, the SFT may be sheared or partially absorbed, or it may form a junction or by completely restored (Osetsky et al., 2006). When an interstitial loop is swept by a gliding dislocation line, it may be dragged when the Burgers vector of the loop belongs to the dislocation slip plane or forming junction when the Burgers vector of the loop intersects the dislocation slip plane. After the junction forms, the interstitial loop may be fully absorbed by the dislocation lines or it may be left with a different Burgers vector (Terentyev et al., 2013). A great degree of success has been achieved using 3D-DDD simulations to capture the near-atomic resolution of dislocation-barrier interactions for radiation defects with dislocations (Cui et al., 2017a; Martinez et al., 2008; Shi et al., 2015). Plastic instability, manifesting in the formation of microscale-localized plasticity zones called cleared channels, has also been studied (Arsenlis et al., 2012; Diaz de la Rubia et al., 2000; Ghoniem et al., 2001).

Nevertheless, irradiation defects generally have a very high density. The number density of irradiation defects N generally ranges from 10^{21} to 10^{23} m^{-3}. Therefore, an investigation of their collective interactions with dislocations obviously requires extensive computation time if these irradiation defects are to be described discretely. On the other hand, if these high-density irradiation defects are described at a coarse scale through a continuous

irradiation defect field, higher computation efficiency and capability will be obtained. That is, the behavior of materials with superhigh-density barriers and a large size can be understood. Based on these considerations, the discrete 3D-DDD framework was coupled with the FEM solution of a continuum field equation for the evolution of dispersed barriers (Cui et al., 2018). Here, we briefly review the process to familiarize readers.

First, one must include information about the hardening or resistance that a defect field imposes on glide dislocations. This is usually described by the irradiation hardening model

$$\tau = k\mu b\sqrt{Na} \tag{7.5.1}$$

where τ is the slip resistance induced by the irradiation defects, N is the irradiation defect number density, and a is the irradiation defect size. k is the dimensionless coefficient, which depends on the barrier strength and distribution. For a specific type of irradiation defect, k can be determined by a statistical analysis of the output from dedicated DDD simulations coupled with discrete irradiation defects (Cui et al., 2018; Monnet, 2015; Sobie et al., 2015).

Then, one needs to determine the kinetics of the irradiation defect density and size evolution to consider the interaction between the dislocations and irradiation defects. Simulations show that irradiation defect clusters are either absorbed or swept away by gliding dislocations within a prescribed capture distance $y/2$ (Ghoniem et al., 2000a; Osetsky et al., 2005; Rodney et al., 2001), as shown in Fig. 7.27. Therefore, the irradiation defect destruction rate induced by gliding dislocations is described as

$$\Delta N - \Delta N_s = -\lambda N \frac{y\Delta A}{V} \tag{7.5.2}$$

where ΔN is the change in the irradiation defect number density. ΔN_s describes the irradiation production caused by concurrent radiation damage. ΔA is the slip area increase in the dislocation line. V is the characteristic volume. λ is the irradiation defect

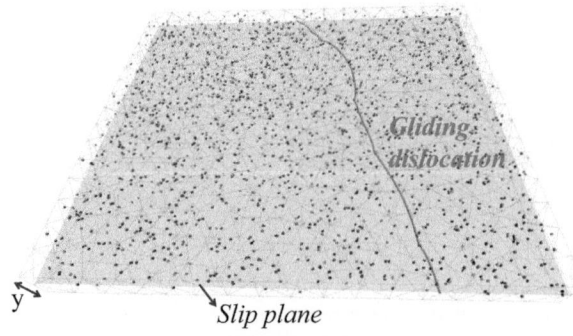

Figure 7.27
Possible influence range of gliding dislocation on irradiation defects.

annihilation fraction. The value of λy can be determined by fitting stress-strain curves (Patra and McDowell, 2016) or directly recording and analyzing the irradiation defect density evolution by DDD simulations (Cui et al., 2018).

The variables transferring process between DDD and the continuum irradiation defect field (IDF) mainly contains two processes. First, the irradiation defect density is passed from the IDF model to the DDD code to calculate the resistance stress field. Then, the local plastic shear strain is transferred from the DDD to the IDF to update the local irradiation defect density, which is similar to the process described in Section 7.3.1. The FEM model is mainly used to calculate the kinetics equation of the irradiation defect (Eq. 7.5.2).

Such a hybrid continuum field-DDD model leads to new possibilities for efficiently investigating plastic instabilities and flow localization behavior in high-dose irradiated materials. For example, it becomes possible to study micrometer-sized irradiated Fe with an irradiation defect density up to 3×10^{22} m^{-3} at a feasible computational cost. The formation of a dislocation channel can be effectively investigated by such a method, such as shown in Fig. 7.28. This method can also be extended to studying flow localization problems in materials with other kinds of dispersed barriers, such as precipitation-hardened alloys, and in materials with hydrogen, etc.

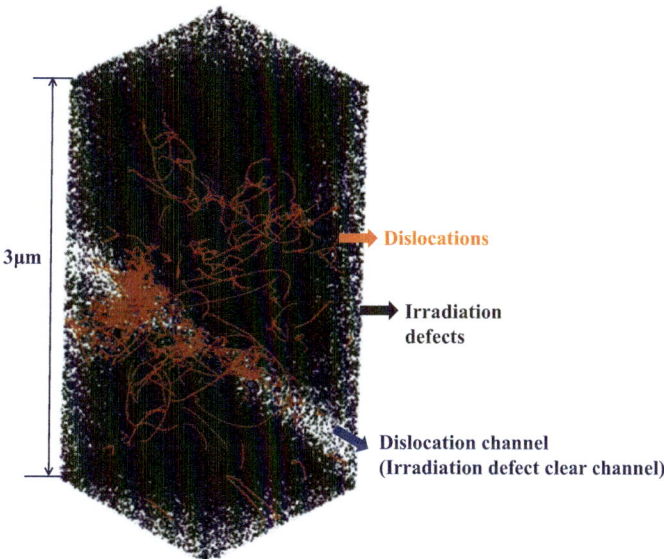

3μm

Dislocations

Irradiation defects

Dislocation channel
(Irradiation defect clear channel)

Figure 7.28

Dislocation configurations and irradiation defect distributions after deformation obtained by coupled field-discrete dislocation dynamics simulations. More details of the simulation setup are given in Cui et al. (2018).

7.6 Summary

This chapter describes the DCM, which couples the DDD with the FEM. Algorithmic details are presented, including how to consider the finite deformation of crystal plasticity, complex boundary conditions, and surface effects. The modified DCM is applied to study dislocation behavior in heteroepitaxial films. Thermoelastic analogical calculations were conducted to calculate the internal stress field induced by the lattice misfit between films and substrates. Dislocation behaviors and corresponding stress fields were analyzed for thin and thick substrates. The DCM computation framework can be extended to investigate the plastic flow localization of materials with dispersed barriers. An irradiated material is taken as a typical example. Use of this tool to reveal the collective interactions between high-density barriers and dislocations is expected to guide the design of stronger and more ductile materials.

Single-Arm Dislocation Source (SAS)-Controlled Submicron Plasticity

Chapter Outline

8.1 Introduction

Over the past decade, submicron crystals with a characteristic size ranging from several hundred nanometers to several micrometers have attracted continuous interest owing to the increasing use and broader potential application of small-scaled devices and materials (Akarapu et al., 2010; Frank and Van der Merwe, 1949; Zhang et al., 2013a,b). There is a considerable body of experimental and simulation evidence showing that the plastic behavior of submicron materials is generally in sharp contrast to that of their bulk counterpart (Kraft et al., 2010). Some typical stress-strain curves and the dislocation density evolution are given in Fig. 8.1, which correspond to the compression tests on Ni single-crystal micropillars with different diameters from 200 to 800 nm. Fig. 8.1 shows that a significant size effect of yield stress is observed even without a strain gradient. The popular notion that small is stronger emphasizes the strength advantage of small-scaled materials, which makes them capable of supporting high stress before permanently deforming or failing (Frank and Van der Merwe, 1949; Greer and De Hosson, 2011; Gu and Ngan, 2013; Kim et al., 2012). However, researchers also noticed that stress-strain curves at small scales are far from being smooth and continuous, but exhibit a step-like (Fig. 8.1) or serrated character (Csikor et al., 2007). The stochastic occurrence of the strain burst makes it difficult to control the plastic formation of microdevices, which poses a challenge for realizing the full potential of emerging submicron technologies.

Dislocation Mechanism-Based Crystal Plasticity. https://doi.org/10.1016/B978-0-12-814591-3.00008-X
Copyright © 2019 Tsinghua University Press. Published by Elsevier Inc.

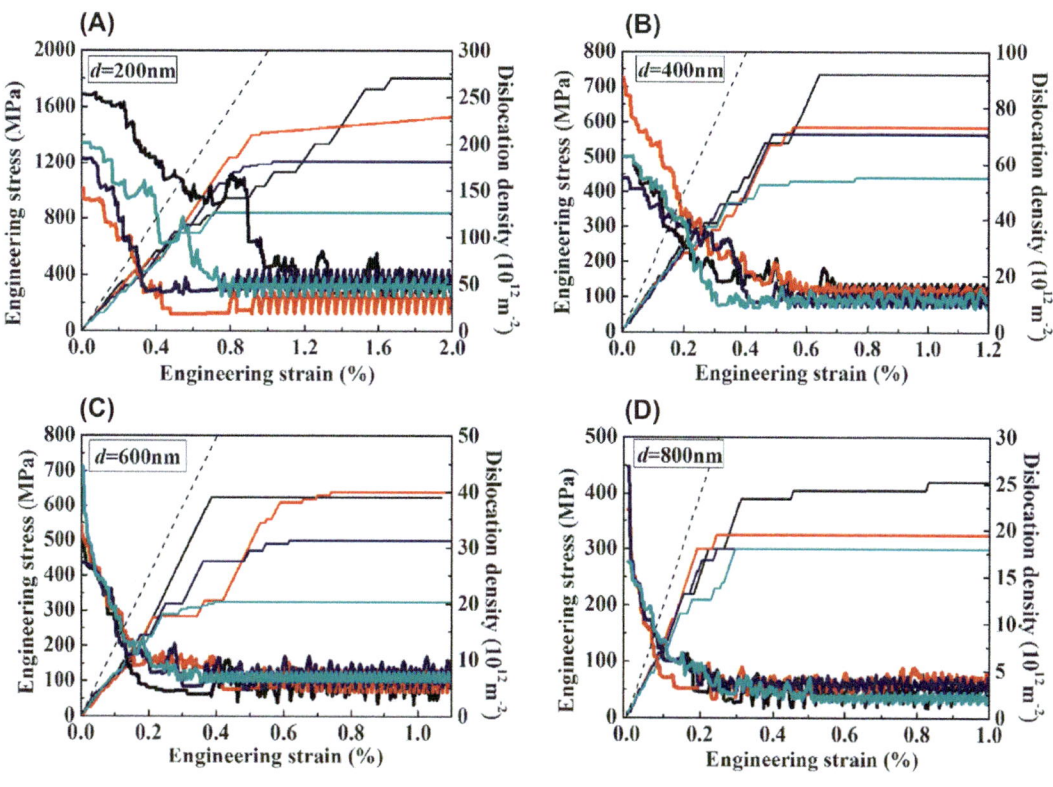

Figure 8.1

Stress (*thin solid line*) and density (*thick solid line*) evolution for Ni micropillars with different diameters: (A) 200 nm, (B) 400 nm, (C) 600 nm, and (D) 800 nm. Different colors denote results for different initial configurations. The *dotted line* reflects the purely elastic response. These were obtained by discrete dislocation dynamic simulations and agreed well with available experimental results (Dimiduk et al., 2005). *Reprinted from Cui, Y., Lin, P., Liu, Z., Zhuang, Z., 2014. Theoretical and numerical investigations of single arm dislocation source controlled plastic flow in FCC micropillars. International Journal of Plasticity 55, 279–292, Copyright 2014, with permission from Elsevier.*

An immense effort has been made to disclose the physical origin of these features of submicron plasticity (Greer and De Hosson, 2011; Ngan and Ng, 2010; Uchic et al., 2009a; Zhu et al., 2008a,b). Among the simulation methods, three-dimensional (3D) discrete dislocation dynamics (DDD) and discrete-continuous models described in Chapter 7 are suggested as a natural and powerful approach to revealing collective dislocation mechanisms at submicron scales (Espinosa et al., 2006; Groh and Zbib, 2009; Po et al., 2014; Shao et al., 2013). The following discussion reveals the main insights obtained from DDD simulations and experiments. Furthermore, some theoretical models are discussed based on these insights to better understand and predict submicron plasticity.

8.2 Single-Arm Dislocation Source Mechanisms at Submicron Scales

The plastic behavior of bulk crystals is usually dominated by collective interactions between dislocations. This can be well-described by the Taylor hardening law, e.g., flow stress exhibits a good correlation with the square root of dislocation density, $\tau \propto \sqrt{\rho}$. However, both theoretical and experimental investigations demonstrate that the Taylor relation ceases to hold for materials at submicron scales (Akarapu et al., 2010; El-Awady et al., 2009a,b; Mayeur and McDowell, 2013; Tang et al., 2008). First, the Taylor relation alone is not enough to consider the size-dependent initial yield strength explicitly in the absence of strain gradients. To analyze the power law relation between strength and size, other mechanisms must be incorporated. For example, some valuable attempts are made to combine the Taylor relation and the fractal-like dislocation network (Gu and Ngan, 2013). Second, the Taylor relation envisages plasticity as a smooth and steady flow in both time and space, which renders it incapable of capturing the discontinuous deformation with intermittent strain bursts. Third, at small scales, the limited slip distance makes the dislocation surface annihilation rate exceed the multiplication rate. Dislocation density always tends to decrease instead of increasing during deformation (Greer, 2006), as is shown in Fig. 8.1. According to the Taylor relation, strain softening should definitely occur after the reduction of dislocation density. However, the stress-strain trend usually does not meet this expectation (Fig. 8.1). Hence, besides the mean-field forest hardening described by the Taylor relation, there must be another key mechanism that controls the plastic flow behavior of submicron crystals.

Considerable pioneering work has been carried out to reveal this mechanism. Currently, there are two generally accepted explanations. One is the dislocation starvation model (Greer and Nix, 2006), which expects that smaller samples contain fewer sources on average. In this model, the increase in flow stress results from a loss of dislocation sources by quick surface annihilation. The other is the dislocation source truncation model, which expects that smaller samples contain shorter sources (Beanland, 1995; Parthasarathy et al., 2007), so higher-flow stress is usually required to activate such shorter sources. One can see that these theories emphasize the role of dislocation sources in controlling plastic behavior at submicron scales.

For submicron crystals, the surface-to-volume ratio is very high and inside dislocations have a high probability of interacting with free surfaces. Hence, substantial dislocation-surface interaction makes dislocation sources result in new features. As revealed by transmission electron microscopy (TEM) observation (Kiener and Minor, 2011a; Oh et al., 2009) and DDD simulations (Akarapu et al., 2010; El-Awady et al., 2009a,b; Tang et al., 2008), the leading dislocation sources change from Frank-Read (FR) source to SAS, with one pinning point located inside the crystal and a free end on the surface (Fig. 8.2). It was shown that activation stress for the weakest SAS can successively explain the size effect of

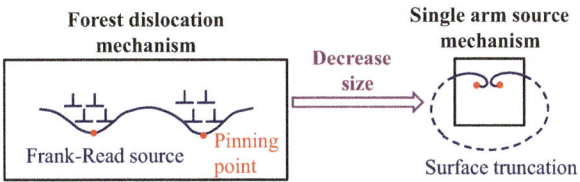

Figure 8.2

Schematic showing the mechanism transition from forest dislocation interaction to single arm source mechanism with the reduction of external size.

the initial yield strength for face-centered cubic (FCC) and body-centered cubic (BCC) crystals (Lee et al., 2009; Lee and Nix, 2012; Ng and Ngan, 2008c). In the following sections, the correlations between the SAS mechanism and the specific features of submicron plasticity will be analyzed in detail.

8.3 Single-Arm Dislocation Source-Controlled Strain Burst and Dislocation Avalanche

Quakes and avalanches are widely observed in many physical systems such as strain bursts and dislocation avalanches in submicron crystals, sandpile slides, snow avalanches, and geologic earthquakes on a much larger length scale. The magnitude and frequency of occurrence of avalanches exhibit wide scatter. However, one astonishing feature of these avalanches spanning different spatial and temporal scales is "scale-free" power law distribution (Papanikolaou et al., 2017a; Uhl et al., 2015; Zaiser, 2006): namely, the probability $P(s)$ of observing an avalanche with magnitude s is proportional to $s^{-\alpha}$. This universality has become a topic of considerable experimental and theoretical attention (Maaß and Derlet, 2018; Papanikolaou et al., 2017a). With respect to submicron plasticity, the statistical analysis of strain burst and dislocation avalanches is summarized in Table 2 in Cui et al. (2017b).

To understand this power law scaling avalanche behavior, the first aspect to be in investigated is the controlling dislocation mechanisms of strain burst and dislocation avalanches. Csikor et al. (2007) proposed that strain burst originates from the collective and avalanche-like motion of dislocations owing to long-range interactions. Hu et al. (2014) investigated strain burst in compressed nanoparticles and stated that it was induced by the robust emissions of numerous pileup dislocations in a particular area as the result of strain gradients. The other widely accepted opinion is that the occurrence and termination of strain bursts are directly controlled by the intermittent operation of SAS, which is supported by several DDD simulation results (Cui et al., 2014, 2016b; Rao et al., 2008; Tang et al., 2008) and experimental observations (Kiener and Minor, 2011b).

One example of the correspondence between SAS operation and strain burst is given in Fig. 8.3 (Cui et al., 2014). This shows the simulation result of a compressed Ni pillar in which $d = 200$ nm. Initially, a number of SAS are arranged in a network (such as in Fig. 8.3A) connected by dislocation junctions. As stress increases, the weakest SAS begins

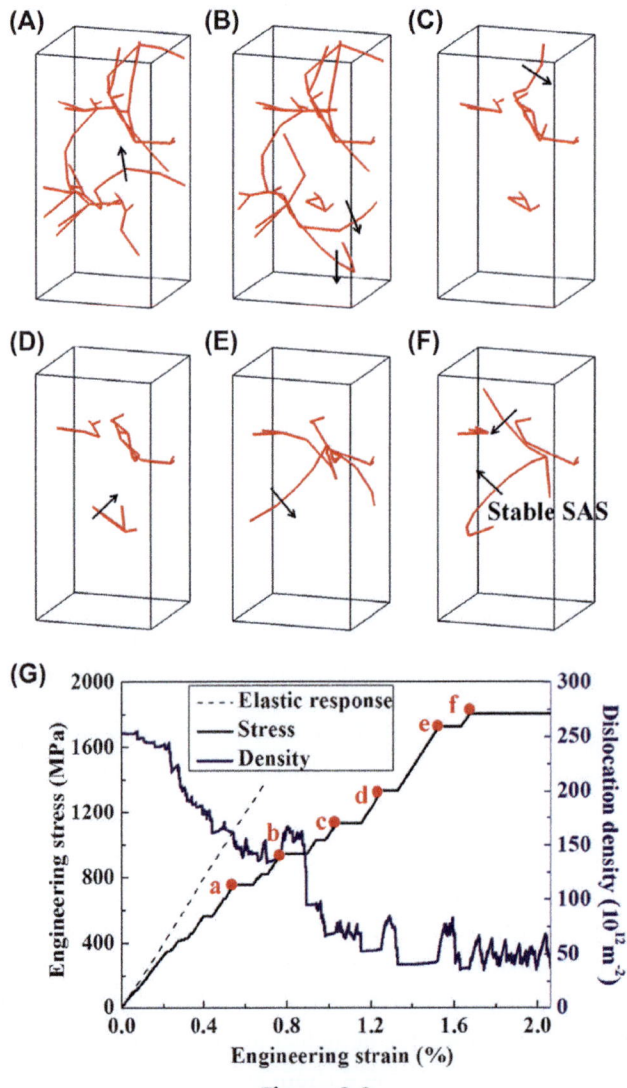

Figure 8.3

(A–F) Snapshots of dislocation configurations during compression for an Ni micropillar with $d = 200$ nm, in which the *arrows* indicate the bowing out directions of the activated single-arm dislocation source; (G) Stress-strain and dislocation density-strain curve. The *marked points* correspond to the dislocation configurations in (A–F). *Reprinted from Cui, Y., Lin, P., Liu, Z., Zhuang, Z., 2014. Theoretical and numerical investigations of single arm dislocation source controlled plastic flow in FCC micropillars. International Journal of Plasticity 55, 279–292, Copyright 2014, with permission from Elsevier.*

to bow out along the direction indicated by the arrows in Fig. 8.3A; meanwhile, a strain burst event takes place. Plastic strain is then generated by the spiral motion of the SAS. However, SAS usually operates only for a short time before they cease to work. For example, the activated sources shown in Fig. 8.3A−C form new junctions by interacting with other dislocations and could no longer be activated under the current stress level. The activated source in Fig. 8.3D disappears owing to the destruction of the pinning point under high stress and the subsequent surface annihilation. The dynamic source in Fig. 8.3E is broken as the pinning point glides out of the crystal in such a limited size. The number of times that SAS sweeps the slip plane can be defined as the lifetime of the SAS. This analysis shows that its lifetime depends on the neighboring dislocation microstructure, the instantaneous stress level, and the relative position to the surface. When dislocation sources cease to operate, the current strain burst terminates. Then, the stress increases nearly elastically to activate new sources. Because of the stochastic distribution of the strength and the lifetime of the SAS, the stress always increases in a discontinuous and stochastic manner during deformation. This also explains why strain burst occurs in a stochastic way. As the strain burst takes places, the dislocation sources are gradually exhausted. Finally, the initial SAS network becomes several isolated sources (as shown in Fig. 8.3F). Similar simulation results were also reported and are discussed in the provided references (Akarapu et al., 2010; Benzerga, 2008; Rao et al., 2008; Tang et al., 2008).

In a micropillar with diameter d, the intersection between the slip plane and the pillar surface is an ellipse with major axis $a = d/cos\beta$. Here, β is the angle between the slip plane normal direction and the compressive axis. When one activated SAS traverses the whole glide plane for one time, the produced plastic shear strain γ can be estimated as

$$\gamma = \frac{bA_{slip}}{V} = \frac{b \cdot \pi d(d/cos\beta)/4}{\pi d^2 h/4} = \frac{b}{hcos\beta} \tag{8.3.1}$$

where b is the Burgers vector, A_{slip} is the slipped area, V is the sample volume, and h is the height of the sample. The corresponding engineering strain ε can be calculated by $\varepsilon = \gamma \cdot M$, where M is the Schmid factor. Given the same lifetime of SAS for both large and small samples, the small samples correspond to larger plastic strain and thus a longer stress plateau and more noticeable strain burst. However, at the same time, small samples correspond to more significant image force, which always tends to drive dislocations out and leads to a shorter lifetime of the SAS. Therefore, the actual strain burst magnitude is a consequence of competition between the amount of generated strain and lifetime of the SAS.

A branching SAS operation model was proposed to explain the avalanche statistics (Cui et al., 2016b). The basic idea of the model is given in Fig. 8.4. First, information is given on the material properties and the statistical distribution of dislocation source. Then, the external load increases, until the weakest source is activated. The discrete plastic deformation is assumed to proceed mainly through the intermittent activation and

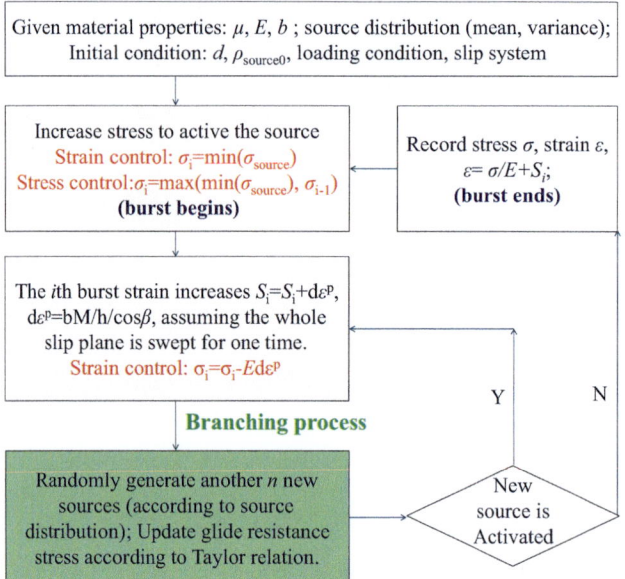

Figure 8.4

Calculation procedure of branching dislocation source operation model (Cui et al., 2016b). Here, μ is the shear modulus, E is the elastic modulus, b is the Burgers vector magnitude, d is the external size, ρ is the dislocation density, σ is stress, ε is strain, S is the burst strain, and subscript i represents the results corresponding to the ith step.

deactivation of the SAS. Each time one SAS operates, the intermittent strain burst magnitude increases according to Eq. (8.3.1). One activated source may lead to the stochastic generation or activation of other sources, similar to a branching process, described in Fig. 8.5. The number of newly generated dislocation sources n_a is taken as the nearest integer of $2 \times$ rand, where rand represents a random value from 0 to 1. The fate of all sources (active or not) is determined by checking whether the instantaneous applied stress can reach the source operation stress. It is remarkable that this simple model successfully captures the power law scaling statistics of strain burst, as shown in Fig. 8.6. The power law exponent is -1.5, which is consistent with previous experimental results (Csikor et al., 2007; Ng and Ngan, 2008c; Zhang et al., 2012). Moreover, statistical analysis on the burst magnitude based on this branching SAS operation model indicates that the loading mode can induce a controllable dynamical regime transition from self-organized criticality avalanche power-law scaling to quasiperiodic strain burst oscillations (Cui et al., 2016b, 2017b). This finding is further verified by 3D DDD simulations in conjunction with the external system. Simulation results show that this transition occurs because the correlated extent of the operated SAS during each burst event is sensitive to the loading modes (Cui et al., 2016b, 2017b). This opens up new possibilities for novel experiments with a faster response rate than is currently obtainable, which can be designed to explore the dynamical regime transition of dislocation systems.

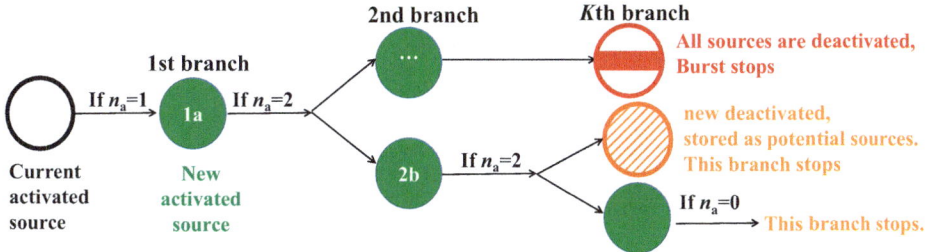

Figure 8.5

Schematic showing the random branching dislocation source generation and activation process; n_a is the number of newly generated dislocation sources. *Green filled circles* represent that a new source is activated, and only the activated source may trigger further branching processes (Cui et al., 2016b).

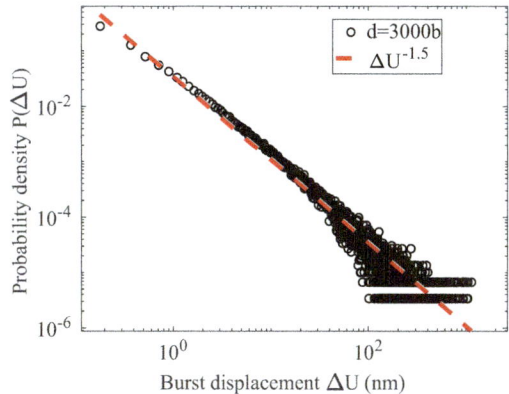

Figure 8.6

Power law avalanche statistics predicted by branching dislocation source operation model (Cui et al., 2016b).

8.4 Description of Single-Arm Dislocation Source-Controlled Plasticity

In the following discussion, a comprehensive, especially quantitative understanding of the direct correlation between the SAS mechanism and flow stress will be presented. In particular, the conventional dislocation density evolution equation and the strain hardening model are modified to consider the SAS operation mechanism.

8.4.1 Single-Arm Dislocation Source-Controlled Dislocation Density Evolution

At the submicron scales, dislocations are close to free surfaces, which serve as sinks for dislocations. The image force further enhances the surface annihilation of dislocations. Therefore, the dislocations escape from the free surface in such limited size at a faster rate than they multiply (Greer and Nix, 2006; Liu et al., 2009b; Shan et al., 2008).

The dislocation density continuously decreases during the early deformation stage (Fig. 8.1). At the steady flow stage, the dislocation densities can fluctuate near a constant value (Zhou et al., 2011). Interestingly, for different diameters d, this constant value is almost the same for different initial dislocation configurations in Fig. 8.1.

The fluctuation of the dislocation density curve directly reflects the operation of the SAS. Once the activated SAS traverses the whole glide plane one time, one peak appears in the dislocation density curve. This is because the dislocation arm length fluctuates when it revolves in the slip plane (Fig. 8.7A). Accordingly, the number of peaks in each dislocation density curve actually reflects the lifetime of the SAS. When the dislocation density finally reaches the stable value, the emergence of many peaks in the dislocation density curve means that a stable SAS forms and continues to operate. At the steady flow stage, there are usually only several stable SAS operating corresponding to this stage, as shown in Fig. 8.7A.

The length of such a stable SAS is called the stable source length and is expressed as λ_s in the following analysis. Next, one simulation result, corresponding to the green curves for $d = 400$ nm in Fig. 8.1B, is taken as an example to show how to calculate the value of λ_s in the 3D DDD method. As shown in Fig. 8.7A, there are two glissile junctions at the

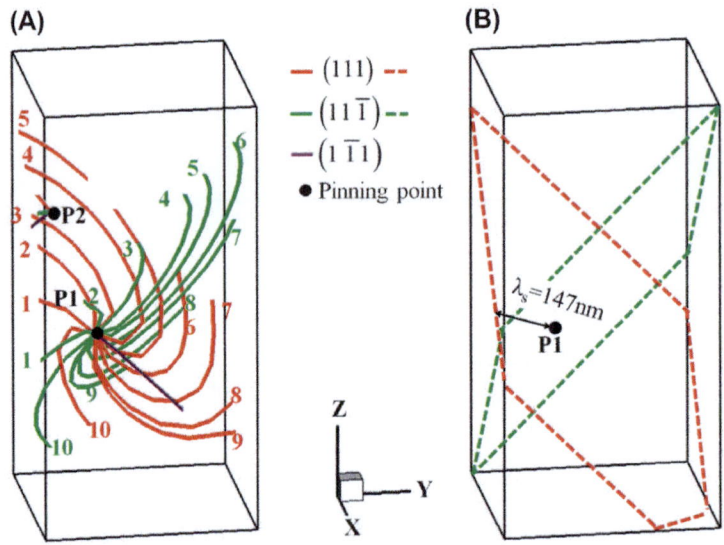

Figure 8.7

(A) Overlay of snapshots at the steady flow stage for a micropillar with $d = 400$ nm (steps 1–10) showing the revolving of dislocation arms; (B) Schematic definition of stable source length λ_s. *Dotted lines* are the boundary lines of the slip plane for considered dislocation arms. *Different colors* reflect slip plane information. *Reprinted from Cui, Y., Lin, P., Liu, Z., Zhuang, Z., 2014. Theoretical and numerical investigations of single arm dislocation source controlled plastic flow in FCC micropillars. International Journal of Plasticity 55, 279–292, Copyright 2014, with permission from Elsevier.*

steady flow stage. The purple dislocation segments are immobile because they belong to immobilized slip systems $[110](1\bar{1}1)$ under uniaxial compression. They provide two stable pinning points P1 and P2 for the operation of the SAS. Here, the SAS with pinning point P2 is not activated because of its short arm length. Attention is focused on the two activated SAS with pinning point P1. The red and green spiral dislocation arms in Fig. 8.7A lie in slip planes (111) and $(11\bar{1})$, respectively. The intersection lines of these two slip planes with the free surface are given by the red and green dotted lines in Fig. 8.7B, respectively. The stable source length λ_s is defined as the shortest distance between pinning point P1 and these intersection lines. According to the coordinates of P1, λ_s is about 147 nm, as indicated in Fig. 8.7B. Following this method, the values of λ_s for all simulations in Fig. 8.1 can be calculated. For each diameter, the average λ_s normalized by pillar diameter d as well as the root-mean-square error are plotted in Fig. 8.8. Surprisingly, the average λ_s/d is essentially constant for different sample sizes. Similar results were also reported in Ryu et al. (2015). This implies that although DDD itself is scale-invariant, a characteristic size length can emerge as a natural outcome of the geometrical constraints on the source operation and motion of dislocations without the need for artificial treatment of the initial dislocation source length. This finding confirms the observation and theoretical analysis made by Dunstan and Bushby (2014).

Because the dislocation density evolution at submicron scales is strongly controlled by the operation of the SAS, the dislocation kinetic equation must explicitly include this effect. Therefore, at least four dislocation density terms should be incorporated for submicron materials: the generation of dislocation from the SAS operation (Malygin, 2010), the

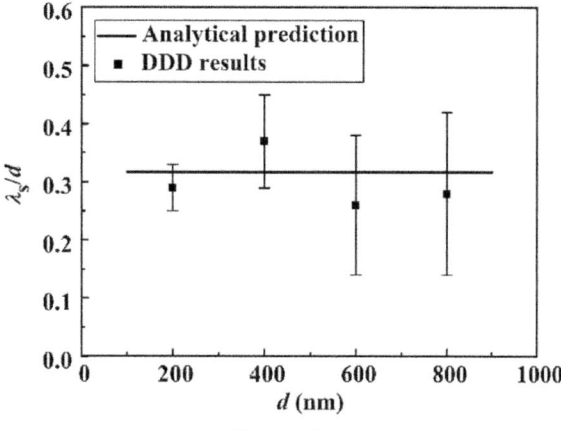

Figure 8.8

Stable source length normalized by the pillar diameter when the dislocation density reaches a stable value. *DDD*, discrete dislocation dynamics. *Reprinted from Cui, Y., Lin, P., Liu, Z., Zhuang, Z., 2014. Theoretical and numerical investigations of single arm dislocation source controlled plastic flow in FCC micropillars. International Journal of Plasticity 55, 279–292, Copyright 2014, with permission from Elsevier.*

escape of dislocation through the free surface (Greer, 2006), dislocation multiplication owing to forest dislocations, and mutual annihilation for closely spaced dislocations of opposite sign.

We begin with building in a traditional kinetic equation of bulk-like dislocation density ρ_{bulk} (Devincre et al., 2008):

$$\frac{d\rho_{\text{bulk}}}{d\gamma} = \frac{1}{b\lambda_{\text{FR}}} - \frac{y}{b}\rho \tag{8.4.1}$$

where ρ is the total dislocation density and λ_{FR} is the mean free path along which the dislocation can move before storage. It can also be considered to be the characteristic length of an FR source in forest dislocations. Generally, $\lambda_{\text{FR}} = 1/(k_f\rho)$, where k_f is a dimensionless constant and is set to 10^{-2} (Malygin, 2012). y is the effective mutual annihilation distance.

Inspired by the dislocation multiplication term in Eq. (8.4.1), the dislocation generation rate by SAS is proposed as

$$\frac{d\rho_{\text{SAS}}}{d\gamma} = \frac{1}{2b\bar{\lambda}} \tag{8.4.2}$$

where $\bar{\lambda}$ is a length of the statistically average effective SAS. The coefficient 1/2 is chosen to consider that SAS has only one pinning point compared with the FR source, and thus the dislocation generation effect reduces by half.

The dislocation surface annihilation term was first built by Greer (2006). It is assumed that dislocations in the subsurface region have 1/2 probability of escaping from the free surface. To incorporate the crystal orientation effect, Zhou et al. (2011) gave an improved formulation:

$$\frac{d\rho_{\text{surf}}}{d\gamma} = -\frac{2\cos^2(\beta/2)}{bd} \tag{8.4.3}$$

Considering these four different terms, the dislocation density evolution equation at submicron scales could be represented as

$$\frac{d\rho}{d\gamma} = \frac{1}{2b\bar{\lambda}} - \frac{2\cos^2(\beta/2)}{bd} + \frac{k_f\sqrt{\rho}}{b} - \frac{y}{b}\rho \tag{8.4.4}$$

Only the first two terms on the right side of Eq. (8.4.4) are related to sample size d. According to simple calculations, the contributions of the last two terms for micropillars with diameters of several hundred nanometers are almost an order of magnitude smaller than those of the first two terms. Thus, the last two terms will be ignored in the following analysis and the SAS multiplication and surface annihilation mechanism have an essential role in determining the dislocation density at submicron scales.

From Fig. 8.1, the dislocation density almost reaches a stable value at the steady flow stage. This means that the generation rate by the SAS and the escape rate from the free surface are balanced. According to Eq. (8.4.4), this can be expressed as

$$\frac{d\rho}{d\gamma} = \frac{1}{2b\lambda_s} - \frac{2\cos^2(\beta/2)}{bd} = 0 \quad \text{or} \quad \frac{\lambda_s}{d} = \frac{1}{4\cos^2(\beta/2)} \tag{8.4.5}$$

It can be seen that λ_s/d, the stable source length normalized by the pillar diameter, is only related to the crystal orientation. When loaded along [001], $\lambda_s/d = 0.317$ for <011> (111) slip systems. As shown in Fig. 8.8, this prediction is quantitatively in accordance with the simulation results, even though the simulation results are relative scattered owing to the sensitivity to the initial dislocation configuration.

8.4.2 Effective Single-Arm Dislocation Source Length

For bulk materials, a number of dislocation sources can form because of the collective interactions between dislocations. Generally, the average length of these sources can be expressed as a function of dislocation density (Malygin, 2010). What happens at submicron scales? Can we infer the stable dislocation density according to the stable source length? In the micropillars, the length of the activated SAS scatters from sample to sample. Thus, it is reasonable to estimate the statistically average length of the activated SAS theoretically during plastic deformation. 3D DDD simulations revealed that the dislocation sources are relatively spatially isolated and interactions between dislocation sources at different slip systems are weak at submicron scales (Cui et al., 2014). These characters make it possible to use some isolated pinning points to describe these SAS sources (Parthasarathy et al., 2007). As described earlier, the SAS length is defined as the shortest distance from the pinning point to the surface in the slip plane of dislocation arms (Fig. 8.4B). When several sources exist, the SAS with the longest length will be activated preferentially and determine instantaneous flow stress. Hence, the longest SAS length is defined as the effective SAS length λ. Bearing these facts in mind, the statistically average length of effective SAS $\bar{\lambda}$ during plastic deformation can be decided by the following statistical model (Parthasarathy et al., 2007; Zhou and LeSar, 2012).

As shown in Fig. 8.9, the slip plane in the pillar is an ellipse with major axis $a = d/\cos\beta$ and minor axis d. Assuming there are n pinning points, all of which randomly distribute in the slip plane, the probability of finding a pinning point in the filled area (distance to free surface is l_1) can be expressed as

$$P(l_1)dl_1 = \frac{\pi[(d/2 - l_1) + (a/2 - l_1)]dl_1}{\pi ad/4} \tag{8.4.6}$$

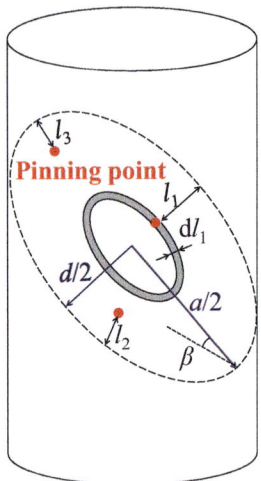

Figure 8.9

Schematic showing the statistical model of calculating the effective single-arm dislocation source length λ, $\lambda = \max(l_1, l_2, l_3, \ldots, l_n)$ (Parthasarathy et al., 2007). The *dashed ellipse* indicates the slip plane of the dislocation arm.

Eq. (8.4.6) describes the probability of finding a source with length l_1. Then, if the total pinning points number is n, the probability of finding an effective SAS with length λ has the following form:

$$P(\lambda)\mathrm{d}\lambda = \left[1 - \frac{\pi(d/2 - \lambda)(a/2 - \lambda)}{\pi ad/4}\right]^{n-1} \times \left(\frac{\pi[(d/2 - \lambda) + (a/2 - \lambda)]}{\pi ad/4}\right) n\mathrm{d}\lambda \qquad (8.4.7)$$

Thus, the average effective SAS length $\bar{\lambda}$ can be determined as

$$\bar{\lambda} = \int_0^{d/2} P(\lambda)\lambda\mathrm{d}\lambda \qquad (8.4.8)$$

This equation gives the relation between n and $\bar{\lambda}$. If either of these two quantities is known during plastic deformation, the other can be obtained correspondingly.

Moreover, the number of pinning points n can be related to sample size d and dislocation density ρ. Given ρ, the total length of dislocation segments L is approximately $\rho\pi d^2 h/4$, where h is the height of the micropillar. Assuming that the average length of each dislocation segment is equal to the pillar diameter and each dislocation segment represents an SAS, n can be estimated as (Jennings et al., 2012)

$$n = \text{Integer}\left[\frac{\rho\pi dh}{4}\right] \qquad (8.4.9)$$

Table 8.1: Stable Dislocation Density for Different Sample Sizes (Cui et al., 2014)

Diameter (nm)		200	400	600	800
Stable dislocation density $(10^{12}\ m^{-2})$	Analytical prediction	48	12	5.3	3.0
	Discrete dislocation dynamics results	46 ± 9	14 ± 3	5.4 ± 2	3.2 ± 1

Following this analysis, we can estimate the dislocation density ρ and statistically average length $\bar{\lambda}$ of an effective SAS during plastic deformation by solving Eqs. (8.4.4), (8.4.8), and (8.4.9) together.

Analytical prediction by Eq. (8.4.5) shows that the average stable source length $\lambda_s = 0.317\ d$ at the steady flow stage. By substituting this value into Eq. (8.4.8), the number of pinning points $n = 3$ can be obtained. This means there are three stable SAS at the steady flow stage. This prediction is close to the DDD simulation results, such as the stable dislocation configurations shown in Figs. 8.3F and 8.7A. Thus, the stable dislocation densities for different sample sizes can be predicted by Eq. (8.4.9); the results are listed in Table 8.1 together with the DDD results. The good agreement further demonstrates the applicability of the dislocation density evolution equation and the statistical model. The stable dislocation density increases with a decreasing sample size, which agrees well with the experimental observations (Norfleet et al., 2008). In addition, both the analytical prediction and the simulation results show that this stable value does not evidently depend on the initial dislocation density. As long as the initial dislocation density is larger than this stable value, the density will decline toward this stable value.

For other loading orientations ($0 < \beta < \pi/2$) with d ranging from 200 to 800 nm, the corresponding normalized stable source lengths and pinning point numbers are given in Fig. 8.10. The pinning point numbers are 2 or 3 for the most loading orientations. Thus, according to Eq. (8.4.9), the stable dislocation density value for most loading cases should be a little bit smaller than the predicted value in Table 8.1 or close to it.

8.4.3 Single-Arm Dislocation Source-Controlled Flow Stress

The next issue is how to estimate flow stress for SAS-controlled plasticity. Generally, the critical resolved shear stress τ to activate the SAS is decided by lattice friction stress τ_0, the elastic interaction stress related to dislocation density ρ, and the line tension stress (Lee and Nix, 2012; Ng and Ngan, 2008a; Norfleet et al., 2008; Parthasarathy et al., 2007), which is a function of average effective source length $\bar{\lambda}$. Following the model proposed by Parthasarathy et al. (2007), the critical resolved shear stress can be expressed as

$$\tau = \tau_0 + \alpha\mu b\sqrt{\rho} + \frac{k\mu b}{\bar{\lambda}} \qquad (8.4.10)$$

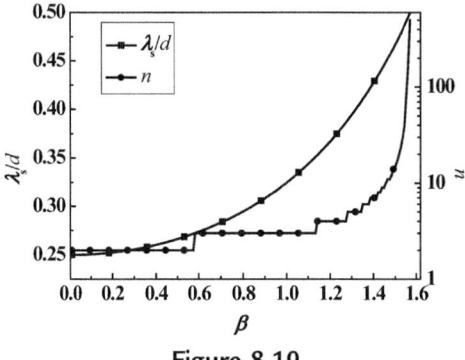

Figure 8.10

Stable source length normalized by pillar diameter λ_s/d and the pinning point number n for different loading orientations. *Reprinted from Cui, Y., Lin, P., Liu, Z., Zhuang, Z., 2014. Theoretical and numerical investigations of single arm dislocation source controlled plastic flow in FCC micropillars. International Journal of Plasticity 55, 279–292, Copyright 2014, with permission from Elsevier.*

where α and k are dimensionless constants, usually α is set at 0.5, and k is taken to be 1.0 (Lee and Nix, 2012; Parthasarathy et al., 2007).

3D DDD simulations reveal that the dislocation configuration is relatively isolated and the flow stress keeps constant at the steady flow stage. These suggest that dislocation sources themselves, the third term on the right side of Eq. (8.4.10), may dominate the resolved shear stress, instead of dislocation interactions presented in the second term of Eq. (8.4.10). In most of the studies, Eq. (8.4.10) is used only to predict critical stress at the onset of yield (Ng and Ngan, 2008a; Parthasarathy et al., 2007). However, as long as the evolutions of ρ and $\bar{\lambda}$ are known, it may be further extended to predict the whole hardening behavior at a submicron scale. Here, the dislocation density evolution law is given by Eq. (8.4.4) and the SAS length evolution can be estimated based on the statistical model given by Eqs. (8.4.8) and (8.4.9). In the calculations, Eqs. (8.4.4) and (8.4.8)–(8.4.10) are solved explicitly to obtain the evolution of flow stress. The whole procedure is schematically shown in Fig. 8.11. First, given the initial dislocation density ρ_0, the initial average effective SAS length can be estimated by Eqs. (8.4.8) and (8.4.9). Then, the dislocation density for the next time step can be obtained according to Eq. (8.4.4). Meanwhile, the average effective SAS length for the next step is updated according to the current dislocation density by the statistical model. Then, the shear stress τ is calculated via Eq. (8.4.10). Meanwhile, axial stress σ and strain ε can be calculated according to Schmid factor M and elastic modulus E.

As shown in Fig. 8.12, once given the diameter d and initial dislocation density ρ_0 of the micropillar, the dislocation density, average effective SAS length $\bar{\lambda}$, stress-strain curve, and stable flow stress all can be predicted theoretically. Here, the initial dislocation densities for different sizes are taken as three times the corresponding stable dislocation density

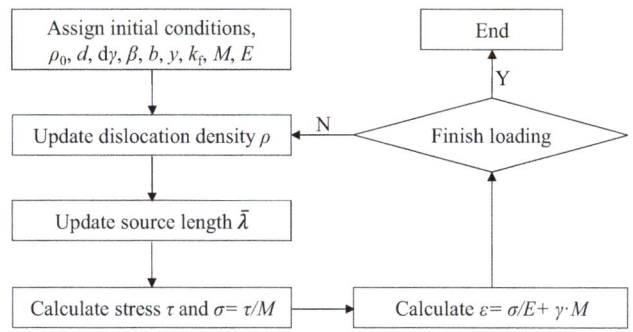

Figure 8.11

Calculation flowchart of the theoretical model.

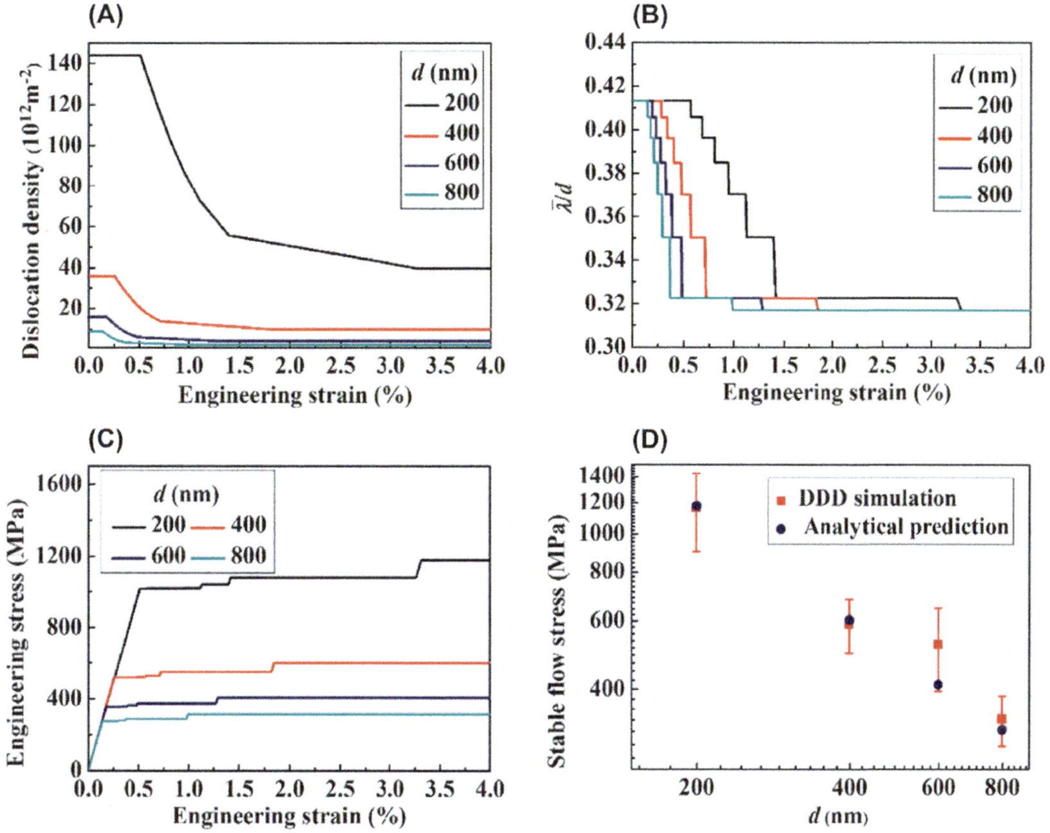

Figure 8.12

Results for theoretical analysis: (A) Evolution of dislocation density; (B) Evolution of average effective single-arm dislocation source length; (C) Evolution of flow stress; (D) Comparison of stable flow stress between analytical and simulation results. *Reprinted from Cui, Y., Lin, P., Liu, Z., Zhuang, Z., 2014. Theoretical and numerical investigations of single arm dislocation source controlled plastic flow in FCC micropillars. International Journal of Plasticity 55, 279–292, Copyright 2014, with permission from Elsevier.*

given in Table 8.1. The dislocation density evolution curves in Fig. 8.12A show the same trend as the DDD results in Fig. 8.1. However, the critical strain, corresponding to which the dislocation density reaches the stable value, is much larger than that in DDD simulations. This is because in DDD simulations, at the early stage of loading, microplasticity usually happens as a result of the motion of free dislocations and the breakup of weak junctions (Maaβ and Derlet, 2018). Thus, dislocation density also decreases, evidently during the microplasticity stage. However, this effect is not taken into account in the analytical model. In Fig. 8.12B, the evolution curve of $\bar{\lambda}$ reveals the intermittent activation feature of the sources, as observed in DDD simulations. During the early stage of deformation, the dislocation density is high, so the dislocation sources have a high probability of interacting with neighboring dislocations to form shorter sources. As a result, the statistically average lifetime of the sources is relatively short. The final constant SAS length physically corresponds to the stable SAS with a long lifetime. Meanwhile, the statistically average stress-strain curves for each diameter are given in Fig. 8.12C. For different diameters, both size-dependent flow stress and strain burst are obtained, as observed in DDD simulations. The smaller the pillar diameter, the more significant the strain burst. In addition, the predicted stable flow stresses for different diameters are plotted in Fig. 8.12D together with the root-mean-square error for the DDD results. It can be seen that the theoretical model successfully predicts the average stable flow stress for different initial configurations.

A further quantitative comparison with the TEM experiment data (Frick et al., 2008) is depicted in Fig. 8.13. The stable flow stress of the analytical and DDD simulation results

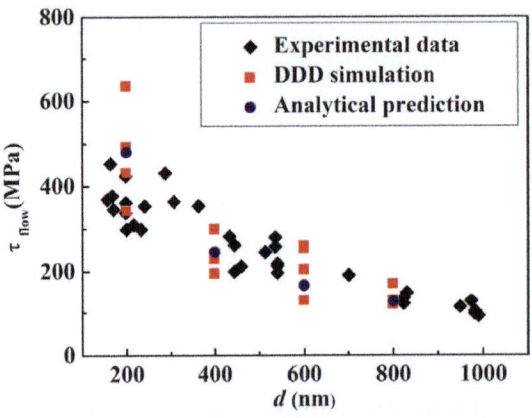

Figure 8.13

Comparison between resolved shear stress (RSS) at 3% strain for Ni micropillar experiments (Frick et al., 2008) and RSS at a steady flow state for simulation and analytical analysis. DDD, discrete dislocation dynamics. *Reprinted from Cui, Y., Lin, P., Liu, Z., Zhuang, Z., 2014. Theoretical and numerical investigations of single arm dislocation source controlled plastic flow in FCC micropillars. International Journal of Plasticity 55, 279–292, Copyright 2014, with permission from Elsevier.*

is transformed into the resolved shear stress τ_{flow} to facilitate comparison. In addition, the experimental resolved shear stress values are taken at 3% strain for [111] orientated compression nickel pillars. A good quantitative match is also achieved.

Least-squares fit shows that stable flow stress σ_{flow} exhibits a power-law type size dependence $\sigma_{flow} \propto d^{-0.85}$, which is consistent with the power-law exponent by experiments, ranging from -0.61 to -0.97 for FCC micropillars (Uchic et al., 2009b). The actual value of the exponent in the power-law relation depends on the initial dislocation density (Rao et al., 2008) and dislocation structure, and thus can be altered by prestrain (Gu and Ngan, 2013). Dunstan and Bushby (2013) pointed out that the dependence on dislocation density actually reflects a dependence on the corresponding bulk strength; they reported a unified exponent of 1 with a clear physical explanation. Interestingly, if the least-squares fit is made according to the average stable flow stress and sample diameter (such as is shown in Fig. 8.12D), the exponent is very close to 1.

In calculating stable flow stress, the third term on the right side of Eq. (8.4.10), which is associated with the activation stress of the SAS, occupies about 80%. In contrast, the second term of Eq. (8.4.10), the Taylor hardening part, contributes less than 20%. This further emphasizes the importance of the SAS operation in submicron plastic flow.

In addition, the SAS mechanism itself is not suitable for explaining hardening behavior for a very small or very large sample size. In deep submicron samples, the image force becomes so large that together with the external applied stress, dislocations can easily be driven out of the crystal. The available dislocation sources become progressively exhausted, even causing dislocation starvation. TEM experiments revealed that a micropillar with a diameter smaller than about 160 nm could achieve a dislocation starvation state by mechanical annihilation (Shan et al., 2008). Thus, stochastic dislocation surface nucleation events control the occurrence of strain burst (Wang et al., 2012a). Significant dislocation starvation hardening may be observed at the same time. If one wants to consider both the starvation of preexisting dislocations and the nucleation of new ones, the method proposed by Jérusalem et al. (2012) can be referenced for smaller pillars. Then, Eq. (8.4.10) needs to be replaced by the expression

$$\tau = \text{Min}\left(\left(1 - \frac{\varepsilon_p}{\varepsilon_p^{starv}}\right)\left(\tau_0 + \alpha\mu b\sqrt{\rho} + \frac{k\mu b}{\bar{\lambda}}\right) + \frac{\varepsilon_p}{\varepsilon_p^{starv}}\tau_{nucl}, \tau_{nucl} \right) \qquad (8.4.11)$$

where ε_p is the engineering plastic strain and ε_p^{starv} is the plastic strain for which dislocation nucleation is more favorable than the SAS operation and at which the critical resolved shear stress for dislocation nucleation τ_{nucl} is reached.

Moreover, for very large samples, the strain hardening is bulk-like and associated with pronounced dislocation interaction. The forest hardening mechanism has a leading role.

Kraft et al. (2010) proposed that the critical length scale for the bulk-like mechanism may be approximately 1000 nm or more, because TEM investigations revealed that the dislocation substructures in samples larger than this size are qualitatively similar to those found in bulk samples. Thus, according to a conservative estimate, a theoretical analysis for a SAS mechanism is more suitable for single-crystal micropillars with a diameter of 200–1000 nm. For larger pillars up to about 40 μm (Parthasarathy et al., 2007), the size effect of the initial yield strength can be well-predicted in terms of SAS activation stress by Eq. (8.4.10). However, the statistically based evolution equation of an effective SAS length must be modified owing to the collective dislocation interactions if this model is applied to a larger sample size.

This discussion reveals that dislocation density ρ and average effective source length $\bar{\lambda}$ are two main links that connect discrete dislocation activities with plastic flow behavior for submicron FCC single crystals with a moderate initial dislocation density. This raises the possibility of developing continuum crystal plasticity theories in which ρ and $\bar{\lambda}$ act as two internal variables in the constitutive relation, to study more complicated plastic deformation for submicron crystals.

8.5 Summary

Discovering the plastic mechanism for submicron crystals is a key issue not only for developing crystal plastic theory but also for designing reliable microdevices. However, atypical plastic behavior at submicron scales cannot be investigated effectively by traditional crystal plastic theory. Based on insights gained by experimental studies, 3D DDD, and a statistically based theoretical model, this chapter describes SAS-controlled plastic behaviors in submicron single crystals. The role of SAS operation and evolution on intermittent strain bursts, the evolution of dislocation density, and plastic flow stress are revealed.

Confined Plasticity in Micropillars

Chapter Outline

Interfaces are commonly introduced to improve the strength of the material, because interfaces can partly block dislocation motion. Examples are found in bimaterials, polycrystals with grain boundaries (Gao et al., 2011b; Li et al., 2009b; Quek et al., 2016) or twin boundaries (Fan et al., 2015), and materials with a coating interface. However, in most cases, enhanced strength is accompanied by a loss of ductility. Therefore, it is necessary to understand confined plasticity to design strong and reliable materials.

In this chapter, we take the coated micropillar as an example to study confined plastic behavior, because crystal devices in microelectromechanical systems (MEMS) often have a protective hard coating. This gives them higher strength but also greatly improves their erosion or wear resistance, prevents stiction or electrical shorting (Hoivik et al., 2003), and improve thermal stability (Zhuang et al., 2006). However, the deposition of coating also leads to ultrahigh local flow stress during plastic deformation (Greer, 2007; Gu and Ngan, 2012; Jennings et al., 2012; Lee et al., 2013; Ng and Ngan, 2009a) and causes mechanical reliability issues for MEMS.

This chapter aims to reveal the underlying dislocation mechanism and predict flow stress in confined plasticity at the microscale. It is organized as follows. Section 9.1 presents the mechanical behavior and underlying dislocation mechanisms in coated pillars and compares them with the uncoated counterpart, mainly based on dynamic causal modeling

Dislocation Mechanism-Based Crystal Plasticity. https://doi.org/10.1016/B978-0-12-814591-3.00009-1

(DCM) simulations. Then, Section 9.2 preliminarily connects the insights obtained by simulation to the continuum crystal plasticity theory for confined plasticity at a microscale. In Section 9.3, a simple stochastic theory model is described that can predict the upper and lower bounds of flow stress in the coated micropillar. Finally, the coating failure mechanism is preliminarily discussed in Section 9.4, including high hoop stress and the transmission effect of dislocation from the interface.

9.1 Insights into Coated Micropillar Plasticity

Compression test experiments for coated micropillars provided a good opportunity to investigate the confined plasticity problem. During compression, numerous dislocations were trapped at the pillar-coating interface. The following discussion compares coated and uncoated micropillars based on a series of DCM simulations and experimental results, to give a clear picture about the coating effect on microscopic deformation mechanisms and mechanical responses. Some theoretical models are discussed based on insights into coated micropillar plasticity.

9.1.1 Stress-Strain Curves in Coated and Uncoated Pillars

By carrying out compression tests for Au pillars with diameters of 500–900 nm, Greer (2007) first reported that the coated pillars displayed much higher flow stress and a significant amount of linear strain hardening, which differed substantially from those of pillars with free surfaces. Then, Ng and Ngan (2009a) pointed out that the overall mechanical response was insensitive to the volume fraction of the coating $V_{coating}$, when $V_{coating}$ varied from 0.07 to 0.32. These results suggested that the load-sharing effect was unimportant in the coated pillars under consideration.

Fig. 9.1A gives typical DCM simulation results of an Ni single crystal micropillar with and without an Al_2O_3 coating (Cui et al., 2015a). The pillar diameters range from 200 to 800 nm. The thickness of the Al_2O_3 coating layer is 5 nm as measured in the experiments (Lee et al., 2013). For uncoated pillars, the stress-strain curves exhibit no evident strain hardening, as shown in Fig. 9.1. For coated pillars, linear strain hardening and higher stress are observed (Greer, 2007; Jennings et al., 2012).

The stress-strain curves for both uncoated and coated pillars are composed of strain bursts under constant stress, separated by elastic segments with a slope similar to the elastic modulus (Fig. 9.1A). The difference is that the strain burst is larger for uncoated than for coated pillars. Similar inhibited strain burst behaviors are also observed in experiments. Ng and Ngan (2009a) found that stress-strain behavior could be smoothened by coating, and strain bursts were effectively suppressed for micropillars with a diameter ranging from 1.2 to 6.0 μm and a $V_{coating}$ larger than about 0.26. Experimental results by

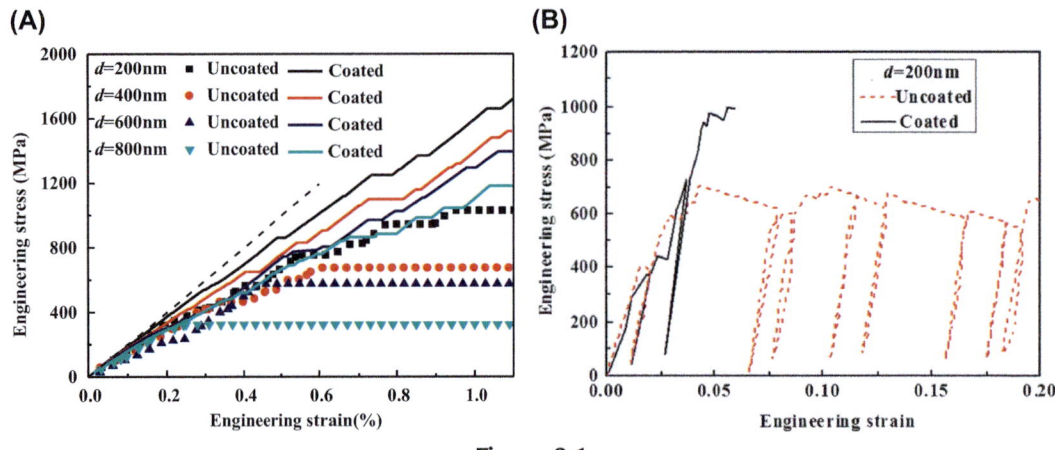

Figure 9.1

Comparison of typical stress-strain curves between uncoated and coated micropillars (A) obtained by dynamic causal modeling simulations, and (B) obtained by experiments (Jennings et al., 2012). The *dashed line* in (A) reflects the purely elastic response. *Reprinted from Cui, Y.N., Liu, Z.L., Zhuang, Z., 2015a. Theoretical and numerical investigations on confined plasticity in micropillars. Journal of the Mechanics and Physics of Solids 76, 127–143, Copyright 2015, with permission from Elsevier.*

Jenning et al. (2012) further showed that the coating could not fully suppress the strain burst in small pillars with a diameter of 200 nm and $V_{coating}$ of about 0.17, and plastic strain recovery occurred during unloading. El-Awady et al. (2011) simulated the dislocation penetration process based on Koehler barrier strength criteria and discussed the influence of barrier strength on strain burst. The initiation of large strain bursts was delayed to higher strain and stress regimes with an increase in the barrier strength. In addition, as the barrier strength increased, the magnitude of the strain bursts decreased.

9.1.2 Dislocation Source Mechanism in Coated Micropillars

As described in Chapter 8, for uncoated micropillars, a single-arm dislocation source (SAS) mechanism can well explain the mechanical response: the size effect is related to the characteristic length of the SAS, the intermittent strain burst is directly caused by the operation and shutdown of the SAS, and the lack of strain hardening is caused by the continuous operation of the stable SAS and weak dislocation interactions (Cui et al., 2014; El-Awady et al., 2009a,b; Parthasarathy et al., 2007). However, whether the SAS mechanism still works for coated micropillars needs to be restudied because a lot of dislocations are observed to pile up at the pillar-coating interface (Zhou et al., 2010b).

By observing the evolution of dislocation microstructures, the researchers found that the plastic deformation of coated micropillars is mainly accommodated by the spiral motion of the SAS for the chosen dislocation density range ($\rho_0 > 47 \times 10^{12} \text{m}^{-2}$) and deformation

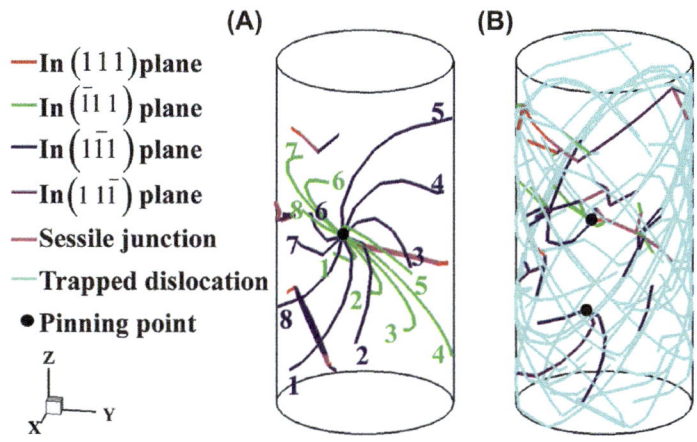

(A)   **(B)**

—In $(\bar{1}\,1\,1)$ plane

—In $(\bar{1}1\,1)$ plane

—In $(1\bar{1}1)$ plane

—In $(1\,1\bar{1})$ plane

—Sessile junction

—Trapped dislocation

● Pinning point

Figure 9.2

With the engineering strain at about 1%: (A) an overlay of images showing the operation of the single-arm dislocation source in the uncoated micropillars with a diameter of 400 nm; (B) The dislocation configuration showing the inner dislocation source character in coated micropillars with a diameter of 400 nm. *Reprinted from Cui, Y.N., Liu, Z.L., Zhuang, Z., 2015a. Theoretical and numerical investigations on confined plasticity in micropillars. Journal of the Mechanics and Physics of Solids 76, 127–143, Copyright 2015, with permission from Elsevier.*

stage ($\varepsilon < 1.2\%$). One example is presented in Fig. 9.2. Dislocation configurations at 1% engineering strain are given for uncoated and coated pillars with a diameter of 400 nm, which have the same initial dislocation configuration. In the uncoated pillars, as shown in Fig. 9.2A, the stable SAS can continuously sweep the slip plane, generating a large amount of plastic strain. However, in the coated pillars, operation of the SAS leads to significant deposition of trapped dislocations, as shown in Fig. 9.2B. The SAS can operate only intermittently owing to back stress induced by the trapped dislocations.

The initial dislocation density, ρ_0, has an important role in influencing the destiny of the internal dislocation sources (Benzerga, 2009; El-Awady et al., 2013; Rao et al., 2008; Zhou et al., 2011). For coated micropillar with a low ρ_0, the exhaustion of SAS may occur because high stress in the coated pillars may easily destroy the internal SAS.

To analyze the coating effect on the activation of SAS quantitatively, the operation processes of individual strong SAS with indestructible pinning points in uncoated and coated micropillars are further presented in Fig. 9.3 (Cui et al., 2015a), which excludes the influence of collective dislocation interactions. Similar to the results with a complex dislocation configuration, the stable SAS in an uncoated sample can operate continuously once the applied stress reaches its operation stress. Nevertheless, in a coated sample, stress needs to increase intermittently to remobilize the SAS, leading to a high strain hardening rate (SHR) close to the elastic modulus, as shown in Fig. 9.3A. In addition, these results

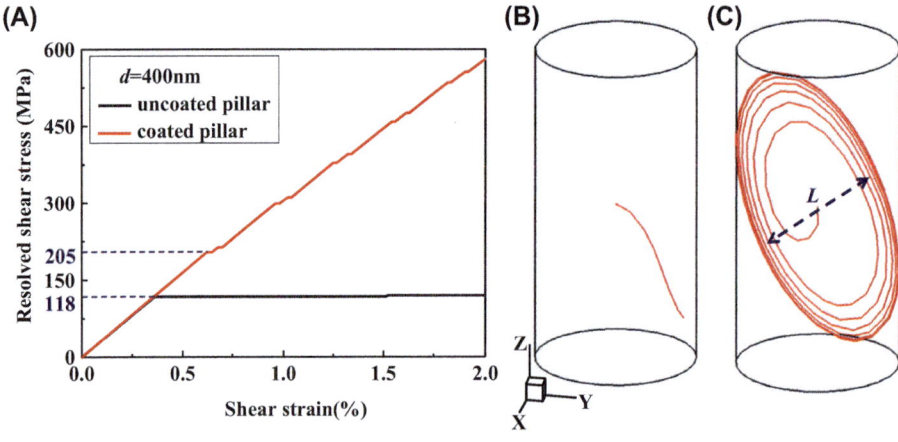

Figure 9.3

(A) Stress-strain curve for micropillar containing individual single-arm dislocation source under compression; (B and C) Dislocation configurations when shear strain is 2.45% in uncoated and coated micropillars, respectively. The *dashed arrow* in (C) indicates the characteristic length in the continuum back-stress model. *Reprinted from Cui, Y.N., Liu, Z.L., Zhuang, Z., 2015a. Theoretical and numerical investigations on confined plasticity in micropillars. Journal of the Mechanics and Physics of Solids 76, 127–143, Copyright 2015, with permission from Elsevier.*

show that for a free pillar, the operation of one stable source is enough to keep stable plastic flow, but for a coated pillar, observable plastic flow requires multiple sources to be activated simultaneously.

As described in Section 8.4.3, critical resolved shear stress τ to activate the SAS in uncoated samples can be estimated as (Lee and Nix, 2012; Parthasarathy et al., 2007)

$$\tau = \tau_0 + \alpha\mu b\sqrt{\rho} + \frac{k_s\mu}{\lambda/b} \tag{9.1.1}$$

where τ_0 is the lattice friction stress, b is the magnitude of the Burgers vector, ρ is the dislocation density, and λ is the effective SAS length. Dimensionless constants α and k_s are set to 0.5 and 1.0, respectively (Lee and Nix, 2012; Parthasarathy et al., 2007). The simulation results in Fig. 9.3A show that the critical resolved shear stress to activate an SAS in an uncoated micropillar is 118 MPa. For the considered case, the second dislocation interaction term on the right side of Eq. (9.1.1) can be ignored because there is no evident forest hardening. The deduced coefficient k_s is very close to 1.0 according to Eq. (9.1.1). In uncoated samples, the operation stress of the SAS depends on the effective source length but also the angle between the initial dislocation segment and the Burgers vector (Rao et al., 2007). Thus, only in a statistical sense is the value of k_s 1.0 for an uncoated sample.

To extend this model to predict the operation stress of the SAS in the coated case, coefficient k_s should be recalibrated to take into account the pinning effect of coating. Meanwhile, the back-stress term τ_b should be introduced:

$$\tau = \tau_0 + \alpha\mu b\sqrt{\rho_{\text{mobile}}} + \frac{k_s\mu}{\lambda/b} + \tau_b \tag{9.1.2}$$

where ρ_{mobile} represents mobile dislocation density. Why ρ_{mobile} is used instead of ρ will be discussed later.

Simulations indicate that the SAS in coated micropillars is similar to the Frank-Read (FR) source, and its operation stress is insensitive to the geometrical orientation. Thus, coefficient k_s for the coated case can be estimated by the simulation results. According to the critical resolved shear stress that initially activates the SAS, $\tau_y = 205$ MPa ($\tau_b = 0$), as shown in Fig. 9.3A, it can be deduced that $k_s \approx 2.0$. The estimation of back stress τ_b is given and expressed as follows.

9.1.3 Back Stress in Coated Micropillars

For the individual strong SAS considered in Fig. 9.3, ρ_{mobile} and λ are substantially unchanged during deformation. Thus, the sum of the first three terms on the right side of Eq. (9.1.2) is equal to the initial operation stress $\tau_y = 205$ MPa. Accordingly, the back-stress term can be obtained by $\tau_b = \sigma \cdot M - \tau_y$, where M is the Schmidt factor. The relations between back stress and instantaneous trapped dislocation density ρ_{trapped} are plotted in Fig. 9.4A. The back stress increases stepwise as the pileup of dislocations.

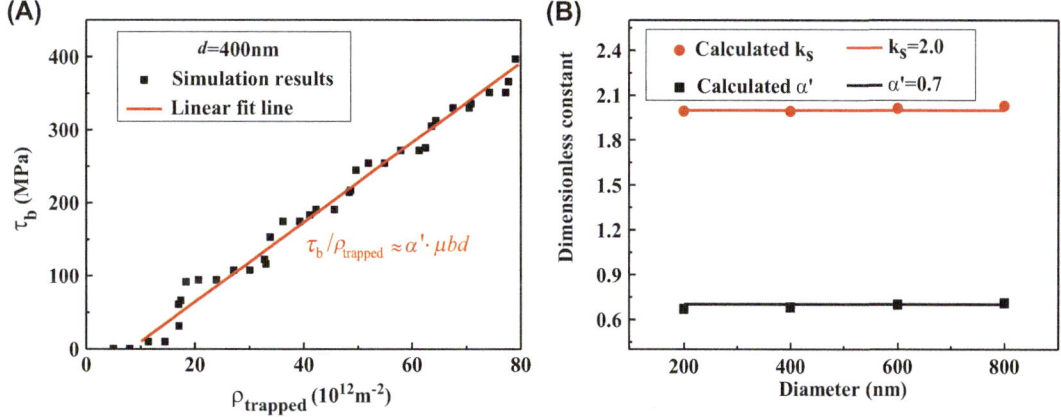

Figure 9.4

(A) Back stress versus trapped dislocation density for micropillars with a diameter of 400 nm; (B) Calculated values of k_s and α' for micropillars with different diameters. *Reprinted from Cui, Y.N., Liu, Z.L., Zhuang, Z., 2015a. Theoretical and numerical investigations on confined plasticity in micropillars. Journal of the Mechanics and Physics of Solids 76, 127–143, Copyright 2015, with permission from Elsevier.*

If ignoring these discrete stress steps, a linear relation can be found between the trapped dislocation density and back stress. In a micropillar with diameter d, the mean diameter of dislocation loops located at the pillar-coating interface can be approximately taken as d. According to the classical dislocation pileup theory, the local resolved back stress acting on a given dislocation source resulting from the n already-emitted dislocation loops can be estimated by $\tau_b \approx n\pi\mu b/d$ (Allain and Bouaziz, 2010; Hirth and Lothe, 1982). This implies that the back stress linearly depends on the trapped dislocation density, because it also linearly depends on the number of dislocation loops, n. Besides, the linear relation between flow stress and geometrically necessary dislocation (GND) was also suggested by Guruprasad and Benzerga (2008a). Accordingly, it is reasonable to assume that the following linear relation exists in a dimensionless form:

$$\tau_b/\mu \approx \alpha' \cdot \rho_{\text{trapped}} \cdot b \cdot d \tag{9.1.3}$$

Here, coefficient α' is approximately 0.7, according to the least-square fitting of simulation results in Fig. 9.4A.

To verify that k_s and α' are constants independent of the sample size, similar simulations were carried out in coated micropillars with other sizes. They also initially contained one individual SAS with the same slip plane, Burgers vector, effective length, and initial orientation. In coated micropillars with a diameter of 200, 600, and 800 nm, critical resolved shear stress to operate the SAS was 399, 142, and 110 MPa, respectively. The corresponding k_s was calculated as shown in Fig. 9.4B. By least-square fitting of τ_b and ρ_{trapped} data, the values of α' were also obtained. Fig. 9.4B shows that both k_s and α' are almost independent of the sample size.

Then, by combining Eqs. (9.1.2) and (9.1.3), the activation stress of the SAS in coated samples can be estimated as

$$\tau = \tau_0 + \alpha\mu b\sqrt{\rho_{\text{mobile}}} + \frac{k_s\mu}{\lambda/b} + \alpha'\rho_{\text{trapped}}bd\mu \tag{9.1.4}$$

9.1.4 Evolution of Mobile and Trapped Dislocation

Obviously, a prerequisite for applying Eq. (9.1.4) is to obtain the evolution law of mobile and trapped dislocation density. In the following discussion, the simulation results for micropillars with different complex dislocation configurations are further analyzed to study the evolution of ρ_{mobile} and ρ_{trapped}. The distinction method between ρ_{mobile} and ρ_{trapped} in DCM simulations is described in detail in Cui et al. (2015a). The evolution of dislocation density in a pillar with a diameter of 400 nm is illustrated in Fig. 9.5. For the coated pillar,

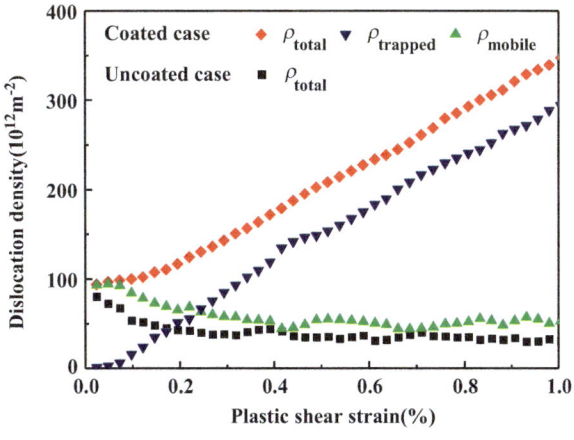

Figure 9.5
Evolution of total dislocation density ρ_{total}, mobile dislocation density ρ_{mobile}, and trapped dislocation density ρ_{trapped} in an uncoated and coated micropillar with a diameter of 400 nm. *Reprinted from Cui, Y.N., Liu, Z.L., Zhuang, Z., 2015a. Theoretical and numerical investigations on confined plasticity in micropillars. Journal of the Mechanics and Physics of Solids 76, 127—143, Copyright 2015, with permission from Elsevier.*

though, the total dislocation density increases during deformation, and the mobile dislocation density first decreases and then tends to remain stable, similar to the uncoated case. Generally, mobile dislocations exist in two forms. One is surface dislocation with both ends terminating at the surface/interface. The other is dislocation sources with one or two anchor points inside the pillar resulting from the formation of dislocation junctions or jogs. The dislocation microstructures observed in the simulations reveal that most surface dislocations quickly glide to the interface region and pile up during the initial microplastic stage. The main form of the mobile dislocations is the SAS. Thus, to a certain extent, the initial gradual decrease in mobile dislocation density reflects the gradual destruction of the SAS.

On the other hand, the trapped dislocation density has a linear dependence on the plastic strain, as indicated in Fig. 9.5. Because the plastic deformation is mainly induced by the operation of the SAS, it is natural to think of deriving the evolution of the trapped dislocation density by an SAS model. When the SAS rotates one circle, slipped area A_{slip} is $\pi d(d/\cos \beta)/4$ (Fig. 8.9). Assuming the slip plane is swept by the SAS n times, the produced plastic shear strain, γ_{p}, can be calculated as

$$\gamma_{\text{p}} = \frac{nbA_{\text{slip}}}{V} = \frac{nb\pi d^2/(4cos\ \beta)}{\pi d^2 h/4} = \frac{nb}{hcos\ \beta} \qquad (9.1.5)$$

where V is the sample volume and h is the height of the sample. Meanwhile, the accumulated dislocation length corresponding to each deposited dislocation loop by the

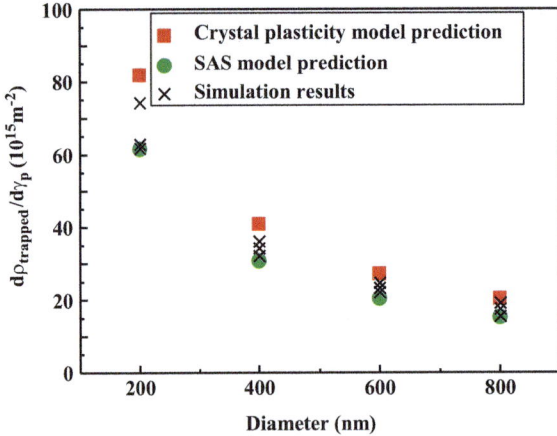

Figure 9.6

Accumulation rate of trapped dislocation for different diameters. *SAS*, single-arm dislocation source. *Reprinted from Cui, Y.N., Liu, Z.L., Zhuang, Z., 2015a. Theoretical and numerical investigations on confined plasticity in micropillars. Journal of the Mechanics and Physics of Solids 76, 127—143, Copyright 2015, with permission from Elsevier.*

SAS operation is $\pi(d + d/\cos \beta)/2$. Therefore, trapped dislocation density ρ_{trapped} after n cycles of rotation of the SAS is

$$\rho_{\text{trapped}} = \frac{n\pi(d + d/\cos \beta)/2}{\pi d^2 h/4} \tag{9.1.6}$$

By substituting Eq. (9.1.5) into Eq. (9.1.6), we have

$$\rho_{\text{trapped}} = \frac{2(1 + \cos \beta)}{bd}\gamma_{\text{p}} \tag{9.1.7}$$

Eq. (9.1.7) shows the linear relationship between ρ_{trapped} and γ_{p}. On the other hand, the linear coefficient between ρ_{trapped} and γ_{p} can be calculated by fitting ρ_{trapped} and γ_{p} data in Fig. 9.5. The initial small plastic strain stage ($\gamma_{\text{p}} = 0-0.02\%$) is not taken into account in linear fitting because plastic deformation at this stage is accommodated by inner dislocation glide and dislocation configuration rebuilding without significant interface dislocation deposition. The simulation results for different pillar diameters are fitted and plotted together with SAS model predictions in Fig. 9.6. The coefficients are well-predicted by Eq. (9.1.7). The accumulation rate of trapped dislocation $d\rho_{\text{trapped}}/d\gamma_{\text{p}}$ decreases as the pillar diameter increases.

9.2 Implications for Crystal Plasticity Model

Within the realm of continuum mechanics, the gradient crystal plasticity theory is often applied to investigate confined plasticity at the microscale. In the strain gradient theory,

the expression of GND and its influence on strain hardening behavior have been extensively studied (Bayley et al., 2006). One typical work is the higher-order crystal plasticity theory developed by Gurtin (2002), in which back stress associated with trapped dislocations is incorporated. The back stress is derived from a defect energy term, which is the quadratic function of GND. This model can capture the size effect successfully in several constraint plastic flow problems (Bittencourt et al., 2003; Gurtin et al., 2007; Nicola et al., 2005a). However, the physical meaning of defect energy and the length parameters in the theory are not clear. For the coated pillar during plastic deformation, the coating introduces a deformation gradient because the coating and micropillar have different mechanical properties. Therefore, the trapped dislocation can be considered a GND, which is automatically maintained to ensure deformation compatibility (Ohno and Okumura, 2007). In addition, it is always expected that the GND are concentrated near the interface (Cleveringa et al., 1997; Evers et al., 2004a). Accordingly, if the evolution of GND and back stress can be obtained by a lower-scale discrete dislocation simulation, the results can be used directly to develop the strain gradient plasticity theory.

DCM simulation results show that the back stress and trapped dislocation density have a linear relation (Fig. 9.4), and so do the trapped dislocation density and plastic shear strain (Eq. 9.1.7). In this section, these results are further verified by comparing them with solutions obtained from higher-order crystal plasticity theory. One of the most difficult challenges in higher-order crystal plasticity theory is how to formulate flow stress through the dislocation density at a microscale. DCM simulation results are preliminarily explored in the hope of shedding some light on this problem.

In the following discussion, the compression of the coated micropillar is analyzed by using the higher-order crystal plasticity model developed by Gurtin (2002). Here, to facilitate comparison with the simulation results given in Section 9.1, it is supposed that only one slip system denoted by "k" is activated. A local coordinate system, OXYZ, is built in the elliptic slip plane, as indicated in Fig. 9.7A, where the origin of the coordinate is the center of the ellipse. The base vectors of axes X, Y, and Z are $\mathbf{e}_1$, $\mathbf{e}_2$, and $\mathbf{e}_3$, respectively. Here, $\mathbf{e}_1$ and $\mathbf{e}_2$ are along the minor and major axis of the ellipse, respectively, so that $\mathbf{e}_3$ is the out-of-plane direction, which is equal to the unit vector indicating the slip plane normal $\mathbf{m}^{(k)}$. Assume that in the kth slip system the angle between $\mathbf{e}_1$ and slip direction is ϕ. Then, the slip direction $\mathbf{s}^{(k)}$ can be expressed as

$$\mathbf{s}^{(k)} = \cos\phi\mathbf{e}_1 + \sin\phi\mathbf{e}_2 \tag{9.2.1}$$

The corresponding tangent line direction of edge dislocation $\mathbf{I}^{(k)}$ can be expressed as

$$\mathbf{I}^{(k)} = \mathbf{m}^{(k)} \times \mathbf{s}^{(k)} \tag{9.2.2}$$

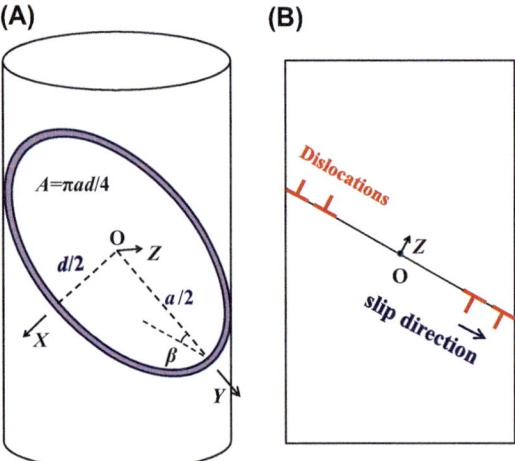

Figure 9.7

Schematic description of dislocations gliding in the slip plane in a coated micropillar (A) in full three-dimensional view, where the *blue elliptical ring* is the coating layer; and (B) in cross-sectional view. *Reprinted from Cui, Y.N., Liu, Z.L., Zhuang, Z., 2015a. Theoretical and numerical investigations on confined plasticity in micropillars. Journal of the Mechanics and Physics of Solids 76, 127—143, Copyright 2015, with permission from Elsevier.*

Confining attention to the small displacement gradient, the total strain rate $\dot{\boldsymbol{\varepsilon}}$ can be decomposed as an elastic part $\dot{\boldsymbol{\varepsilon}}_e$ and a plastic part $\dot{\boldsymbol{\varepsilon}}_p$:

$$\dot{\boldsymbol{\varepsilon}} = \dot{\boldsymbol{\varepsilon}}_e + \dot{\boldsymbol{\varepsilon}}_p, \quad \dot{\boldsymbol{\varepsilon}}_p = \sum_{k=1}^{N_S} \dot{\gamma}_p^{(k)} \mathbf{P}^{(k)}, \quad \mathbf{P}^{(k)} = \frac{1}{2}\left(\mathbf{s}^{(k)} \otimes \mathbf{m}^{(k)} + \mathbf{m}^{(k)} \otimes \mathbf{s}^{(k)}\right) \tag{9.2.3}$$

where $\dot{\gamma}_p^{(k)}$ is the slip rate on the kth slip system, N_S denotes the number of activated slip system ($N_S = 1$ here), and $\mathbf{P}^{(k)}$ is the Schmid tensor. Stress rate tensor $\dot{\boldsymbol{\sigma}}$ is related to the elastic strain by

$$\dot{\boldsymbol{\sigma}} = \mathbf{C}^e : \dot{\boldsymbol{\varepsilon}}_e \tag{9.2.4}$$

The microforce balance equation is expressed as

$$\dot{\tau}^{(k)} = \dot{\pi}^{(k)} - \nabla \cdot \dot{\boldsymbol{\xi}}^{(k)} \tag{9.2.5}$$

where $\tau^{(k)}$ is the resolved shear stress in the kth slip system. $\pi^{(k)}$ is the slip resistance, which corresponds to the first three terms on the right side of Eq. (9.1.2). $-\nabla \cdot \dot{\boldsymbol{\xi}}^{(k)}$ corresponds to the back-stress term in Eq. (9.1.2). Higher-order stress $\boldsymbol{\xi}^{(k)}$ can be expressed by the partial derivative of defect energy $\psi^{(k)}$ with respect to the slip gradient (Gurtin, 2002):

$$\boldsymbol{\xi}^{(k)} = \frac{\partial \psi^{(k)}}{\partial \nabla \gamma_p^{(k)}}, \quad \psi^{(k)} = \frac{1}{2}S_0 L^2 \sum_k \left[\left(\nabla \gamma_p^{(k)} \cdot \mathbf{s}^{(k)}\right)^2 + \left(\nabla \gamma_p^{(k)} \cdot \mathbf{I}^{(k)}\right)^2\right] \tag{9.2.6}$$

where S_0 is a material constant with the dimension of stress. L is a characteristic length, which is determined later. Because we restrict attention to the uniaxial compression test for a micropillar, only the axial components of stress and strain tensor, such as σ, ε, and ε_p, are discussed next. According to Eqs. (9.2.3) and (9.2.4), axial stress rate $\dot{\sigma}$ can be expressed as

$$\dot{\sigma} = E(\dot{\varepsilon} - \dot{\varepsilon}_p) = E\left(\dot{\varepsilon} - \sum_{k=1}^{N_s} \dot{\gamma}_p^{(k)} M^{(k)}\right) \tag{9.2.7}$$

By taking only the kth slip system into account, the axial stress rate can be expressed as $\dot{\sigma} = E\left(\dot{\varepsilon} - \dot{\gamma}_p^{(k)} M^{(k)}\right)$. The resolved shear stress rate is $\dot{\tau}^{(k)} = M^{(k)}\dot{\sigma}$. Combined with Eqs. (9.2.5) and (9.2.6), the general solution for the plastic shear strain rate can be obtained:

$$\dot{\gamma}_p^{(k)} = \dot{\varepsilon}\Big/ M^{(k)} + C_1 e^{\sqrt{Q}Y} + C_2 e^{-\sqrt{Q}Y} + C_3 e^{\sqrt{Q}X} + C_4 e^{-\sqrt{Q}X},$$
$$Q = E\left(M^{(k)}\right)^2 \Big/ \left(S_0 L^2\right) \tag{9.2.8}$$

where C_1, C_2, C_3, and C_4 are constants to fulfill the boundary conditions. Here, the microclamped conditions at the coated interface are given by

$$\gamma_p^{(k)}(X, Y) = 0, \ \ if \ \ \frac{X^2}{(d/2)^2} + \frac{Y^2}{(a/2)^2} = 1 \tag{9.2.9}$$

To obtain the analytical solution of Eq. (9.2.8), the first term is kept unchanged and Taylor series are used around original point O for the other terms, which is given by

$$\dot{\gamma}_p^{(k)} = \dot{\varepsilon}\Big/ M^{(k)} + C_1\left(1 + \sqrt{Q}\,Y + \left(\sqrt{Q}\,Y\right)^2\Big/2\right) + C_2\left(1 - \sqrt{Q}\,Y + \left(\sqrt{Q}\,Y\right)^2\Big/2\right) +$$
$$C_3\left(1 + \sqrt{Q}\,X + \left(\sqrt{Q}\,X\right)^2\Big/2\right) + C_4\left(1 - \sqrt{Q}\,X + \left(\sqrt{Q}X^2/2\right)\right) \tag{9.2.10}$$

According to the boundary conditions, the solution is obtained by

$$\dot{\gamma}_p^{(k)} = P\left(4d^2 Y^2 + 4a^2 X^2 - a^2 d^2\right), \ \ P = \frac{-\dot{\varepsilon} E M^{(k)}}{8 S_0 L^2 (a^2 + d^2) + E\left(M^{(k)}\right)^2 a^2 d^2} \tag{9.2.11}$$

Because the mixed dislocations can be decomposed into edge and screw components, the corresponding GND density is calculated (Ohno and Okumura, 2007):

$$\dot{\rho}_{GND}^{(k)} = \sqrt{\left(\dot{\rho}_{GND,edge}^{(k)}\right)^2 + \left(\dot{\rho}_{GND,screw}^{(k)}\right)^2}$$
$$= \frac{1}{b}\sqrt{\left(\nabla\dot{\gamma}_p^{(k)} \cdot s^{(k)}\right)^2 + \left(\nabla\dot{\gamma}_p^{(k)} \cdot \mathbf{I}^{(k)}\right)^2} = \frac{8|P|}{b}\sqrt{a^4 X^2 + d^4 Y^2} \tag{9.2.12}$$

Here, the dislocation cross-slip and climb are not taken into account. Assuming dislocation glide only, the pillar is considered to be composed of many elliptic cylinders, each of which corresponds to a slip plane. Thus, the volume-averaged GND is thought to be equivalent to the average GND density through the slip plane:

$$\left\langle \dot{\rho}_{\text{GND}}^{(k)} \right\rangle = \frac{1}{A} \int_A \left| \dot{\rho}_{\text{GND}}^{(k)} \right| dA = \frac{4|P|a^2 d}{3\pi b} C_5, \quad C_5 = \int_0^{2\pi} \sqrt{\cos^2 \theta + \cos^2 \beta \sin^2 \theta} \, d\theta \quad (9.2.13)$$

where C_5 is a dimensionless constant, which depends only on the loading orientation. Correspondingly, the average plastic shear strain caused by dislocation slip is

$$\left\langle \dot{\gamma}_{\text{p}}^{(k)} \right\rangle = \frac{1}{A} \int_A \left| \dot{\gamma}_{\text{p}}^{(k)} \right| dA = \frac{|P|a^2 d^2}{2} \quad (9.2.14)$$

According to Eqs. (9.2.13) and (9.2.14), the relation between the average GND density and average plastic shear strain can be deduced as

$$\left\langle \dot{\rho}_{\text{GND}}^{(k)} \right\rangle = \frac{8C_5}{3\pi bd} \left\langle \dot{\gamma}_{\text{p}}^{(k)} \right\rangle \quad (9.2.15)$$

Eq. (9.2.15) reflects the linear dependence of $\left\langle \dot{\rho}_{\text{GND}}^{(k)} \right\rangle$ on $\left\langle \dot{\gamma}_{\text{p}}^{(k)} \right\rangle$, and $\left\langle \dot{\rho}_{\text{GND}}^{(k)} \right\rangle \big/ \left\langle \dot{\gamma}_{\text{p}}^{(k)} \right\rangle = 4.2/(bd)$ for the [001] loading orientation. As discussed previously, the trapped dislocation can be thought as GND. According to the SAS model prediction in Section 9.1.4, Eq. (9.1.7) also gives a linear relation between the trapped dislocation density and volume-averaged plastic shear strain. For the slip plane considered here, $d\rho_{\text{trapped}}/d\gamma_{\text{p}}$ is $3.2/(bd)$. The relations predicted by both Eq. (9.1.7) and Eq. (9.2.15) agree well with the simulation results shown in Fig. 9.6.

Furthermore, the average back stress can be obtained as

$$\langle \dot{\tau}_b \rangle = \frac{1}{A} \int_A \left| -\nabla \cdot \dot{\xi}^{(k)} \right| dA = 8S_0 L^2 |P| \left(a^2 + d^2 \right) \quad (9.2.16)$$

Combining Eqs. (9.2.16) and (9.2.13), the following relation can be derived:

$$\langle \dot{\tau}_b \rangle / \mu = \left(\frac{6\pi \left(1 + \cos^2 \beta \right) S_0 L^2}{C_5 \mu d^2} \right) \left\langle \dot{\rho}_{\text{GND}}^{(k)} \right\rangle bd \quad (9.2.17)$$

Comparing Eq. (9.2.17) with Eq. (9.1.3), the characteristic parameter $S_0 L^2$ can be determined to be $0.14\mu d^2$. According to work by Liu et al. (2011), $S_0 = \mu/8(1-\nu)$, which is an elastic constant. Then, L is estimated as $0.88d$, as denoted in Fig. 9.3C. The characteristic length L just corresponds to the size of the typical dislocation configuration that is influenced by the extrinsic characteristic length; here, d is the pillar diameter. This provides a reference for determining the length parameter in a higher-order back stress model. The material length parameter reflects the region of influence of short-range dislocation interactions.

In gradient-based plasticity formulations, identification of a certain dislocation as being statistically stored or geometrically necessary remains unclear (Evers et al., 2004a). Guruprasad and Benzerga (2008a,b) reported some inspiring 2-5-dimensional (2.5D)-*discrete dislocation* dynamics (DDD) work to analyze the local GND density in a free micropillar using the net Burgers vector based on Nye's tensor. However, in 3D-DDD, a strict distinction of GND is difficult. Consistency between the crystal plasticity theory and simulation results here further suggests a correspondence between GND and trapped dislocation and a linear relation between back stress and trapped dislocation density, which may shed some light on intuitively understanding the GND.

Many studies also consider the contribution of GND to slip resistance by the Taylor interaction term (Fleck et al., 1994; Han et al., 2005a). However, Mayeur and McDowell (2013) found that adding the GND density to the Taylor relation would overestimate flow stress. Because the back-stress term already considers the contribution of GND, including the GND density in the Taylor relation will double-count its contribution. The simulation results presented here show that the trapped dislocations (GND) contribute most to the flow stress and the trapped dislocation density increases linearly with plastic strain in the coated pillars. Therefore, if the contribution of trapped dislocation is introduced by the Taylor hardening law, a square root dependence of flow stress on plastic strain is obtained. However, linear strain hardening is clearly observed in both simulations and experiments (Greer, 2007) in coated pillars at the microscale. This means that the trapped dislocations contribute to an increase in flow stress in terms of back stress hardening instead of Taylor hardening. That is why only the back-stress term in Eqs. (9.1.3) and (9.1.4) considers the trapped dislocation density, which is also consistent with previous work (Guruprasad and Benzerga, 2008a). The slip system resistance is mainly influenced by mobile dislocations and the internal dislocation source operation. In particular, at such a small scale, the source operation has a more crucial role compared with the Taylor interactions in relation to the mobile dislocation density.

9.3 Theoretical Models for Coated Micropillars

Some theoretical studies were also performed to develop a continuum model describing confined plasticity in coated pillars. Lee et al. (2013) used a simple numerical model to illustrate the coating effect on the source operation. He introduced an additional stress term, $\Delta\sigma_{coating}$, to the SAS model to consider dislocation pinning and the pile-up effect. However, the value of $\Delta\sigma_{coating}$ was directly calculated from the experimental sample strength and its evolution was not provided. Thus, it was difficult to use it to predict the mechanical response for other samples. On the other hand, some researchers try to correlate high flow stress with the total dislocation density in coated pillars based on the Taylor hardening theory (Jennings et al., 2012; Ng and Ngan, 2009a). For a large coated pillar, a good correlation was obtained (Gu and Ngan, 2012). However, for a small coated

pillar ($\sim$ <1 μm), the Taylor relation failed. Generally, the Taylor relation worked well when forest dislocation hardening prevailed. Nevertheless, studies (Gu and Ngan, 2012) indicated that the coating could not store the forest dislocations effectively in a small sample. Internal mobile dislocations were scarce and most dislocations were trapped at the interface. In the following discussion, based on the operation stress equation of SAS and the linear back-stress model obtained from DCM simulations, a theoretical model is presented that can conveniently predict the stress-strain curve of a coated sample.

9.3.1 Dislocation Density Evolution Model

From Fig. 9.1, it is reasonable to assume that for the considered sample size, all of the plastic strain is composed of discrete strain bursts and the stage between each strain burst is a pure elastic response. Accordingly, as long as the stress value at which strain burst occurs and the magnitude of strain burst are known, the stress-strain curve can be predicted. According to the simulation results in Section 9.1, the plasticity of the coated pillar is still controlled by the operation of the SAS, so each detectable strain burst is assumed to be caused by the operation of the SAS, similar to that in the uncoated pillar (Cui et al., 2014; Rao et al., 2008). Therefore, the key issue is to calculate the operation stress of the SAS and their evolution.

First, given initial dislocation density ρ_0 and the pillar's geometrical size, the number of SAS can be estimated by n = integer[$\rho\pi dh/4$], according to Eq. (8.4.9). Based on the statistical model described in Section 8.4.2, the statistically averaged effective length of the SAS can be estimated but only one slip system can be considered. Here, the dislocation activities for all the available slip systems are considered by a stochastic method inspired by Ng and Ngan's work (2008a,b). We assumed that each SAS is randomly assigned to one of 12 slip systems in a *face-centered cubic* (FCC) crystal. The pinning point for each SAS distributes randomly in the corresponding elliptic slip plane, with the same probability of locating at any position in the slip plane. The shortest distance of each pinning point from the elliptic perimeter is calculated as the effective length of SAS $\lambda^{(j)}$. Here, superscript "(j)" means the variables corresponding to the jth SAS.

In the micropillar under consideration, the number of SAS is limited. For the early deformation stage ($\varepsilon < 0.2\%$), let us assume that there is no interaction between SAS and no coupling between slip systems. The SAS can be activated one by one independently of each other. Thus, the stress plateau for each strain burst corresponds to the operation stress of one SAS, which can be deduced from Eq. (9.1.4) as

$$\sigma^{(j)} = \tau^{(j)} \Big/ M^{(j)} = \left[\tau_0 + \alpha\mu b\sqrt{\rho_{\text{mobile}}} + \frac{k_s\mu}{\lambda^{(j)}\big/b} + 2\alpha'(1 + \cos\beta)\gamma_{\text{p}}^{(j)}\mu\right]\Big/ M^{(j)} \quad (9.3.1)$$

where Schmid factor $M^{(j)}$ for the jth SAS can be calculated according to its slip system information and the loading orientation.

Obviously, to obtain the evolution of $\sigma^{(j)}$, the evolutions of ρ_{mobile}, $\lambda^{(j)}$, and $\gamma_p^{(j)}$ must first be calculated. For simplicity, the change of ρ_{mobile} is ignored because its influence on the source operation stress is weak compared with the other two terms in Eq. (9.3.1). $\gamma_p^{(j)}$ can be calculated according to Eq. (9.1.5) by taking n as the operation times of the jth SAS. The most difficult one is to estimate the evolution of $\lambda^{(j)}$, which reflects the magnitude of strain burst and corresponds to the lifetime of the SAS (i.e., the number of times that SAS can sweep the slip plane before it ruptures). In an uncoated micropillar, weak dislocation interactions and the small amount of SAS make it possible to predict the lifetime of the SAS statistically according to the instantaneous dislocation density and sample size (Cui et al., 2014). Nevertheless, strong interactions between the SAS and trapped dislocations in the coated case make it difficult to predict the lifetime of the SAS. For simplicity, two extreme cases are considered. From the simulation results shown in Fig. 9.3A, after the SAS sweeps the glide plane one time, back stress leads to the complete shutdown of this source. Therefore, one extreme case assumes that the SAS ruptures once it operates and sweeps the slip plane one time. Namely, $\lambda^{(j)}$ become 0 in the subsequent steps after one operation. This can be considered the shortest lifetime of the SAS and the maximum effect of SAS exhaustion hardening. Therefore, the predicted mechanical response represents the upper bound of stress. The other assumes that all SAS have an unlimited lifetime and $\lambda^{(j)}$ is unchanged during the whole deformation stage. This assumption considers back-stress hardening but ignores SAS exhaustion hardening. Thus, the predicted results represent a lower bound of stress. The physical process behind the extreme case that ignores the failure of the SAS is similar to the DDD simulation in which a population of FR sources is initially put in the crystal using indestructible pinning points and a presumed source length distribution (El-Awady et al., 2011; Zhou et al., 2010b).

More details about the calculation procedure of this theoretical model are schematically described in Fig. 9.8. Under uniaxial compression, the applied stress elastically rises until the ath SAS with the minimum operation stress $\sigma_{min} = min(\sigma^{(1)}, \sigma^{(2)}, ..., \sigma^{(N)})$ activates. The operation stress of the weakest ath SAS can be considered the initial yield stress, which is also the stress value for the first strain burst. Accordingly, it begins from $\sigma = \sigma^{(a)}$, $\varepsilon = \sigma/E$ and ends with $\sigma = \sigma^{(a)}$, $\varepsilon = \sigma/E + \gamma_p M^{(a)}$, where E is the Young's modulus and γ_p is the plastic shear strain produced by the activated SAS, calculated according to Eq. (9.1.5) by setting n as 1. Afterward, the operation stress for the activated SAS needs to be updated. Then, the applied stress elastically increases again until another SAS with the minimum operation stress is activated. Correspondingly, the second strain burst occurs. This activated SAS also sweeps the glide plane one time and produces γ_p. This procedure is repeated until the expected loading is reached, as schematically described in Fig. 9.8. To plot the stress-strain curve, the stress and strain values are

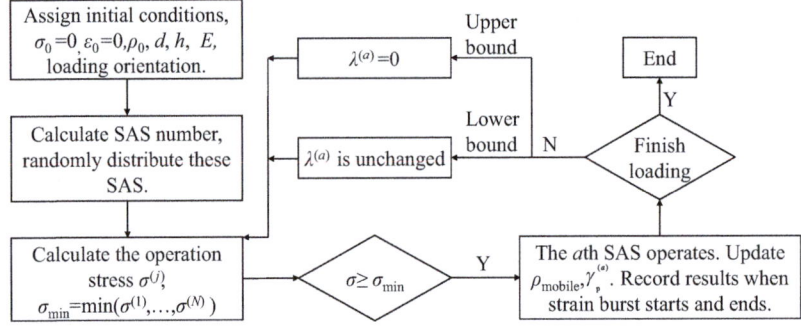

Figure 9.8

Calculation flowchart of theoretical model. *SAS*, single-arm dislocation source. *Reprinted from Cui, Y.N., Liu, Z.L., Zhuang, Z., 2015a. Theoretical and numerical investigations on confined plasticity in micropillars. Journal of the Mechanics and Physics of Solids 76, 127–143, Copyright 2015, with permission from Elsevier.*

recorded at the start and end of the ith strain burst, respectively. The ith strain burst starts when $\sigma_{2i} = \sigma_{min}$ and $\varepsilon_{2i} = \varepsilon_{2i-1} + (\sigma_{2i} - \sigma_{2i-1})/E$, and ends when $\sigma_{2i+1} = \sigma_{min}$ and $\varepsilon_{2i+1} = \varepsilon_{2i} + \gamma_p M^{(a)}$, respectively.

9.3.2 Prediction of Stress-Strain Curve

Some stress-strain responses for these two extreme cases obtained by the theoretical model prediction are given in Fig. 9.9A for the coated micropillar under uniaxial compression.

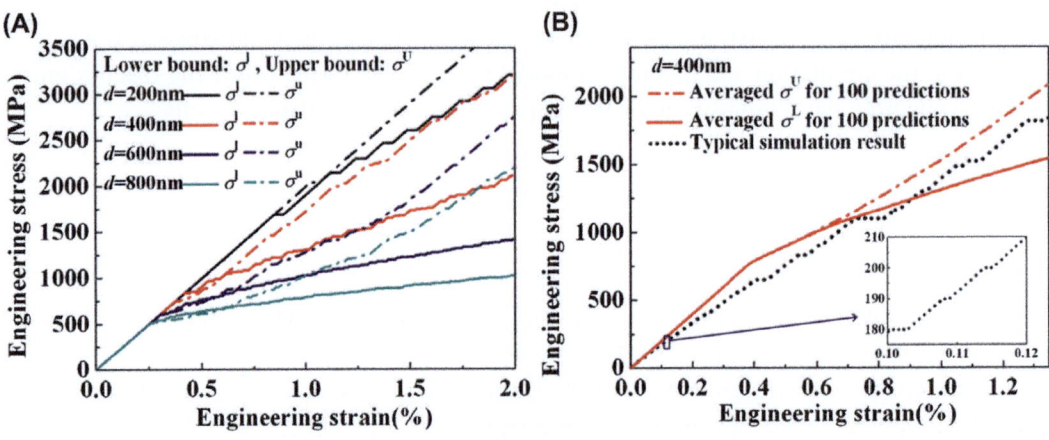

Figure 9.9

(A) Predicted upper-bound and lower-bound stress-strain curves for coated micropillar under uniaxial compression; (B) Comparison between theoretical model predictions and simulation results. The *inset* is a higher-magnification image of the microplasticity region during the early deformation stage. *Reprinted from Cui, Y.N., Liu, Z.L., Zhuang, Z., 2015a. Theoretical and numerical investigations on confined plasticity in micropillars. Journal of the Mechanics and Physics of Solids 76, 127–143, Copyright 2015, with permission from Elsevier.*

Here, the loading axis is taken as [001] and the initial dislocation density is 10^{14} m^{-2}. The results for two extreme cases share some similarities: (1) The initial yield stresses are almost the same for the same diameter, although some slight differences arise from the stochastic distribution of sources. (2) The smaller micropillar exhibits higher flow stress, which agrees with the simulation results shown in Fig. 9.1A. Interestingly, although the stress-strain curves are composed of collective discrete strain bursts, the obtained stress-strain curves are smooth for large micropillars. This is consistent with a previous investigation (Csikor et al., 2007). Moreover, for a micropillar with a diameter of 200 nm, the hardening moduli for both extreme cases are close to the Young's modulus, in accord with simulation results shown in Fig. 9.1A and previous experimental data in Fig. 9.1B (Jennings et al., 2012).

The theoretical predictions and simulation results were directly compared for the pillar with a diameter of 400 nm, as shown in Fig. 9.9B. Here, the initial dislocation density for simulation results was close to 10^{14} m^{-2}. The results of the theoretical model were the average over 100 separate realizations to give a sense to the statistic representation. The strain hardening stage of the simulation results are well-captured by the theoretical model, whereas stress from the theoretical model predictions at the early stage of loading is higher than that in the simulation results. This is because the theoretical model assumes that the initial deformation is purely elastic, whereas in the simulations, microplasticity usually occurs owing to dislocation motion and the breakup of weak junctions (Motz et al., 2009; Ni et al., 2017), as shown in the inset of Fig. 9.9B. Moreover, the existence of microplasticity and the stochastic strain burst cause difficulty in defining the initial yield stress (Ispánovity et al., 2010) for the simulation results. Hence, the yield stress is not compared in the simulation and theoretical model predictions. In the following discussion, the predicted strain hardening behavior is compared with the simulation results.

The SHR is calculated by

$$\text{SHR} = d\sigma_T/d\varepsilon_T (\sigma > \sigma_c), \sigma_T = \sigma(1 + \varepsilon), \varepsilon_T = \ln(1 + \varepsilon) \qquad (9.3.2)$$

where σ_T and ε_T are the averaged true stress and true strain, respectively, which are obtained according to engineering stress and strain. The averaged SHR is derived by least-square fitting the σ_T-ε_T curve. For the theoretical model, cutoff stress σ_c is the critical stress that activates the weakest SAS. For the simulation results, σ_c is taken to be the stress when the first detectable strain burst occurs (the burst extent $\Delta\varepsilon > 0.02\%$). The analytical prediction and simulation results for SHR are plotted in Fig. 9.10, in which the initial dislocation densities ρ_0 used in the simulations are indicated. The simulation results fall into the region bounded by the upper and lower bounds.

The SHR values exhibit some kind of size dependence. The smaller the sample size is, the higher the SHR is. In addition, the simulation results indicate that the micropillar with a

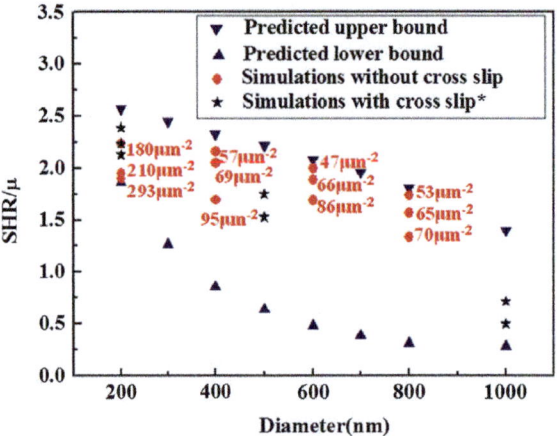

Figure 9.10

Comparison of strain hardening rate (SHR) between theoretical model predictions and simulation results. The initial dislocation density ρ_0 for the simulated micropillar is numerically labeled. The ρ_0 for analytical results is 100 μm^{-2}. *Reprinted from Cui, Y.N., Liu, Z.L., Zhuang, Z., 2015a. Theoretical and numerical investigations on confined plasticity in micropillars. Journal of the Mechanics and Physics of Solids 76, 127—143, Copyright 2015, with permission from Elsevier. *Results marked are from previous studies Lee, S.W., Jennings, A.T., Greer, J.R., 2013. Emergence of enhanced strengths and Bauschinger effect in conformally passivated copper nanopillars as revealed by dislocation dynamics. Acta Materialia 61, 1872—1885; Zhou, C., Biner, S., LeSar, R., 2010b. Simulations of the effect of surface coatings on plasticity at small scales. Scripta Materialia 63, 1096—1099.*

lower initial dislocation density exhibits a higher SHR because it is easier for the SAS to exhaust, which is consistent with previous studies (Benzerga, 2009; Gu and Ngan, 2012). On the other hand, the activation of cross-slip is thought to have an important role in promoting dislocation multiplication and decreasing the SHR (Zhou et al., 2010b). However, the effect of cross-slip is not considered in either theoretical analysis or numerical simulations here. To discuss this, the SHRs are calculated according to simulation results in the available references that consider cross-slip, e.g., uniaxial compression tests on a Cu micropillar coated with TiO_2/Al_2O_3 with a diameter of 200 nm and an initial dislocation density of about 100 μm^{-2} (Lee et al., 2013), and coated Ni with a diameter of 500/1000 nm and an initial dislocation density of about 10—20 μm^{-2} (Zhou et al., 2010b), respectively. As evidenced by the data in Fig. 9.10, the incorporation of cross-slip does not dramatically decrease the SHR for such a small coated micropillar.

9.4 Brief Discussion on Coating Failure Mechanism

According to experimental observations, a coated micropillar is usually damaged by failure of the coating followed by delamination at the interface (Jennings et al., 2012). Failure of the coating may occur in two different ways. One is as a result of local high stress.

The other is caused by penetration of the dislocations. These two mechanisms are preliminarily discussed next (Cui, 2016).

9.4.1 High Hoop Stress of Coated Layer

Previous postdeformation scanning electron microscopy observations show that the coating usually fails as a result of axial cracks in the coating (Jennings et al., 2012). Thus, high hoop stress may be the reason for coating failure. As shown in Fig. 9.11A, lots of dislocations are trapped along the circumference of the pillar. The corresponding hoop stress distribution is obtained by DCM simulation, as shown in Fig. 9.11B and C. Even when the trapped dislocation density is not high, the hoop stress value of the coating is significantly higher than that of the micropillar. Therefore, it is easy for the coating to crack as a result of high local hoop stress. Then, a large deformation is triggered where the crack occurs in the coating layer (Ng and Ngan, 2009a).

9.4.2 Transmission Effect of Dislocations Across Coating

On the other hand, a ceramic coating is opaque to dislocations and acts as a dislocation sink. Because of its limited ability to absorb matrix dislocations, the transmission of dislocation can lead to brittle-type failure of the coating. Whether a dislocation segment

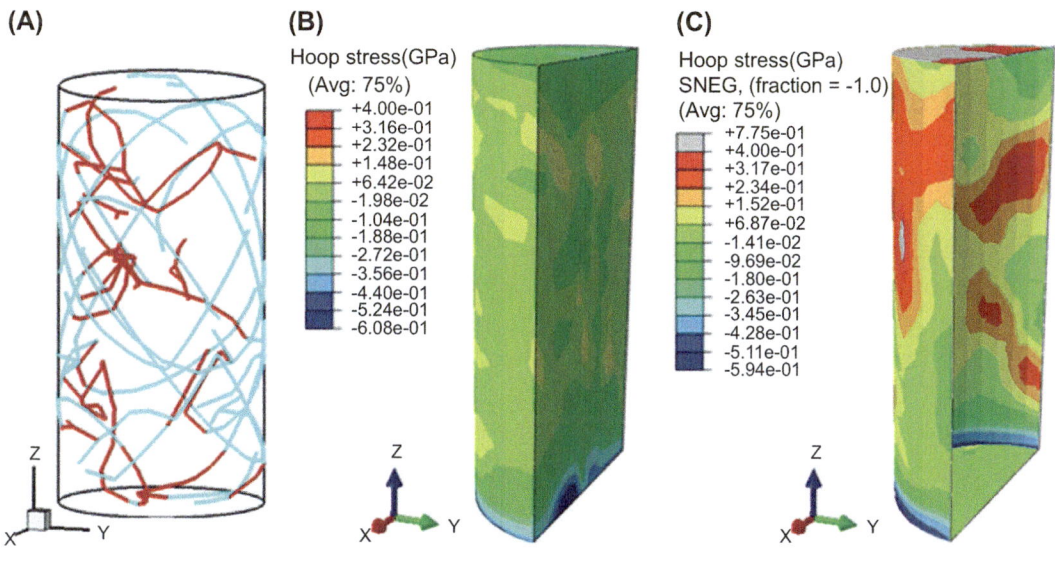

Figure 9.11

A micropillar with a diameter of 400 nm in which the engineering strain is 0.6%: (A) The corresponding dislocation configuration, in which the blue dislocation lines are trapped and the others are the mobile; (B) a longitudinal cross-sectional view of hoop stress in the micropillar; and (C) a longitudinal cross-sectional view of hoop stress in the coating (Cui, 2016). *SNEG* is used to describe the orientation of surfaces in ABAQUS software.

will pile up or transmit across the interface mainly depends on the stress acting upon it. As discussed earlier, when the dislocation line approaches the interface, it will be strongly repelled by the image forces. The dislocation transmission can occur only when the other stress acting on the dislocation overcomes the high image force. In DCM methods, it is difficult to calculate the stress field of dislocations close to the interface with accuracy because very fine meshes are needed to capture the singularity. To estimate the critical image stress value at which the transmission happens, the analytical solution for the image force of the bimaterial is used. The available analytical solution has been found only for some specific cases such as an edge-screw dislocation near a surface layer or a dislocation loop in an anisotropic biomaterial (Chu et al., 2012; Head, 1953a,b; Weeks et al., 1968). Generally, the analytical image force diverges as the distance between the dislocation and the interface nears 0. However, this divergence is an artifact of treating the dislocation as an infinitely thin filament. When the distance to the interface is equal to the core radius, $r = 2b$, the repulsion reaches the maximum possible value (Koehler, 1970). For the cases considered here, the resolved shear stress for image force τ_c is 0.7—0.8 GPa. Thus, τ_c is taken to be 0.8 GPa in our simulation. To study the effect of the τ_c value, τ_c is also calculated when the cutoff radius is taken to be the magnitude of the Burgers vector. Here, $\tau_c = 1.4$—1.7 GPa and τ_c is set as 1.5 GPa during the simulation. Then, dislocations can transmit if the following criterion is met:

$$\mathbf{b} \cdot (\boldsymbol{\sigma} + \boldsymbol{\sigma}_{self} + \boldsymbol{\sigma}_{inter}) \cdot \mathbf{n} > \tau_c \qquad (9.4.1)$$

where $\boldsymbol{\sigma}_{self}$ is the line tension, $\boldsymbol{\sigma}_{inter}$ is the interaction stress caused by the other dislocations, and $\mathbf{n}$ is the normal direction vector of the slip plane. Such a treatment is consistent with the dislocation penetration criteria (El-Awady et al., 2011). τ_c corresponds to the Koehler barrier strength (Koehler, 1970), which depends on the thickness of the coating and the material properties of the micropillar and coating layer.

To investigate the dislocation transmission process and exclude the influence of collective dislocation interactions, the simulated samples contain only one SAS with indestructible pinning points. The parameters for the SAS are the same as those described in Section 9.1.3. Fig. 9.12A indicates that the strain hardening behavior can practically vanish when dislocation penetration is allowed, even if there is only one activated SAS. The final stable flow stress scales proportionally to the value of τ_c. The stable resolved shear stress is about 509 and 670 MPa when τ_c is 0.8 and 1.5 GPa, respectively. This observation is consistent with previous simulation work (El-Awady et al., 2011). The dislocation configurations when the engineering strain is 1% are shown in Fig. 9.12B and C, in which the local penetration sites are denoted by arrows.

The hoop stress distributions are shown in Fig. 9.13, in which the dislocations begin to transmit from the interface. In the case of $\tau_c = 800$ MPa, the maximum hoop stress in the coating σ_{hoop}^{max} is 588 MPa, as indicated by the gray region. In the case of $\tau_c = 1500$ MPa,

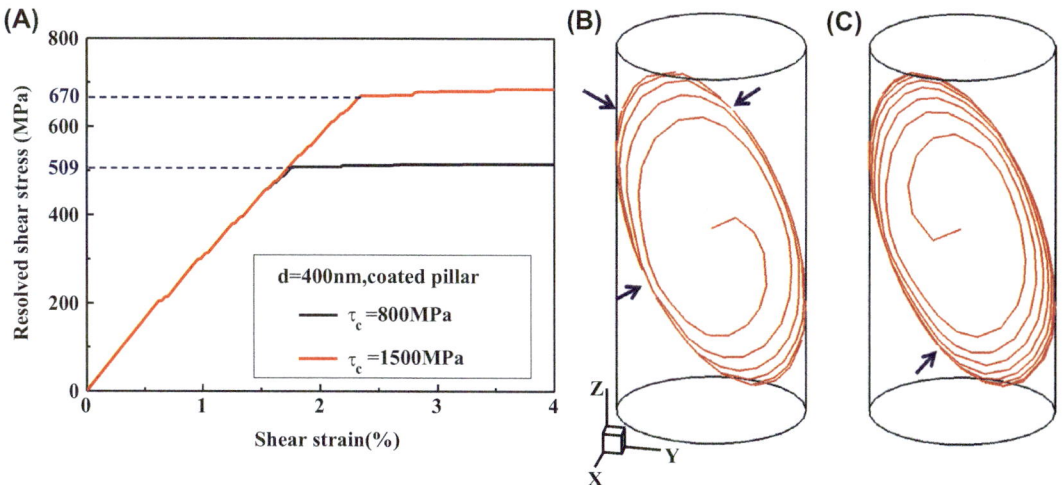

Figure 9.12
(A) Stress-strain curve for micropillar containing individual single-arm dislocation source under compression; (B and C) Dislocation configuration when shear strain is 2.45% in coated micropillar with τ_c of 800 and 1500 MPa, respectively (Cui, 2016).

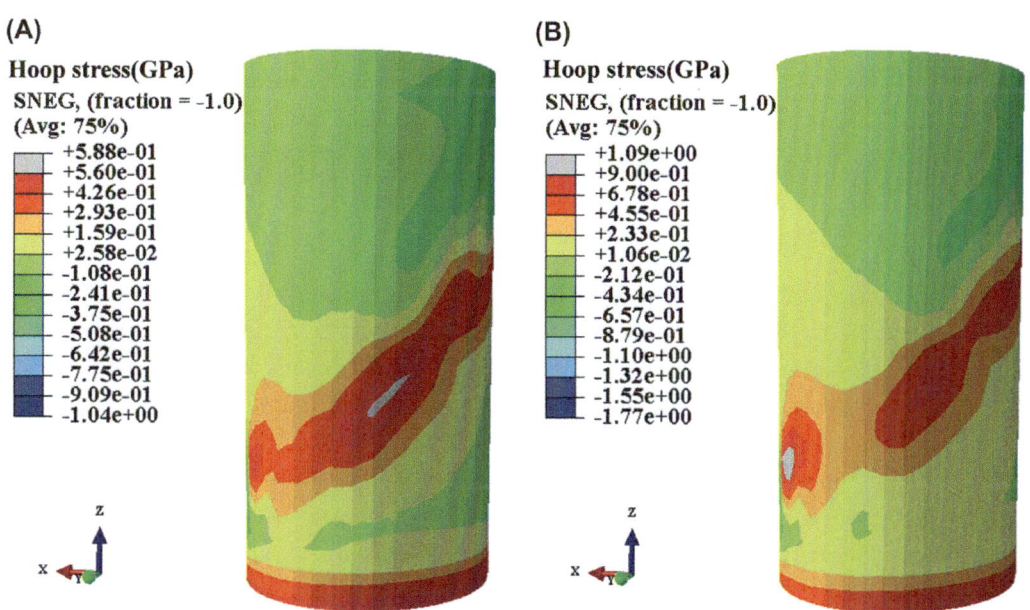

Figure 9.13
For a micropillar with a diameter of 400 nm, hoop stress of the coating is shown when dislocations begin to transmit from the interface (A) in the case of $\tau_c = 0.8$ GPa; and (B) in the case of $\tau_c = 1.5$ GPa (Cui, 2016). *SNEG* is used to describe the orientation of surfaces in ABAQUS software.

$\sigma_{\text{hoop}}^{\text{max}}$ is 1090 MPa. Because the tensile strength of Al_2O_3 is only about 267 MPa (Munro, 1997), the coating will crack as a result of high hoop stress before dislocation transmission.

9.5 Summary

In this chapter, the stress-strain behavior and underlying dislocation mechanisms of coated micropillars are presented to understand confined plasticity at the microscale. The evolution of mobile and trapped dislocations is separately discussed. The exhaustion of mobile dislocation reflects the destruction of the SAS; it contributes to flow stress through the Taylor hardening law. The trapped dislocation density exhibits a linear relation with plastic strain and induces an increase in flow stress in terms of back-stress hardening. Back stress contributes most to flow stress and exhibits a linear dependence on the trapped dislocation density. This relation correlates well with the derivation obtained by the higher-order crystal plasticity theory and is used directly to determine the material parameters in the continuum back-stress model. The theoretical model of predicting the upper and lower bounds of the stress-strain curve for coated compression micropillars is given. Finally, the basics of two kinds of coating failure mechanisms are discussed.

Mechanical Annealing Under Low-Amplitude Cyclic Loading

Chapter Outline

10.1 Introduction

Clearing out preexisting dislocations is important to obtain multifunctional microdevices that have high-strength (Bei et al., 2008), improved dielectric properties (Alpay et al., 2004) and electrical and thermal conductivity (Watling and Paul, 2011). A considerable body of evidence shows that dislocation density tends to decrease in small submicron single crystals when it is subjected to monotonic loading (Greer and Nix, 2006; Shan et al., 2008; Uchic et al., 2009b). This process, called mechanical annealing (Shan et al., 2008), is usually attributed to massive dislocation escape from the free surface (Greer, 2006; Greer and Nix, 2006; Wang et al., 2012a). Generally, the occurrence of this phenomenon depends on the sample size (several hundred nanometers) and applied stress level (several *gigapascals*). A smaller sample size corresponds to a higher attractive image force and more significant surface annihilation, whereas higher stress means that more dislocation junctions can be broken. However, relatively high applied monotonic loading (as high as several *gigapascals*) also triggers a significant amount of dislocation multiplication. The decreasing dislocation density requires the annihilation of both preexisting and multiplied dislocations. This leads to evident changes in the specimen shape, which is not anticipated in practical applications. Therefore, it would be attractive if a new loading method existed (e.g., with relative low amplitude) to drive out dislocations and also keep the specimen shape without significant changes.

Dislocation Mechanism-Based Crystal Plasticity. https://doi.org/10.1016/B978-0-12-814591-3.00010-8

This raises the question of whether nonmonotonic loading performed with a relative low amplitude can disengage complex preexisting dislocation structures without causing significant dislocation multiplication. Generally, under cyclic loading conditions, dislocations accumulate by forming pattern structures such as persistent slip bands (Erel et al., 2017) and well-ordered veins and walls (Aifantis, 1987), even at the micron scale (Zhang et al., 2006). The accumulation of such defects gradually contributes to cyclic strain localization and the initiation of fatigue crack in bulk samples (Mughrabi, 2010). Therefore, there seems to be much consensus that cyclic loading should lead to dislocation accumulation and material fatigue.

Thus, the question is whether defect healing can be observed if the external size further decreases to several hundred nanometers. In situ low-amplitude cyclic loading experiments were carried out for a pure single-crystalline Al pillar inside a Transmission electron microscopy (TEM) (Wang et al., 2015). The nominal dimensions of the cuboid-shaped sample were 300 nm in thickness and 500 nm in width. Surprisingly, the initial high dislocation density significantly decreased within cycles. There was no such pronounced shape change compared with relatively high stress monotonic loading. In addition, the yield strength after such cyclic loading significantly improved as expected. This phenomenon makes great promising for applications in obtaining high strength crystals with low-density dislocations.

Naturally, this intriguing observation raises several questions. First, how could low cyclic stress contribute to the decline in dislocation density? How will cyclic loading affect the dislocation annihilation and multiplication process? In addition, both the line tension model (Dupuy and Fivel, 2002) and atomic-level analysis (Rodney and Phillips, 1999) indicate that failure of the dislocation junction often requires a high-enough applied stress. How can dislocation junctions be destroyed without high stress under cyclic loading? Second, how can dislocation density evolution be predicted under low cyclic stress? Finally, determining the critical conditions for mechanical annealing is especially interesting. Because both monotonic and cyclic loadings can contribute to dislocation annihilation, a comparison can be made of aspects of the critical size, to discuss conditions under which dislocation starvation occurs.

This chapter mainly addresses these problems and clarifies the underlying mechanism under low-amplitude cyclic loading conditions, based on efforts in three-dimensional discrete *dislocation* dynamics (3D-DDD) simulations and theoretical analysis.

10.2 Cyclic Behavior of Collective Dislocations

This section gives several sets of 3D-DDD simulations results to give a basic idea of the cyclic behavior of collective dislocations in a micropillar (Cui et al., 2016a). Similar to

previous experiments (Wang et al., 2015), cuboid-shaped Al pillars are considered. Initial equilibrium dislocation configurations are generated by relaxing randomly distributed straight dislocation lines before applying a load. The dislocation junctions emerge as a natural outcome of the dislocation interaction without requiring initial fixed pinning points. The obtained initial dislocation density is within the range of the experimentally measured dislocation density for a focused ion beam (FIB) fabricated micropillar (10^{13}–10^{14} m^{-2}). Two kinds of small-strain cyclic loadings are imposed on top of the pillar. The maximum strain ε_{max} is twice the minimum one ε_{min}. In the one case, $\varepsilon_{max} = 2\varepsilon_{min} = 0.1\%$; in the other case, $\varepsilon_{max} = 2\varepsilon_{min} = 0.2\%$. The corresponding peak stress is lower than 140 MPa.

As revealed in Fig. 10.1A, the dislocation density decreases within cycles for a micropillar with a cross-section of 400 nm × 400 nm and height of 800 nm, which is consistent with the experimental sample. During a loading stage of the first cycle, the dislocation annihilation rate exhibits an almost linear relation to the applied strain. Thus, the higher the peak strain is, the lower the dislocation density is after the first cycle. Actually, this corresponds to the dislocation density evolution trend during the initial microplastic stage of monotonic loading (Motz et al., 2009). Fig. 10.1B and C show that the decrease in dislocation density under low-amplitude cyclic strains is mainly induced by the surface annihilation of mobile dislocations during the first several cycles. The comparison of dislocation configuration in Fig. 10.1B and D shows that in the case of a larger peak strain, broken, weak dislocation junctions also contribute to the decrease in dislocation density.

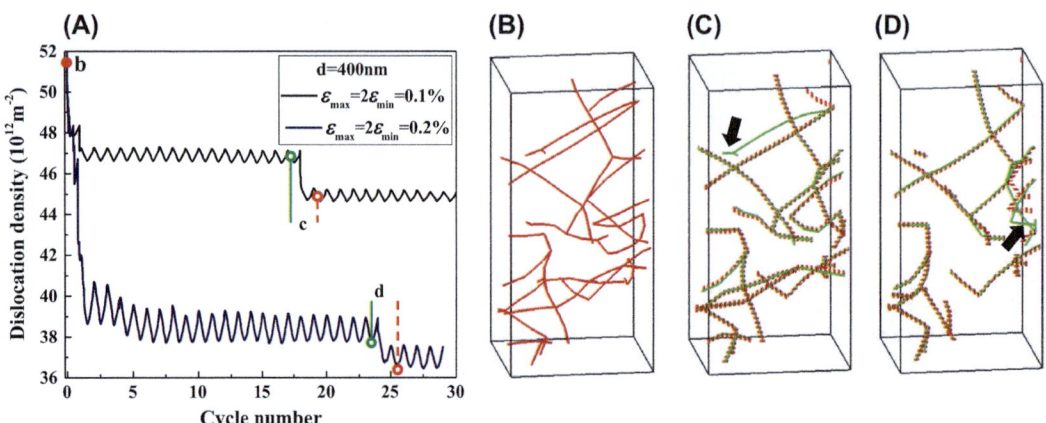

Figure 10.1

(A) Evolution of dislocation density during low-amplitude cyclic loading for micropillar with diameter of 400 nm; (B) Initial dislocation configuration; (C, D) Snapshots of dislocation configuration corresponding to *circles* of the same color as in (A). *Arrows* indicate junctions destroyed during cyclic loading. *Reprinted from Cui, Y.N., Liu, Z.L., Wang, Z.J., Zhuang, Z., 2016a. Mechanical annealing under low-amplitude cyclic loading in micropillars. Journal of the Mechanics and Physics of Solids 89, 1–15, Copyright 2016, with permission from Elsevier.*

Fig. 10.2A shows that the dislocation density seems to change little from cycle number $N = 3$ to $N = 18$, especially for loading condition $\varepsilon_{max} = 2\varepsilon_{min} = 0.1\%$. This is because the dislocations, which are easy to annihilate under the current stress level, are gradually exhausted during the first two cycles. In such a small sample, the characteristic length of dislocation source is short, which means that these sources have high activation stress. Thus, the low applied stress is not high enough to trigger operation of the dislocation source and no substantial dislocation multiplication occurs. Most dislocation segments bow out upon loading and shrink backward during unloading from the second cycle. This leads to a first increase and a subsequent decrease in dislocation density during each cycle (Fig. 10.1A). The recovery of dislocation density and the reverse plasticity during unloading are also observed in other DDD simulations (Déprés et al., 2008; Erel et al., 2017). Reverse dislocation motion does not fully recover the dislocation configuration back to that at the beginning of the cycle. With such a limited size, incomplete reverse dislocation motion provides opportunities for annihilation at a nearby free surface, leading to a gradual decrease in dislocation density, as shown in Fig. 10.2. This exactly reflects the advantage of cyclic loading upon monotonic loading when dislocation annihilation is triggered. Although the dislocation density decreases by only a small amount after each cycle owing to the small strain amplitude, the cumulative decline with the increasing number of cycles can still be large and ultimately may lead to a significant

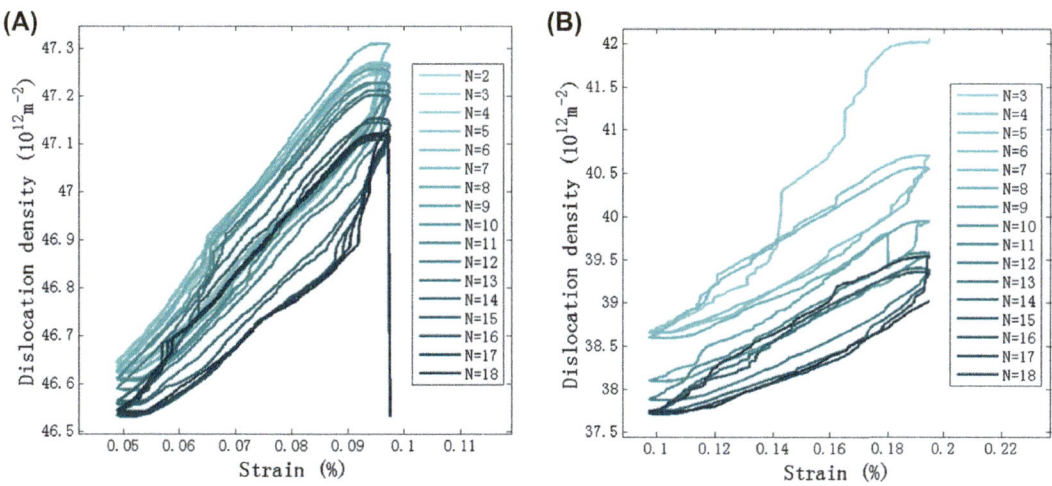

Figure 10.2

Evolution of dislocation density for a micropillar with a diameter of 400 nm during the first several cycles under conditions of (A) $\varepsilon_{max} = 2\varepsilon_{min} = 0.1\%$ and (B) $\varepsilon_{max} = 2\varepsilon_{min} = 0.2\%$. *Reprinted from Cui, Y.N., Liu, Z.L., Wang, Z.J., Zhuang, Z., 2016a. Mechanical annealing under low-amplitude cyclic loading in micropillars. Journal of the Mechanics and Physics of Solids 89, 1−15, Copyright 2016, with permission from Elsevier.*

decrease in dislocation density. This phenomenon is fundamentally different from cyclic behavior for bulk materials.

After multiple cycles, the dislocation junction can be broken and expelled from the crystal, leading to a sudden decline in dislocation density at some cycle. This phenomenon is represented by the green solid and red dotted dislocation configurations in Fig. 10.1C and D. Similar behavior is observed in a micropillar with a cross-section of 800 nm × 800 nm and a height of 1600 nm under the loading condition of $\varepsilon_{max} = 2\varepsilon_{min} = 0.1\%$ (Fig. 10.3). By way of contrast, for a larger sample, there is a large possibility that the junction will interact with other dislocations before directly failing by annihilating from the free surface. Therefore, the dislocation configuration is only rearranged after the failure of some dislocation junction in a large sample.

In addition, the evolution of accumulated plastic strain in Fig. 10.4 illustrates that from the second cycle, plastic strain during the unloading stage is almost comparable to that during loading. Rajagopalan et al. (2007) proposed that the driving force for recoverable plastic deformation arises from residual internal stresses caused by inhomogeneous deformation. Moreover, accumulated plastic strain after multiple cycles remains low during the whole low cyclic loading process, which means that this kind of loading mode causes only a change in small shape.

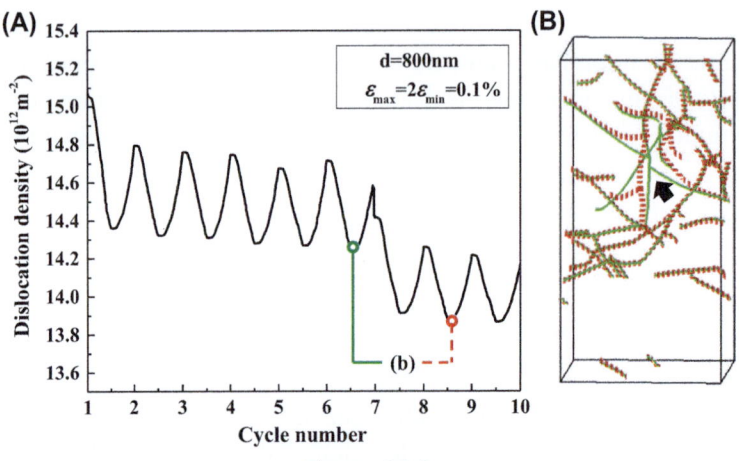

Figure 10.3

(A) Evolution of dislocation density during low-amplitude cyclic loading for a micropillar with a diameter of 800 nm; (B) Images of dislocation configuration corresponding to *circles* of the same color as in (A). *Arrows* indicate that junctions are rearranged during cyclic loading. *Reprinted from Cui, Y.N., Liu, Z.L., Wang, Z.J., Zhuang, Z., 2016a. Mechanical annealing under low-amplitude cyclic loading in micropillars. Journal of the Mechanics and Physics of Solids 89, 1–15, Copyright 2016, with permission from Elsevier.*

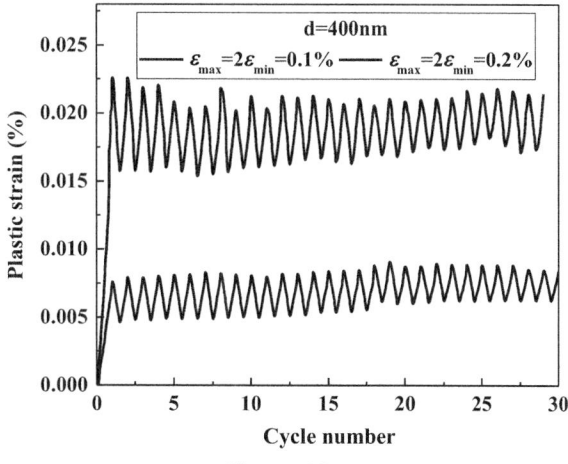

Figure 10.4
Evolution of plastic strain within cycles for a micropillar with a diameter of 400 nm. *Reprinted from Cui, Y.N., Liu, Z.L., Wang, Z.J., Zhuang, Z., 2016a. Mechanical annealing under low-amplitude cyclic loading in micropillars. Journal of the Mechanics and Physics of Solids 89, 1–15, Copyright 2016, with permission from Elsevier.*

10.3 Cyclic Instability of Dislocation Junction

These simulation results present an interesting phenomenon revealing that the dislocation junction can fail under low-amplitude cyclic stress. Most studies focused on the stability of a dislocation junction under monotonic loading (Dupuy and Fivel, 2002; Picu and Soare, 2010). Little attention has been paid to the cyclic stability of dislocation junctions. This section discusses its underlying mechanism. For complex dislocation configurations, the dynamic evolution of dislocations is blurred by the difficulty of distinguishing the role of cyclic loading from collective dislocation interactions. Thus, a simple but illustrative case is considered that contains only one dislocation junction.

The glissile and Lomer-Cottrell (LC) junctions are widely observed in a face-centered cubic structure (FCC) crystal. They are determined to be the strongest junctions and very important (Dupuy and Fivel, 2002; Franciosi and Zaoui, 1982), with representative glissile and sessile characters, respectively. Thus, they are chosen to be discussed to understand the cyclic failure mechanism of the dislocation junction. By comparing their response under monotonic loading and peak stress relaxation, dislocation annihilation mechanisms under different loading modes are also analyzed.

10.3.1 Glissile Dislocation Junction

Generally, two pinning points emerge after a junction forms. At a small scale, there is a high probability that one of the pinning points escaped from the surface. Thus, a glissile

junction with only one pinning point is first considered, as shown in Fig. 10.5A. Initially, Arm 1 is along $\zeta_1 = [1\ -1\ 2]$, with Burgers vector $\mathbf{b}_1 = [\bar{1}10]$, and slip plane normal vector $\mathbf{n}_1 = (11\bar{1})$; Arm 2 is along $\zeta_2 = [-1\ 1\ 2]$, with $\mathbf{b}_2 = [10\bar{1}]$, and $\mathbf{n}_2 = (111)$; and the junction segment is along $\zeta_{jun} = [-1\ 1\ 0]$, with $\mathbf{b}_{jun} = [01\bar{1}]$, $\mathbf{n}_{jun} = (111)$.

Three kinds of strain-controlled tests are carried out: (1) A monotonic tension test is first applied to evaluate the strength of the junction, defined as the instantaneous stress value when the pinning point disappears owing to unzipping or surface annihilation. (2) A low cyclic tension strain test is applied to examine the cyclic stability of the junction. The maximum normal strain is twice the minimum one, $\varepsilon_{max} = 2\varepsilon_{min} = 0.2\%$. (3) The third one results in the stress relaxation test, i.e., total normal strain first increases to 0.2% and then remains constant. This test is designed to check whether the junction failure under cyclic deformation is a time-dependent process.

Once the pinning point disappears as a result of unzipping or surface annihilation, the junction fails to lock the dislocations inside the sample. Hence, attention is focused on the position of the pinning point, to discuss junction stability. Considering that the pinning point can move only along the intersection line between two slip planes $\mathbf{n}_1$ and $\mathbf{n}_2$, the distance to the free surface is defined as length x between the pinning point and the free surface along the intersection line, as shown in Fig. 10.5A.

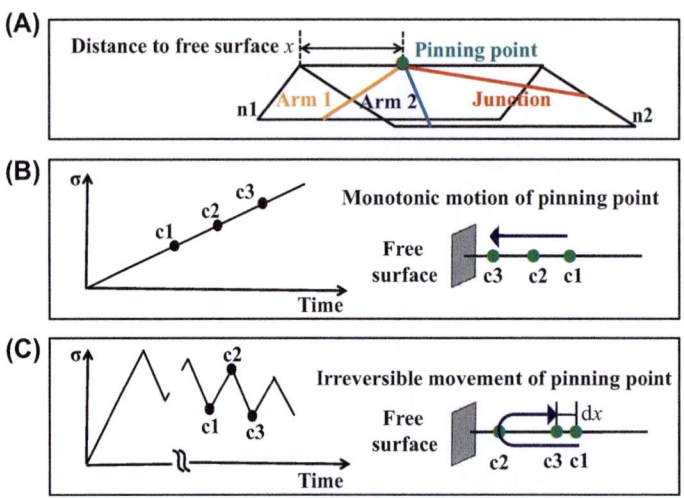

Figure 10.5

(A) Schematic definition of distance of pinning points to free surface (x) for a glissile junction. Schematic diagram for junction destruction process under (B) monotonic deformation and (C) small strain cyclic deformation. *Reprinted from Cui, Y.N., Liu, Z.L., Wang, Z.J., Zhuang, Z., 2016a. Mechanical annealing under low-amplitude cyclic loading in micropillars. Journal of the Mechanics and Physics of Solids 89, 1–15, Copyright 2016, with permission from Elsevier.*

The simulation result for the monotonic tension test is given in Fig. 10.6. For this considered junction, by simple calculation, the initial Peach-Koehler (PK) force on Arm 2 is along [−1 1 0] and the PK force on the junction segment is along [1 1 −2]. Thus, the increasing PK force first makes dislocation Arm 2 and the junction segment meet and react with each other (see the purple dislocation image in Fig. 10.6B). Then, dislocations gradually glide out of the crystal upon higher stress. Finally, the pinning point annihilates from the surface at 265 MPa, which is considered to be the strength of the junction.

During the cyclic tension test, the cyclic peak stress (140 MPa) is much lower than the strength of the junction obtained under monotonic tension. However, Fig. 10.7 shows that the junction also gradually fails. The detailed evolution of the pinning point position is given in Fig. 10.7A. After each loading cycle, the pinning point does not move back to its previous position but shifts away from its previous position by dx. The shift direction after each strain cycle is random but the accumulated effect is that the pinning point moves toward a free surface. In addition, the general trend for the magnitude of the shift distance after each straining cycle is to increase with increasing cycle numbers. Fig. 10.7B shows the configuration evolution of the dislocation junction that corresponds to cycles marked with hollow circles with different colors. When the pinning point becomes close enough to the surface, the unstable dislocation junction is destroyed by fatal attraction to the free surface. At the same time, the dislocation segments escape quickly from the free surface and leave behind a nearly perfect crystal.

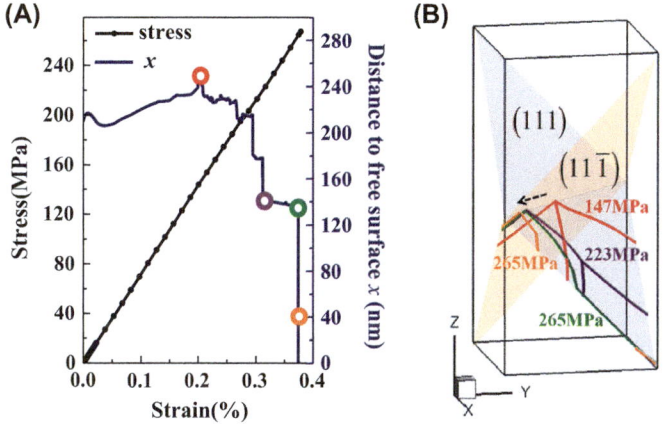

Figure 10.6

(A) Evolution of stress and distance of pinning point to free surface versus applied strain under monotonic tension for a glissile junction; (B) Images of dislocation configuration corresponding to *circles* of the same color as in the curve in (A). *Reprinted from Cui, Y.N., Liu, Z.L., Wang, Z.J., Zhuang, Z., 2016a. Mechanical annealing under low-amplitude cyclic loading in micropillars. Journal of the Mechanics and Physics of Solids 89, 1−15, Copyright 2016, with permission from Elsevier.*

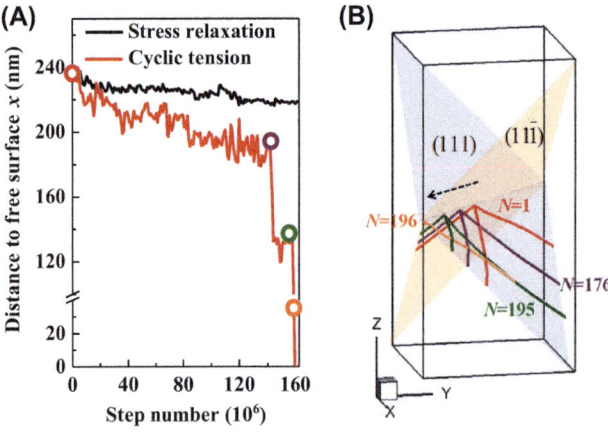

Figure 10.7

(A) Evolution of distance of pinning point to free surface under cyclic tension and stress relaxation for a glissile junction. The total step number corresponds to 200 cycles; (B) Images of dislocation configuration corresponding to circles of the same *color* as in the curve in (A). *Reprinted from Cui, Y.N., Liu, Z.L., Wang, Z.J., Zhuang, Z., 2016a. Mechanical annealing under low-amplitude cyclic loading in micropillars. Journal of the Mechanics and Physics of Solids 89, 1—15, Copyright 2016, with permission from Elsevier.*

In contrast to Tests 1 and 2, the junction stays at an almost stable configuration for stress relaxation Test 3, as shown in Fig. 10.7A. Because the total loading times for stress relaxation and cyclic deformation are the same, their totally different responses illustrate that the cyclic effect is the key factor in triggering dislocation starvation and junction destruction instead of the time accumulation. Compared with the process under monotonic loading, it takes much longer time to fail the junction, because the irreversibility part of dislocation slip occupies only a small fraction of the total amount of slip for each cycle, especially for such low-amplitude cyclic loading.

Similar cyclic instability processes of junction failure and cyclic enhanced mobile dislocation annihilation are observed when the cyclic tension loading condition is changed to a cyclic compression condition. During the uniaxial compression test, as shown in Fig. 10.8, the pinning point directly annihilates from the free surface without an interaction between the dislocation junction and arms. The strength of the junction under uniaxial compression is about 153 MPa. In the low cyclic compression strain test, $\varepsilon_{max} = 2\varepsilon_{min} = -0.156\%$, the cyclic peak stress is about 112 MPa and the pinning point annihilates from the free surface after 18 cycles, as shown in Fig. 10.9. During subsequent cycles, mobile dislocation gradually annihilates from the free surface with the aid of an incomplete reversible slip (Fig. 10.9B). This case not only verifies the cyclic instability of the dislocation junction, it clearly illustrates how cyclic stress contributes to the gradual annihilation of mobile dislocation.

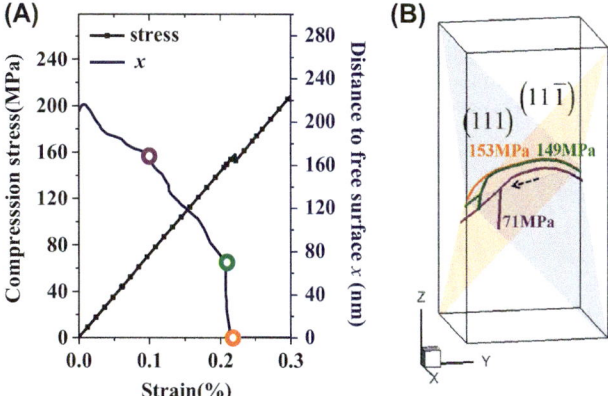

Figure 10.8

(A) Evolution of stress and distance of pinning point to free surface versus applied strain under monotonic compression for a glissile junction; (B) Images of dislocation configuration corresponding to circles of the same *color* as in the curve in (A). *Reprinted from Cui, Y.N., Liu, Z.L., Wang, Z.J., Zhuang, Z., 2016a. Mechanical annealing under low-amplitude cyclic loading in micropillars. Journal of the Mechanics and Physics of Solids 89, 1–15, Copyright 2016, with permission from Elsevier.*

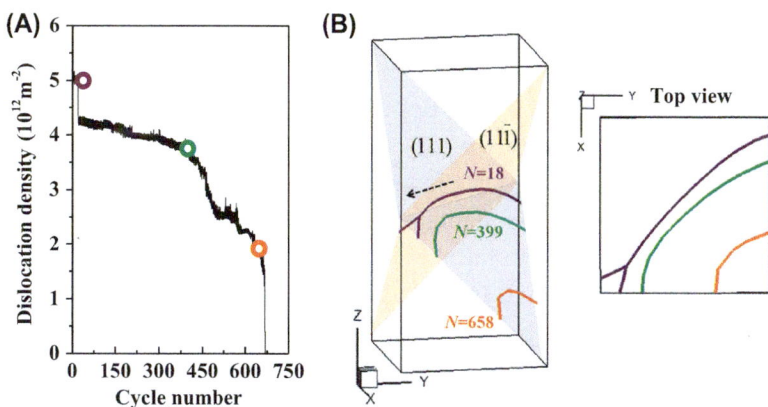

Figure 10.9

(A) Evolution of dislocation density under cyclic compression for a glissile junction. Cyclic peak stress is about 112 MPa; (B) Images of dislocation configuration corresponding to circles of the same *color* as in the curve in (A). *Reprinted from Cui, Y.N., Liu, Z.L., Wang, Z.J., Zhuang, Z., 2016a. Mechanical annealing under low-amplitude cyclic loading in micropillars. Journal of the Mechanics and Physics of Solids 89, 1–15, Copyright 2016, with permission from Elsevier.*

10.3.2 Sessile Dislocation Junction

In light of these inspiring results for glissile junction, the sessile dislocation junction is also considered under monotonic, cyclic and relaxation loading conditions. The loading

parameters are the same as those for the glissile junction. Here, an LC lock is naturally formed by the interaction of two straight surface dislocation lines crossing each other at their midpoints. One dislocation line is initially along $\zeta_1 = [-1\ 1\ 0]$, with Burgers vector $\mathbf{b}_1 = [10\bar{1}]$ and slip plane normal vector $\mathbf{n}_1 = (111)$. The other one is initially along $\zeta_2 = [0\ 1\ 1]$, with $\mathbf{b}_2 = [0\ 1\ 1]$ and $\mathbf{n}_2 = (11\bar{1})$.

The results for monotonic tension are given in Fig. 10.10. Here, the distance to the free surface is defined as the length between the pinning point and Point A. Fig. 10.10B shows that both pinning points are subsequently annihilated from the left-side surface. The junction finally fails at about 215 MPa.

The results for cyclic tension and stress relaxation results are shown in Fig. 10.11. Because the first pinning point is annihilated in the first cycle, only the results for the second pinning point are given. The distance to the free surface is defined as the length between the pinning point and Point B, because the second pinning point progressively escapes from the right-side surface (Fig. 10.11B), which differs from that under monotonic tension (Fig. 10.10B). Cyclic failure for the LC lock is similar to that for the glissile junction. However, because the LC lock segment cannot move in the crystal slip plane, an immobilized lock segment, AB (Fig. 10.11B), leaves along the intersection line. Moreover, during stress relaxation, the dislocation junction remains almost stable, which suggests that junction cyclic failure is time-insensitive.

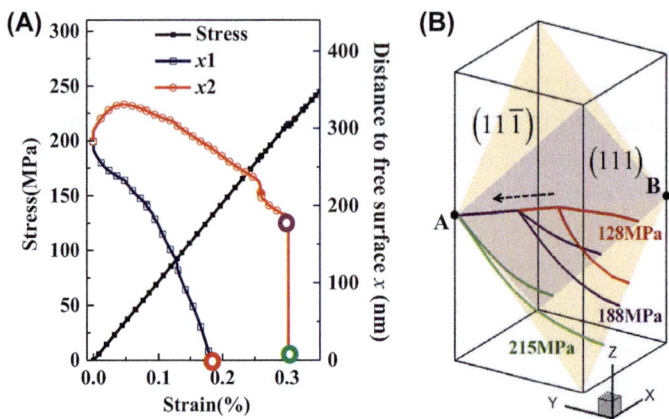

Figure 10.10

(A) Evolution of stress and distance of pinning point to free surface under monotonic tension for an LC junction. *x*1 and *x*2 represent the distance for the first and second pinning points, respectively; (B) Images of dislocation configuration corresponding to *circles* of the same color as in the curve in A after the first pinning point is annihilated. *Reprinted from Cui, Y.N., Liu, Z.L., Wang, Z.J., Zhuang, Z., 2016a. Mechanical annealing under low-amplitude cyclic loading in micropillars. Journal of the Mechanics and Physics of Solids 89, 1—15, Copyright 2016, with permission from Elsevier.*

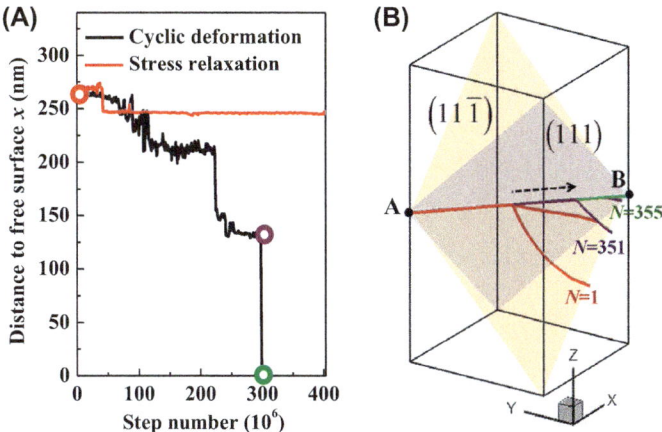

Figure 10.11

(A) Evolution of distance of pinning point to free surface under cyclic tension and stress relaxation for an LC junction. The total step number corresponds to 480 cycles; (B) Images of dislocation configuration corresponding to *circles* of the same color as in the curve in (A). *Reprinted from Cui, Y.N., Liu, Z.L., Wang, Z.J., Zhuang, Z., 2016a. Mechanical annealing under low-amplitude cyclic loading in micropillars. Journal of the Mechanics and Physics of Solids 89, 1–15, Copyright 2016, with permission from Elsevier.*

10.4 Cyclic Enhanced Dislocation Annihilation Mechanism

These simulation results clearly show that cyclic loading enhances junction destruction and surface annihilation. To indicate the underlying dislocation mechanism, different junction failure processes under monotonic loading and cyclic loading are further schematically presented in Fig. 10.5B and C. Under monotonic deformation, the relative high stress leads to the failure of the dislocation junction. The pinning point monotonously moves toward the free surface upon increased loading. Under cyclic deformation, the pinning point always glides to and fro between two equilibrium positions corresponding to maximum strain ε_{max} and minimum strain ε_{min} in one cycle, as shown in Fig. 10.5C. The equilibrium position of the pinning point is determined by minimizing the total potential energy of the entire structures (Picu and Soare, 2010). This can be simplified for solving equations by making sure forces exerted on the pinning point by three intersecting segments equal 0 along the intersection line (Dupuy and Fivel, 2002). Because externally applied stress is always tension stress during deformation, its corresponding PK stress always makes dislocations glide toward a given direction and thus makes the pinning point move toward a given direction. In contrast, the line tension force (Fig. 10.7B) and image force will assist motion toward the free surface but opposite the returning motion. Therefore, the reverse motion in each cycle does not entirely cancel out the forward motion, as shown in Fig. 10.5C. This irreversible movement gradually accumulates and

makes the pinning point move toward surface. When the pinning point gets close enough to the free surface, the image force will have a significant role in promoting motion toward the surface. Finally, the dislocation segments glide out of the crystal. The cyclic annihilation process of mobile dislocation is similar to the cyclic instability process of the dislocation junction, as shown in Section 10.3.1.

Accordingly, the cumulative irreversible slip has a crucial role in dislocation self-organization and surface annihilation. Generally, the main origin of cyclic slip irreversibility includes four parts (Mughrabi, 2009): surface annihilation, the mutual annihilation of opposite sign dislocations, cross-slip of screw dislocations, and random to-and-fro glide of dislocations. All of these slip irreversibilities will manifest themselves in one form or another during dislocation structure evolution. At small scales, they can favor the destruction of dislocation junctions and provide the conditions for surface annihilation.

Based on the insights revealed by the results of the simulation, how cyclic slip irreversibility contributes to dislocation density evolution and the critical sizes of mechanical annealing in submicron single crystals are preliminarily addressed in the following sections.

10.5 Dislocation Density Influenced by Cyclic Slip Irreversibility

Here, the degree of cyclic slip irreversibility is implicitly characterized by dislocation density evolution equations. Theoretical investigations indicate that the change in total dislocation density $d\rho$ under monotonic loading consists of four parts: dislocation multiplication owing to forest dislocations $d\rho_{mult}$, dislocation surface annihilation $d\rho_{surf}$, dislocation mutual annihilation for closely spaced dislocations of opposite sign $d\rho_{inner}$, and dislocation generation by single-arm source operation $d\rho_{SAS}$, as discussed in Section 8.4.1:

$$d\rho = d\rho_{mult} + d\rho_{surf} + d\rho_{inner} + d\rho_{SAS} \qquad (10.5.1)$$

Because low peak stress is not high enough to trigger the complete activation of a single-arm source, the fourth term is ignored in the following discussions. Among the first three terms on the right side of Eq. (10.5.1), the multiplication term is not fully reversible. For example, the bowing dislocations may interact with other dislocations or cross-slip, or be strongly influenced by the image force induced by the free surface. All of these processes prevent them from returning to their original shape. Under low-amplitude cyclic loading conditions, the dislocation segments will bow out during the loading stage, leading to an increase in dislocation density. Then, during the unloading stage, the shrinkage of dislocation segments will decrease the dislocation density. Accordingly, the recovery part of dislocation multiplication during unloading stage $d\rho_{recovery-mult}$ will suppress the increase in dislocation density.

For simplicity, a dimensionless parameter, k_{rm}, is proposed to reflect the effect of reversible dislocation slip on the multiplication rate:

$$k_{rm} = \frac{\sum\limits_{d\varepsilon<0} d\rho_{recovery-mult}}{\sum\limits_{d\varepsilon>0} d\rho_{mult}}$$ (10.5.2)

At the same time, the dislocation surface annihilation and mutual annihilation parts are entirely irreversible but will be enhanced by incomplete reversible dislocation slip, as indicated by the simulation results in Sections 10.2 and 10.3. Thus, the cyclic slip-irreversible part will promote the decrease in dislocation density. Assuming $d\rho_{irrever-anni}$ denotes the dislocation surface and mutual annihilation enhanced by the irreversible dislocation slip, a dimensionless parameter k_{ira} is defined to reflect the effect of irreversible dislocation slip on the annihilation rate:

$$k_{ira} = \frac{\sum\limits_{d\varepsilon>0} d\rho_{irrever-anni}}{\sum\limits_{d\varepsilon>0} d\rho_{surf} + d\rho_{inner}}$$ (10.5.3)

The values of k_{rm} and k_{ira} change during deformation and mainly depend on the dislocation configuration, cyclic number, and amplitude of applied strain. During the first several cycles, the bowing dislocations have a high possibility of interacting with the other dislocations, so the recoverable multiplication coefficient k_{rm} is low. However, there is a higher possibility of finding mobile dislocations that are attractive and tend to annihilate each other, or are close to the free surface but have not yet glided out of the crystal. Therefore, the coefficient of annihilation enhanced by irreversible slip k_{ira} is high. When the cyclic number is large, a relative stable dislocation configuration is formed. The recoverable multiplication coefficient, k_{rm}, is high. However, irreversibility mainly manifests itself through irreversible slip induced by image force and cross-slip, so k_{ira} is low. As for the strain amplitude, in general, small to-and-fro dislocation displacements are more reversible than larger ones (Mughrabi, 2009). Thus, the smaller the amplitude of applied strain is, the larger the value of k_{rm} and the smaller the value of k_{ira} are. From a quantitative standpoint, it is difficult to develop a mathematical expression for these two parameters. Given the material and loading conditions, k_{rm} and k_{ira} can be estimated by fitting experimental data or simulation results.

Eq. (10.5.1) gives the dislocation density evolution law for monotonic deformation. Based on the previous analysis, the unique partial reversible dislocation multiplication and irreversible slip-enhanced dislocation annihilation should be taken into account to analyze the mechanical annealing phenomenon. In addition, a dimensionless parameter is

introduced to consider the different dislocation annihilation ability approximately during loading and unloading stages:

$$k_{\mathrm{u}} = \frac{\sum\limits_{d\varepsilon<0} d\rho_{\mathrm{surf}} + d\rho_{\mathrm{inner}}}{\sum\limits_{d\varepsilon>0} d\rho_{\mathrm{surf}} + d\rho_{\mathrm{inner}}} \tag{10.5.4}$$

Because dislocation annihilation occurs dominantly during loading, k_{u} is less than 1. Besides, the cyclic enhanced annihilation effect is considered during the loading stage. Accordingly, the following expression for dislocation density evolution is developed to distinguish loading and unloading:

$$\rho_i = \begin{cases} \rho_{i-1} + d\rho_{\mathrm{mult}} + (1 + k_{\mathrm{ira}})(d\rho_{\mathrm{surf}} + d\rho_{\mathrm{inner}}) & (d\varepsilon > 0) \\ \rho_{i-1} - k_{\mathrm{rm}}d\rho_{\mathrm{mult}} + k_{\mathrm{u}}(d\rho_{\mathrm{surf}} + d\rho_{\mathrm{inner}}) & (d\varepsilon < 0) \end{cases} \tag{10.5.5}$$

In the following, unless specified, the subscript i and $i-1$ denote variables at the ith and $(i-1)$th time increments, respectively.

According to earlier work by Gilman (1969), the multiplication term $d\rho_{\mathrm{mult}}$ can be written as a function of breeding coefficient δ, which is defined as the inverse of the mean free path along which the dislocation should move before storage. Generally, δ is set as a function of dislocation density (Ungar et al., 2011; Zhou et al., 2011):

$$d\rho_{\mathrm{mult}} = \rho\delta v dt, \quad \delta = f(\rho) = k_{\mathrm{f}}\sqrt{\rho} \tag{10.5.6}$$

where v is the dislocation velocity and k_{f} is a dimensionless proportionality coefficient accounting for the details of dislocation structure (Pantleon, 2004) and loading orientations (Devincre et al., 2008).

The dislocation surface annihilation term $d\rho_{\mathrm{surf}}$ is derived by assuming that the probability of dislocation escape from the surface is 1/2 if a dislocation segment is located in the region within a distance $v dt$ away from the surface (Greer, 2006):

$$d\rho_{\mathrm{surf}} = -\rho\left(\frac{\pi(a/2 + d/2)v dt}{\pi a d/4}\right) = -\rho\frac{2\cos^2(\beta/2)v dt}{d} \tag{10.5.7}$$

where β is an angle between the primary slip plane normal and the axial direction of the pillar (Zhou et al., 2011). d is the diameter of the pillar. $a = d/\cos\beta$, which is a major axis of the ellipse slip plane in a cylindrical sample. Inner mutual annihilation term $d\rho_{\mathrm{inner}}$ is described as a function of effective annihilation distance y (Ungar et al., 2011):

$$d\rho_{\mathrm{inner}} = -\rho y \cdot d\varepsilon^p/(bM) \tag{10.5.8}$$

where b is the Burgers vector with a magnitude of 0.29 nm for aluminum. y is specified as $6b$ (Cleveringa et al., 2000; Kubin et al., 1992) and M is the Schmid factor. Combined with

the relation between the axial plastic strain increment $d\varepsilon^P$ and the dislocation slip amount $d\varepsilon^P = M\rho bvdt$, Eq. (10.5.5) can be expressed as

$$\rho_i = \begin{cases} \rho_{i-1} + \left(k_f\sqrt{\rho_{i-1}} - (1 + k_{ira})\dfrac{2\cos^2(\beta/2)}{d} - (1 + k_{ira})\rho_{i-1}y\right)\dfrac{d\varepsilon^P}{Mb} & (d\varepsilon > 0) \\[3mm] \rho_{i-1} + \left(-k_{rm}k_f\sqrt{\rho_{i-1}} - k_u\dfrac{2\cos^2(\beta/2)}{d} - k_u\rho_{i-1}y\right)\dfrac{d\varepsilon^P}{Mb} & (d\varepsilon < 0) \end{cases}$$

$$(10.5.9)$$

10.6 Critical Size for Mechanical Annealing

One important issue is the critical size for mechanical annealing. For low-amplitude cyclic deformation, $d\varepsilon^P$ periodically changes sign. Therefore, the dislocation density law varies with it. If the sum of dislocation density change is negative during loading and unloading for each cycle, the dislocation density can gradually drop to 0 within cycles. This condition can be expressed as $d\rho(d\varepsilon < 0) + d\rho(d\varepsilon > 0) < 0$. According to the simulation results in Fig. 10.4, the amount of plastic deformation during forward loading is approximately equal to that during reverse loading in one cycle. Therefore, this condition is approximated by $d\rho/d\varepsilon^P(d\varepsilon < 0) + d\rho/d\varepsilon^P(d\varepsilon > 0) < 0$. Combining with Eq. (10.5.9), the critical size for mechanical annealing is estimated by the relation

$$\begin{aligned} d_{crit\_cyclic} &= \frac{2(1 + k_{ira} + k_u)\cos^2(\beta/2)}{(1 - k_{rm})k_f\sqrt{\rho} - (1 + k_{ira} + k_u)\rho y} \\[3mm] &= \frac{2\cos^2(\beta/2)}{\sqrt{\rho}((1 - k_{rm})k_f/(1 + k_{ira} + k_u)) - \rho y} \geq \frac{2\cos^2(\beta/2)}{\sqrt{\rho} - \rho y} \end{aligned}$$

$$(10.6.1)$$

Only when the sample size is smaller than this critical value can the dislocation annihilation within cycles occur. For a large sample, the crystal exhibits the bulk-like dislocation accumulation trend during continued cyclic deformation, which suggests that the value of k_f varies from 10^{-2} to 1 (Malygin, 2012; Zhou et al., 2011). At the same time, the analyses in Section 9.4 indicate that the values of k_r, k_{ira}, and k_u range from 0 to 1. Therefore, $(1 - k_r)$ $k_f/(1 + k_{ira} + k_u)$ in Eq. (10.6.1) varies from 0 to 1. The minimum value of critical diameter can be estimated according to Eq. (10.6.1). For cyclic loading experiments (Wang et al., 2015), the loading orientation is along the $[\bar{1}11]$ direction and the initial density of the long dislocation line is 20×10^{12} m^{-2}. The critical diameter for mechanical annealing under low amplitude cyclic loading should be larger than 300 nm. This explains why mechanical annealing is observed in the sample with a cross-section size of 300 nm $\times$ 500 nm (Wang et al., 2015) but not in other experiments with a larger sample size.

Similarly, the critical size for thorough mechanical annealing under monotonic deformation can be predicted according to its dislocation density evolution equation. If $d\rho/d\varepsilon^p < 0$ and $d\varepsilon^p > 0$, the dislocation density can gradually decrease to 0 during deformation. Therefore, the critical size for mechanical annealing under monotonic deformation can be estimated as $d\rho/d\varepsilon^p = 0$.

For a micropillar with a moderate dislocation density and a diameter of 200–1000 nm, the operation of a single-arm source also contributes to the generation of dislocations. If strong single-arm sources form and operate continuously, the dislocation density will reach a stable value after a sharp decrease (Cui et al., 2014). Thus, the dislocation starvation state can only be reached if a micropillar with an initial dislocation density is lower than the stable dislocation density. Combining Eqs. (10.5.1), (10.5.6), (10.5.7), and (10.5.8), the critical condition for thorough mechanical annealing under monotonic deformation can be expressed as

$$\frac{d\rho}{d\varepsilon^p} = \frac{1}{bM}\left(\frac{1}{2\bar{\lambda}} - \frac{2\cos^2(\beta/2)}{d} + k_f\sqrt{\rho} - y\rho\right) = 0 \qquad (10.6.2)$$

where $\bar{\lambda}$ is the length of the statistically average effective single-arm sources, which can be expressed as a function of the dislocation density and pillar diameter (Jennings et al., 2012; Parthasarathy et al., 2007; Zhou and LeSar, 2012). The critical diameter for thorough dislocation annihilation can be estimated as

$$d_{\text{crit\_monotonic}} = \frac{2\cos^2(\beta/2)}{\dfrac{1}{2\bar{\lambda}(d_{\text{crit\_monotonic}},\rho)} + k_f\sqrt{\rho} - \rho y} < d_{\text{crit\_cyclic}} \qquad (10.6.3)$$

Comparing Eq. (10.6.3) and Eq. (10.6.1), it is easy to see that thorough mechanical annealing can take place in a larger size under cyclic deformation than that under monotonic deformation. If k_f is set to 10^{-2} (Malygin, 2012) and monotonic loading orientation is along [001], for the FIB fabricated micropillar with an initial dislocation density of 10^{14} m^{-2}, the critical pillar diameter for mechanical annealing is calculated to be about 130 nm, according to Eq. (10.6.3). This is close to the sample size (the top diameter of the pillar is 160 nm), for which dislocation starvation is observed via in situ TEM under a monotonic compression test (Shan et al., 2008). It also explains why dislocation starvation is rarely observed in a larger pillar under monotonic loading.

According to the experimental data and this analysis, a comparison is made for dislocation starvation behavior under monotonic loading and cyclic loading in Table 10.1. To achieve mechanical annealing by imposing external loading, surface annihilation has an important role. Thus, both phenomena are apt to occur in a smaller size, which means a more pronounced image force, a shorter mean free path, and a larger specific surface area for surface annihilation. During monotonic loading, high stress destroys the dislocation

Table 10.1: Comparison of Dislocation Annihilation Under Monotonic Loading and Cyclic Loading for Pillar With Initial Dislocation Density of 10^{14} m^{-2}

	Monotonic Compression	Low Amplitude Cyclic Loading
Dislocation junction failure and surface annihilation mechanism	Induced by high stress	Assisted by cumulative irreversible slip
Predicted critical size for dislocation starvation	About 130 nm	>300 nm
Critical size for dislocation starvation in experiments	Small, about 160 nm (Shan et al., 2008)	Large, >387 nm (Wang et al., 2015)
Stress level	High, about 1 GPa (Shan et al., 2008)	Low, about 200–400 MPa
Residual deformation	Large	Very small

junction and increases the probability of making dislocations move to the near surface region. During low-amplitude cyclic loading, both dislocation junction failure and surface annihilation are enhanced by cumulative irreversible slip. The difference in the dislocation mechanism makes it possible for mechanical annealing under cyclic loading to occur in a larger sample under lower applied stress and to produce a smaller change in shape.

10.7 Summary

This chapter presents the finding that cyclic loading with a low stress amplitude can drive dislocations out of submicron single crystals without notably changing the shape of the structure, which is significantly different from traditional dislocation accumulation observed during the cyclic deformation of bulk materials. Cumulative irreversible slip is a key factor in promoting junction destruction and dislocation annihilation at the free surface under low-amplitude cyclic loading at small scales. A dislocation density evolution model is given that introduces this mechanism and can predict critical conditions for mechanical annealing under cyclic and monotonic loadings. Low-amplitude cyclic loading, which strengthens the single crystal without disturbing the structure considerably, has potential applications in the manufacture of defect-free nanodevices.

Strain Rate Effect on Deformation of Single Crystals at Submicron Scale

Chapter Outline

11.1 Introduction

An understanding of the dynamic plastic behavior of metals is important to many engineering applications including high-speed machining and impact loading. Numerous experiments have demonstrated that the mechanical properties of materials, such as yield stress and strength, depend on the applied rate.

In Section 11.2, discrete dislocation dynamics (DDD) simulations were carried out to investigate the strain rate effect systematically on finite-sized copper single-crystalline samples under uniaxial compression and hydrostatic compression. In Section 11.3, a combined finite element (FE) simulation and DDD approach was introduced to investigate the effects of the uniaxial tension loading rate on yield stress and deformation patterning of single-crystal copper at the micron scale. Section 11.4 focuses on the homogeneous

Dislocation Mechanism-Based Crystal Plasticity. https://doi.org/10.1016/B978-0-12-814591-3.00011-X

nucleation (HN) of dislocations under shock loading and dynamic mechanical behavior under different impact speeds. Finally, a summary is provided in Section 11.5.

11.2 Strain Rate Effect on Flow Stress in Single-Crystal Copper Under Compression Loading

It is well-known that various aspects of mechanical behavior at submicron scales are different from those at the bulk scale. In addition to the widely known size effect (Greer et al., 2005; Greer and Nix, 2006; Parthasarathy et al., 2007), strain rate sensitivity at the submicron scale also attracts lots of attention (Liu et al., 2008; Zhang et al., 2013a; Khan et al., 2015). In particular, Jennings et al. (2011) demonstrated a strain rate effect emerging in single-crystalline copper pillars at submicron scales. By computing the activation volumes as a function of the pillar diameter at each strain rate, they postulated a plasticity mechanism transition from dislocation multiplication by operating single-arm sources to surface dislocation nucleation when the pillar size became smaller. The effects of both the strain rate and sample size on the compressive strength of single-crystalline copper were reported although quantitative investigations of dislocation sources and structures were limited. Gurrutxaga-Lerma et al. (2015a) investigated the dominant dislocation mechanism at different strain rates ranging from 10^1 to 10^{10} s^{-1} and illustrated a strain rate effect on both the activation time and the source strength of Frank-Read (FR) sources. They discussed FR sources and HN processes, which are essential in bulk deformation. However, in submicron and nanosized samples, dislocation nucleation from the free surface has been observed to be a potent mechanism (Jennings et al., 2011; Ryu et al., 2015).

DDD simulations reported in the literature provide important insights into mechanical behavior under different strain rates. They serve as an effective method to study dislocation-based plasticity under low strain rates (Zbib and Diaz de la Rubia, 2002), high strain rates (Roos et al., 2001a,b; Wang et al., 2008b; Agnihotri and Van der Giessen, 2015), and extremely high strain rates as shock compression (Shehadeh et al., 2005b; Shehadeh, 2012; Kattoura and Shehadeh, 2014). The calculation procedure always follows the same general form. Forces acting on all of the dislocation segments are evaluated at each time step, dislocation velocities are calculated by solving the equations of motion, and dislocation positions are updated for the next time step. In the most applications of DDD simulations, the equation of dislocation motion involves a linear force-velocity law instead of the full dynamical formula, which includes inertia. This assumption seems sufficient to describe dislocation motion at low deformation rates, but it may not be appropriate for high strain rate deformation. Wang et al. (2007a) employed a full dynamical equation of dislocation motion for comparison with the commonly assumed linear force-velocity dynamics. Through comparison and analysis, they indicated that the inertial effect could not be ignored when the strain rate was 10^4 s^{-1} and larger.

In this section, DDD simulations are carried out to investigate the effect of the strain rate on finite-sized copper single-crystalline samples under uniaxial compression and hydrostatic compression. Mechanical behavior and microstructures are systematically discussed.

11.2.1 Strain Rate Effect of Submicron Copper Pillars Under Uniaxial Compression

The tree-dimensional DDD method used here was described in detail in our previous work (Cui et al., 2014). Compared with the conventional DDD model, two new features are introduced. A full dynamical equation of dislocation motion including an inertial effect is adopted and a surface nucleation process is introduced.

The full dynamical equation of dislocation motion is

$$m_e \dot{v} + Bv = f \tag{11.2.1}$$

where m_e is the effective mass of dislocation per unit length, and $m_e = \rho_c b^2$, where ρ_c is the material density and b is the magnitude of the Burgers vector. B is the viscous drag coefficient and f is the total Peach-Koehler force of the applied stress and interaction with other defects, line tension, and image force by the free surface.

The introduction of dislocation sources associated with surface nucleation in DDD computations was motivated by the atomistic model of Zhu et al. (2008b). Their results suggested that the free surface acts as an effective source of dislocations and the nucleation stress provides an upper bound to the strength of compressive nanopillars. Their probabilistic investigations revealed that the probability of dislocation nucleation from the free surface within a time span dt is given by

$$P = v_0 \exp\left[-\frac{Q(\sigma, T)}{k_B T}\right] \times \left(\frac{S}{b^2}\right) \times dt \tag{11.2.2}$$

Here, v_0 is the attempt frequency, $Q(\sigma, T)$ is the activation free energy of dislocation nucleation from the free surface, $k_B T$ is the thermal energy, and $\frac{S}{b^2}$ is the number of nucleation sites at the surface. In DDD simulations, we expand possible nucleation sites from a sample corner to the whole surface; dislocation can nucleate at any point on the surface of the pillar and stress $\sigma = |\sigma_{33}|$ is used in simulations of the compression test. A first approximation of the temperature effect on the activation free energy is introduced as

$$Q(\sigma, T) = \left(1 - \frac{T}{T_m}\right) \cdot Q_0(\sigma) \tag{11.2.3}$$

where T_m is the surface disordering temperature and is chosen to be 700K. By taking the function form $Q_0(\sigma) = A(1 - \sigma/\sigma_{ath})^{\alpha}$ to fit the calculated activation energies at different stresses, $A = 4.8$ eV, $\sigma_{ath} = 5.2$ GPa, and $\alpha = 4.1$ are obtained (Zhu et al., 2008b).

The probability of surface nucleation is calculated at each time span. The number of nucleated dislocations N is

$$N = \begin{cases} \text{Integer}[P], & \text{if } \int_{t_m}^{t_n} P \geq 1 \\ 0, & \text{if } \int_{t_m}^{t_n} P < 1 \end{cases} \tag{11.2.4}$$

Here, the probability of dislocation nucleation is accumulated from t_m to t_n because that once a dislocation nucleation process occurs, the probability starts from 0 again and an arc-shaped dislocation loop on a random slip system of all 12 typical *face-centered cubic* (FCC) slip systems is introduced at a stochastic site on the surface.

In the simulations, the cross-sections of single-crystalline copper pillars are set to be square. Side lengths vary from 200 to 800 nm whereas the ratio of pillar height H to side length D is fixed at 2. These cuboid cells are used to mimic cylindrical specimens used in the experiment (Rao et al., 2008) because it is easier to deal with the image force and it is known that the cross-section shape has only a weak effect (Kiener et al., 2011). All pillars are loaded along the $(00\bar{1})$ direction in a displacement-controlled manner and the lateral surfaces are traction-free. A wide range of strain rates are performed in DDD simulations, from 10^4 to 10^6 s^{-1}. The initial dislocation densities are around 4×10^{13} m^{-2}. Simulations for each pillar size are carried out for five different initial dislocation distributions to illustrate stochastic scatter. The material properties of copper are: shear modulus $\mu = 48$ GPa, Poisson's ratio $\nu = 0.34$, the density of Cu $\rho_c = 8.96 \times 10^3$ kg/m^3, and the viscous drag coefficient $B = 2 \times 10^{-5}$ Pa s. The attempt frequency ν_0 in Eq. (11.2.2) is set at 10^{13} s^{-1} (Ryu et al., 2011) and temperature $T = 300$K. The time step used in the simulations is $\triangle t = 10^{-12}$ s, which ensures that the numerical results are converged.

The simulation results for pillars with different side lengths are shown in Fig. 11.1A. The stress-strain curves exhibit three typical characteristics. For type I, which is generally observed in small pillars ($D < 400$ nm), as marked in Fig. 11.1A, stress increases in an elastic way at the initial stage, and then early on, yield events of a short duration occur at various stress levels (varying between ~ 500 and ~ 1200 MPa), followed by an extensive elastic period, after which massive yield takes place at a stress level of around 1600 MPa. Despite the small oscillation, flow stress is approximately stable at the end. Distinct from this, in type III plasticity, yield is massive right away and stress remains roughly stable without obvious strain hardening. As a result, flow stress is relatively low compared with that in type I. Type II falls between types I and III; it seems that stress remains stable after initial yielding but strain hardening can start unexpectedly.

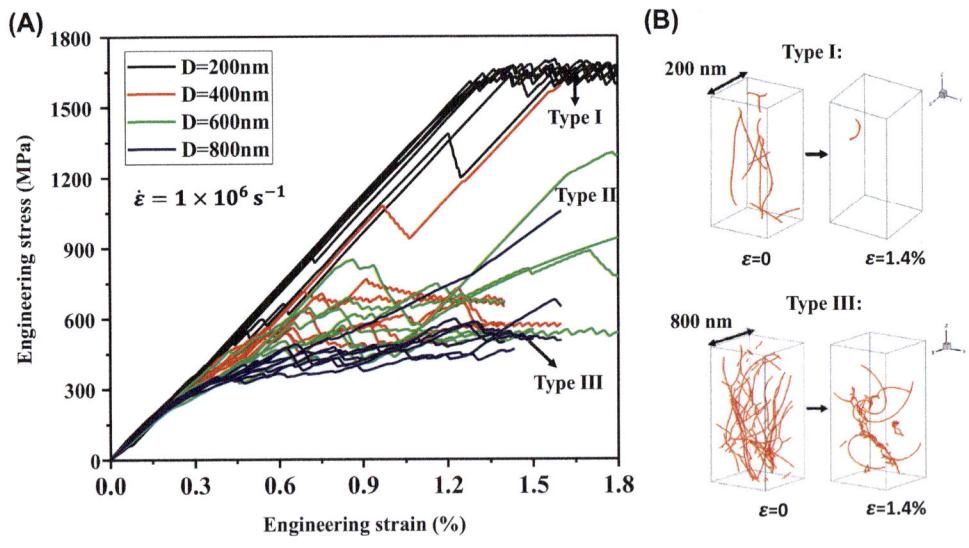

Figure 11.1

(A) Stress-strain curves for pillars with different sizes at an applied strain rate of 10^6 s^{-1}; (B) Typical dislocation structures for types I and III. *Reprinted from Hu, J.Q., Liu, Z.L., Van der Giessen, E. Zhuang, Z., 2017a. Strain rate effects on the plastic flow in submicron copper pillars: Considering the influence of sample size and dislocation nucleation. Extreme Mechanics Letters 17, 33–37. Copyright 2017, with permission from Elsevier.*

The evolution of representative dislocation structures corresponding to types I and III is shown in Fig. 11.1B. For pillars as small as 200 nm, it is difficult for dislocations to form stable pinning points and the initial dislocations will be driven out quickly by subsequent external loading. After that, new dislocations emerge from the free surface and immediately escape out of the pillar. The small oscillation of flow stress is attributed to the repeated nucleation and escape of surface dislocations. Therefore, it is a surface nucleation-dominant mechanism in small pillars characterized by very high flow stress. In large pillars ($D = 800$ nm) such as type III, in contrast, internal dislocations can self-organize initially to form a jammed configuration. With the increase in external loading, the jammed network will continuously evolve until internal dislocation sources with stable pinning points are formed. Both single-arm sources and FR sources exist inside large pillars at the moment. Operation of these internal sources can generate enough plasticity to keep the flow stress stable. In this case, the dislocation multiplication mechanism has a leading role featuring relative low stress levels. The mechanical behavior of medium-sized pillars ($D = 400-600$ nm) is sensitive to the initial dislocation configuration. For samples without enough dislocation junctions and tangles, after the break of weak links between dislocations, only a few single-arm sources are formed inside the pillars and their operation leads to a provisional yielding stage. However, the

dislocation sources can become progressively exhausted, causing dislocation starvation hardening, shown as type II in Fig. 11.1A.

In addition to the sample size, the strain rate has a significant effect on plastic behavior. To investigate the effects of strain rate imposed by different displacement rates at the pillar ends, stress-strain curves for pillars with side lengths of 200 and 800 nm, respectively, are shown in Fig. 11.2. Two distinct types of plastic behavior are observed, distinguished by different stress levels. For small pillars ($D = 200$ nm), flow stresses are generally high within the investigated range of strain rates ($\geq 10^4$ s^{-1}). As in type I shown in Fig. 11.1B, all initial dislocations tend to escape owing to external loading. Then, surface nucleation takes over as the dominant dislocation mechanism because there no preexisting dislocations are left. Moreover, dislocation starvation and surface nucleation are more likely to happen with the increase in strain rate. Nevertheless, in a small number of pillars with a side length 200 nm, one or a few stable single-arm sources may persist at relative low strain rates. Operation of these sources can also result in relatively low flow stress such as in large pillars. In contrast, as shown in Fig. 11.2B, flow stress in large pillars ($D = 800$ nm) is generally not high enough for dislocation nucleation from the surface to be activated.

The exhaustion of internal sources leads to an increase in elastic stress and eventually results in the surface nucleation process. Stable flow stresses that are involve in nucleation, characterized by the absence of a hardening process, can be predicted by the surface nucleation formula. In the case of the dislocation multiplication-dominated

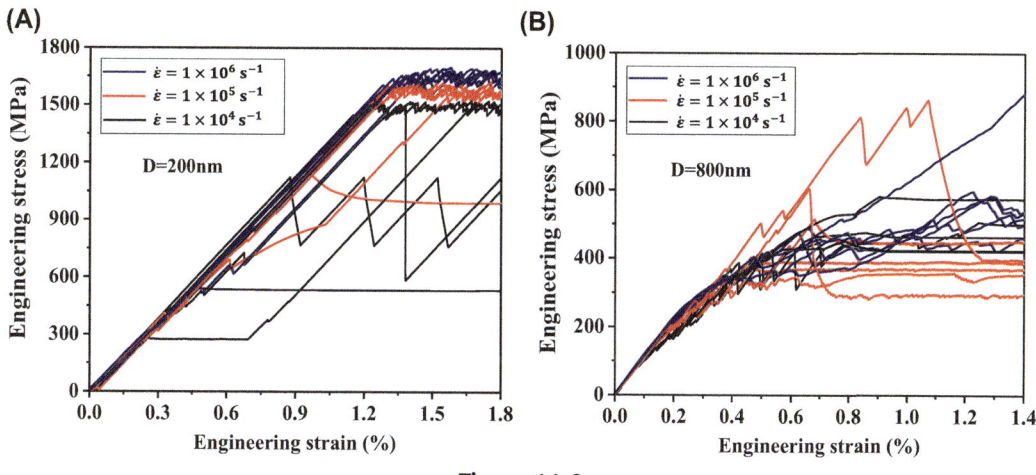

Figure 11.2

Stress-strain curves for pillars with different sizes at strain rates from 10^4 to 10^6 s^{-1}: (A) $D = 200$ nm; (B) $D = 800$ nm. *Reprinted from Hu, J.Q., Liu, Z.L., Van der Giessen, E. Zhuang, Z., 2017a. Strain rate effects on the plastic flow in submicron copper pillars: Considering the influence of sample size and dislocation nucleation. Extreme Mechanics Letters 17, 33—37. Copyright 2017, with permission from Elsevier.*

process, stable flow stresses vary greatly even in the case of the same strain rate and sample size, as shown in Fig. 11.2B. We systematically investigated the evolution of internal dislocation structures and found that the internal source-controlled yield is controlled solely by the number of internal dislocation sources. Consequently, stable flow stress depends on the number of dislocation sources in addition to the sample size and strain rates. Following Eq. (8.4.10) in Chapter 8 and referencing works Parthasarathy et al. (2007), Lee and Nix (2012) and Cui et al. (2014), we propose a formula for stable shear flow strength associated with three factors:

$$\tau = \underbrace{\tau_0 + \alpha\mu b\sqrt{\rho} + \frac{k\mu b}{\lambda}}_{\tau_{\text{static}}} + \frac{B\dot{\varepsilon}V}{Mb^2\lambda \cdot n} \qquad (11.2.5)$$

Here, the first three items on the right-hand side are the internal source-controlled critical resolved shear stress under quasistatic loading, as systematically studied by Cui et al. (2014); τ_0 is the lattice friction stress; α and k are dimensionless constants, where usually α is set at 0.5 and k is taken as 1.0 in previous literature (Parthasarathy et al., 2007; Lee and Nix, 2012); ρ is the dislocation density; and λ is the average effective source length. The last term is an additional part owing to the strain rate effect, in which $\dot{\varepsilon}$ is the strain rate, V is the sample volume, n is the number of stable single-arm sources, and M is the Schmid factor. Obviously, the contribution of the last term increases when the strain rate becomes higher. In particular, for pillars with adequate internal dislocation sources, the last item in Eq. (11.2.5) can be neglected as $n \to \infty$ (except when the strain rate $\dot{\varepsilon} \to \infty$) and the stable flow strength reduces to a quasistatic one. It possesses a physical foundation in Orowan's law. Furthermore, for a certain strain rate loading, the stress obtained in Eq. (11.2.5) could become high enough to trigger surface nucleation if the source number n is limited. Once the shear flow strength is obtained by Eq. (11.2.5), flow stress can be calculated by $\sigma = \tau/M$, where M is the Schmid factor.

To test this proposition, especially the effect of the number of dislocation sources, the results for two pillars with a side length of 800 nm are investigated in detail. For each case, the dislocation configuration at three strain levels well into the plastic regime are considered (Fig. 11.3). During the analysis, stable flow stress is calculated by averaging all stress values within the strains ranging from 1.2% to 1.6% in the simulation to eliminate the influence of stress oscillation. More dislocation sources in Pillar 2 result in relatively low flow stress. The stable flow stresses are given in Table 11.1 for two pillars by numerical simulation and an analytical solution. The analytical solutions are consistent with the results of the simulation.

In addition, the analytical solution we achieved is based on the single-arm source dominant mechanism, which is suitable at submicron scales but not for a very small or very large sample size. Transmission electron microscopy (TEM) experiments (Shan et al., 2008)

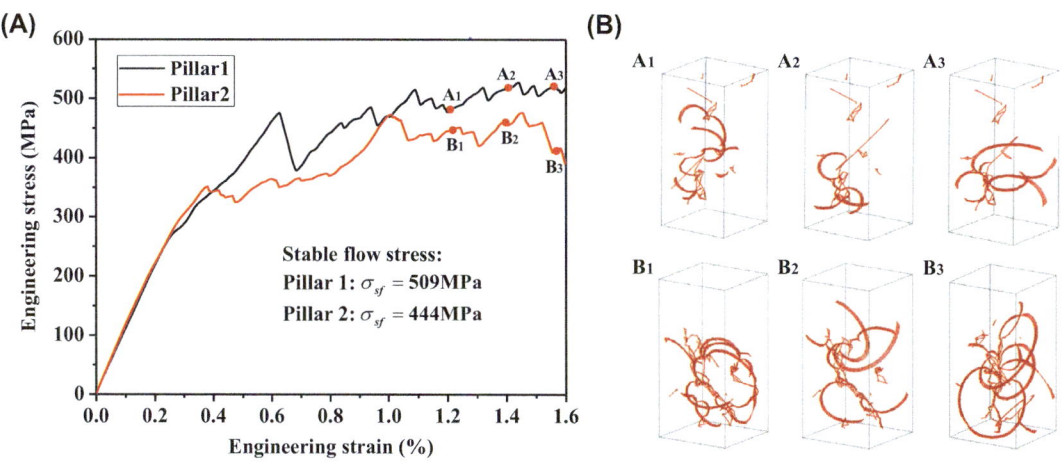

Figure 11.3

(A) Stress-strain curves for pillars with the same size ($D = 800$ nm) and strain rate ($\dot{\varepsilon} = 1 \times 10^6$ s^{-1}); (B) Dislocation structures corresponding to each marked point in (A). *Reprinted from Hu, J.Q., Liu, Z.L., Van der Giessen, E. Zhuang, Z., 2017a. Strain rate effects on the plastic flow in submicron copper pillars: Considering the influence of sample size and dislocation nucleation. Extreme Mechanics Letters 17, 33—37. Copyright 2017, with permission from Elsevier.*

Table 11.1: Comparison of Simulation and Analytical Solution Between Two Pillars

	Number of Single-Arm Sources n at points labeled in Fig. 11.3			Simulated Flow Stress (MPa)	Analytical Prediction (Value of n)
	A_1 or B_1	A_2 or B_2	A_3 or B_3		
Pillar 1	8	7	9	509	534 MPa (8)
Pillar 2	13	9	12	444	451 MPa (11)

revealed that pillars with a diameter smaller than 160 nm can achieve dislocation starvation by mechanical annihilation. According to this analysis, Fig. 11.4 presents a schematic diagram of the governing dislocation mechanisms in plastic deformation at submicron scales that unifies the three combined effects, in which the mechanisms are divided into three zones depending on sample size D, strain rate $\dot{\varepsilon}$, and stable dislocation density ρ_s or source number n. The three zones correspond to the three types in Fig. 11.1. Several characteristic values for transitions extracted from computations in this work are also marked in Fig. 11.4.

In small pillars (~ 200 nm), dislocation pinning points are not formed and the initial dislocations run across the sample and easily escape. Dislocation starvation triggers the nucleation of dislocations from the surface, which causes very high stable flow stress. In contrast, at the same initial dislocation density, dislocations in large pillars have a smaller

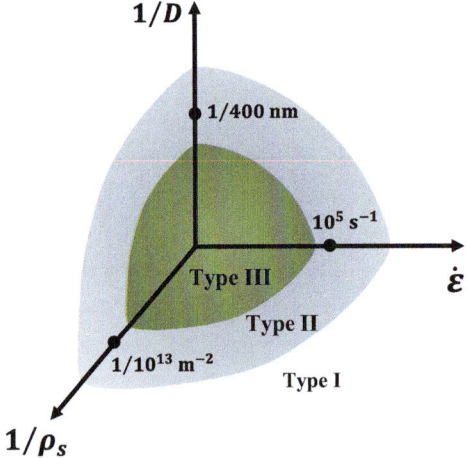

Figure 11.4

Dislocation mechanism map for plastic flow at submicron scales: nucleation of surface dislocations (Type I), cooperation of dislocation multiplication and surface nucleation (Type II), and stable internal single-arm dislocation multiplication (Type III). *Reprinted from Hu, J.Q., Liu, Z.L., Van der Giessen, E. Zhuang, Z., 2017a. Strain rate effects on the plastic flow in submicron copper pillars: Considering the influence of sample size and dislocation nucleation. Extreme Mechanics Letters 17, 33—37. Copyright 2017, with permission from Elsevier.*

tendency to escape from the free surface and more probability of developing stable pinning points. As a result, the plastic behavior of large pillars is always dominated by dislocation multiplication, featuring relatively low stress levels. At higher strain loading rates, the flow stress tends to increase because the limited number of internal sources cannot generate enough plasticity to resist the increase in external loading: that is, $\dot{\varepsilon}^p < \dot{\varepsilon}$. In turn, these internal dislocation sources may be broken owing to the increasing stress. Finally, to illustrate the central role of stable dislocation sources, an analytical expression is proposed for comparison with the simulation results and further explains why the stable flow stress varies even in pillars with the same size and strain rate. We emphasize the role of stable dislocation sources but not the initial dislocation density, because it directly translates into the probability of achieving stable internal dislocation sources.

11.2.2 Strain Rate Effect of Dislocation Evolution in Copper Cubes Under Hydrostatic Pressure

Most of investigations in the open literature have concentrated on mechanical behavior and its corresponding dislocation evolution under uniaxial compression (Shan et al., 2008; Liu et al., 2009a,b; Ryu et al., 2015). However, mechanical behaviors along other loading paths might exhibit some different features. For instance, the effect of size on the response to hydrostatic pressure, which is rarely studied, might have valuable potential applications

under extreme environments such as deep-sea operations and the three-dimensional printing of metallic parts. Investigations into mechanical behavior under hydrostatic pressure are also valuable for a better understanding of the plate impact loading process in which a hydrostatic state of stress becomes dominant (Shehadeh et al., 2006; Kattoura and Shehadeh, 2014; Hu et al., 2017b). In addition, Yuan and Wu (2014) investigated the interplay between pressure and grain size with a series of large-scale molecular dynamics (MD) simulations that showed different dislocation behaviors below and above a critical hydrostatic pressure. Work by Chen et al. (2014) explored the effect of the hydrodynamic compression rate on crystal formation from a noble fluid that contained a Cu inclusion. Hence, an important issue arises regarding whether Cu inclusion of different sizes might exhibit different mechanical behaviors under hydrostatic compression. Because the mechanical responses of metallic samples are insensitive to confining pressure at the macroscale, von Mises elastoplasticity has commonly been used to model metals in engineering practice. To clarify the applicability of von Mises elastoplasticity at the microscale, we used DDD in this work to explore the mechanical response and corresponding dislocation evolution under hydrostatic pressure. The focus of this section is on the effects of hydrostatic pressure, the loading rate, and the sample size on the evolution of the dislocation microstructure.

A series of DDD simulations of the hydrostatic compression test are performed on finite-sized Cu single-crystalline samples. Each sample is set to a cube with the same edge length, ranging from 200 to 1200 nm in three dimensions. These cubes are set with the x-, y-, and z-axes along [100], [010], and [001], respectively. As a reference, the uniaxial compression test is also carried out on all Cu samples to investigate the effect of the loading mode. All simulations are performed under stress-controlled compression loading. The material properties are: shear modulus $\mu = 48$ GPa, Poisson's ratio $\nu = 0.34$, mass density $\rho_c = 8.96 \times 10^3$ kg/m^3, magnitude of the Burgers vector $b = 0.256$ nm, and viscous drag coefficient $B = 2 \times 10^{-5}$ Pa s.

Initial equilibrium dislocation configurations are generated via a relaxation procedure that approximates a real thermal annealing process (Zhou et al., 2010a,b, 2011). This process starts by generating randomly created straight dislocation lines and internal dislocation loops spreading through all 12 FCC slip systems. The ends of straight dislocation lines terminate at the free surface. All the internal dislocations then evolve dynamically without external loading until the structures stabilize. More details can be found in Cui et al. (2014). According to experimental observations (Lee et al., 2009; Jennings et al., 2010), the mechanical response of micropillars with the same size also varies. This motivates us to use different initial dislocation configurations for each sample size D. As a result, the initial dislocation structure and density are different from each other in the Cu samples. After the initial relaxation procedure, hydrodynamic compression loading rates ranging from 0.1 to 1.0 MPa/ps are applied on all three dimensions of cubes, i.e., along the [100],

[010], and [001] crystal orientations. For the case of uniaxial compression, however, the same compression loading rates are applied on the top of the cube along the [001] crystal orientation and the lateral surfaces are traction-free. The time step used in the simulations is $\triangle t = 1$ ps, which ensures that the numerical results are converged. To compare responses between hydrostatic compression and uniaxial compression in the simulation results, the stress in hydrostatic compression is defined as $\sigma = -(\sigma_{xx} + \sigma_{yy} + \sigma_{zz})/3$ and the one in uniaxial loading is $\sigma = -\sigma_{zz}$, for which the corresponding strains are both defined as $\varepsilon = -(\varepsilon_{xx} + \varepsilon_{yy} + \varepsilon_{zz})$.

Copper cubes with an edge length of 400 nm are loaded under two different modes, i.e., uniaxial and hydrostatic compression, respectively. The stress-strain curves and evolution of dislocation density under the two loading modes are plotted for the Cu samples with the same size and three different initial configurations, as shown in Fig. 11.5. In particular, different curve types (i.e., solid, dotted, and dashed lines) are used to distinguish the different initial dislocation configurations and indicate the correspondence between the evolution of dislocation density and mechanical response. Fig. 11.5A shows that the stress-strain curves of Cu cubes are almost the same under hydrostatic compression when the edge length is 400 nm, for which the stress increases approximately linearly with the strain. Hence, the initial dislocation structures have a limited effect on mechanical behavior and the dislocation density gradually decreases with an increase in hydrostatic pressure. In the case of uniaxial compressive loading, as shown in Fig. 11.5B, however, the stress-strain curves exhibit a conventional pronounced

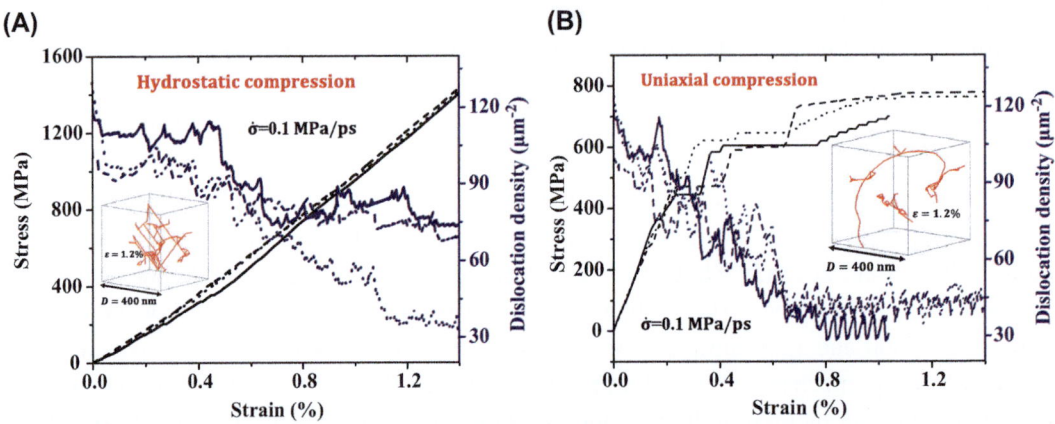

Figure 11.5

Stress (*black line*) and dislocation density (*blue line*) evolution for Cu cubes with different initial structures under (A) hydrostatic compression test, and (B) uniaxial compression test. *Reprinted from Hu, J.Q., Chen, Z., Liu, Z.L., Zhuang, Z., 2018. Pressure sensitivity of dislocation density in copper single crystals at submicron scale. Materials Research Express 5, 016504. Copyright 2018, with permission from IOP Publishing.*

step-like character (Kiener and Minor, 2011a,b), which suggests that the discrete strain burst events occur in an uncontrollable and stochastic way. After that, the stress reaches a critical stable value at which the yield strength is obtained. The dislocation density is fluctuant, which is related to the operation and suppression of single-arm dislocation sources (Cui et al., 2014).

Another feature of dislocation evolution under uniaxial compression is that the dislocation density decreases more rapidly to a stable state because dislocation located on the slip plane is activated via effective resolved shear stress on the plane. For an FCC single crystal with an orientation of <001>, the resolved shear stress is much larger in the case of uniaxial compression. By contrast, the dislocations can hardly be activated and tend to remain stable under hydrostatic pressure, which leads to insensitivity of the mechanical response to the confining pressure. Dislocation microstructures at the strain of 1.2% in the Cu sample with the same initial dislocation configuration are also illustrated in Fig. 11.5. All dislocations are trapped on the slip plane in the hydrostatic compression test even though the stress is much higher ($\sim$ 1200 MPa), whereas operation of the single-arm dislocation source continues in the case of uniaxial loading, which leads to massive yielding. Next, the dislocation density and microstructure under hydrostatic pressure will be further analyzed.

The previous analysis shows that the mechanical responses appear to be insensitive to confining pressure at small scales; this is further confirmed by analyzing the stress-strain curves for Cu samples of different sizes. However, evolution of the dislocation shows some different characteristics. The following investigation focuses on the evolution of the dislocation density and microstructures. Cu samples with edge lengths from 200 to 1200 nm are loaded under hydrostatic pressure with a loading rate 0.5 MPa/ps. Evolution of the dislocation density with increasing pressure is shown in Fig. 11.6A−D. It exhibits size dependency under hydrostatic compression. For Cu samples with edge lengths of 200 and 400 nm, the dislocation density is relatively stable at the beginning and then gradually decreases with increasing stress, exhibiting pressure sensitivity. This may be because the operation of dislocation sources requires larger critical shear stress for a smaller sample (Zhou et al., 2010a,b; Zhou et al., 2011) and dislocations tend to move out owing to the image force by the free surface when they are activated (Cui et al., 2014). For larger samples ($D = 800$ and 1200 nm), after the initial decrease in dislocation density, which corresponds to the mechanical annealing process (Shan et al., 2008), the dislocation density can increase with an increase in hydrostatic pressure and finally achieve a relatively stable value. This process exhibits a characteristic and insensitivity similar to those of the pressure, featuring a relatively stable dislocation density although the pressure keeps increasing.

As a reference, the dislocation evolution in Cu samples with edge lengths of 400 and 800 nm under uniaxial loading are also presented in Fig. 11.7. The evolution of dislocation

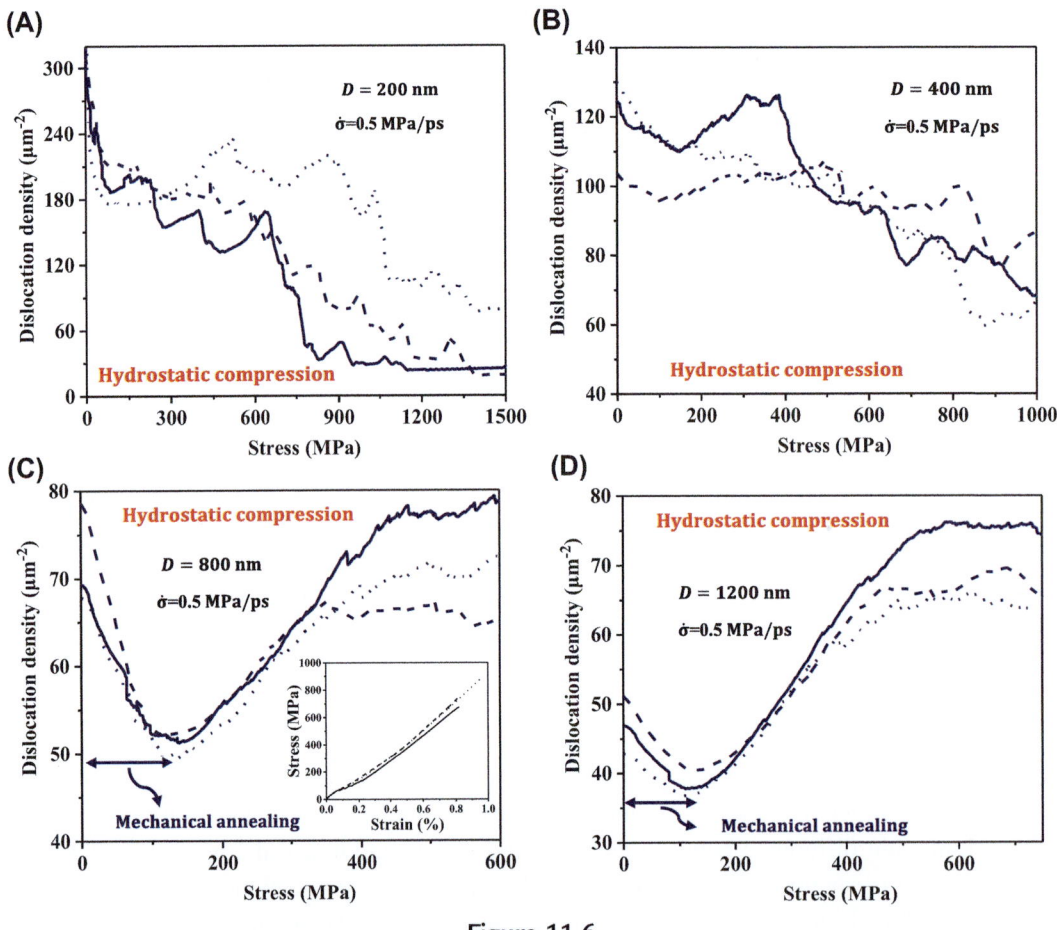

Figure 11.6

Evolution of dislocation density in Cu cubes with different edge lengths: (A) $D = 200$ nm, (B) $D = 400$ nm, (C) $D = 800$ nm, and (D) $D = 1200$ nm under hydrostatic loading. *Reprinted from Hu, J.Q., Chen, Z., Liu, Z.L., Zhuang, Z., 2018. Pressure sensitivity of dislocation density in copper single crystals at submicron scale. Materials Research Express 5, 016504. Copyright 2018, with permission from IOP Publishing.*

density is different from that under hydrostatic pressure loading. In the early stage of plastic deformation, stress is relatively low when the massive dislocation multiplication occurs, as characterized by the appearance of stress steps in the stress-strain curves. Vertical lines in the curves of dislocation density evolution correspond to dramatic dislocation activity indicating that dislocations are more easily activated and generate plenty of plasticity in this case. The dislocation density continues to increase after stress reaches the yield strength, resulting in a sample that is generally unable to withstand large loads under uniaxial loading conditions.

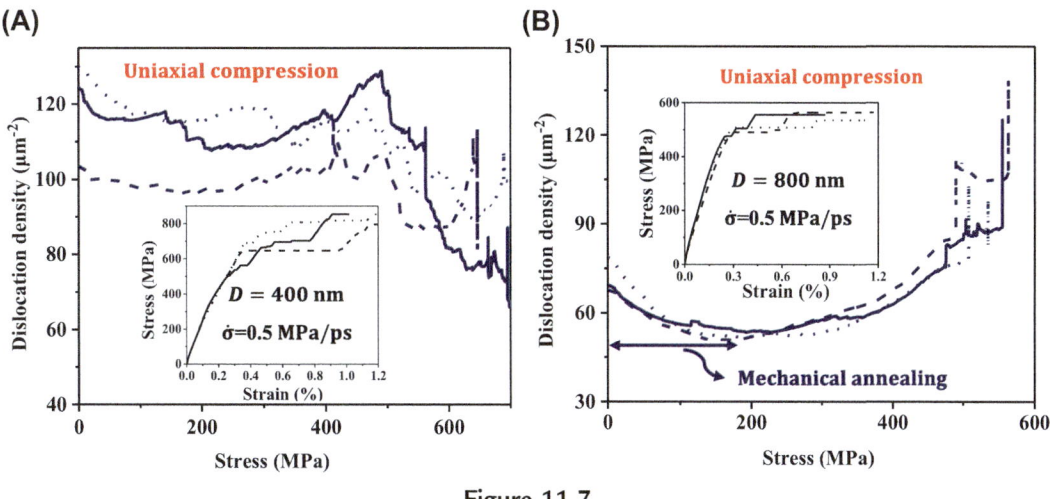

Figure 11.7

Evolution of dislocation density for Cu cubes with edge lengths of (A) $D = 400$ nm and (B) $D = 800$ nm under uniaxial compression. *Reprinted from Hu, J.Q., Chen, Z., Liu, Z.L., Zhuang, Z., 2018. Pressure sensitivity of dislocation density in copper single crystals at submicron scale. Materials Research Express 5, 016504. Copyright 2018, with permission from IOP Publishing.*

Thus, two questions arise: why does the sample size have an important role in the evolution of dislocation density under hydrostatic compression, especially when sample size becomes smaller ($D \leq 400$ nm), and why does the evolution of dislocation density show a significant difference between cases of hydrostatic and uniaxial compression? The dislocation microstructures are further analyzed to answer these questions. As shown in Fig. 11.8A and B, in the case of hydrostatic compression, for Cu samples with an edge length of 400 nm, the dislocations are intermittently activated and the links between dislocations are destroyed as the pressure increases. At the same time, it is difficult for new stable dislocation links to form before these dislocations are driven out of the samples. The remaining dislocations are more stable and require a higher activation stress. As a result, the dislocation density gradually decreases with the increase in stress, leading to pressure sensitivity in the evolution of the dislocation density. By contrast, for a Cu sample as large as 800 nm, as shown in Fig. 11.8B, many of the generated dislocations can form new junctions and tangles as new sources rather than slip out immediately. Therefore, the dislocation density can increase again after the initial mechanical annealing process occurs. The reconstruction of a jammed dislocation network is stable, which is responsible for the pressure insensitivity.

Moreover, dislocation microstructures for Cu samples with edge lengths of 400 and 800 nm under uniaxial compression are presented in Fig. 11.8C and D, respectively. For a Cu sample with an edge length of 400 nm (Fig. 11.8C), some stable single-arm dislocation

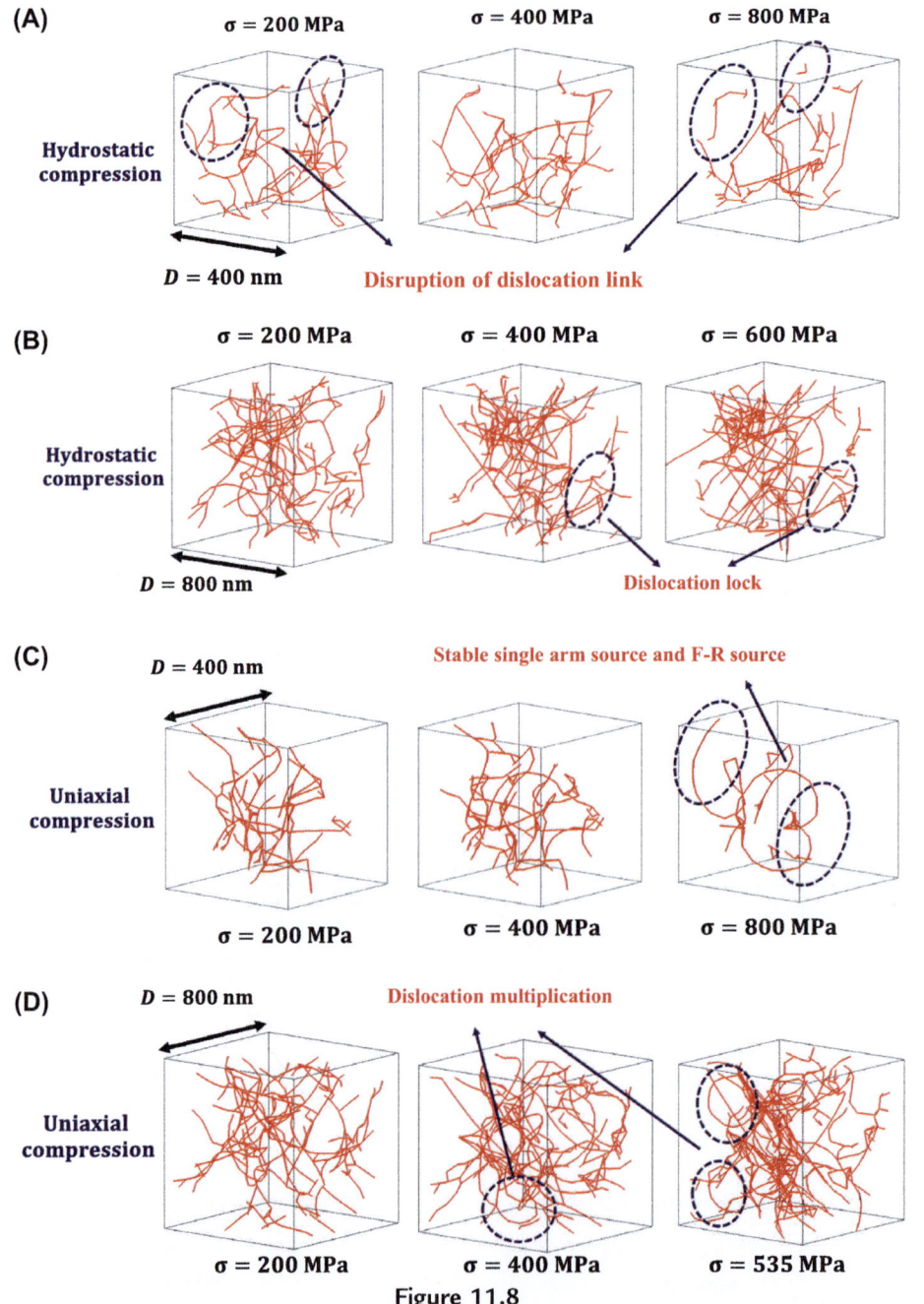

Figure 11.8

Evolution of dislocation microstructures for Cu cubes with different edge lengths: (A) *D* = 400 nm and (B) *D* = 800 nm under hydrostatic compression, and (C) *D* = 400 nm and (D) *D* = 800 nm under uniaxial compression. *F-R*, Frank-Read. *Reprinted from Hu, J.Q., Chen, Z., Liu, Z.L., Zhuang, Z., 2018. Pressure sensitivity of dislocation density in copper single crystals at submicron scale. Materials Research Express 5, 016504. Copyright 2018, with permission from IOP Publishing.*

sources and FR sources can eventually form. These stable dislocation sources continuously sweep the slip plane and generate a large amount of plastic deformation to maintain a steady plastic flow. Similarly, in a Cu sample with an edge length of 800 nm, with an increase in the external load, more dislocation sources are activated, which leads to dislocation multiplication. When the stress reaches a yield strength of 535 MPa, dislocation multiplication continues. For the hydrostatic loading considered here, the internal dislocations are barely activated under the confining pressure. Thus, a number of dislocation locks are formed and are stable, which can serve as an obstacle to dislocation motion. This may why materials can always reach a much higher pressure level under shock loading, in which the region of the material that has been deformed will normally enter a hydrostatic state (Hu et al., 2017a,b,c), so that the movement of internal dislocation is suppressed and stress relaxation can barely be achieved.

To study the rate effect, the mechanical responses for Cu samples with an edge length of 800 nm as well as the evolution of dislocation density under different pressure loading rates are presented in Fig. 11.9. The results in Fig. 11.9A further confirm that the mechanical responses are insensitive to the confining pressure, even under different pressure loading rates, which is different from the rate-dependent mechanical behavior under uniaxial compression (Jennings et al., 2011). However, the evolution of the dislocation density has a significant rate effect. After the initial mechanical annealing process, the dislocation density starts to increase as a result of the activation of dislocation sources and finally becomes nearly stable. The stable dislocation density increases with an increase in the pressure loading rates.

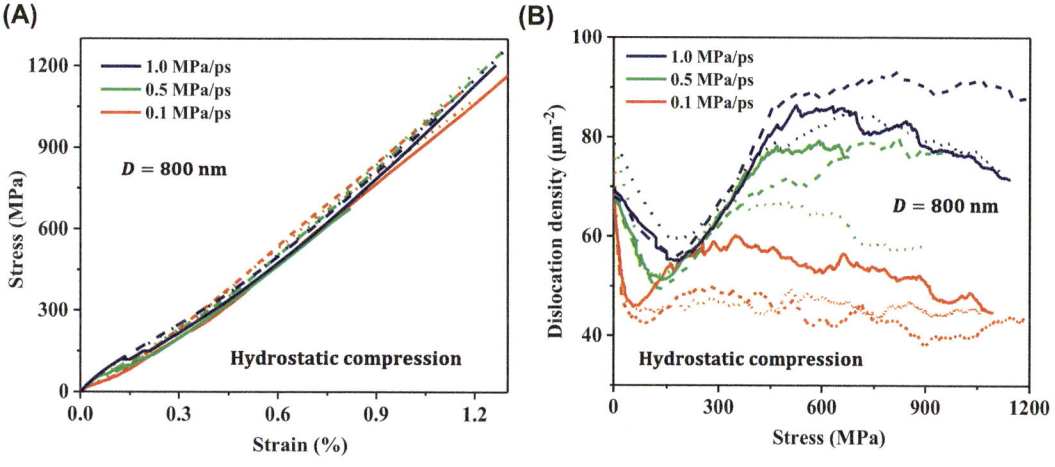

Figure 11.9

(A) Stress-strain curves for Cu samples under different pressure loading rates, and (B) the evolution of dislocation density. *Reprinted from Hu, J.Q., Chen, Z., Liu, Z.L., Zhuang, Z., 2018. Pressure sensitivity of dislocation density in copper single crystals at submicron scale. Materials Research Express 5, 016504. Copyright 2018, with permission from IOP Publishing.*

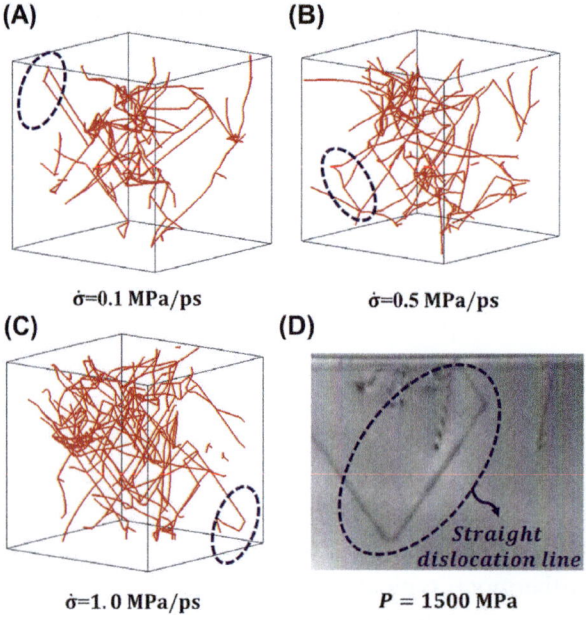

(A) $\dot{\sigma}=0.1$ MPa/ps

(B) $\dot{\sigma}=0.5$ MPa/ps

(C) $\dot{\sigma}=1.0$ MPa/ps

(D) $P = 1500$ MPa

Straight dislocation line

Figure 11.10

Dislocation structures in Cu sample with an edge length of 800 nm at a hydrostatic pressure of 1000 MPa and loading rates of (A) 0.1 MPa/ps, (B) 0.5 MPa/ps, and (C) 1.0 MPa/ps; (D) Typical dislocation microstructure in silicon at a pressure of 1500 MPa (Wzorek et al., 2007). *Reprinted from Hu, J.Q., Chen, Z., Liu, Z.L., Zhuang, Z., 2018. Pressure sensitivity of dislocation density in copper single crystals at submicron scale. Materials Research Express 5, 016504. Copyright 2018, with permission from IOP Publishing.*

Correspondingly, dislocation microstructures in the Cu sample with the same initial dislocation configuration at a hydrostatic pressure of 1000 MPa are shown in Fig. 11.10. First, the typical structural characteristics of dislocations are different from those of uniaxial compression, in which most dislocation lines are curved when the operation of single-arm or FR dislocation sources occurs. Conversely, dislocation structures under hydrostatic pressure are always straight with sharp edges and corners, which is consistent with the microstructures observed in the silicon sample under 1500 MPa pressure (Wzorek et al., 2007), as shown in Fig. 11.10D. Furthermore, by analyzing the evolution of dislocation structures under different loading rates, the rate effect on the dislocation density can be attributed to the transformation of the dislocation structures from curved lines to straight ones that are the final stable morphological structures of dislocation under hydrostatic pressure. A faster-increasing rate in pressure makes it more probable that it will form straight dislocation lines before the activated curved dislocations slip out from the samples. As a result, the stable dislocation structures tend to concentrate in the central region. A high hydrostatic pressure can effectively eliminate the effect of the image force and keep dislocations trapped inside the Cu samples.

With an increase in hydrostatic pressure, however, forces acting on the dislocations are still insufficient to break the strong links between the internal dislocations so that the dislocations cannot generate enough plasticity under hydrostatic loading. As a result, the mechanical response does not exhibit an obvious rate effect.

The dislocation density of Cu samples and their corresponding microstructures were examined under hydrostatic compression and compared with the results under uniaxial compression. The effects of the loading mode, sample size, and loading rate were explored on the mechanical responses. The dislocation structures were less activated under hydrostatic compression compared with uniaxial compression. A size-dependent transition existed in the evolution of the dislocation density, which exhibits pressure sensitivity as the sample size became smaller than a specific value ($D \leq 400$ nm). Although a rate-dependent stress-strain relationship was not observed under hydrostatic compression, the evolution of dislocation density exhibited a significant rate dependence. This is because a more rapid increase in stress made it quicker to suppress the effect of image forces and easier to keep dislocations trapped inside the samples. Furthermore, typical characteristics of dislocation structures under hydrostatic compression were analyzed. Straight dislocations with sharp edges and corners were stable under the confining pressure. These findings may provide a better understanding of the elastoplastic responses of metallic samples at the microscale.

11.3 Strain Rate Effect on Dynamic Deformation of Single-Crystal Copper Under Tensile Loading

The response of crystal metals to high strain rate deformation is a complicated phenomenon involving many physical mechanisms such as dislocation dynamics, twinning, displacive phase transformations, grain size, stacking fault, and solution hardening effects. Several experimental techniques have been used to study the response of materials to a high strain rate, such as plate impact, split Hopkinson pressure bar, explosives, and pulsed laser loading. In particular, high-intensity laser facilities coupled with X-ray diffraction techniques have been used to study dynamic deformation in crystalline materials (Meyers et al., 2003).

In addition to these experimental techniques, computer simulation methodologies have been used to study various dynamic deformation behaviors of materials. In FE simulations, constitutive laws are critical for investigating the dynamic mechanical behavior of crystalline materials in applications spanning from the microscale to the macroscale, especially when materials are sensitive to the strain rate. MD simulations have also been performed to explore the mechanical properties of nanoscale materials. With the use of atomistic methods, dislocation behavior in the plastic flow has been numerically studied (Lee et al., 2015; Gunkelmann et al., 2016; Wen et al., 2016). As can be seen from the

earlier literature, much research has been conducted to investigate the rate dependence of material properties at the atomic scale and macroscale, respectively. However, little has been done to understand loading rate effects on the microscale, because many important failure forms such as shear instability and fracture take place on the microscale (Needleman, 2000).

In this section, a multiscale model is developed by coupling three-dimensional DDD code with FEM. In this method, the DDD code is substituted for the constitutive form traditionally used in FE calculation. A three-dimensional DDD code solves the dynamic and local reactions of discrete dislocation lines and computes the plastic strain generated by dislocation glide. On the other hand, an FE code computes the stress and displacement field, providing a solution to the boundary value problem by using the plastic strain field developed by the DDD simulation. The loading rate effects on yield stress and the deformation patterning of single-crystal copper are investigated.

11.3.1 Resolution of Discrete Dislocation Dynamics

DDD is a microscale approach that simulates the dynamic behavior of a large number of curved dislocations on a finite length. Simulations based on discrete dislocations are mechanism-based, allowing the effects of dislocation motion and dislocation interactions to be analyzed. In three-dimensional DDD, dislocations are discretized into segments of mixed character. The self-stress field of a straight dislocation segment in an infinite solid serves as the basis of DDD code. Because the stress field of dislocation varies with the inverse of distance from the dislocation core, dislocations interact among themselves over long distances. As the dislocation moves, it has to overcome internal drag and local barriers such as Peierls stress. The dislocation may encounter local obstacles such as defect clusters and vacancies, which interact with the dislocation at short ranges and affect its local dynamics. The Peach—Koehler force acting on each dislocation is calculated at the center of each dislocation segment, and the effective force on segment i is given by

$$\mathbf{f}_i = (\boldsymbol{\sigma}_i \cdot \mathbf{b}_i) \times \boldsymbol{\xi}_i + \mathbf{f}_{\text{self}} \tag{11.3.1}$$

where $\mathbf{b}_i$ is the Burgers vector of segment i, $\boldsymbol{\xi}_i$ is the line sense unit vector describing its direction, $\mathbf{f}_{\text{self}}$ is the self-force caused by the nearest neighbor segments, and $\boldsymbol{\sigma}_i$ is the total stress tensor estimated at the center of segment i and can be decomposed into

$$\boldsymbol{\sigma}_i = \boldsymbol{\sigma}^{\text{applied}} + \boldsymbol{\sigma}^{\text{interaction}} + \boldsymbol{\sigma}^{\text{other}} \tag{11.3.2}$$

where $\boldsymbol{\sigma}^{\text{applied}}$ is the externally applied stress, which is computed by the FE method (FEM) in our multiscale model; $\boldsymbol{\sigma}^{\text{interaction}}$ is the contribution of the stress field of the other dislocation segments; and $\boldsymbol{\sigma}^{\text{other}}$ arises from the interaction of dislocations with other defects or free surfaces, which can also be computed by FEM.

The velocity, v, of a dislocation segment is governed by a first-order differential equation consisting of an inertia term, a drag term, and a driving force such as

$$m^*(v)\dot{v} + Bv = F = \tau b \tag{11.3.3}$$

where m^* is the effective mass in the dislocation segment, B is the viscous drag coefficient, b is the magnitude of the Burgers vector, and τ is the resolved effective stress. Because the initial acceleration of dislocation is high, the dislocations can reach a stable velocity in a short time and the corresponding moving distance is so small (less than 1 nm) that it can be neglected. Therefore, we assume that dislocation immediately reaches a stable velocity (Gillis and Kratochvil, 1970) such as

$$Bv = \tau b \tag{11.3.4}$$

Gillis et al. (1969) suggested that Eq. (11.3.4) might be extended to higher velocities by replacing B with

$$B = \frac{B_0}{1 - v^2/v_s^2} \tag{11.3.5}$$

where B_0 is the static viscous drag coefficient and v_s is the speed of shear wave. Substituting Eq. (11.3.5) into Eq. (11.3.4), we can get the stable velocity

$$v = \frac{v_s}{2} \left(\sqrt{\left(\frac{v_s B_0}{\tau b}\right)^2 + 4} - \frac{v_s B_0}{\tau b} \right) \tag{11.3.6}$$

with $v_s = 2.92 \times 10^3$, $b = 2.55 \times 10^{-10}$, and $B_0 = 5 \times 10^{-5}$; the velocity of dislocation versus resolved shear stress is plotted in Fig. 11.11. As the resolved shear stress increases, the velocity approaches the speed of the shear wave. Then, the evolving dislocation structure can be simulated, such as bowing out, expansion or shrinkage of loops, and pileup.

11.3.2 Coupling Dislocation Dynamics Plasticity With Finite Element

The kinematics and constitutive relations of micron plasticity that follow those of Peirce et al. (1983) are used in this section. The velocity gradient can be expressed as

$$\mathbf{L} = \mathbf{D} + \boldsymbol{\Omega} \tag{11.3.7}$$

where $\mathbf{D}$ is the symmetric part of the deformation rate and $\boldsymbol{\Omega}$ is the skew symmetric spin tensor. $\mathbf{D}$ and $\boldsymbol{\Omega}$ are also decomposed into elastic parts ($\mathbf{D}^e$ and $\boldsymbol{\Omega}^e$) and plastic parts ($\mathbf{D}^p$ and $\boldsymbol{\Omega}^p$), respectively:

$$\mathbf{D} = \mathbf{D}^e + \mathbf{D}^p \tag{11.3.8}$$

$$\boldsymbol{\Omega} = \boldsymbol{\Omega}^e + \boldsymbol{\Omega}^p \tag{11.3.9}$$

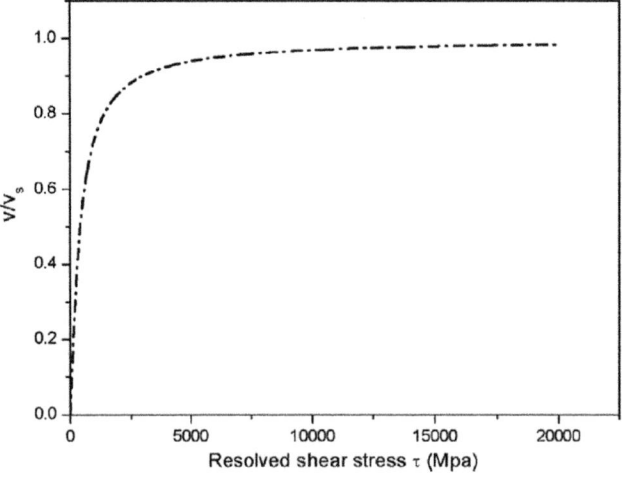

Figure 11.11

Variation of dislocation velocity (normalized by the speed of the shear wave) with resolved shear stress. *Reprinted from Liu, Z.L., et al., 2008. A mesoscale investigation of strain rate effect on dynamic deformation of single-crystal copper. International Journal of Solids and Structures 45 (13), 3674–3687. Copyright 2008, with permission from Elsevier.*

Assuming that the crystal elasticity is unaffected by slip, the elastic constitutive equation is given by

$$\overset{\nabla}{\boldsymbol{\sigma}}{}^{e} = \mathbf{C}^{e} : \mathbf{D}^{e} \tag{11.3.10}$$

where $\mathbf{C}^{e}$ is the tensor of elastic moduli and $\overset{\nabla}{\boldsymbol{\sigma}}{}^{e}$ is the Jaumann rate of Cauchy stress $\boldsymbol{\sigma}$, which is the corotational stress rate in terms of the coordinate system that rotates with the lattice. $\overset{\nabla}{\boldsymbol{\sigma}}{}^{e}$ is determined by

$$\overset{\nabla}{\boldsymbol{\sigma}}{}^{e} = \overset{\nabla}{\boldsymbol{\sigma}}{}^{e} + \boldsymbol{\Omega}^{\mathbf{p}} \cdot \boldsymbol{\sigma} - \boldsymbol{\sigma} \cdot \boldsymbol{\Omega}^{\mathbf{p}} \tag{11.3.11}$$

where $\overset{\nabla}{\boldsymbol{\sigma}}$ is the corotational stress rate on the coordinate system that rotates with the material. In: small strain

$$\boldsymbol{\Omega}^{\mathbf{p}} = 0, \quad \overset{\nabla}{\boldsymbol{\sigma}} = \overset{\nabla}{\boldsymbol{\sigma}}{}^{e} = \mathbf{C}^{e} : \mathbf{D}^{e} \tag{11.3.12}$$

Erik van der Giessen and Needleman (1995) combined two-dimensional dislocation dynamics with FE to solve the boundary value problem of finite crystals using the superposition principle. The solution to the boundary value problem is obtained as the sum of two contributions. The first represents the solution for dislocations in an unbounded

continuum and the other is the complementary elastic solution needed to satisfy equilibrium at the external and internal boundary. The last contribution is computed by an FE code.

Based on the superposition principle, the boundary value problem of finite crystals can be cast into a standard FE framework and has the general form:

$$\mathbf{Ma} + \mathbf{f}^{\text{int}} = \mathbf{f}^{\text{ext}} \tag{11.3.13}$$

where

$$\mathbf{f}^{\text{ext}} = \mathbf{f}^{\text{a}} + \mathbf{f}^{\text{im}} + \mathbf{f}^{\text{dis}} \tag{11.3.14}$$

where $\mathbf{f}^{\text{int}} \int_V \mathbf{B}^T \boldsymbol{\sigma} dv$ is internal force vector; $\boldsymbol{\sigma}$ is the total stress tensor and can be computed by Eq. (11.3.11); $\mathbf{f}^{\text{a}} \int_S \mathbf{N}^T \mathbf{t}^{\text{a}} ds$ is applied force vector; $\mathbf{t}^{\text{a}}$ is traction applied on the stress boundary; $\mathbf{f}^{\text{im}} \int_S \mathbf{N}^T \mathbf{t}^{\text{im}} ds$ is the image force vector; $\mathbf{t}^{\text{im}}$ is traction arising from the interaction of dislocations with the free surface; the image field can be accurately solved by Boussinesq–Cerruti formalism or Lothe-force approximation (Liu and Schwarz, 2005); $\mathbf{f}^{\text{dis}} \int_V \mathbf{B}^T \boldsymbol{\sigma}^{\text{dis}} dV$ is a body force (Zbib and Diaz de la Rubia, 2002) in the crystal; and $\boldsymbol{\sigma}^{\text{dis}}$ is the stress tensor of dislocations in an infinite crystal. $\mathbf{N}$ is the shape function vector and $\mathbf{B} = \text{grad}[\mathbf{N}]$.

The key to this method is to connect the observed macroscopic parameters (used in continuum models), such as stress and strain at material points, to the elementary shears produced by a constitutive relationship with dislocations. A direct linkage to continuum codes will be too inefficient to make computation practical. Because FE calculations typically involve a large number of material points, each with its own stress-strain trajectory, running a statistically representative DDD simulation for each material point simultaneously will remain prohibitively expensive for years to come (Bulatov, 2002). So, within the continuum framework, we choose a representative volume element (RVE), in which the deformation field is assumed to be homogeneous (Zbib and Diaz de la Rubia, 2002). The dislocation dynamics model is presumably capable of describing the heterogeneous nature of the deformation field. With a proper homogenization theory, or field averaging, one can couple the DDD model with continuum mechanics.

Macroscopic plastic strain $\mathbf{D}^{\text{p}}$ and plastic spin $\boldsymbol{\Omega}^{\text{p}}$ are evaluated at the Gauss points and can be related to the motion of each dislocation segment:

$$\mathbf{D}^{\text{p}} = \frac{\Delta \gamma}{2} (\mathbf{n}_i \otimes \mathbf{b}_i + \mathbf{b}_i \otimes \mathbf{n}_i), \quad \boldsymbol{\Omega}^{\text{p}} = \frac{\Delta \gamma}{2} (\mathbf{n}_i \otimes \mathbf{b}_i - \mathbf{b}_i \otimes \mathbf{n}_i)$$

$$\Delta \gamma = \rho bv = \sum_{i=1}^{N} \frac{l_i b v_{gi}}{V} \tag{11.3.15}$$

where V is the volume of RVE that includes the integration point, N is the number of segments that slip through the element in Δt (the time step in the FE simulation), l_i is the length of segments i, v_{gi} is the dislocation glide velocity, $\mathbf{n_i}$ is the unit normal to the slip plane, and $\mathbf{b_i}$ is the Burgers vector.

Eqs. (11.3.11), (11.3.14), and (11.3.15) can be implemented in ABAQUS/Standard by a user-defined material subroutine, UMAT, and a user-defined load, DLOAD, respectively. Eq. (11.3.13) can be solved with the Newmark β integration method in ABAQUS, and then the stress and displacement fields are obtained in the finite single crystal. Once the stress field in the crystal is known, the Peach-Koehler force can be estimated for all dislocation segments. A new plastic step is repeatedly performed by the DDD code.

11.3.3 Model Description and Simulation Results

An isotropic copper single crystal with the following properties (relevant to both the continuum model and the DDD model) is considered: shear modulus, $\mu = 42$ GPa; Poisson's ratio, $v = 0.347$; density, $\rho = 8900$ kg/m^3; magnitude of Burgers vector, $b = 2.5525$A; and static viscous drag coefficient, $B_0 = 5.5 \times 10^{-5}$ Pa/s. The temperature is 300K.

As illustrated in Fig. 11.12, the simulation setup consists of a $5 \times 5 \times 10$-μm block that is divided into an FE mesh using eight-node brick elements. The bottom surface of the sample is fixed and the upper surface is moved at controlled rate so that the strain rate,

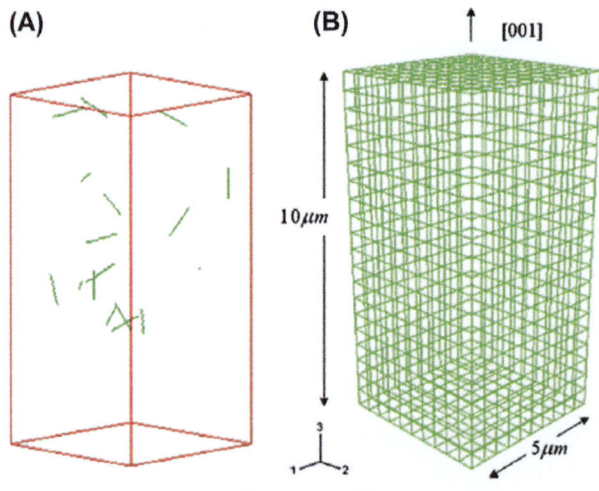

Figure 11.12
(A) Initial dislocation line distribution in single-crystal copper; (B) Dimension of simulation cell and finite element mesh. *Reprinted from Liu, Z.L., et al., 2008. A mesoscale investigation of strain rate effect on dynamic deformation of single-crystal copper. International Journal of Solids and Structures 45 (13), 3674–3687. Copyright 2008, with permission from Elsevier.*

ranging from 10^2 to 10^5 s^{-1}, is made constant (displacement of the upper surface increases in time). All other surfaces are assumed to be free, in which traction resulting from the interaction of dislocations with free surfaces is computed at four integration points.

Fig. 11.12A shows the initial state of dislocations in the simulation block. The mobile dislocation density is 0. Sixteen dislocation lines pinned at two ends are distributed randomly in the crystal and the length of each dislocation line is 1 μm. They are located in the four different slip systems: $(111)[10\bar{1}]$, $(1\bar{1}1)[\bar{1}01]$, $(11\bar{1})[011]$, and $(\bar{1}11)[110]$, and Schmid factors are 0.408, 0.408, 0.408, and 0, respectively. Each slip system possesses four dislocation sources and the dislocation density is about 6.4×10^{10} m^{-2}.

Various FE meshes were examined ranging from very coarse ($5 \times 5 \times 10$) to very fine ($20 \times 20 \times 40$). The results for $10 \times 10 \times 20$ are presented, which is fine enough to reflect the deformation patterning of a dislocation structure. The time step of DDD computation is chosen to be from 10^{-9} to 10^{-11} with an increase in the strain rate, which is one order lower than that of FE simulations.

The plastic flow of the metal single crystal is determined by many factors, including crystal orientation, temperature, applied strain rate, specimen size, deformation path, and the microstructure of the material. The focus of the current work is on the microstructure and strain rate effects. The stress-strain curves are computed by average values of the elements. Four different strain rates are plotted in Fig. 11.13, from which we can see that

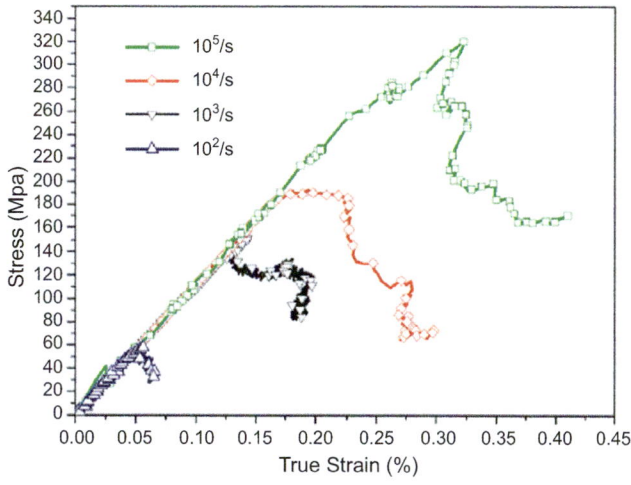

Figure 11.13

Stress-strain curves of a $5 \times 5 \times 5$-μm^3 single-crystal copper block at different strain rates.
Reprinted from Liu, Z.L., et al., 2008. A mesoscale investigation of strain rate effect on dynamic deformation of single-crystal copper. International Journal of Solids and Structures 45 (13), 3674–3687. Copyright 2008, with permission from Elsevier.

the curves can be distinctly divided into three segments: an elastic part, a strain hardening part, and a strain softening part. The initial transient peak in Fig. 11.13 is caused by an artifact related to the initial configuration: Initially, the mobile dislocation density is too small to achieve the imposed strain rate because the pinned segment sources cannot be activated. In cases in which the initial configuration has a higher density of dislocation that can accommodate the strain rate more rapidly, the transient peak is absent. The slight oscillations and a stress drop-off in Fig. 11.13 are the result of the heterogeneous nature of the activation and progression of dislocations.

The transition point between the elastic and strain hardening parts can be regarded as the stress yield point. Fig. 11.13 shows that the yield stress increases with an increase in the strain rate. This phenomenon was discovered by Edington (1969) in uniaxial compression tests of copper specimens with a size of several millimeters, by Horstemeyer et al. (2001) in simple shear atomistic simulations of single-crystal nickel, and by Guo et al. (2007) in simple shear MD simulations of single-crystal copper, as shown in Fig. 11.14.

To understand the rate effect of single-crystal copper at the microscale, we carried out a quantitative analysis based on our multiscale simulation results. The flow stress τ of crystal material is composed of two components:

$$\tau = \tau_\mu + \tau^* \tag{11.3.16}$$

where τ^* is the effective stress on the dislocation, which governs the thermally activated motion of dislocations through localized obstacles such as Peierls barriers (Hirth J P 1982);

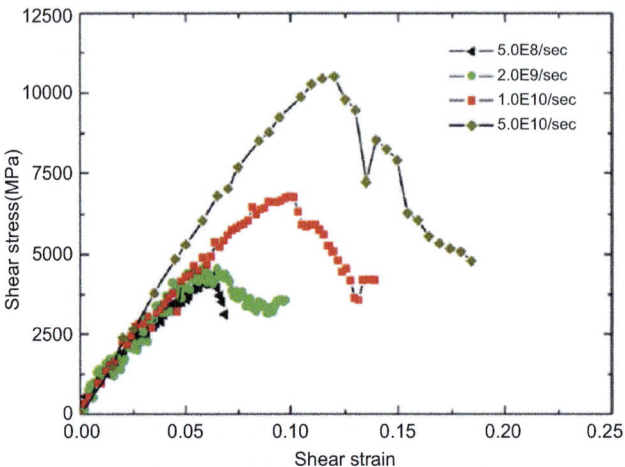

Figure 11.14

Shear stress-strain curves of $32.35 \times 10.98 \times 2.40$-nm^3 single-crystal copper block at different strain rates (Guo et al., 2007). *Reprinted from Liu, Z.L., et al., 2008. A mesoscale investigation of strain rate effect on dynamic deformation of single-crystal copper. International Journal of Solids and Structures 45 (13), 3674—3687. Copyright 2008, with permission from Elsevier.*

and τ_μ is athermal stress on the dislocation, which stems from the long-range elastic interaction with all other dislocations and short-range dislocation-dislocation interactions.

At a normal temperature and strain rate, for an FCC single crystal, thermally activated resistance is small: are mainly Peierls barriers, just about 0.5 MPa for single-crystal copper. It can be easily overcome, so the contribution of τ^* to flow stress can be neglected. In this case, flow stress τ is mainly determined by the athermal part, τ_μ. In other words, short-range reactions between dislocations are responsible for most of the flow stress. In our simulation, it is mainly decided by the critical stress, τ_{FR}, for the operation of an FR source, and the nonlocal contribution typically is on the order of $0.1-0.15\ \tau_{FR}$ for a segment length in the micrometer range (Devincre and Condat, 1992); the influence of surface correction is neglected. The critical resolved stress for a pinned dislocation line can be expressed as

$$\tau_{FR} = 2\alpha \frac{\Gamma}{bl} \approx 2\alpha \frac{\mu b}{l}, \quad \tau_\mu \approx \tau = 1.15\tau_{FR}, \tag{11.3.17}$$

where $\Gamma \approx \mu b^2$ is the line tension of the dislocation, μ is the shear modulus, b is the magnitude of the Burgers vector, α is a parameter that quantifies the strength of the obstacle for an FR source $\alpha \approx 1$; and $l = 1\ \mu m$ is the length of the dislocation line. We get

$$\tau_\mu = 5.88 \times 10^{-4} \quad \mu = 24.73 \text{ MPa} \tag{11.3.18}$$

As can be seen from Eq. (11.3.17), when the average length of the dislocation line, l, is governed by one specimen dimension, a size effect arises.

Once the critical resolved stress is obtained, initially pinned dislocation lines are activated and begin to bow out (Fig. 11.15). Yielding begins when there is a balance between the increase in the dislocation-induced plastic strain and the boundary-imposed strain rate, so that when the yield stress is achieved, more and more FR sources become operative to accommodate high strain rate deformation with the increasing strain rate, as shown in Fig. 11.16. Assuming flow stress τ is the maximum shear stress, we can estimate static yield stress $\sigma_s = 2\tau = 49.96$ MPa $= 0.001176\mu$ for our simulation, which accords well with the value obtained under a strain rate of 100 s^{-1}, as shown in Fig. 11.13.

When the strain rate is high, according to Orowan's law, which relates the total strain rate to the dislocation flux:

$$\dot{\gamma} = \rho b v \tag{11.3.19}$$

where v is the average velocity of dislocations. Increasing the strain rate is equivalent to increasing the average velocity, according to Eq. (11.3.19). We have the expression of effective stress τ^* and the velocity for the viscous type used in pure FCC metals:

$$\tau^* = \frac{Bv}{b} = \frac{B\dot{\gamma}}{b^2\rho}, \quad B = \frac{B_0}{1 - v^2/v_s^2} \tag{11.3.20}$$

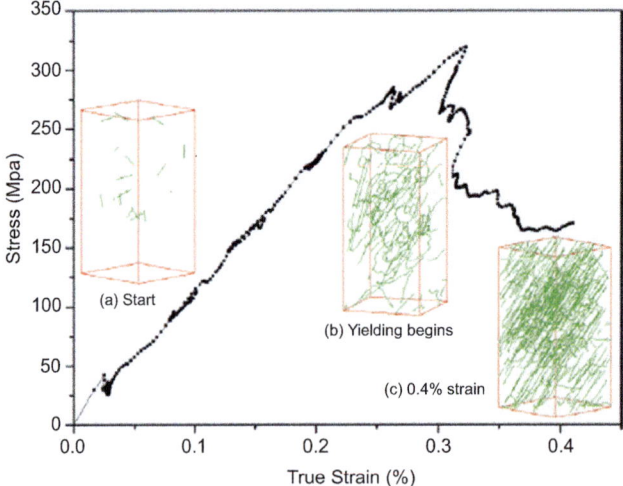

Figure 11.15

Stress-strain curve under a strain rate of $10^5\,\text{s}^{-1}$ and corresponding dislocation states illustrating activation and motion of dislocation lines at different strains. *Reprinted from Liu, Z.L., et al., 2008. A mesoscale investigation of strain rate effect on dynamic deformation of single-crystal copper. International Journal of Solids and Structures 45 (13), 3674–3687. Copyright 2008, with permission from Elsevier.*

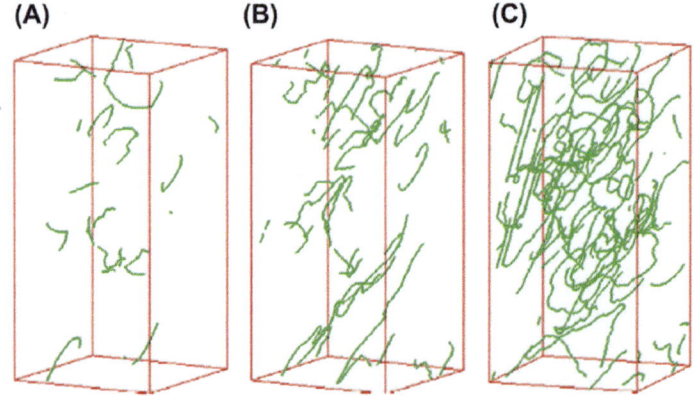

Figure 11.16

Dislocation microstructure at the yield point under strain rates of (A) $10^3\,\text{s}^{-1}$, (B) $10^4\,\text{s}^{-1}$, and (C) $10^5\,\text{s}^{-1}$. More and more Frank-Read sources are operating to accommodate high strain rate deformation with an increase in the strain rate when yielding occurs. *Reprinted from Liu, Z.L., et al., 2008. A mesoscale investigation of strain rate effect on dynamic deformation of single-crystal copper. International Journal of Solids and Structures 45 (13), 3674–3687. Copyright 2008, with permission from Elsevier.*

Coefficient B accounts for electron and phonon drag; it also impedes the motion of dislocations with an increase in the strain rate. Assuming that dislocation density ρ is not significantly modified, a high stress level is necessary to achieve the imposed velocity, and it will become larger than what is needed to cross forest obstacles τ_μ, with an increase in the strain rate. In other words, increasing the strain rate is equivalent to increasing the thermally activated glide resistance of dislocations. The same effect can be obtained by decreasing the temperature. They all make the movement of dislocations more difficult.

So, according to Eq. (11.3.20), when $\dot{\gamma}$ is larger than some critical value, $\dot{\gamma}_c$, τ^* will dominate the value of the flow stress, and then the yield stress will be strongly sensitive to the strain rate. According to $\tau^* > \tau_\mu$, $\dot{\gamma}_c$ is controlled by the value of $\frac{\alpha\mu b^3}{B}\left(\frac{\rho}{7}\right)$, which is decided by the dislocation density, the size of the dislocation source (the distance between the two pinned ends of the source), the size of the specimen, or simply the average size of the dislocation mean free path and the strength of the obstacles. In our simulation, the crystal strain rate is 10^2-10^3 s^{-1}. Yield stress normalized by the shear modulus is plotted as a function of strain rate in Fig. 11.17, which yields the following linear relation with a correlation coefficient of $r^2 = 0.99$:

$$\frac{\sigma_s}{\mu} = 0.00162 \cdot \log\dot{\varepsilon} - 0.00186 \qquad (11.3.21)$$

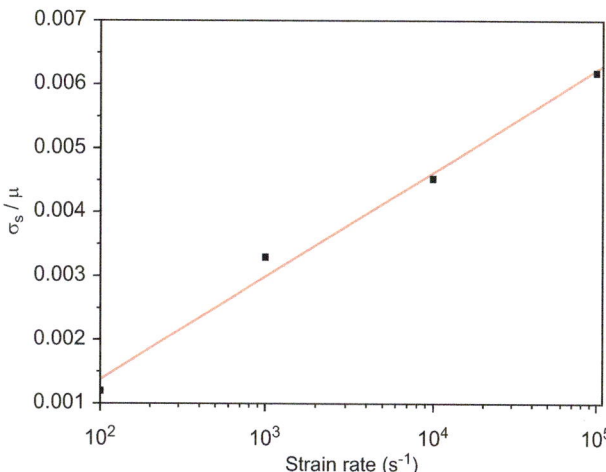

Figure 11.17

Normalized resolved yield stress compared with four different strain rates. The linear regression line has a correlation coefficient of $r^2 = 0.99$. *Reprinted from Liu, Z.L., et al., 2008. A mesoscale investigation of strain rate effect on dynamic deformation of single-crystal copper. International Journal of Solids and Structures 45 (13), 3674–3687. Copyright 2008, with permission from Elsevier.*

Above the critical strain rate, yield stress normalized by the shear modulus has a linear relation to $\log \dot{\varepsilon}$. A similar conclusion was drawn from experiments in tantalum (Chen et al., 1999) and copper (Follansbee and Gray, 1991).

Yield stress is also sensitive to the applied loading rate under a high strain rate at the atomic scale (Guo et al., 2007), where the effects of dislocation inertia and phonon drag are important. A higher strain rate requires a higher acceleration and velocity of dislocations nucleated in the specimen to accommodate quick deformation. Therefore, a higher yield stress is required to overcome the inertial force and phonon drag force acting on the dislocations for larger specimens under a high strain rate. On the microscale, the initial dislocation density, the average size of the dislocation mean free path, and the strength of barriers to dislocation also affect yield stress considerably under a high strain rate besides the specimen size.

Dislocation patterning is also determined by the ratio of τ_{μ}/τ^*: When τ_{μ} dominates flow stress, dislocation-dislocation interaction is strong and the multibody interaction effect is evident. One can observe dislocation patterning in crystals; when τ^* is dominant, dislocation-dislocation interaction is comparatively weak and each dislocation moves as if it were alone in the crystal. The dislocation microstructure exhibits no patterning, as can be seen in Fig. 11.18A–C. Dislocation microstructures change from nonuniform to uniform.

In single-crystal copper, there are 12 different slip systems that can contribute to the deformation process. The slip systems are listed in Table 11.2 as combinations of the slip planes and slip direction. The contribution of each slip system to plastic deformation at different strain rates was investigated by plotting the dislocation density distribution of each slip system, as shown in Fig. 11.19. Dislocation tends to become localized in one slip system $(1\bar{1}1)\left[\bar{1}01\right]$ and in the areas of the simulated crystal where dislocation multiplication is most active (Fig. 11.18); the number of initial dislocation sources in each slip system is the same. This may be attributed to the interaction process between dislocations in different systems.

Under a high strain rate, the presence of active dislocation sources in a rapidly deforming crystalline material can serve as the focus of localized plastic flow and associated energy concentrations that in turn determine the mechanical response of the material. Most of these localized regions often manifest in the form of a shear band (Wright, 2002). In our simulation, a deformation band can be clearly observed in Fig. 11.18A–C. The simulation results show that deformation is mostly localized in the bands along the most active slip plane $(1\bar{1}1)$, and with an increase in the strain rate, the width of the band also increases from 2 μm to several tens of micrometers, and is restricted by the size of the simulated crystal. In experiments with bulk crystal materials, the width of the shear band is often hundreds of micrometers. In our simulation, few

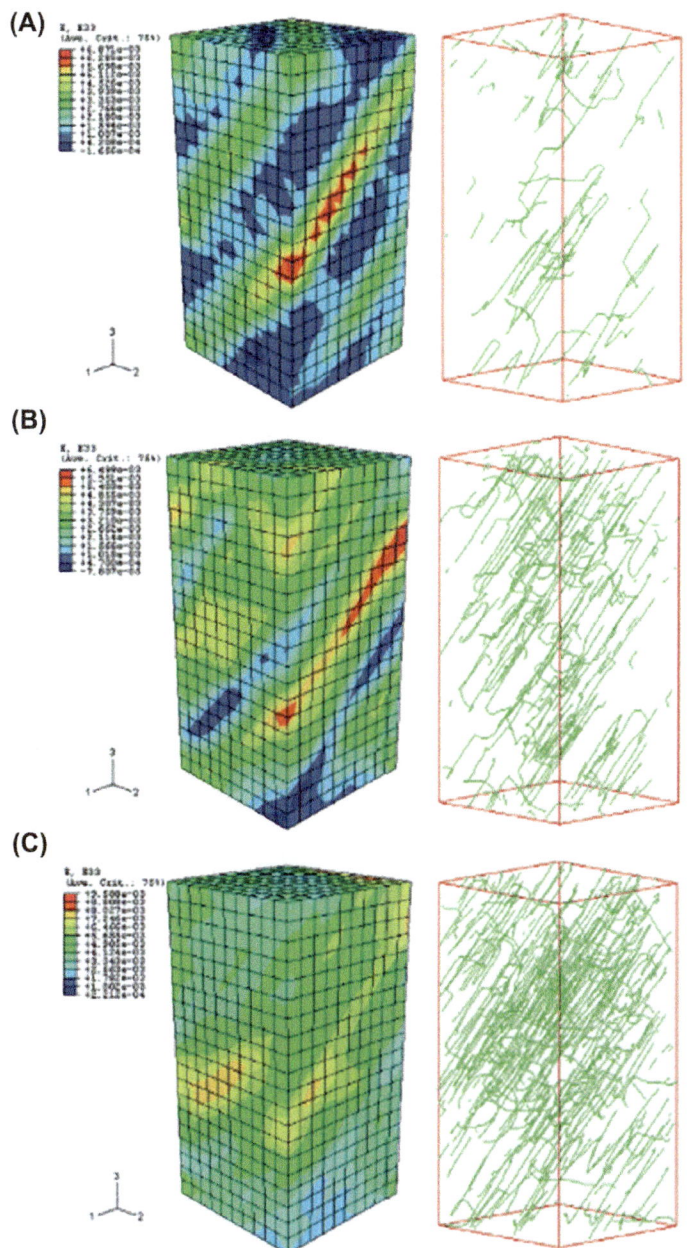

Figure 11.18

Distribution of strain ε_{33} in the crystal and corresponding dislocation microstructures after yield under strain rates of (A) $10^3 \, s^{-1}$, (B) $10^4 \, s^{-1}$, and (C) $10^5 \, s^{-1}$. Deformation is mostly localized in the bands along the most active slip plane $(1\bar{1}1)$, and with an increase in the strain rate, the width of the band also increases. *Reprinted from Liu, Z.L., et al., 2008. A mesoscale investigation of strain rate effect on dynamic deformation of single-crystal copper. International Journal of Solids and Structures 45 (13), 3674–3687. Copyright 2008, with permission from Elsevier.*

Table 11.2: Slip Systems in Single-Crystal Copper

Slip System	Burgers Vector	Slip Plane
1	$[10\bar{1}]$	(111)
2	$[\bar{1}01]$	$(1\bar{1}1)$
3	$[011]$	$(1\bar{1}1)$
4	$[011]$	$(11\bar{1})$
5	$[\bar{1}10]$	$(11\bar{1})$
6	$[110]$	$(\bar{1}11)$
7	$[1\bar{1}0]$	(111)
8	$[110]$	$(1\bar{1}1)$
9	$[0\bar{1}1]$	(111)
10	$[01\bar{1}]$	$(\bar{1}11)$
11	$[101]$	$(11\bar{1})$
12	$[101]$	$(\bar{1}11)$

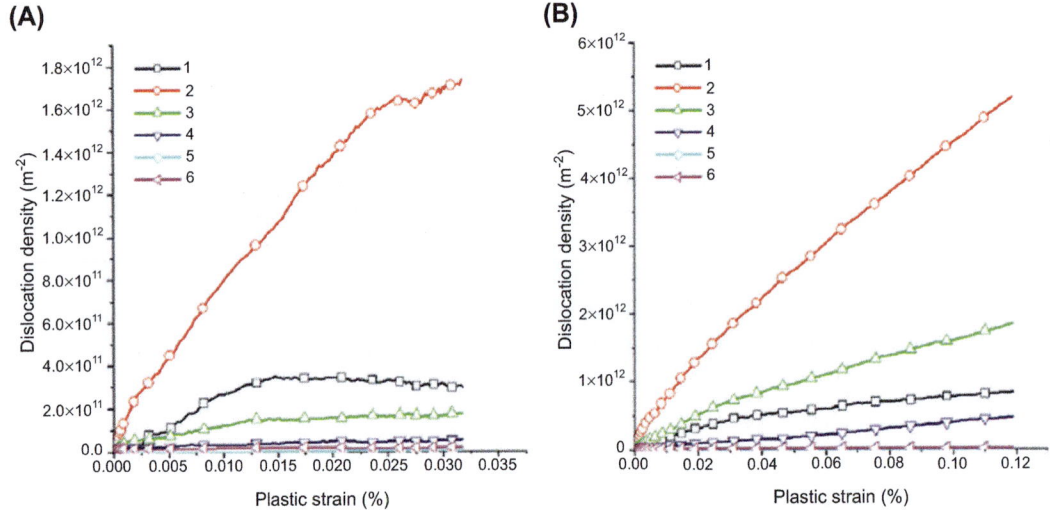

(A) **(B)**

Figure 11.19

Evolution of dislocation density in different slip systems under strain rates of (A) 10^4 s^{-1} and (B) 10^5 s^{-1}. *Reprinted from Liu, Z.L., et al., 2008. A mesoscale investigation of strain rate effect on dynamic deformation of single-crystal copper. International Journal of Solids and Structures 45 (13), 3674–3687. Copyright 2008, with permission from Elsevier.*

dislocation sources are responsible for creating the dislocations that form the shear band. Under an extremely high strain rate, such as during shock loading (up to 10^{10} s^{-1}), because the classical dislocation sources of the FR type are slow-acting, they no longer supply a sufficient quantity of dislocations for an appreciable plastic strain rate during a

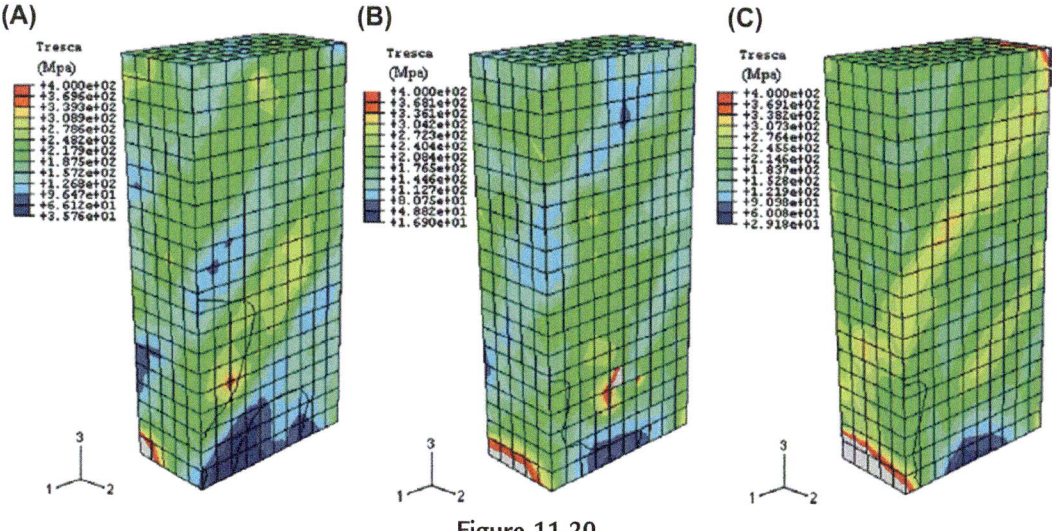

Figure 11.20

Distribution of Tresca stress in crystal after yield under strain rates of (A) 10^3 s^{-1}, (B) 10^4 s^{-1}, and (C) 10^5 s^{-1}. *Reprinted from Liu, Z.L., et al., 2008. A mesoscale investigation of strain rate effect on dynamic deformation of single-crystal copper. International Journal of Solids and Structures 45 (13), 3674–3687. Copyright 2008, with permission from Elsevier.*

time interval, and the new physical mechanisms of dislocation multiplication or generation are still lacking (Bringa et al., 2006; Shehadeh et al., 2006).

The stress distributions in single-crystal copper under different strain rates were also investigated. Tresca stress distributions are plotted in Fig. 11.20. The stress value is a relatively higher in the band compared with those in neighboring regions, and which will induce materials in the bands to lose the ability to transfer shear stress, leading to shear instability of the material. As also can be seen in Fig. 11.20, some stress and strain concentration regions exist in the band because of the heterogeneity of plastic deformation, and they are often the locations where further failure occurs, such as the formation of cracks and the growth of voids (Wright, 2002). Shear bands must be given more attention in crystalline materials under high strain rate deformation.

With an increase in the strain rate, the yield stress of single-crystal copper increases and a critical strain rate exists for each single-crystal copper block for a given size and dislocation source that is determined by the initial dislocation density, the average size of the dislocation mean free path, and the strength of the barriers to dislocation. Below the critical strain rate, the yield stress is relatively insensitive to the strain rate; above it, the yield stress will rapidly increase with an increase in the strain rate. It has a linear relation with log$\dot{\varepsilon}$. Under a high strain rate, dislocation structure patterning changes from nonuniform to uniform and dislocations are localized in one slip system. Deformation

bands are also observed in the simulations, and the bandwidth increases with the strain rate. Shear stress in the band is higher than that in the neighboring region, which may induce shear instability in the crystalline materials. Some material points in the band will be the location where further failures occur because of the concentrations of stress and strain.

Moreover, with an increase in the strain rate, such as takes place under shock loading, the temperature of the material will arise considerably, which will affect the properties of the material and dislocation mobility. These effects will be taken into account in future research.

11.4 Shock-Induced Deformation and Dislocation Mechanisms in Single-Crystal Copper

11.4.1 Dynamic Mechanical Behavior Corresponding to Dislocation Microstructure

The mechanical response of a material under shock loading has been a topic of interest for more than a century. The exploration of material dynamic deformation is valuable and has a wide range of applications, such as in explosive forming, shale oil and gas horizontal well perforation, and shock synthesis. Various experiments (Dekel et al., 1998; Moshe et al., 1998; Damian and Swift, 2006) have been carried out to investigate the dynamic mechanical behavior of materials, especially the equation of state and the strength of materials under shock loading. A uniaxial strain state of deformation is achieved in materials under shock conditions and a high-pressure wave propagates along the loading direction. With an increase in the impact strength, plastic deformation occurs when the applied pressure exceeds the Hugoniot elastic limit, accompanied by an amount of dislocation nucleation and multiplication.

In the case of high strain rate loading, dislocation has an important role in determining the behavior of a material. Many researchers contributed their talents to exploring the connection between microscopic dislocation structures and the dynamic mechanical behavior of materials through a series of experiments (Cao et al., 2005, 2010; Zhao et al., 2016). Simultaneous shock compression and X-ray diffraction experiments were introduced by Johnson and Mitchell (1972), who offered an attractive method of observing distortions in the lattice as it was being compressed. By creating laser-induced shock compression in monocrystalline copper, Meyers et al. (2003) found that the number of dislocation loops was much larger than that observed in undeformed copper. They suggested that loop nucleation was an essential event of laser-induced shock compression at pressures of 12 and 20 GPa. Ye et al. (2011) revealed that there was a close relation between laser shock processing-induced nano-crystallization and dislocation generation. In general, the dislocation microstructures observed in the recovered samples consisted of dislocation cells, shear bands, and deformation twins (Mogilevskii and Bushnev, 1990; Rivas et al., 1995). Moreover, there was a threshold pressure at which the deformation

mechanism changed from dislocation glide to twinning (Schneider et al., 2005) and the transition pressure depended on the crystal orientation. These studies provide significant information to enable an understanding of the evolution of microstructures under shock loading. However, to our best knowledge, a higher-resolution observation of the simultaneous evolution of dislocations is still lacking, because much of what we know is inferred from recovery experiments or the analysis of wave profiles.

Various simulations were carried out to investigate how simultaneous microstructures affect the dynamic mechanical behavior of materials (Shehadeh et al., 2005b; Austin and McDowell, 2011; Liu et al., 2015a; Gunkelmann et al., 2016), especially shock-induced dislocation plasticity. Cao et al. (2007) analyzed dislocation structures under shock loading by MD simulations. The density of microstructures in the simulations was much larger than that in the experiments, which they attributed to the difference in the recovery process and the much longer stress rise time (~ 1 ns) in flyer-plate shock experiments. To eliminate the space-time gap between MD simulations and experiments, multiscale DDD simulations (Cheng and Shehadeh, 2006) were conducted to investigate dislocation structures and stress distribution in silicon crystals during the laser shock process. HN and heterogeneous nucleation of dislocations in copper single crystals were investigated using the same model (Shehadeh et al., 2006). The results suggested that the HN of dislocations would be dominant as shock rise time decreased. This conclusion is consistent with results from the simulation by Gurrutxaga-Lerma et al. (2015a). The HN would become the dominant dislocation generation mechanism when the strain rates exceeded $5 \times 10^7 \, \text{s}^{-1}$ because the FR sources were unable to operate before HN relaxed elastic stresses. In those studies, the HN of dislocation was generally introduced at the position where critical shear stress exceeded a specific value. The density of nucleated dislocation sources was always artificially defined and not discussed (Shehadeh et al., 2006; Gurrutxaga-Lerma et al., 2015a). However, the density of dislocation sources has a significant effect on the dynamic mechanical behavior of materials, especially stress relaxation at the front of the shock wave. How to introduce HN processes properly, i.e., the dislocation shape, location, and nucleation density, still needs critical testing.

In the next section, the dynamic mechanical behavior of single-crystalline copper pillars and their corresponding dislocation microstructures are investigated by coupling DDD with explicit FE analyses. Dislocation motion and discrete plasticity are determined by DDD modeling. The basic laws of continuum mechanics including momentum balance and the conservation of energy equation are solved by FEM. The HN of dislocation is introduced into the DDD-FE calculation by a coarse-grained model developed from systematic MD simulations. Shock-induced plasticity and the dynamic mechanical behavior of single-crystalline copper pillars are investigated. Typical dislocation microstructures and their connection with dynamic mechanical responses are analyzed. The results are compared with experimental findings available in the literature.

11.4.2 Dynamic Multiscale Discrete Dislocation Plasticity Model

A multiscale discrete dislocation plasticity model that links discrete and continuum analysis has achieved remarkable research results (Verdier et al., 1998; Zbib and Diaz de la Rubia, 2002; Vattré et al., 2013). Within the framework, DDD was presented in detail in our previous work (Liu et al., 2009a,b; Cui et al., 2014). Dislocation curves are discretized into straight segments. At each time step, forces acting on all dislocation segments are evaluated and the dislocation velocities are calculated by solving the equations of motion. Then, the dislocation positions are updated for the next time step.

To investigate impact problems, the DDD-FE coupling procedure is different from those static issues (Wallin et al., 2008; Vattré et al., 2013). First, under shock loading, a large amount of plastic work is generated in a short time as a result of dislocation motion. The plastic work can be converted into thermal energy, which leads to a significant rise in local temperature. The conservation of energy equation is expressed as

$$\rho c \dot{T} = \lambda_t \nabla^2 T + \gamma \cdot (\boldsymbol{\sigma} : \dot{\boldsymbol{\varepsilon}}^p), \tag{11.4.1}$$

where ρ is the mass density, T is the temperature, and c and λ_t are specific heat and thermal conductivity. γ is a thermal scaling factor representing the transformation from plastic work to thermal energy. $\boldsymbol{\sigma}$ is the stress tensor and $\dot{\boldsymbol{\varepsilon}}^p$ is the plastic strain rate owing to dislocation motion.

Then, the thermoelastic response of metal materials can be expressed using the rate form of Hooke's law for large deformation such that

$$\dot{\boldsymbol{\sigma}} = \boldsymbol{C}^e : (\dot{\boldsymbol{\varepsilon}} - \dot{\boldsymbol{\varepsilon}}^p) + \beta_t \dot{T} \cdot \boldsymbol{I}, \tag{11.4.2}$$

Here, $\dot{\boldsymbol{\sigma}}$ is the corotational stress rate, $\boldsymbol{C}^e$ is the elastic stiffness tensor, and $\dot{\boldsymbol{\varepsilon}}$ and $\dot{\boldsymbol{\varepsilon}}^p$ are total strain rate tensor and plastic strain rate tensor. $\beta_t = -\frac{E}{1-2v} \cdot \alpha$, in which E, v, and α are the Young's modulus, Poisson's ratio, and line expansion coefficient, respectively. $\dot{T}$ is the temperature change rate and $\boldsymbol{I}$ is a unit matrix.

The momentum equation is solved using the dynamic FE method such that

$$M\ddot{U} + C\dot{U} + KU = F, \tag{11.4.3}$$

where M, C, and K are mass matrix, damping matrix, and stiffness matrix, respectively. U is the nodal displacement. F is the force vector, which includes the external nodal force and the thermal force resulting from the temperature change. The coupled thermal-stress analysis is implemented by the commercial software ABAQUS. An explicit integration scheme is used to solve the displacement vector. Determination of the time step in the FE procedure is related to two processes, i.e., the stable time increase of solid deformation and the stability limit for thermal diffusion.

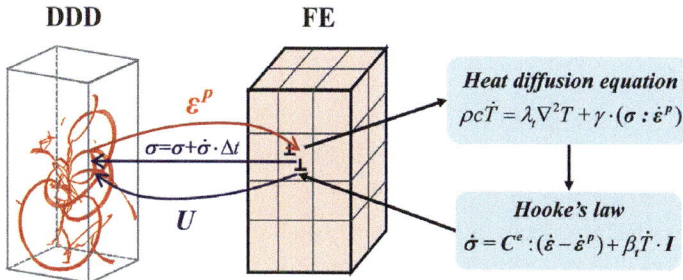

Figure 11.21

Schematic of discrete dislocation dynamics (DDD)-finite element (FE) coupling procedure. *FEM, finite element modeling. Reprinted from Hu, J.Q., Liu, Z.L., Chen, K.G., Zhuang, Z., 2017b. Investigations of shock-induced deformation and dislocation mechanism by a multiscale discrete dislocation plasticity model. Computational Materials Science 131, 78–85. Copyright 2017, with permission from Elsevier.*

The DDD-FE coupling procedure is illustrated in Fig. 11.21 and mainly contains the following information transfer processes: (1) Calculate plastic strain ε^p generated by dislocation slip in the DDD simulation. Then, the discrete plastic strain is localized to continuum material points by a new regularization method proposed by Cui et al. (2015b). (2) Solve the heat conduction equation by taking the plastic work as the inner heat source. The dynamic stress field associated with the thermal effect is calculated by FEM under a specific boundary condition. Stress field σ is then transferred to DDD and serves as the applied force to drive dislocation motion. (3) Displacement field U of the FEM cell is also transferred into the DDD cell to update the geometry configuration.

11.4.3 Coarse-Grained Homogeneous Nucleation Model

For metals under shock loading, an elastic precursor wave propagating at the longitudinal speed of sound precedes a slower plastic wave front. The dislocation activity is generally known as a dominant contributor at the onset of the plastic front (Armstrong et al., 2007; Armstrong and Li, 2015). As emphasized in the introduction, when the strain rate exceeds 5×10^7 s^{-1}, the HN process of dislocation is extremely important because the FR sources are too slow with respect to the shock front's rise time (Gurrutxaga-Lerma et al., 2015a). Thus, a new issue emerges regarding how to introduce the HN process in a DDD-FE model, especially the trigger conditions, shape, and number of incipient dislocation sources.

First, we carry out a series of MD simulations to investigate the HN process. A significant space-time gap exists between the MD and DDD-FE simulations. A typical timescale in MD is femtoseconds or picoseconds whereas for DDD-FE it is nanoseconds or microseconds. At the macroscale, the whole sample exhibits a dynamic mechanical behavior and the shock wave propagates along the loading direction. In contrast, the

nanosized local region just behind the elastic precursor has been fully deformed where the HN of dislocation originates. Thus, MD simulations are performed under uniform deformation instead of a dynamic impact to obtain the details of dislocation nucleation.

The Large-Scale Atomic/Molecular Massively Parallel Simulator (Plimpton, 1995) is employed for MD simulations. The embedded atom method potential (Mishin et al., 2001) is adopted to describe interactions among copper atoms. This potential is widely used and demonstrated accurately in shock and nonshock simulations of copper (An et al., 2008; Lin et al., 2012; Li et al., 2014; Wen et al., 2016). We tested different specimens with x-, y-, and z-axes along [100], [010], and [001], respectively. Periodic boundary conditions were applied in all directions. The simulated systems contain approximately $10^5 - 10^7$ Cu atoms, depending on their size. Before loading, the initial structures are fully relaxed at 300K and 0 GPa with a constant pressure-temperature ensemble and a time step of 10^{-15} s. After the specimen reaches thermal equilibrium, uniaxial compression is performed along the z-axis. The system is loaded in a displacement-controlled manner by imposing compressive displacements to atoms along the z-direction that vary linearly from 0 at the bottom to a maximum value on top. The strain rate in the simulation is set to $1-20$ ns^{-1}. Tracking of dislocations is performed by the dislocation extraction algorithm (Stukowski and Albe, 2010) and Ovito (Stukowski, 2010) is used to visualize the defect structures.

In MD simulations, monocrystalline copper is loaded under different strain rates and the smallest Cu system has a dimension of $\sim 11 \times 11 \times 54$ nm^3 ($30 \times 30 \times 150$ FCC unit cells), which is sufficiently large to capture the HN events of dislocation. In the analysis, shear stress is defined as $\tau = -\frac{1}{2}\left[\sigma_{zz} - \frac{1}{2}\cdot(\sigma_{xx} + \sigma_{yy})\right]$ to make it a positive number under compressive loading. The shear stress-strain curves under different strain rates are shown in Fig. 11.22A. After the initial rise in elastic loading, there is a drop in the shear stress that reveals dislocation nucleation. Three shear stress-strain curves have almost no difference when the nucleation event occurs. It demonstrates that the strain rate has little effect on the initial dislocation nucleation processes. The subsequent evolution of dislocations is rate dependent because the three curves exhibit different characteristics. Furthermore, the effect of the system size was investigated, as shown in Fig. 11.22B. There is almost no difference among the three curves when dislocation nucleation occurs. Interestingly, stress can increase transiently after the initial softening, which may be the result of the formation of strong cross-links between dislocations after the rapid multiplication of incipient dislocations.

Correspondingly, the details of incipient dislocation structures in systems of different sizes are presented as shown in Fig. 11.22C−E. All of the shapes of incipient dislocations are loops that exhibit the characteristics of HN. All loops are Shockley partial dislocations with a Burgers vector of 1/6<112> distributed on the {111} slip planes. The nucleation

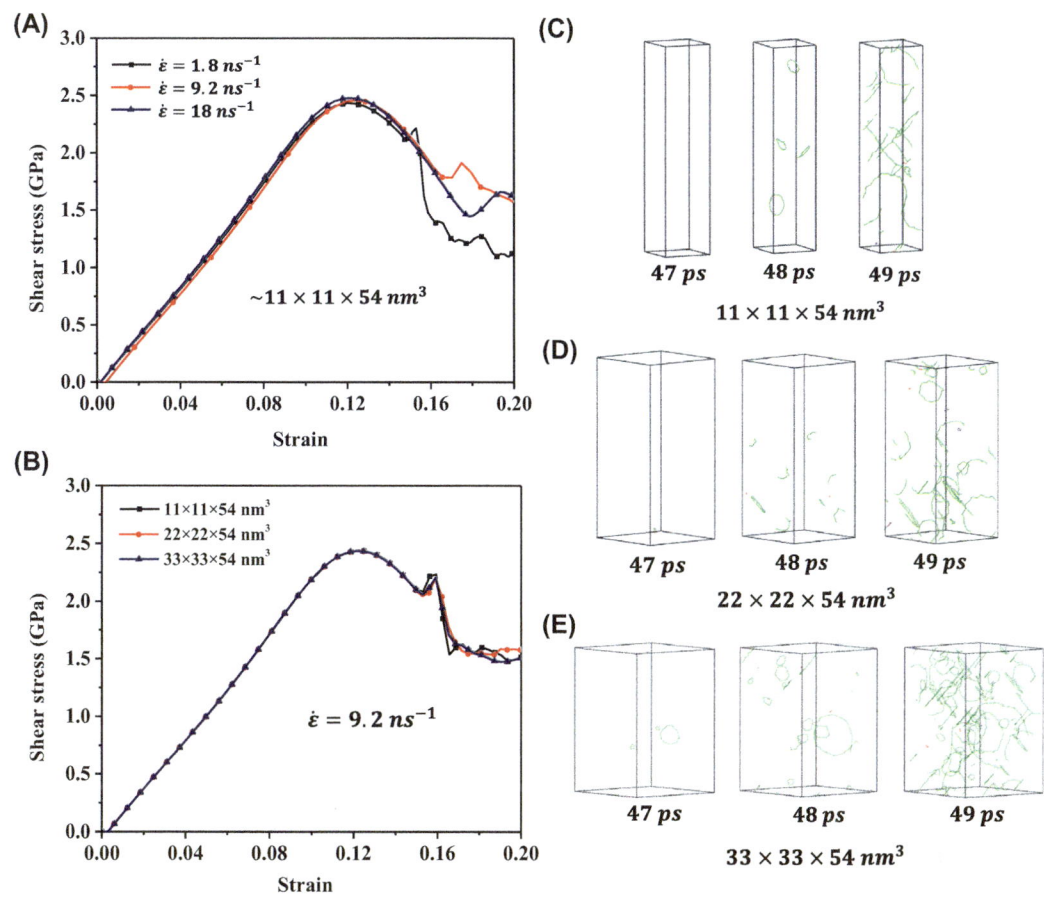

Figure 11.22

(A) Stress-strain curves for specimen at different strain rates; (B) Stress-strain curves for different system sizes; (C—E) Initial nucleated dislocation structures corresponding to different system sizes. *Reprinted from Hu, J.Q., Liu, Z.L., Chen, K.G., Zhuang, Z., 2017b. Investigations of shock-induced deformation and dislocation mechanism by a multiscale discrete dislocation plasticity model. Computational Materials Science 131, 78—85. Copyright 2017, with permission from Elsevier.*

condition is calculated by analyzing the stress state at the softening point in Fig. 11.22B; the local resolved shear stresses on the four {111} slip planes are: 1.948 GPa on the (111) slip plane, 1.951 GPa on $(-1-11)$, 1.957 GPa on (-111), and 1.958 GPa on $(1-11)$, respectively. As a consequence, a local resolved shear stress of approximately 2 GPa is considered to be the stress criterion for HN on all slip planes in the simulations.

Furthermore, by analyzing the incipient dislocation structures as well as their subsequent evolution, it can be found that the HN of dislocation triggers to saturation on the

picosecond timescale, which is comparable to the time step in the DDD-FE simulation. Dislocation multiplication will be before the nucleation events once plenty of HN loops exist. Then, it is important to determine how many dislocation loops should be introduced once the HN condition is satisfied. By counting the number of dislocation loops during the nucleation period, it is found that the density of dislocation loops, i.e., the number of dislocation loops N per unit volume V, becomes stable with an increase in the MD system size, which is approximately 180 loops per 10^6 nm^3. It is reasonable that the loop density converges with the increase in simulation size. Both HN and heterogeneous nucleation of dislocations exhibit some probabilistic nature (Zhu et al., 2008b). Decreasing the volume size in the MD simulation makes the HN process of dislocation more stochastic.

Next, we focus attention on how to introduce the HN information obtained from the MD simulation into the DDD-FE model in detail. As analyzed in MD simulations, the criterion of locally resolved shear stress for HN is used in the DDD-FE simulation and the value is defined as 2 GPa. Because the density of dislocation loops is calculated and approximately stable in MD simulations, we try to configure the same source density in the DDD-FE calculations when the HN event is triggered. However, the sample size and loading duration time are much larger in the DDD-FE calculation, so it is nearly impossible to introduce fully congruent dislocation sources from the MD to the DDD-FE simulation owing to the limited computational capability. To solve this conflict, a coarse-grained parameter, β, representing spatial resolution, is proposed to transmit the information of HN from MD to DDD-FE calculations. For a nucleated dislocation loop within the RVE in the FE calculation, as illustrated in Fig. 11.23, it can be taken as a dislocation cluster or a superdislocation at the MD scale. Here, the coarse-grained model with the incipient HN events has the same loop density in MD and DDD-FE simulations, expressed as

$$\frac{N_{MD}}{V_{MD}} = \beta \cdot \frac{N_{FE}}{V_{FE}}, \tag{11.4.4}$$

where N_{MD} is the number of incipient nucleated dislocation loops obtained in MD simulations. N_{FE} is the number of loops introduced into one RVE in DDD-FE calculations. V_{MD} and V_{FE} are the system size of MD simulations and the size of the RVE in FE calculations, respectively.

Coarse-grained parameter β is highly effective for bridging the gap between different temporal and spatial scales. In the MD simulation, the occurrence and saturation of HNs are on the picosecond timescale, which is comparable to the time step in the DDD-FE calculation. Thus, as shown in Fig. 11.23, several dislocation loops with different slip systems are introduced into an RVE in the DDD-FE model just at the initial time step when the local resolved shear stress exceeds 2 GPa. After that, as described in MD

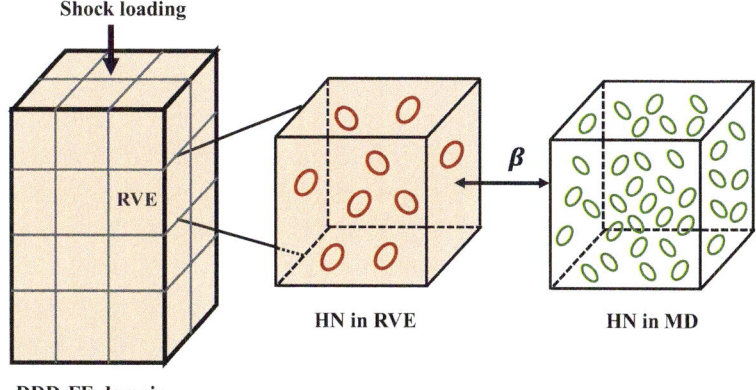

Figure 11.23

Schematic of homogeneous nucleation (HN) connection between molecular dynamics (MD) and DDD-FE (discrete dislocation dynamics- finite element) modeling using a coarse-grained model. *Reprinted from Hu, J.Q., Liu, Z.L., Chen, K.G., Zhuang, Z., 2017b. Investigations of shock-induced deformation and dislocation mechanism by a multiscale discrete dislocation plasticity model. Computational Materials Science 131, 78—85. Copyright 2017, with permission from Elsevier.*

simulations, multiplication of these sources will be dominant during subsequent deformation. Therefore, the HN of dislocation is suppressed in the RVE once the incipient HN event happens. The plastic deformation rate generated by dislocation slip is also enhanced by coarse-grained parameter β in the DDD-FE calculation, which is different from conventional DDD simulations. In particular, for the case of $\beta = 1$, Eq. (11.4.4) represents a one-to-one correspondence of HN from an MD to a DDD-FE simulation. The coarse-grained model can physically connect the HN of dislocations at two different spatial scales and remarkably improve the computational efficiency without the loss of typical dislocation characters. More important, it avoids the artificial introduction of dislocation sources.

11.4.4 Shock-Induced Plasticity at the Submicron Scale

In this section, single-crystalline copper pillars were loaded under different impact speeds to investigate their dynamic mechanical behavior at the submicron scale. The simulation domain is a prismatic bar with an $0.8 \times 0.8\text{-}\mu m^2$ cross-section and a height of 8 μm. The copper pillars are oriented in the [001] direction. In the DDD-FE procedure, the time step is 2 ps and the shock wave is generated by applying a velocity-controlled boundary condition on the upper surface such that the velocity increases linearly to its maximum value at a 0.5-ns rise time. Then, the velocity is held constant for a 2-ns shock holding time. The lateral surfaces are fixed in transverse directions to achieve the

uniaxial strain condition of shock waves. The loading and boundary conditions may be summarized as:

$$
u_z = u_z =
\begin{cases}
-at^2, & 0 \le t \le t_{\text{rise}} \\
-at_{\text{rise}}^2 - V_p(t - t_{\text{rise}}), & t > t_{\text{rise}}
\end{cases}
, \quad \text{at } z = \frac{L_z}{2}
$$

$$
u_z(t) = 0, \quad \text{at } z = -\frac{L_z}{2}
$$

$$
u_x(t) = 0, \quad \text{at } x = \pm\frac{L_x}{2}
$$

$$
u_y(t) = 0, \quad \text{at } y = \pm\frac{L_y}{2},
$$

(11.4.5)

Here, L_x, L_y, and L_z are the lengths of the simulation domain in the x, y, and z directions, respectively. u_x, u_y, and u_z are the displacement components and a is the acceleration making the shock velocity V_p during the rise time period t_{rise}.

For the small scale of interest here, no initial dislocation source is preset in the simulation domain. The HN of dislocation is introduced by the coarse-grained model when the local resolved shear stresses at the wave front attain or exceed 2 GPa. Coarse-grained parameter β given by Eq. (11.4.4) is associated with both the FE element size and the computational domain volume in the MD simulation. The mesh size in the FE is set as $\sim 160 \times 160 \times 160$ nm^3 and nine dislocation loops that are not in contact with each other at the initial time are randomly introduced into each element when the nucleation conditions are satisfied. These nine loops are all Shockley partial dislocations with a Burgers vector pf 1/6<112> distributed on the {111} slip planes, the same as the MD simulation results. This configuration can provide the plasticity generated by dislocation motion on different slip systems in an RVE. Then, the value of coarse-grained parameter β is set to 80 from Eq. (11.4.4). As mentioned earlier, the density of dislocation loops is slightly different when the size of the MD system changes. This uncertainty comes from the probabilistic nature of HN. The effect of the coarse-grained parameter will be further discussed in the following section. In this study, initial radius R of the nucleated dislocation loops is set to be as small as 50 times the magnitude of the Burgers vector. When a loop is formed, its contribution to the plastic strain and spin is given by the relations

$$
\boldsymbol{\varepsilon}^p = \beta \cdot \frac{\boldsymbol{b} \otimes \boldsymbol{n} + \boldsymbol{n} \otimes \boldsymbol{b}}{2V_0} A_0
$$

$$
\boldsymbol{\omega}^p = \beta \cdot \frac{\boldsymbol{b} \otimes \boldsymbol{n} - \boldsymbol{n} \otimes \boldsymbol{b}}{2V_0} A_0,
$$

(11.4.6)

Table 11.3: Material Properties and Parameters for Copper

Material and Parameters	Value	Dimension
Shear modulus (G)	48	GPa
Poisson's ratio (ν)	0.34	
Burgers vector length (b)	0.256	nm
Initial temperature (T_0)	300	K
Specific heat (c)	390	J/(kg·K)
Thermal conductivity (λ_t)	380	W/(m·K)
Line expansion coefficient (α)	17.7×10^{-6}	K^{-1}
Thermal scaling factor (γ)	0.9	
Mass density (ρ)	8.9×10^3	kg/m^3
Drag coefficient (B)	2.0×10^{-5}	Pa·s

Here, b is the Burgers vector and n is the normal vector of the slip plane. V_0 is the volume of the representative cell and A_0 is the initial area of the loop.

For the DDD-FE simulations, the material properties and controlling parameters of copper are listed in Table 11.3. This simulation mainly focuses on studying typical dislocation microstructures during shock wave propagation and shock-induced plasticity.

In particular in the following analyses, we emphasize that the pressure is given by $P = -\sigma_{zz}$ as usually defined in the shock wave community (Shehadeh et al., 2005b), and shear stress is also calculated as $\tau = -\frac{1}{2}\left[\sigma_{zz} - \frac{1}{2}\cdot(\sigma_{xx} + \sigma_{yy})\right]$; these are different from those in continuum mechanics. Strain is taken as $\varepsilon = -(\varepsilon_{xx} + \varepsilon_{yy} + \varepsilon_{zz}) = -\varepsilon_{zz}$ because deformation under shock loading conditions undergoes uniaxial compression without lateral strains in the simulations, and here ε_{zz} is longitudinal strain.

Fig. 11.24A shows shock wave characteristics and the corresponding dislocation structures at an impact speed of 500 m/s. The solid line is the distribution of pressure P along the loading direction in the current HN process. As a reference, the wave profiles of the simulation using a perfectly elastic constitutive law are also given as a short dashed line in Fig. 11.24A. By comparing two sets of curves, it can clearly be seen that there is a significant attenuation in stress level. As the shock wave advances in the material, the value of peak stress decreases. Four typical features marked in the wave profile are observed: (I) a leading wave front in which the pressure increases from 0 to a peak value is approximately 1 μm wide; (II) a sharp plasticity relaxation in stress occurs just behind the elastic precursor; (III) the local stress yield is followed by a relatively smooth increase in stress; and (IV) there are some release parts where the stress level decreases.

By carefully analyzing the corresponding microstructures shown in Fig. 11.24A, it can be seen that there is no dislocation in the leading wave front, which is an elastic precursor. Just behind this region, the abrupt decrease in the stress level is related to

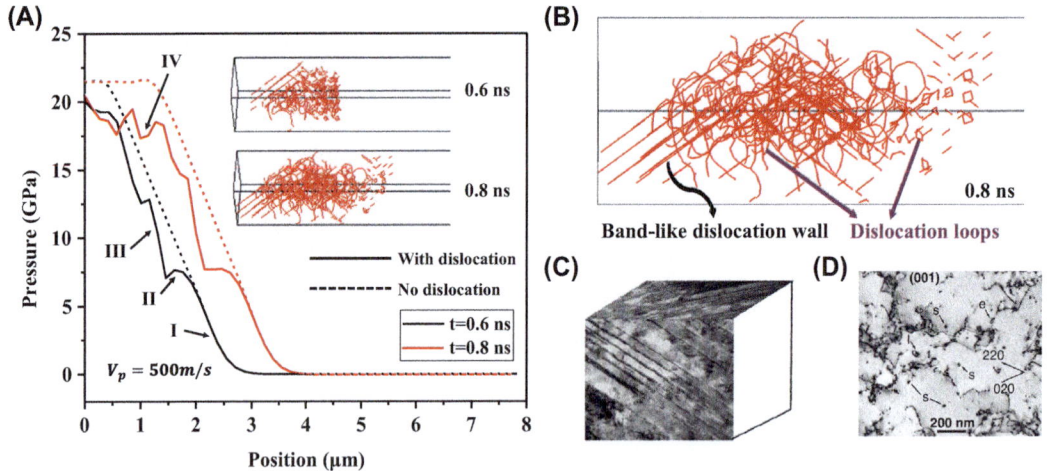

Figure 11.24

(A) Pressure profiles and dislocation structures at different times; (B) Typical microstructures in discrete dislocation dynamics-finite element simulation; (C, D) Experimental photographs from Meyers et al. (Meyers et al., 2003; Cao et al., 2007). *Reprinted from Hu, J.Q., Liu, Z.L., Chen, K.G., Zhuang, Z., 2017b. Investigations of shock-induced deformation and dislocation mechanism by a multiscale discrete dislocation plasticity model. Computational Materials Science 131, 78—85. Copyright 2017, with permission from Elsevier.*

the HN of dislocations and numerous dislocation loops exist in a limited region. As the shock wave propagates in the copper pillars, nucleation events continue to burst when the stress increases to a high value. One interesting feature is that once a dislocation loop expands, the adjacent dislocation loops located on different slip planes tend to be suppressed and even annihilated. Then, the extended dislocations generate a strain localization on the parallel slip planes and result in band-like dislocation walls or shear bands, as shown in Fig. 11.24B, which conform well with the micromorphology observed in the experiment (Cao et al., 2007), as shown in Fig. 11.24C. It seems there is a shielding effect of the band-like walls in the evolution of dislocation microstructures. That is, once the parallel shear bands are formed, dislocations between these walls are trapped and suppressed. The generation rates of dislocations on the dislocation walls and between them are different. As a result, the stress level on the shear band is relatively low owing to a higher rate of dislocation multiplication. Moreover, after the incipient HN processes, the deformation microstructures consist of both dislocation line defects caused by the multiplication and unexpanded dislocation loops from HN processes, which are consistent with the TEM observations (Meyers et al., 2003), as shown in Fig. 11.24D.

The evolution of local dislocation density in a slice and shear stress at the impact speed of 500 m/s are further investigated in this section. The slice is chosen to be a cube with a

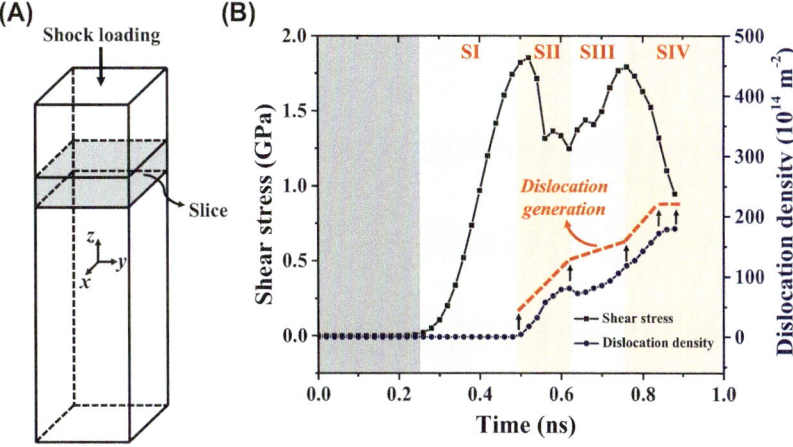

Figure 11.25

(A) Schematic of the slice; (B) Evolution of shear stress and dislocation density in a slice. *Reprinted from Hu, J.Q., Liu, Z.L., Chen, K.G., Zhuang, Z., 2017b. Investigations of shock-induced deformation and dislocation mechanism by a multiscale discrete dislocation plasticity model. Computational Materials Science 131, 78–85. Copyright 2017, with permission from Elsevier.*

dimension of $800 \times 800 \times 320$ nm^3 in the copper pillar, as shown in Fig. 11.25A. During the analysis, the nominal local dislocation density is evaluated by

$$\rho_s = \beta \cdot \frac{\sum l_i}{V_s}, \tag{11.4.7}$$

where l_i is the length of dislocation segment i inside the slice and V_s is the slice volume.

Fig. 11.25B shows the evolution of shear stress and local dislocation density in the slice. The initial undeformed state results from the fact that the shock wave has not yet reached the location of the slice. After that, a slice in the sample typically undergoes four stages under impact loading: Stage I (SI) is elastic rise without dislocation activity because there is no preset dislocation source in the simulation and HN has not started. Stage II (SII) corresponds to the HN of dislocations. The increase in dislocation density mainly comes from the HN process and the initial expansion of dislocation loops, resulting in local yielding in the sample. At Stage III (SIII), dislocation multiplication occurs in the slice and the dislocation density increases at a slower rate. Shear stress begins to increase again and the local dislocation density reaches as high as $\sim 10^{16}$ m^{-2}, which falls in the range between recovery experimental observations (Meyers et al., 2003) and MD simulations (Bringa et al., 2006; Cao et al., 2007). Finally, in Stage IV (SIV), interactions between dislocations and shock waves continue until the relaxation process takes place. The dislocation density saturates rapidly at a steady-state value, which is similar to the findings (Shehadeh et al., 2005b, 2006).

Two distinct dislocation mechanisms are proposed to explain stress relaxation by further analyzing the evolution of dislocation density in the slice, as shown by the red dashed line in Fig. 11.25B. Both the dislocation density, ρ_s, and its generation rate, $\dot{\rho}_s$, are discussed because the plastic shear strain rate can be written as (Austin and McDowell, 2012)

$$\dot{\gamma} = b\rho_s\bar{v} + b\dot{\rho}_s\bar{x}, \tag{11.4.8}$$

where $\bar{v}$ is the average velocity of dislocation segments and $\bar{x}$ is a displacement during the formation of a generated dislocation segment. The first item on the right-hand side of Eq. (11.4.8) is the conventional Orowan's relation; the second item corresponds to the differential area swept out during the formation of the new dislocation segment. This generalized relation provides a natural transition in the slip kinematics of different dislocation mechanisms. The initial HN of dislocations induces a local yield of the slice as SII, which is a nucleation-controlled plasticity because the dislocation generation rate related to the HN is sufficiently large (i.e., $\dot{\rho}_s\bar{x} > \rho_s\bar{v}$). Thereafter, as in the MD simulation results, the shear stress will increase again as SIII after the initial softening, which may occur because of the formation of strong cross-links between dislocations after the rapid expansion of incipient dislocation loops. As a result, the dislocation generation rate ($\dot{\rho}_s$) begins to slow down and the dislocation density (ρ_s) is also not as high as that in SIV. In the early period of SIV, parts of the strong cross-links between dislocations are destroyed with an increase in stress. Then, the dislocation density increases rapidly as a result of avalanche-like dislocation multiplication, which leads to shear stress relaxation. However, a saturated dislocation density is achieved soon afterward and stress relaxation continues, so it can be inferred that the conventional Orowan's relation is recovered because the generation term is negligibly small (i.e., $\rho_s\bar{v} \gg \dot{\rho}_s\bar{x}$). We attribute the second stress relaxation to a multiplication-controlled plasticity because the HN of dislocation is suppressed during this period.

In addition, the dynamic mechanical behavior of the slice in the copper pillar mentioned earlier was investigated under different impact speeds. The shear stress-strain curves of the slice under a wide range of loading velocities are presented in Fig. 11.26. Similarly, the shear stress exhibited an elastic increase for up to approximately 4% strain, followed by an abrupt drop of stress, except for the case of an impact speed of 200 m/s. For a shock loading of 200 m/s, the shear stress exhibited unrelaxed elastic behavior and immediately became stable with no attenuation when the slice was fully deformed. This suggests that no dislocation activity is detected at this impact speed and no plasticity relaxation takes place as a result. For a medium speed of loading, i.e., 500 and 800 m/s, the incipient abrupt drop in shear stress is related to the HN process. The shear stress started to increase again after the initial drop. Both annihilation and expansion of the nucleated loops occurred at this moment. However, the stress finally dropped to a shear-free state in which a hydrostatic state of stress $\sigma_{xx} \sim \sigma_{yy} \sim \sigma_{zz}$ was achieved. Dislocation multiplication is an

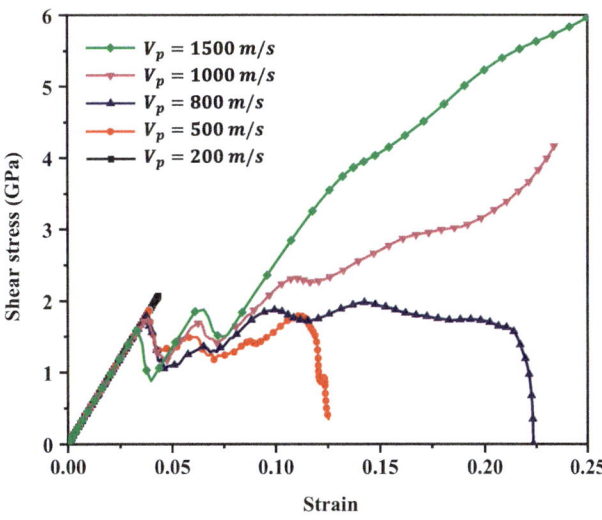

Figure 11.26

Shear stress-strain curves of the slice under various loading velocities. *Reprinted from Hu, J.Q., Liu, Z.L., Chen, K.G., Zhuang, Z., 2017b. Investigations of shock-induced deformation and dislocation mechanism by a multiscale discrete dislocation plasticity model. Computational Materials Science 131, 78—85. Copyright 2017, with permission from Elsevier.*

effective relaxation mechanism during this period. A higher impact speed requires more dislocations to relax external loading completely. For the extremely high-speed impact, i.e., 1000 and 1500 m/s, however, shear stress continued to increase again after the initial yielding. This indicates that the HN and subsequent dislocation multiplication cannot provide enough plastic relaxation. Therefore, a hydrostatic state of stress cannot be achieved with this situation.

These findings seem to conflict with results from previous reported simulations (Shehadeh et al., 2006; Kattoura and Shehadeh, 2014), which always achieved a shear-free state under various shock loadings. The contradiction is that most previous simulations introduced various dislocation sources immoderately. As a result, the dislocation-shock interaction could generate enough plasticity to achieve complete stress relaxation, whereas in our simulations, the HN process, and especially the dislocation source density, was introduced into the DDD-FE modeling by a coarse-grained method. Then, once the HN of dislocation was involved, dislocations in the slice interacted with the shock wave until a saturated dislocation density was achieved. The response of plastic relaxation owing to dislocation activity could not catch up with the increase in stress at the extremely high impact speed, resulting in the hardening behavior. Furthermore, the shock experiment with copper by Meyers et al. (2003) indicates that the microstructure primarily consists of dislocation cells at pressures of 12 and 20 GPa; twinning of stacking-fault bundles are the principal defects at a pressure of 40 GPa, and the structure shows microtwinning

and thermal recovery at a pressure of 55–60 GPa. The related pressure levels at different impact speeds in our simulations are 8.5 GPa at 200 m/s, 20 GPa at 500 m/s, 31.5 GPa at 800 m/s, and 48 GPa at 1500 m/s. The simulation results match the experimental observations. For a shock loading of 1500 m/s, the dislocation mechanism becomes invalid because the pressure is as high as 48 GPa. Thus, a threshold speed around 1000 m/s for the dislocation-dominant mechanism may be proposed from Fig. 11.26, beyond which the effect of other defects such as the stacking fault and twinning would be prominent (Mackenchery et al., 2016).

11.4.5 Discussion and Conclusion

Coarse-grained parameter β representing the spatial resolution of defects is proposed to bridge the gap between MD and DDD-FE simulations. As illustrated in Section 11.4.2, the value of β is slightly different and will converge to a certain value when the MD system becomes larger. In this section, the effect of β is discussed. The shear stress histories in the slice with different values of β are shown in Fig. 11.27A. The shear stress exhibits the typical behavior of dislocation plasticity. After the initial abrupt drop in shear stress, a temporary increase in stress is achieved, followed by a thorough drop in shear stress arising from avalanche-like dislocation multiplication. It seems that the values of β have a limited effect on the shear stress histories. However, the temperature histories of the slice vary greatly with different β, as shown in Fig. 11.27B. It is reasonable that thermal-related plastic work depends on the values of β. For the case of $\beta = 80$, the temperature can reach as high

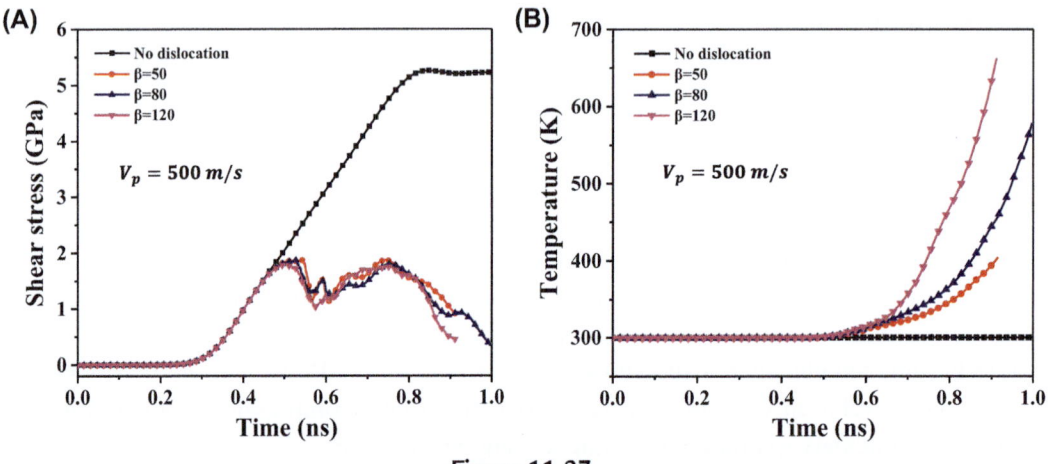

Figure 11.27

(A) Shear stress versus time; (B) Temperature versus time with different values of β at an impact speed of 500 m/s. *Reprinted from Hu, J.Q., Liu, Z.L., Chen, K.G., Zhuang, Z., 2017b. Investigations of shock-induced deformation and dislocation mechanism by a multiscale discrete dislocation plasticity model. Computational Materials Science 131, 78–85. Copyright 2017, with permission from Elsevier.*

as several hundred Kelvin. However, because the temperature has the same influence on three normal stresses, the shear stress histories are nearly the same. For the investigated range of β, the values have a limited effect on the mechanical behaviors of concern.

In this section, a coarse-grained parameter β is proposed to bridge the spatial gap between MD and DDD-FE simulations effectively, which introduces the HN without artificial dislocation sources. The thermal effect caused by dislocation motion and the thermoelastic constitutive law is also considered in the DDD-FE coupling model. The simulation results show the characteristics of shock wave profiles and dislocation microstructures. Two typical features of the microstructures, i.e., band-like dislocation walls and the collaboration of HN and dislocation multiplication, are analyzed and compared with the experimental observations. The shielding effect from the dislocation walls is proposed to explain the parallel shear bands. The evolution of shear stress and local dislocation density in a slice is investigated. Four typical stages of a dislocation-free material are elaborated under impact loading. Both rapid HN and avalanche-like multiplication of dislocations can lead to the softening of shear stress. Furthermore, by analyzing the stress-strain curves of the slice under different impact speeds, the dislocation mechanism is proposed to be dominant only when the impact speed is below 1000 m/s. Beyond this shock velocity, the response of dislocation is not quick enough to generate sufficient plasticity to relax the stress fully, and other defects such as twinning and stacking faults will be present. Finally, the influence of coarse-grained parameter β is discussed. A higher value of β leads to a more pronounced increase in temperature but has a limited effect on the evolution of shear stress.

11.5 Summary

In this chapter, the effect of the strain rate on the deformation of single crystals at the submicron scale was numerically studied. The main conclusions are:

1. DDD simulations were carried out to investigate the strain rate effect on finite-sized copper single crystalline samples under uniaxial compression and hydrostatic compression. The dislocation structures were less activated under hydrostatic compression compared with uniaxial compression. A size-dependent transition existed in the evolution of dislocation density that exhibited pressure sensitivity as the sample size became smaller than a specific value ($D \leq 400$ nm). Although a rate-dependent stress-strain relation was not observed under hydrostatic compression, the evolution of dislocation density exhibited a significant rate dependence.
2. With an increase in the strain rate, the yield stress of single-crystal copper increased rapidly. A critical strain rate existed in each single-crystal copper block for the given size and dislocation sources, below which the yield stress was relatively insensitive to the strain rate. The dislocation patterning changed from nonuniform to uniform under

high strain rate. The shear stresses in the bands were higher than those in neighboring regions, which formed shear bands in the crystal. The bandwidth increased with the strain rate, which often took place where the damage occurred.

3. Interactions between shock waves and dislocations were investigated by a multiscale analysis that coupled DDD with explicit FE. Band-like dislocation walls and their shielding effect on other dislocations were observed during shock wave propagation. The simulation results were in good agreement with experimental findings and were able to capture the typical microstructure of dislocation under shock loading. Both fast HN and avalanche-like dislocation multiplication became involved and led to the softening of shear stress. By comparing dynamic behavior under different impact speeds, a threshold speed for a dislocation-dominant mechanism was proposed from the computations in this work at around 1000 m/s, beyond which the effects of other defects such as stacking faults and twinning would be prominent.

Glide-Climb Coupling Model and Temperature Effect on Microscale Crystal Plasticity

Chapter Outline

Dislocation Mechanism-Based Crystal Plasticity. https://doi.org/10.1016/B978-0-12-814591-3.00012-1

12.1 Introduction

In previous chapters, we mainly focused on dislocation glide-dominated plastic flow in microscale crystals. In a thermally activated process, dislocations can be translated perpendicular to their slip planes, called dislocation climb, which has a significant role in relaxing dislocation pileups and promoting dislocation annihilation between different slip planes. Thus, dislocation climb motion cannot be ignored in thermally activated events. Dislocation climb is different from dislocation glide. First, the climb is a nonconservative motion; atoms must be added or removed from the dislocation core by a diffusion process, so it is a diffusion-controlled motion (Geers et al., 2014). Second, the diffusion process occurs over timescales that are much larger than those of dislocation glide, which creates a great time gap between glide and climb (Keralavarma et al., 2012). Thus, dealing with dislocation climb and glide together is a complex problem involving the coupling of multifield and multitimescale. Although plenty of studies have been carried out to elucidate how the combined movements of dislocation climb and glide influence plastic flow (Ayas et al., 2012; Bakó et al., 2011; Geers et al., 2014; Huang et al., 2014a,b; Keralavarma and Benzerga, 2015; Keralavarma et al., 2012), especially at high temperatures, key issues that remain about the coupling problem involve (1) how to couple vacancy diffusion and dislocation climb precisely to surpass phenomenological characteristics (Fedelich, 2002; Hartmaier et al., 2003; Hiratani et al., 2003; Lebensohn et al., 2010; Pontes et al., 2006) or treating it roughly as a drag-type movement like dislocation glide (Argaman et al., 2001; Bakó et al., 2006; Danas and Deshpande, 2013; Roters et al., 1996; Xiang and Srolovitz, 2006); and (2) how to tackle the timescale separation of about several orders of magnitude between glide and climb (Bakó et al., 2011; Keralavarma and Benzerga, 2015; Keralavarma et al., 2012).

Early works on the interaction between dislocation climb and vacancy diffusion were conducted using the molecular dynamics approach (Ortega, 2002) and atomistic simulations (Clouet, 2006; Kabir et al., 2010; Lau et al., 2009). With approaches such as these, the migration of vacancies and the interaction between microstructures can be characterized in detail at an atomistic level, but the real diffusion process occurs over timescales that are out of reach using atomistic methods and are limited by the length

scales. More studies have improved the climb model by incorporating vacancy diffusion theory into discrete dislocation dynamics (DDD) (Bakó et al., 2011; Mordehai et al., 2008, 2009), which provides an effective computational technique to simulate the collective behaviors of dislocation climb. Gao et al. (2011a) extended this method to pipe diffusion-controlled dislocation climb and enriched it to couple bulk diffusion and pipe diffusion in their work (Gao et al., 2013). Within these methods, the climb rates are derived from analytical formulations of a steady-state diffusion equation for a single infinite straight dislocation, which limits the models to be at the local level; this means that diffusion fields of different dislocation segments do not interact with each other. Although this treatment is not necessarily accurate in general for dislocation lines with high dislocation densities, as demonstrated by Gu et al. (2015), it is precise enough for low-dislocation density cases. Because of their concise and high efficiency, these methods are intensively applied in numerous works to describe climb-related problems qualitatively (Ayas and Deshpande, 2015; Ayas et al., 2012, 2014; Davoudi et al., 2012; Hafez Haghighat et al., 2013; Huang et al., 2014a). For quantitative analyses, however, we need to break up the local limitation and solve the diffusion equation with appropriate boundary conditions in three dimensions to build a global model. This is a complex and time-consuming work, and several attempts have been made to address this issue (Geers et al., 2014; Geslin et al., 2014, 2015; Gu et al., 2015; Ke et al., 2014; Keralavarma et al., 2012).

To bridge the timescales, several schemes have been proposed. Bakó et al. (2011) settled this matter by magnifying the climb rate and superimposing it with the glide rate, which is helpful for quantitative investigation but enlarged the role of dislocation climb. Keralavarma et al. (2012) put forward an adaptive scheme to address this issue. In their method, dislocation climb happens only at jammed or quasiequilibrium dislocation states. Once a dislocation has climbed to a new slip plane, it will glide again. It is an efficiency and fairly accurate scheme for large-timescale separation cases. However, for fast-climb cases in which the dislocation climb rate is high and cannot be neglected even at non-jammed states, the adaptive scheme would weaken the role of dislocation climb. A more general scheme is required to settle this matter.

In this chapter, methods coupling dislocation glide and climb in a unified three-dimensional (3D)-DDD model are developed to figure out the intrinsic formation mechanism of helical dislocations and high-temperature anneal hardening in Au submicron pillars; both multifield and multitimescale issues are addressed. This chapter is organized as follows. In Section 12.2, a bulk diffusion-controlled dislocation climb model is established by coupling vacancy diffusion and dislocation climb based on 3D-DDD. Then, in Section 12.0, the formation mechanism of the helical dislocation and factors influencing helical configuration are studied using this coupled glide-climb model. A modified 3D discrete-continuous method (DCM) is proposed in Section 12.4 to deal with thermally activated dislocation climb, in which the vacancy diffusion equation is solved by the finite

element method (FEM) and the timescale separation between glide and climb is handled by an adaptive scheme. Based on this coupled glide-climb model, the intrinsic mechanism of high-temperature anneal hardening at a submicron scale is investigated in Section 12.5. Finally, concluding remarks are presented in Section 12.6.

12.2 Coupled-Dislocation Glide-Climb Model-Based Analysis
12.2.1 Development of Vacancy Diffusion-Based Dislocation Climb Model

Dislocation climb is a nonconservative motion. Atoms must be added or removed at the dislocation core by a diffusion process that usually takes place at high temperature. As shown in Fig. 12.1, two different processes generate an equivalent climb motion: (1) emission of a vacancy or (2) absorption of an interstitial atom. Thus, the present work takes vacancy as an example. The same laws are valid for interstitials but in an opposite direction.

Because of the severe local distortion at the dislocation core, vacancies are prone to aggregating around the dislocation core and being absorbed or emitted by jogs. The dislocation lines could be assumed to be perfect sinks or sources of vacancies as long as the vacancy emission process is rapid enough compared with the long-range vacancy diffusion rate (Bakó et al., 2011; Mordehai et al., 2008, 2009). The two basic diffusion patterns of vacancies are pipe diffusion and bulk diffusion. Although the former has a smaller activation energy and vacancies can spread much faster than that of bulk diffusion (Gao et al., 2011a; Legros et al., 2008; Tang et al., 2003), bulk diffusion is the dominant diffusion mode when the temperature is high enough to guarantee a high jog density and the mean free path is larger than the jog spacing (Caillard and Martin, 2003). Therefore, only bulk diffusion is considered in our computational model.

Classically, vacancy diffusion follows Fick's first law:

$$J = -D_v \nabla c \qquad (12.2.1)$$

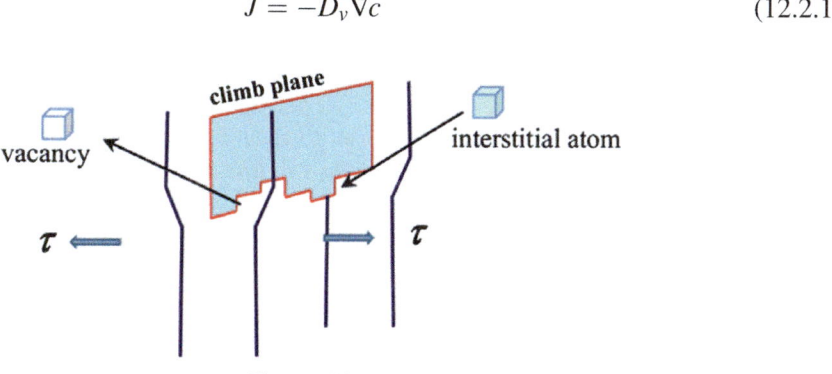

Figure 12.1

Climb of an edge dislocation. *Reprinted from Liu, F.X., Liu, Z.L., Zhuang, Z., 2017a. Numerical investigations of helical dislocations based on coupled glide-climb model. International Journal of Plasticity 92, 2—18. Copyright 2017, with permission from Elsevier.*

where c indicates the vacancy concentration, $D_v = D_v^0 \cdot \exp(-U_v^m/kT)$ denotes the diffusion coefficient, U_v^m equals the migration energy, k is the Boltzmann's constant, and T is the absolute temperature in Kelvin. J is the vacancy flux. The divergence of this flux, J, can be described by the diffusion governing equation (Fick's second law):

$$\frac{\partial c}{\partial t} = -\nabla \cdot J = -\nabla \cdot (D_v \nabla c) \tag{12.2.2}$$

Provided that D_v is constant at a given temperature, and a steady-state diffusion assumption is employed when the climb rate is slowly enough compared with the vacancy diffusion rate, Eq. (12.2.2) can be simplified to

$$\Delta c = \nabla^2 c = 0 \tag{12.2.3}$$

12.2.2 Incorporating the Dislocation Climb Model Into Three-Dimensional Discrete Dislocation Dynamics

Dislocation climbs as a result of vacancies flowing into or out of the dislocation core (Gao and Cocks, 2009). Denoting a positive climb velocity v_c when a dislocation climbs upward after emitting vacancies, we have a volume change in vacancies in the dislocation core resulting from climb, which is $v_c b$. Then, for a dislocation line of unit length experiencing a volumetric vacancy flux ϕ, the matter conservation requires that

$$v_c b = \phi \tag{12.2.4}$$

At high temperature and low stress, the jogs are dense enough to guarantee that the mean free path is larger than the jog spacing. The diffusion process of vacancies from the dislocation to the crystal has a cylindrical symmetry (Caillard and Martin, 2003). Thus, the corresponding volumetric flux across the lateral surface of a cylinder of radius r can be expressed as

$$\phi = 2\pi r D_v \frac{\partial c_v^r}{\partial r} \tag{12.2.5}$$

where r is a distance away from the dislocation core and c_v^r is the vacancy concentration at the corresponding location. Then, the climb rate is

$$v_c = \frac{2\pi r}{b} D_v \frac{\partial c_v^r}{\partial r} \tag{12.2.6}$$

To get the climb rate v_c in Eq. (12.2.6) precisely, Eq. (12.2.3) should be solved with appropriate boundary conditions to obtain the vacancy concentration field. Because the vacancy emission and absorption process is rapid compared with a long-range vacancy diffusion rate (Bakó et al., 2011; Mordehai et al., 2008, 2009), the dislocation lines could act as perfect sources or sinks of vacancies. However, dislocation lines are always in

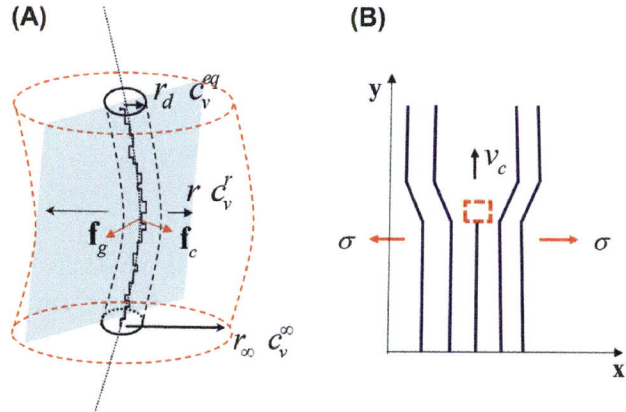

(A)   **(B)**

Figure 12.2

(A) Cylindrical bulk diffusion of vacancies; (B) Edge dislocation in a crystal under external stress. *Reprinted from Liu, F.X., Liu, Z.L., Zhuang, Z., 2017a. Numerical investigations of helical dislocations based on coupled glide-climb model. International Journal of Plasticity 92, 2—18. Copyright 2017, with permission from Elsevier.*

motion during evolution, which causes great difficulties in solving Eq. (12.2.3) in a full 3D domain. To circumvent the computational complexity, an isolated straight dislocation with a cylindrical bulk diffusion model, as shown in Fig. 12.2A, is adopted to derive the analytical solution of Eq. (12.2.3). This assumption is reasonable as long as the dislocation density is small enough in the calculation model (Gu et al., 2015).

On an inner surface of the cylinder, radius r_d is small enough to maintain equilibrium vacancy concentration c_v^{eq}. Because the equilibrium concentration of vacancies c_v^{eq} is determined by the free energy of formation G (Hirth and Lothe, 1982), it is given as

$$c_v^{eq} = c_v^0 \exp\left(-\frac{W_e}{kT}\right) \tag{12.2.7}$$

where c_v^0 is the reference-state or standard-state concentration of vacancies, $c_v^0 = \exp\left(-U_v^f/kT\right)$, where U_v^f denotes the vacancy formation energy. W_e is the extra external work done in creating the vacancies.

Now, consider the simplest case, an edge dislocation shown in Fig 12.2B, with stress σ acting on its core. By neglecting the relaxation volume, external work W_e done in creating a vacancy can be expressed as

$$W_e = -\sigma\Omega \tag{12.2.8}$$

Substituting Eq. (12.2.8) into Eq. (12.2.7), we can get

$$c_v^{eq} = c_v^0 \exp\left(\frac{\sigma\Omega}{kT}\right) \tag{12.2.9}$$

When a tensile stress, $\sigma > 0$, is applied, $c_v^{eq} > c_v^0$. The dislocation lines will emit vacancies and climb in a negative direction, and vice versa.

Now, generalizing to the case where $\sigma = f_c/b_e$, the equilibrium vacancy concentration is

$$c_v^{eq} = c_v^0 \exp\left(\frac{f_c\Omega}{b_e kT}\right) \tag{12.2.10}$$

where f_c denotes the mechanical climb force and $b_e = b\sin\theta$ denotes the edge component of the Burgers vector. θ is the character angle between the unit vector of dislocation line direction ζ and Burgers vector b. Considering that the screw dislocation does not climb, we enforce 0 climb velocity for dislocations with character angle θ less than 10^{-6} to avoid singularity (Mordehai et al., 2009).

According to the principle of 3D-DDD, dislocation lines are discretized into a chain of individual segments connected to each other head to tail by nodes. The evolution of dislocation configuration is represented by the movement of nodes (Bulatov and Cai, 2006; Schwarz, 1999a,b; Zbib et al., 1998). Mechanical climb force f_c is the climb component of nodal force:

$$f_c = \left(f^{PK} + f^{tension} + f^{image}\right)\cdot\mathbf{n} = \left[(\boldsymbol{\sigma}\cdot\boldsymbol{b})\times\zeta + f^{tension} + f^{image}\right]\cdot\mathbf{n} \tag{12.2.11}$$

where $\mathbf{n}$ is the slip plane normal and $(\boldsymbol{\sigma}\cdot\boldsymbol{b})\times\zeta$ denotes the Peach-Koehler force caused by externally applied stress and the long-range interaction of the other dislocation segments. $f^{tension}$ is the line tension, which is analogous to the surface tension of the soap bubble, and tends to straighten the line and reduce the total energy. $f^{tension}$ is defined as the increase in energy per unit in a length of the dislocation line (Hull and Bacon, 2001), which is computed in the same way as in conventional discrete dislocation methodology (Bulatov and Cai, 2006). f^{image} is the image force acting on the dislocation line of per-unit length, which has an important role in attracting dislocations toward the free surface because the material is more compliant there and the dislocation energy is lower (Hull and Bacon, 2001). Hirth and Lothe's analytical solution (1982) regarding a force acting on a straight dislocation segment near the free surface is adopted to add to the infinite-body stress field to capture the image force in the near-surface region. Details regarding the image force in the DDD code were described in previous publications (Cui et al., 2014; Gao et al., 2011a; Liu et al., 2009b).

On the outer surface of the cylindrical control volume, radius r_∞ is large enough to obtain the average vacancy concentration for the whole calculation model and is expressed as c_v^∞. Here, c_v^∞ depends on the spatial distribution of the vacancy sources and sinks. For thin foils or small bulk materials, the surfaces are perfect vacancy sinks, because $c_v^\infty = c_v^0$; for bulk materials, the surface or interface effect can be neglected. Thus, climbing of the dislocation line is the only way to generate or consume vacancies,

for simplicity (Mordehai et al., 2008). The evolution of c_v^∞ can be determined by the swept area of dislocation climb:

$$\frac{dc_v^\infty}{dt} = \frac{b}{V} \frac{dS}{dt} \tag{12.2.12}$$

where V is the volume of the computation cell and S is the swept area. If Δt is the time step and l is the length of the dislocation line, then the area swept by dislocations during climb in corresponding time step Δt can be expressed as $\Delta s = v_c \cdot l \cdot \sin\theta \cdot \Delta t$. It follows that the sign of climb velocity, v_c, determines the sign of the swept area, Δs, and therefore determines whether the dislocation segments are absorbing vacancies, $\Delta s > 0$, or emitting vacancies, $\Delta s < 0$.

In practice, the inner radius, r_d, has the same order of magnitude with a dislocation core, and the outer radius r_∞ is equal to the average distance between dislocation lines, $r_\infty = 1/(2\sqrt{\rho})$, where ρ is a dislocation density in the computational model (Caillard and Martin, 2003). Then, the vacancy concentration at any position can be obtained by the difference between concentrations on the inner and outer boundaries, c_v^{eq} and c_v^∞, respectively. The climb rate can be expressed as (Bakó et al., 2011)

$$v_c = -\frac{2\pi D_v}{b \sin\theta \ln(r_\infty/r_d)} \left(c_v^{eq} - c_v^\infty \right) \tag{12.2.13}$$

Climb direction **n** can be determined by Burgers vector **b** and dislocation line vector **ζ**:

$$\mathbf{n} = \frac{\boldsymbol{b} \times \boldsymbol{\zeta}}{|\boldsymbol{b} \times \boldsymbol{\zeta}|} \tag{12.2.14}$$

After the nodal climb velocity is determined, the topological evolution of dislocation climb can be simulated by our 3D-DDD code, in which the curved dislocation lines are discretized into straight dislocation segments with mixed characters (Bulatov and Cai, 2006; Schwarz, 1999a,b; Zbib et al., 1998). The dislocation nodes and segments can take an arbitrary position in space unrelated to any underlying lattice. Topological changes, including merging and splitting of the dislocation segment and nodes, as well as remeshing of the dislocation networks, are all taken into account and updated at every time step. Details regarding the DDD code are described in our previous publications (Cui et al., 2014; Gao et al., 2011a; Liu et al., 2009). Two typical examples are introduced to validate the developed bulk diffusion-based climb model.

12.2.3 Validation of Dislocation Climb Model

When a metal is irradiated or rapidly cooled from a high temperature, the supersaturation of vacant lattice sites will form and aggregate into small prismatic

loops that evidently change the physical properties. Recovery of these properties occurs when subsequent annealing takes place at elevated temperature (Semenov and Woo, 2003; Silcox and Whelan, 1960).

In the current work, the cases are optimally focused on the evolution of prismatic dislocation loops constructed of closed edge-dislocations with their Burgers vectors perpendicular to the loop planes. The loops can change their sizes only by climbing on the loop plane and modifying their location with a prismatic slip on the cylindrical surface. The configuration of forces acting on these loops mainly contain three parts: (1) line tension caused by curvature; (2) climb component of the Peach-Koehler force; and (3) osmotic force arising from the change in free energy of vacancies owing to the evolution of pressure during the dislocation climb (Lothe, 1967), which is also called the chemical force. It depends on the vacancy concentration gradient and the temperature and can be expressed as (Bakó et al., 2011; Caillard and Martin, 2003)

$$f^{\text{osmotic}} = \frac{kTb\sin\theta}{\Omega}\left(1 - \frac{c_v^\infty}{c_v^0}\right) \tag{12.2.15}$$

Many experiments and analytical methods for loop-vacancy systems have offered clues about the role vacancies may have in the evolution of loop sizes. Silcox and Whelan (1960) found that the lifetime of a loop decreases exponentially with temperature and loop areas decrease linearly with time in thin foils, whereas in bulk materials, big loops grow at the expense of smaller ones. Gao et al. (2011a) and the other researchers (Bakó et al., 2011; Gao et al., 2011a; Mordehai et al., 2008, 2009) validated these time- and temperature-dependent phenomena using analytical methods. In the current work, both thin foil and bulk material conditions are considered for comparison qualitatively and quantitatively with the experimental data.

Face-*centered cubic* (FCC) aluminum is chosen as the object material. The parameters used in the simulation are given in Table 12.1.

12.2.3.1 Case Study 1: Annihilation of Single Prismatic Loop in Thin Foils

In thin foils, the osmotic force generated from the vacancy concentration gradients is negligible owing to effective vacancy sinks provided by the foil surfaces. This is the case at elevated temperatures when vacancies move so fast that they can be removed by these sinks faster than they are produced by the prismatic loops. The average vacancy concentration is $c_v^\infty = c_v^0$. Without externally applied stress and neglecting the image force caused by the free surfaces, the only force acting on the segment of a loop is the line tension caused by the curvature, which tends to shrink the loop.

A 36-segment, discretized, circular, prismatic, vacancy-type loop 200 nm in radius is introduced as an initial configuration in a thin foil computational cell

Table 12.1: Parameters of Face-*Centered Cubic* Al Used in Discrete Dislocation Dynamics Simulation

Symbol	Description	Value
E^a	Young's modulus	72.7 GPa
ν^a	Poisson's ratio	0.347
a^b	Lattice constant	0.404
Ω^b	Atomic volume	16.3×10^{-30} m^3
b^c	Value of Burgers vector	0.256 nm
D_v^{0b}	Diffusion coefficient preexponential	1.18×10^{-5} m^2/s
U_v^{fb}	Vacancy formation energy	0.67 eV
U_v^{mb}	Vacancy migration energy	0.61 eV

[a]Data from Hirth and Lothe (1982).
[b]Data from Bakó et al. (2011).
[c]Data from Gao et al. (2013).

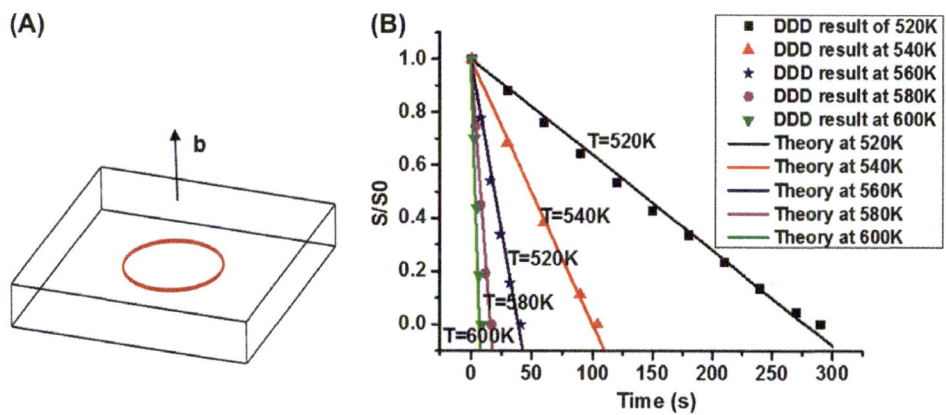

Figure 12.3
(A) Prismatic loops in thin foils; (B) Evolution of normalized loop area. *DDD*, discrete dislocation dynamics. *Reprinted from Liu, F.X., Liu, Z.L., Zhuang, Z., 2017a. Numerical investigations of helical dislocations based on coupled glide-climb model. International Journal of Plasticity 92, 2—18. Copyright 2017, with permission from Elsevier. s is the loop area, s₀ is the initial loop area.*

4000 nm × 4000 nm × 400 nm in size, as shown in Fig. 12.3A. The loop has the usual plane in the [011] direction and a Burgers vector in the <011> direction. These loops are the only vacancy sources. The temperatures range from 500 to 620K and the time step is 0.01 s. Calculation results show that the loop shrinks until annihilation and the loop area decreases linearly with time at all studied temperatures, as observed in the experiment. Silcox and Whelan (1960) also put forward a theoretical solution for the time evolution of radius r for the prismatic loop. The detailed derivation process is briefly introduced as follows.

For a loop lying in the (111) plane, the rate of shrinkage will then be given by

$$\frac{dr}{dt} = \frac{1}{2}v_c \tag{12.2.16}$$

where r is the radius of the loop at time t.

In thin foils at high temperatures, the osmotic force is negligible, because $c_v^{\infty} = c_v^0$. Thus, for rounded loops with no applied stress, the only mechanical force acting on the loop is line tension, $f^{\text{tension}} = \Gamma/rb$. Substituting $f_c = f^{\text{tension}}$ and Eq. (12.2.13) into Eq. (12.2.16) yields

$$\frac{dr}{dt} = \frac{\alpha D_v c_v^0}{b} \left[\exp\left(\frac{\Omega\Gamma}{kTrb} \right) - 1 \right] \tag{12.2.17}$$

At a low stress and high temperature, $\Omega\Gamma/rb \ll kT$, the function in square brackets can be simplified through Taylor expansion, and Eq. (12.2.17) can be rewritten as

$$\frac{dr}{dt} = \frac{\alpha D_v c_v^0}{b} \frac{\Omega\Gamma}{kTb} \cdot \frac{1}{r} \tag{12.2.18}$$

Eq. (12.2.18) can be integrated to give

$$r = r_0(1 - t/\tau)^{1/2} \tag{12.2.19}$$

where r_0 is the initial radius of the loop at t = 0. $D_v c_v^0$ is the self-diffusion coefficient D_v^{sd}, with $D_v^{sd} = \exp(U_v/kT)$, where U_v is the activation energy which is the summation of formation energy and migration energy $\tau = \tau_0\exp(U_v/kT)$ with $\tau_0 = 2r_0^2\frac{\Omega\Gamma}{b^2kT}$. Then, we can derive the evolution of loop area from Eq. (12.2.19):

$$\frac{s}{s_0} = \left(\frac{r}{r_0} \right)^2 = 1 - t/\tau \tag{12.2.20}$$

The evolutions of the normalized loop area obtained from the DDD simulation at different temperatures are given in Fig. 12.3B. Scatterplots in different colors in the DDD simulation results correspond to different temperatures, whereas the solid lines are the theoretical solutions of Eq. (12.2.12). They are in good agreement quantitatively. Moreover, the variation in curve slope indicates that the shrink rate increases with temperature, which agrees well with the experimental data (Silcox and Whelan, 1960).

12.2.3.2 Case Study 2: Evolution of Prismatic Loop Group in Bulk Material

Different from the thin foil case, osmotic force caused by vacancy supersaturation effects is expected to be important because vacancies form around the loops much faster than they move to sinks caused by surfaces. In a more ideal case, the loops are assumed to be

the only vacancy sources and the total vacancy number is conserved within a certain period of time. Average vacancy concentration c_v^∞ is time-dependent:

$$c_v^\infty = c_v^0 + \frac{b \cdot (v_c \cdot l \cdot \sin\theta \cdot \Delta t)}{V} \tag{12.2.21}$$

where Δt is the time increment. Without externally applied stress, the coarsening kinetics of these loops will be driven by competition between the line tension and osmotic force. The climb rate can be derived by inserting Eq. (12.2.21) into Eq. (12.2.13).

The initial configuration is composed of 20 randomly distributed circular prismatic loops in a cube computational cell 4000 nm × 4000 nm × 4000 nm, with no overlap or intersecting between them, as shown in Fig. 12.4A. The diameters of these loops range from 160 to 600 nm with a Burgers vector in the <011> crystal direction and their habit plane in the {011} crystallographic planes. The loops can only grow or shrink on their habit planes by climbing. The temperature is 600K.

The initial average vacancy concentration is set to $c_v^\infty = c_v^0$. It evolves according to the number of vacancies absorbed or emitted by climbing of these loops, because the computational cell is assumed to be a conserved system. The normalized average concentration, c_v^∞ / c_v^0, is used to denote the vacancy supersaturation. Its variation with time can be divided into three typical stages, as demonstrated in Fig. 12.4B. At the beginning, denoted as Stage I, all loops shrink because of the line tension and emit vacancies, inducing an increase in average concentration c_v^∞. At Stage II, larger loops with larger curvatures have smaller line tensions, leading to lower equilibrium concentration c_v^d

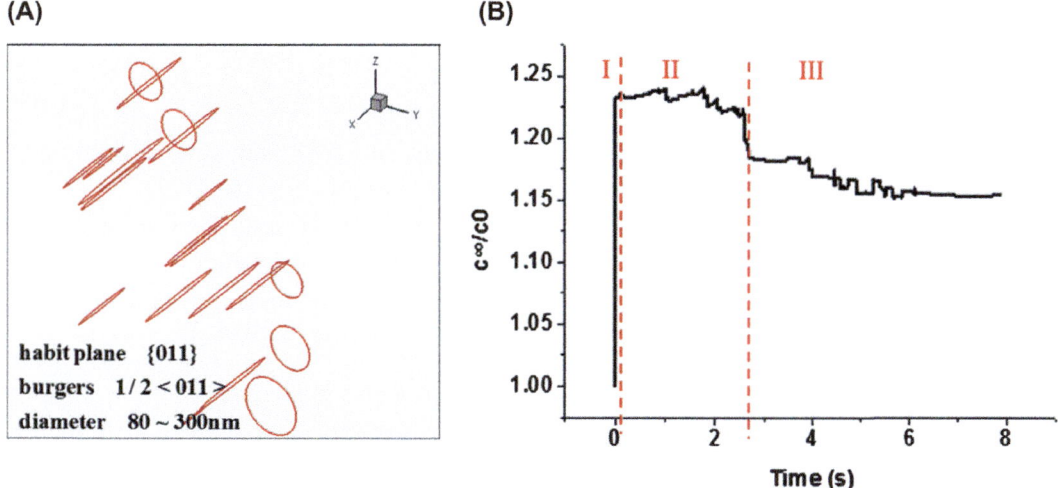

Figure 12.4

(A) Prismatic loops in bulk material; (B) Evolution of normalized average concentration. *Reprinted from Liu, F.X., Liu, Z.L., Zhuang, Z., 2017a. Numerical investigations of helical dislocations based on coupled glide-climb model. International Journal of Plasticity 92, 2−18. Copyright 2017, with permission from Elsevier.*

and then c_v^∞, whereas the smaller ones act otherwise. Thus, the large loops grow by absorbing vacancies emitted by the smaller ones, generating temporary stable Stage II. After annealing of the small loops, the large loops absorb vacancies from the crystal at Stage III, leading to a decrease in the supersaturated concentration until an equilibrium state is reached between average concentration c_v^∞ and equilibrium concentration c_v^d. They are qualitatively consistent with experimental observations (Silcox and Whelan, 1960) and analytical results (Mordehai et al., 2009).

So far, the applicability and accuracy of this vacancy-driven climb model have been validated qualitatively and quantitatively.

12.2.4 Coupled Glide-Climb Model Based on Three-Dimensional Discrete Dislocation Dynamics

A coupled glide-climb model is developed in this section by incorporating dislocation climb into the glide model and a practical time scheme is adopted to bridge the large time separation.

Different from diffusion-controlled climb, dislocation glide is mainly controlled by viscous phonon-drag mechanisms in pure FCC crystals (Marian et al., 2004; Nabarro and Duesbery, 2002), which are assumed to follow a simple linear drag law:

$$v_g = \frac{f_g}{B_g(T)} \tag{12.2.22}$$

where $B_g(T) = 1.4 \times 10^{-5} \cdot T/300$ is the time-dependent drag coefficient, which can be obtained from molecular dynamic calculations (Kuksin et al., 2008). Although the screw segments possess slightly larger glide resistance compared with that of the edge ones, the difference is negligible compared with the difference between the mobility of glide and climb. Thus, in the current simulation, we use a unified drag coefficient for the glide of both edge and screw segments, for simplicity. f_g is the glide component of nodal force:

$$
\begin{aligned}
f_g &= \left(f^{PK} + f^{tension} + f^{image}\right) \times (n \times \zeta) \\
&= \left[(\sigma \times b) \times \zeta + f^{tension} + f^{image}\right] \times (n \times \zeta)
\end{aligned}
\tag{12.2.23}
$$

Because the climb rate is diffusion-limited, it is usually slow. On the other hand, dislocations can glide at a velocity up to the speed of sound (Hirth and Lothe, 1982). The timescale separation between these two types of motion is as much as 10 orders of magnitude (Bakó et al., 2011; Keralavarma et al., 2012). To incorporate them into a unified model, an appropriate scheme is needed to bridge the timescales. Keralavarma et al. (2012) put forward an adaptive scheme to deal with this issue: Let dislocations glide with a smaller time step first, until the overall plastic strain rate approaches 0, which

happens in the case of dislocation pileups. Then, climb-related events are activated with a much larger time step. Once the dislocations are detected as having climbed to new slip planes, the time step will revert to smaller ones. In this scheme, only one type of motion happens at a time, which is efficient and suitable for jammed dislocation states such as power-law creep (Mirzadeh et al., 2015). However, it is not necessarily accurate for fast climb cases, in which the dislocation climb rate is high and cannot be neglected even for non-jammed states. The adaptive scheme would weaken the role of dislocation climb. To solve this problem, a sequentially coupled scheme is proposed in the current work.

In this scheme, dislocations glide and climb naturally with a smaller time step for about 0.1 ns, as shown in Fig. 12.5. Because the climb distance per time step is small compared with the glide distance, the small increments of dislocation climb per time step will accumulate until a climb step of at least one glide plane distance is reached. Then, the cumulative climb distance S_2 will be superimposed onto the total displacement as long as S_2 is greater than or equal to the distance between adjacent slip planes D, where D is set to be equal to b, which is the value of the Burgers vector: here, $b = 0.256$ nm. Hence, displacement increment $\Delta \mathbf{S}$ per time step is

$$\Delta \mathbf{S} = \begin{cases} \mathbf{S}_1, & |\mathbf{S}_2| < D \\ \mathbf{S}_1 + \mathbf{S}_2, & |\mathbf{S}_2| \geq D \end{cases} \tag{12.2.24}$$

where $\mathbf{S}_1$ is the slip distance in each time step Δt and equals $v_g \Delta t$, where v_g is the glide velocity of the current step.

A numerical example is provided to validate the applicability of this method. As schematically shown in Fig. 12.6A, a pure edge dislocation fixed at two ends is driven against an impenetrable obstacle with a small amount of applied stress. The dislocation line cannot bypass the obstacle if it can only glide on the slip plane, whereas with climb mobility that can translate dislocations in a perpendicular direction, this dislocation line may bypass the obstacle. This is often the case in dislocation creep. As shown in Fig. 12.6B and C, the blue dotted circle is the position of the obstacle and the red dash-dot line is the initial configuration of the dislocation line. With the dislocation climb, edge

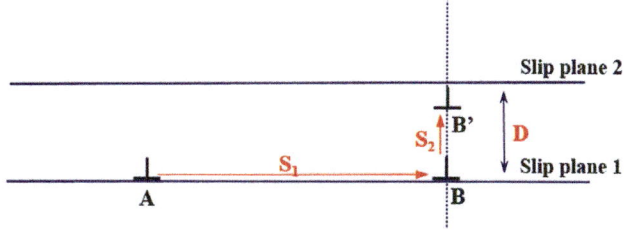

Figure 12.5

Sequentially coupled scheme. *Reprinted from Liu, F.X., Liu, Z.L., Zhuang, Z., 2017a. Numerical investigations of helical dislocations based on coupled glide-climb model. International Journal of Plasticity 92, 2−18. Copyright 2017, with permission from Elsevier.*

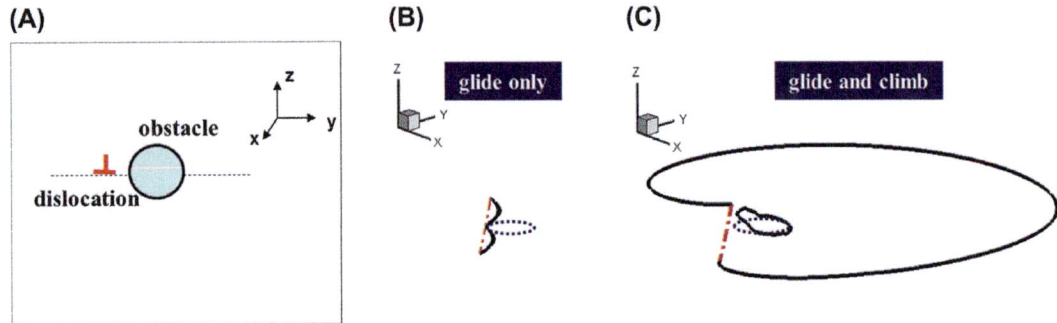

Figure 12.6
An edge dislocation climbs over an impenetrable obstacle: (A) Calculation model; (B) Dislocation configuration only for the glide case; (C) Dislocation configuration for the glide and climb case. *Reprinted from Liu, F.X., Liu, Z.L., Zhuang, Z., 2017a. Numerical investigations of helical dislocations based on coupled glide-climb model. International Journal of Plasticity 92, 2—18. Copyright 2017, with permission from Elsevier.*

dislocation can climb along the obstacle, which lowers the threshold stress to bypass the obstacle, whereas for the glide-only case, the dislocation line is blocked by the obstacle, which indicates that the sequentially coupled scheme is suitable for bridging the timescale separation.

In fact, validation of this sequentially coupled scheme can also be proved by two limiting cases. In one such case, when the climb motion is extremely slow compared with glide motion, it takes a long time before cumulative climb distance S_2 reaches one glide plane distance, D, which will be longer than reaching a jammed state. Then, our time scheme will be equivalent to the adaptive scheme proposed by Keralavarma et al. (2012). In another limiting case, when climb motion is as fast as glide motion, glide distance S_1 and climb distance S_2 will be superimposed directly onto each time step. Then, our time scheme will be more accurate than the adaptive scheme, in which climb motion happens unless glide motion is blocked.

As long as the climb and glide velocities v_c and v_g are derived, respectively, from Eqs. (12.2.13) and (12.2.22) by importing the velocities into a 3D-DDD model (Cui et al., 2014; Gao et al., 2010; Liu et al., 2009) as nodal velocities, a coupled glide-climb DDD model can be established to incorporate two types of dislocation movements into a unified 3D-DDD model. That is, the dislocation segments can climb perpendicular to the slip plane driven by the vacancy concentration gradient, and glide simultaneously on the slip plane driven by nodal forces.

12.3 Study of Helical Dislocations

Helical dislocations have been observed in metallic, ionic, and covalent crystals and even alloys by numerous researchers (Caslavsky and Gazzara, 1971; Kayo et al., 2016; Liao

et al., 2013; Mizuno et al., 2014; Serrano et al., 2006; Wagner and Gottschalch, 2006) since they were first reported by Bontinck and Amelinckx (1957), especially after radiation damage, cold work, or quenching from high temperatures, which can introduce an excess of point defects. Such helical dislocations are believed to have an important role in forming clear bands (Nogaret et al., 2008; Rodney, 2004) or prismatic loops (Mizuno et al., 2014; Munday et al., 2016; Xiang and Srolovitz, 2006), generating a type of whisker on the crystal surface (Amelinckx et al., 1957), and reducing point defects (Kayo et al., 2016), which would have significant impacts on crystal plasticity. Moreover, helical dislocations have poor motility and can act as barriers to other dislocations (Jumel et al., 2005; Liu and Biner, 2008; Nogaret et al., 2008), resulting in remarkable effects on plastic behavior. Three general mechanisms for the formation of helical dislocation have been proposed: (1) A screw dislocation acquires a helical turn by interacting with an interstitial Frank loop with the same Burgers vector (Rodney, 2004); (2) a screw dislocation bends during thermal vibration and obtains edge components in the opposite direction, and then helical turns take shape when they climb in the opposite direction (Frank, 1958); (3) when a mixed dislocation pinned at two ends interacts with supersaturated vacancies, the edge component climbs and the screw part slips. After further movement, the dislocation line will spiral around the pinned points and wind in a helical shape (Amelinckx et al., 1957; Kayo et al., 2016). The first one demands a large number of dislocation loops, whereas the last two require a combination of dislocation glide and climb, which is a nonconservative motion that mostly occurs at elevated temperatures. Although the formation of helical dislocations has been discussed experimentally (Kayo et al., 2016), theoretically (Amelinckx et al., 1957; Weertman, 1957), and even numerically (Liu and Biner, 2008; Munday et al., 2016; Rodney, 2004), there still no comprehensive and coherent mechanism to account for the formation processes in current research.

12.3.1 Formation of Helical Dislocation

To study the formation mechanism of helical dislocation, 3D-DDD simulations in an aluminum micropillar are carried out, as demonstrated in Fig. 12.7A. The size of pillar is 2000 nm × 640 nm × 640 nm. A longitudinal axis is along the [110] direction. A mixed straight dislocation line with a strong component of screw character is introduced, which is pinned at two ends. The Burgers vector is parallel to the longitudinal axis and the slip plane normal is in the [−111] direction. Material properties are obtained from Table 1. The relaxation volume for forming a vacancy is neglected.

Compression with a value of 200 MPa in the [110] direction is applied on top of the pillar. Immobile dislocation lines are placed randomly at the pillar bottom of the DDD model, as shown in Fig. 12.7A. Nodal force $\mathbf{f}$ in our calculation model is

$$f = f^{\mathrm{PK}} + f^{\mathrm{tension}} + f^{\mathrm{image}} + f^{\mathrm{osmotic}} + f^{\mathrm{interact}} \qquad (12.3.1)$$

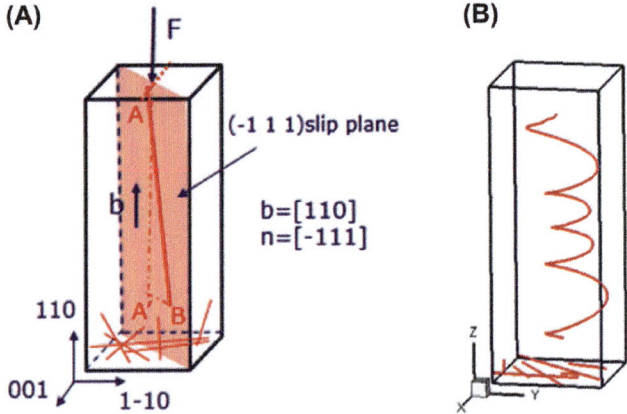

Figure 12.7

(A) Calculation model; (B) Helical dislocation configuration derived from the discrete dislocation dynamics simulation. *Reprinted from Liu, F.X., Liu, Z.L., Zhuang, Z., 2017a. Numerical investigations of helical dislocations based on coupled glide-climb model. International Journal of Plasticity 92, 2–18. Copyright 2017, with permission from Elsevier.*

where f^{PK} is the Peach-Koehler force, $f^{tension}$ is the line tension, f^{image} is the image force caused by the free surfaces, $f^{osmotic}$ is osmotic force caused by the vacancy concentration gradient, and $f^{interact}$ is the interaction generated from dislocations placed at the bottom.

The excess of vacancies results in the climb of edge component A′B in the direction of slip plane normal **n**, whereas screw component AA′ will glide on the surface of the cylinder with an axis parallel to Burgers vector **b**, which is determined by line vector **ξ** and Burgers vector **b**. Because the two ends are pinned, the dislocation lines would wind in spirals after further movement. A helical dislocation would take shape with the combination of prismatic slip and climb.

In the DDD simulation, spirals expand perpendicular to the longitudinal axis during compression, which is consistent with the conclusion adopted by Amelinckx et al. (1957). It can be concluded that dislocation climb brings about winding up and prismatic glide ensures spaces between spirals. In addition, the magnitude of order of the vacancy concentration is 10^{-2}.

12.3.2 Comparison With Theoretical Solution

Since pioneer work by Bontinck and Amelinckx (1957), plenty of theoretical analyses have been conducted revealing the intrinsic formation mechanism of such helical dislocations (Amelinckx et al., 1957; de Wit, 1959; Weertman, 1957). By solving the equilibrium equation of a dislocation subjected to external stress, an osmotic force, and a line tension, Weertman (1957) found that the equilibrium form of such a mixed dislocation line was indeed a helix, but the stable equilibrium configuration contained only one spiral,

which was different from the DDD results shown in Fig. 12.7B. To clarify this issue, the governing equations of theoretical and computational model are compared.

In the theoretical model, a mixed dislocation line in an infinite field is adopted and interaction forces between different segments or image forces are neglected. The governing equation is simplified as

$$f^{PK} + f^{tension} + f^{osmotic} = 0 \qquad (12.3.2)$$

By solving Eq. (12.3.2), the final dislocation configuration is derived regardless of the evolution processes: A mixed dislocation, AB, pinned at two ends, is plotted in Fig. 12.8A. Let $\mathbf{r}(\varphi)$ be a vector whose origin O is in the middle of AB. By changing parameter φ, vector $\mathbf{r}$ would sweep over configurations formed by AB. The general solution of Eq. (12.3.2), which gives rise to a helix passing through A and B in Fig. 12.8A, is given by (Weertman, 1957):

$$\mathbf{r} = a[\cos\varphi\mathbf{i} + (\sin\varphi - \sin\overline{\varphi})]\mathbf{j} + \frac{Ll_2}{a}\left(\frac{\varphi - 1/2m\pi}{\overline{\varphi} - 1/2m\pi}\right)\mathbf{k} \qquad (12.3.3)$$

where a is a radius of the cylinder, $\mathbf{k}$ is a unit vector parallel to Burgers vector $\mathbf{b}$, $\mathbf{i}$ is a unit vector normal to the plane formed by $\mathbf{k}$ and $\mathbf{l}$, $\mathbf{l}$ is a unit vector along direction AB, and $\mathbf{l} = l_1\mathbf{i} + l_2\mathbf{k}$. $\overline{\varphi} = \cos^{-1}(Ll_1/a)$, and $\overline{\varphi} \leq \varphi \leq m\pi - \overline{\varphi}$ with $m = 1, 3, 5, 7, \cdots$. It seems that the number of spirals increases linearly with m, but only the curve with $m = 1$ gives a stable equilibrium. The higher values of m denote positions of unstable equilibrium. In general, this numerical method concludes that the final configuration of a mixed dislocation at stable equilibrium state is a helix with one spiral, whereas the different stresses only influence radius a.

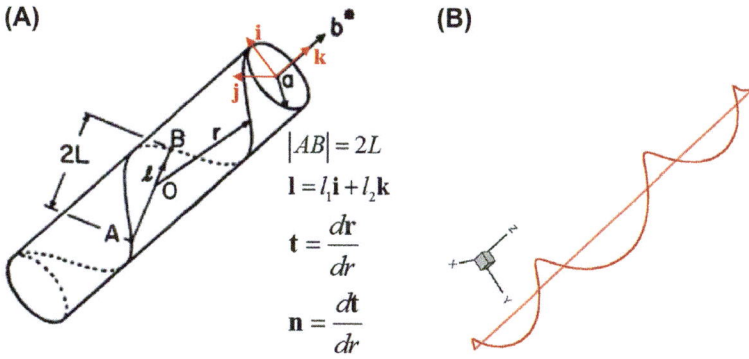

Figure 12.8

Dislocation helix: (A) theoretical solution (Weertman, 1957); (B) Numerical solution. *Reprinted from Liu, F.X., Liu, Z.L., Zhuang, Z., 2017a. Numerical investigations of helical dislocations based on coupled glide-climb model. International Journal of Plasticity 92, 2—18. Copyright 2017, with permission from Elsevier.*

In the DDD model, after neglecting the image force and interaction force corresponding to the theoretical model, the governing equations of Eqs. (12.2.22) and (12.2.13) can be rewritten as

$$\left(f^{PK} + f^{tension}\right) \cdot \mathbf{b}_g = m_g \cdot v_g \tag{12.3.4a}$$

$$\left(f^{PK} + f^{tension}\right) \cdot \mathbf{b}_c + f^{osmotic} = m_c \cdot v_c \tag{12.3.4b}$$

where m_g and m_c may be given a simple physical interpretation as the corresponding damping coefficients of glide and climb, and $\mathbf{b}_g$ and $\mathbf{b}_c$ are the glide and climb components of the Burgers vector, respectively. Eqs. (12.3.4a) and (12.3.4b) can describe the dynamic evolution process of the helical dislocation with an overdamping assumption, whereas the glide and climb velocities are neglected in the theoretical model. By superposing Eqs. (12.3.4a) and (12.3.4b),

$$f^{PK} + f^{tension} + f^{osmotic} = m_c \cdot v_c + m_g \cdot v_g \tag{12.3.5}$$

Terms on the left side of Eq. (12.3.5) are the same as those of Eq. (12.3.2), whereas they are different on the right side. As the evolution proceeds, the values of v_g and v_c tend to be 0, and then Eq. (12.3.5) would be equivalent to Eq. (12.3.2). However, the different equilibrium configurations may exist when the system is unstable. The dislocation may be driven to different equilibrium configurations under different perturbations, which is well-captured in our computational model, as shown in Fig. 12.8B. This is similar to the stability analysis of a structure, which deformed following different equilibrium paths with different initial perturbations when it is unstable.

From this analysis, we can explain the formation of helical dislocations in two ways. From the aspect of kinematics, dislocation climb results in winding up and prismatic glide determines the helix pitches. From the aspect of kinetics, we could reasonably ascribe the formation of helices to the joint action of applied stress, osmotic force, and line tension, which drives the dislocation line into a helix. Glide and climb velocities break the stable equilibrium state and drive the helix into a configuration with multiple turns. In this respect, our numerical calculation is more appropriate for explaining the formation mechanisms of helical dislocations by considering natural evolution.

12.3.3 Influential Factors for Helical Dislocation Configuration

From the numerical simulations, we find that the final configuration of the helix changes under different conditions. In this section, three main factors, including externally applied stress σ, the initial configuration, and vacancy concentration c_v^0, are discussed in detail.

For a straight edge dislocation line with two pinned ends, once the shear stress is large enough to drive the dislocation lines bend past the semicircular equilibrium state,

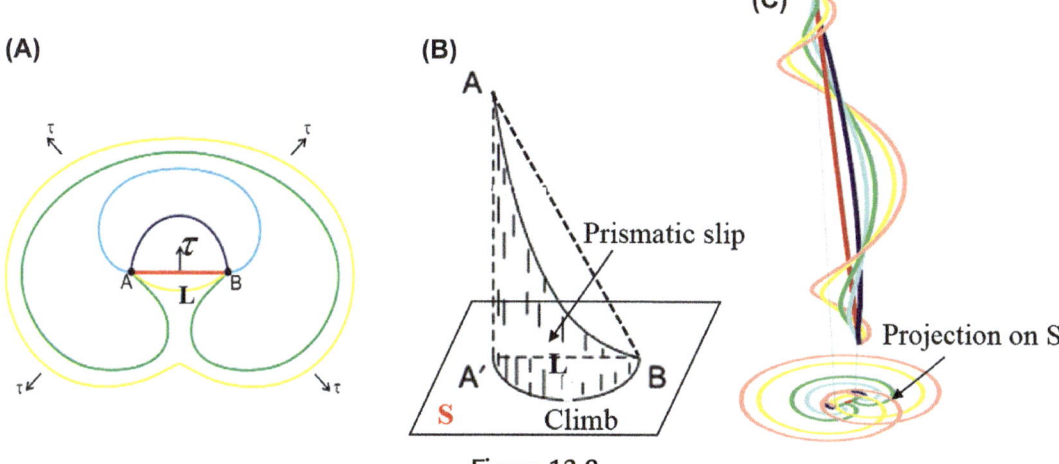

Figure 12.9

(A) Evolution of dislocation sources: Bardeen-Herring source or Frank-Read source; (B, C) Evolution of helical dislocations. *Reprinted from Liu, F.X., Liu, Z.L., Zhuang, Z., 2017a. Numerical investigations of helical dislocations based on coupled glide-climb model. International Journal of Plasticity 92, 2–18. Copyright 2017, with permission from Elsevier.*

it will spontaneously continue to grow and spiral around pinning points A and B, until the segments collide and annihilate each other at the intersection, as the yellow line demonstrates in Fig. 12.9A. This results in a flow of dislocation loops when the dislocations move in the same plane, which is called the Frank-Read source (FRS) for pure glide or the Bardeen-Herring source (BHS) for pure climb (Bakó et al., 2011; Bardeen et al., 1952). The critical resolved shear stress is $\sigma_t = T/br$ for FRS and $\sigma_t = T/br + f^{\text{osmotic}}/b$ for BHS, respectively, where r equals a half-length of source L. It means that a critical shear stress exists to active the source for a dislocation line with given length L.

Whereas for a mixed dislocation line with two pinned ends, A and B, it can glide on a cylindrical prismatic plane and climb on the normal plane, when an appropriate shear stress is applied, it may also spiral around pinning points A and B (Fig. 12.9B). The projection on climb plane S can be seen as a BHS, as shown in Fig. 12.9C. However, different from the BHS or FRS, the helix is a 3D configuration and dislocation segments can move in different planes. It cannot collide with itself or annihilate into dislocation loops. Thus, a critical shear stress does not exist to active such a mixed dislocation to produce spirals constantly, as the dislocation source does. As the spiral number increases, neighboring portions of successive turns would repel each other and source length r would become smaller. A much larger force is needed to drive the helix. With this perspective, the applied external stress and length L of edge component $A'B$ in initial configuration AB would mainly influence the number of spirals in the helical dislocation.

12.3.3.1 Influential Factor 1: Externally Applied Stress

To characterize the final dislocation configuration, we define the number of half-spirals as N. As shown in Fig. 12.10A, we take arbitrary plane P that contains the Burgers vector as the reference plane. The projection of helical dislocations on plane P is the red thick curve. The number of intersections between the projection and the z-axis counts as M, which are black dots in Fig. 12.10A. Then, N is taken to be M-1. For example, if the helical dislocation has one spiral, N = 2.

The cyclic load drives dislocations to move back and forth, providing more time for movement and creating more opportunities for vacancies to interact with dislocation lines at a certain extent. It can change the track of the dislocations, but no essential differences are found in the final configuration. Therefore, unidirectional compression is adopted for simplicity in the following simulations.

The final configurations corresponding to different external applied stresses of 100, 200, 300, and 400 MPa are shown in Fig. 12.10B–E, respectively. All have the same initial configuration with edge component $L = 60\sqrt{2}$ nm and screw component $H = 1800$ nm. Fig. 12.10B–E indicate that the number of spirals increases with external stress: that is, N increases with σ. Interestingly, this is similar to the buckling of a column under axial compression loading, in which the buckling wavenumber increase as the critical load increases.

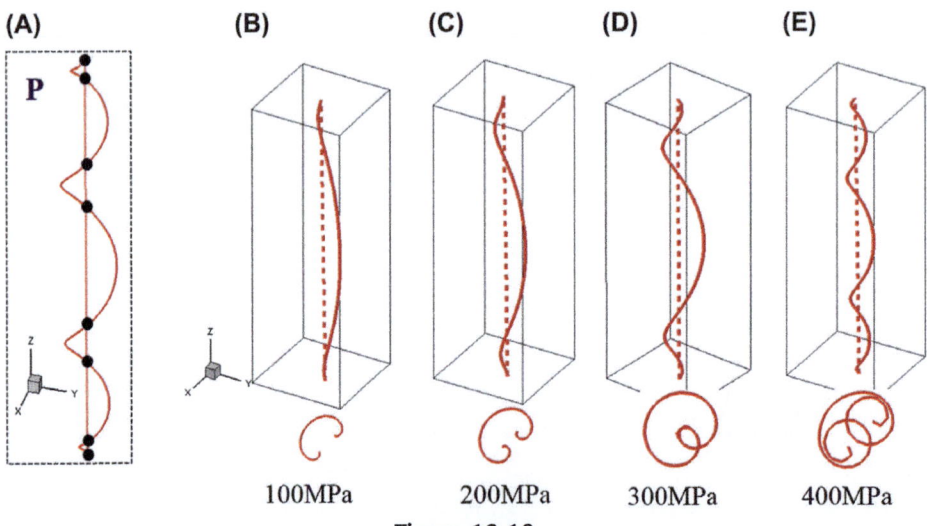

Figure 12.10

(A) Definition of N; (B–E) Final configurations at different applied stresses with the edge component of initial configuration $L = 60\sqrt{2}$ nm. *Reprinted from Liu, F.X., Liu, Z.L., Zhuang, Z., 2017a. Numerical investigations of helical dislocations based on coupled glide-climb model. International Journal of Plasticity 92, 2–18. Copyright 2017, with permission from Elsevier.*

12.3.3.2 Influential Factor 2: Initial Configuration

As stated earlier, the projection on climb plane S can be seen as a BHS, and edge component L of the initial configuration can be compared with the length of the source. Therefore, a smaller stress is required to activate the dislocation with a larger L. The final configurations of different edge components L with the same screw component $H = 1800$ nm at the same stress of 400 MPa are demonstrated in Fig. 12.11. They show that the number of spirals increases with the edge component of the initial configuration: that is, N increases with L.

For different edge components L of the initial configuration, spiral number N cannot keep increasing with externally applied stress, as shown in Fig. 12.12. Beyond a stress value, a horizontal stage of the curve occurs. This is because N increases as the helical dislocation winds and expands; long-range elastic interaction becomes stronger and neighboring successive turns repel each other. Consequently, a much larger stress is needed to drive the helix into more spirals.

12.3.3.3 Influential Factor 3: Vacancy Concentration

As described previously, in a defect-free crystal without external stress, the equilibrium vacancy concentration is $c_v^0 = \exp\left(-U_v^f/kT\right)$. Because the H-trapping effect could decrease the formation energy of vacancies U_v^f, c_v^0 may significantly increase. In our calculation model, the dislocation density is as low as about $10^{13}/m^2$. It is appropriate to

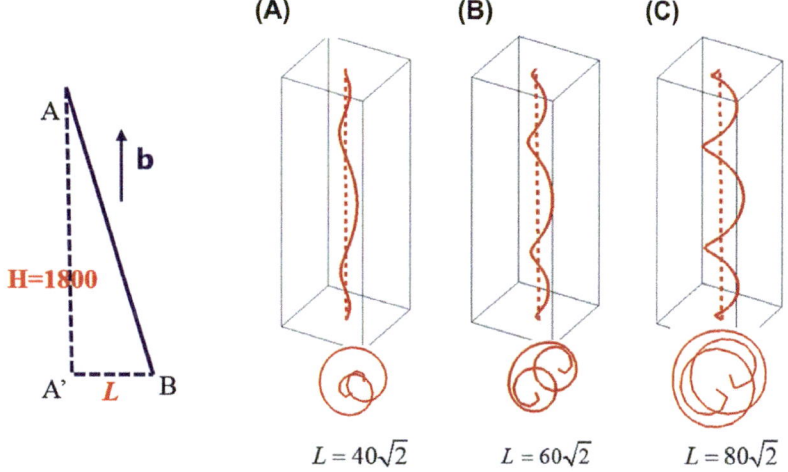

Figure 12.11

Final configurations of different edge components L (nm) at the same applied stress of 400 MPa. *Reprinted from Liu, F.X., Liu, Z.L., Zhuang, Z., 2017a. Numerical investigations of helical dislocations based on coupled glide-climb model. International Journal of Plasticity 92, 2–18. Copyright 2017, with permission from Elsevier.*

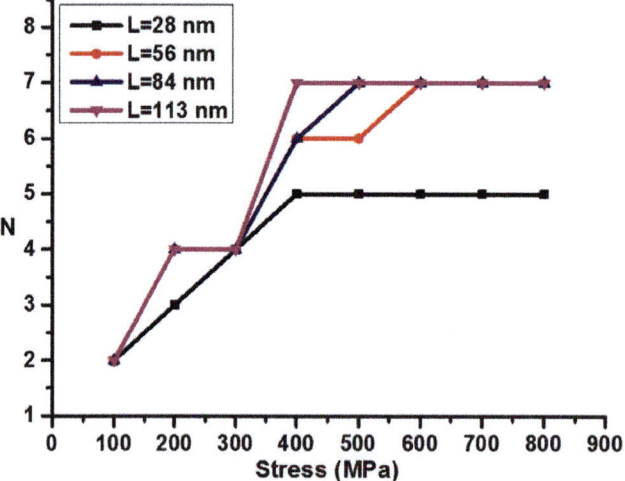

Figure 12.12

Influence of external stress and edge component *L* on initial configuration. *Reprinted from Liu, F.X., Liu, Z.L., Zhuang, Z., 2017a. Numerical investigations of helical dislocations based on coupled glide-climb model. International Journal of Plasticity 92, 2—18. Copyright 2017, with permission from Elsevier.*

assume that average vacancy concentration c_v^∞ is roughly equal to c_v^0. With this assumption, Eq. (12.2.13) can be converted into the form

$$v_c = -\frac{2\pi D_v^0 c_v^0}{b \sin\theta \ln(r_\infty/r_d)} \left(\exp\left(\frac{F_c\Omega}{KTb \sin\theta}\right) - 1 \right) \tag{12.3.6}$$

Vacancy concentration c_v^0 affects the climb velocity directly, which would cause a significant change in the helical configuration. Fig. 12.13 illustrates the dislocation

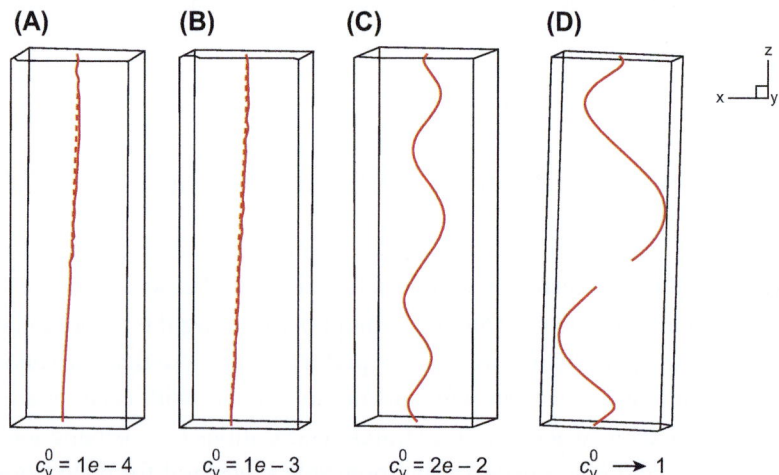

Figure 12.13

Configurations at different vacancy concentrations. *Reprinted from Liu, F.X., Liu, Z.L., Zhuang, Z., 2017a. Numerical investigations of helical dislocations based on coupled glide-climb model. International Journal of Plasticity 92, 2—18. Copyright 2017, with permission from Elsevier.*

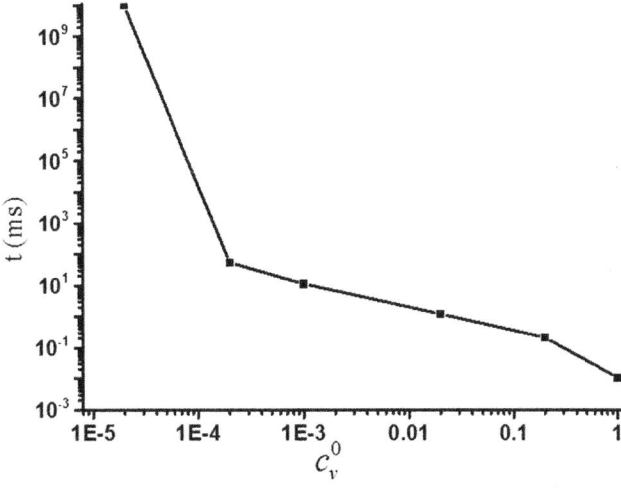

Figure 12.14

Time required to form the first spiral at different vacancy concentrations. *Reprinted from Liu, F.X., Liu, Z.L., Zhuang, Z., 2017a. Numerical investigations of helical dislocations based on coupled glide-climb model. International Journal of Plasticity 92, 2–18. Copyright 2017, with permission from Elsevier.*

configurations corresponding to different vacancy concentrations varying from 1×10^{-4} to 1.0 during evolution at the same time.

The purpose of this simulation is to describe quantitatively the influence of vacancy concentrations on dislocation configurations. The time required to form the first spiral is defined as t, which is plotted as a function of the vacancy concentration in Fig. 12.14. It indicates that the formation time of the first spiral t decreases exponentially as the vacancy concentration increases. The numerical result shows that the formation time is as large as 1×10^{10} ms at a vacancy concentration of 2×10^{-5}. Thus, when the vacancy concentration is lower than 1×10^{-5}, the formation time tends to be infinite, which means that it is hard to form a helix at such a low vacancy concentration at room temperature because the dislocation climb velocity is negligibly small.

12.4 Discrete-Continuous Method for Coupling Dislocation Glide-Climb

In the dislocation climb model in Section 12.2, the climb rate is derived from analytical formulations of a steady-state diffusion equation for the special case of a single straight edge dislocation with a cylindrical diffusion assumption, limiting the model to be at the local level, which means that dislocations act as perfect vacancy sources or sinks and diffusion fields of different dislocation segments do not interact with each other. This treatment has been proved to work only for straight dislocations or prismatic loops with a low dislocation density and is not necessarily applicable in general for dislocation lines with arbitrary shapes or high dislocation densities (Gu et al., 2015).

To break up the local level limitation, the actual diffusion equation with appropriate boundary conditions needs to be solved considering long-range interactions between different diffusion fields, which is complex and time-consuming, especially for a high dislocation density. To provide new insight into this type of problem, Geers et al. (2014) proposed a global-local approach to assess the influence of local diffusion on a global plasticity problem, which is incorporated into an extended strain gradient crystal plasticity model. It is a 2D framework in which 3D dislocation lines are simplified to be 2D infinite long dislocations, and in which realistic dislocation geometrics and complicated dislocation interactions such as junction formation or the cross-slip of 3D dislocation lines cannot be captured. Keralavarma and Benzerga (2015) and Keralavarma et al. (2012) proposed an extended 2D discrete dislocation plasticity framework to address this issue, in which a superposition method proposed by Giessen and Needleman (1995) is used to obtain the dislocation mechanical fields and solve vacancy diffusion equations on the same finite element grid. However, moving core-level boundary conditions are avoided by invoking the analytical solution with a cylindrical diffusion assumption. Similar approaches are used to study other climb-related problems (Ayas and Deshpande, 2015; Ayas et al., 2012, 2014; Danas and Deshpande, 2013; Davoudi, 2017; Gao et al., 2017). Gu et al. (2015) circumvented computational complexity by incorporating a Green's function formulation of dislocation climb into a 3D-DDD model, which offers an ideal nonlocal manner to consider the long-range effect of vacancy diffusion. Other alternative approaches, such as the phase-field method (Geslin et al., 2014, 2015), have been proposed. Despite intense study, it is rare for a general solution to deal with coupling between dislocation climb and vacancy diffusion with arbitrary boundary conditions and diffusion modes (steady-state or transient-state).

To address this issue, a modified 3D discrete-continuous model (DCM) is developed in this section. The DCM was proposed by pioneering researchers (Cui et al., 2015b; Gao et al., 2010; Lemarchand et al., 2001; Liu et al., 2009; Zbib and Diaz de la Rubia, 2002), in which 3D-DDD and continuum mechanics are combined directly. The DDD method is outstanding for capturing the details of dislocation evolution during plastic deformation but it is powerless for dealing with complex boundary conditions or surface effects as well as finite deformation, so it is a good complement to continuum mechanics at the submicron scale. In this DCM, a DDD simulation is used to calculate plastic strain generated by dislocation glide and substitutes for the constitutive form used in FEM. However, unlike previous research, in the current work, a mechanical field is calculated in the DDD simulation whereas FEM is employed to solve the vacancy concentration field to derive the dislocation climb velocity, which is the component of dislocation velocity in the normal direction of the slip plane. An intelligent localization method is adopted to guarantee the efficient transmission of variables between DDD and FEM. The benchmark simulation shows that the localization method works well in reflecting vacancy

distributions around dislocation cores and is feasible enough to guarantee computational efficiency and accuracy. By incorporating dislocation glide into this modified DCM, a coupled glide-climb model is established to investigate the intrinsic mechanisms of annealing hardening and the evolution of detailed defects during annealing at the submicron scale. Numerical results suggest that annealing hardening can be ascribed to: (1) dislocation annihilation during annealing, which significantly reduces the mobile dislocation density; and (2) the formation of dislocation jogs, which possess poor motilities and act as barriers to dislocation motion, leading to mobile dislocation starvation and resulting in higher flow strength compared with the prestrained one.

12.4.1 Dislocation Climb Model in Discrete-Continuous Method

The dislocation climb results from vacancy diffusion between the core and boundary. Dislocation lines may be assumed to be perfect sinks or sources of vacancies as long as vacancy emission is rapid compared with the long-range vacancy diffusion rate (Bakó et al., 2011; Mordehai et al., 2008, 2009). Consider a 3D model containing arbitrarily shaped dislocation lines, as illustrated in Fig. 12.15, in which B denotes the continuous domain excluding internal boundaries from Γ_1 to Γ_n and outer boundaries ∂B, $\mathbf{T}^0$, and $\mathbf{u}^0$ are the force and displacement boundary conditions on ∂B, respectively. Small closed contour Γ_i is one of the internal boundaries, which is cut off from the ith dislocation core. Its size and shape can be determined by atomic simulations (Clouet, 2006).

The instantaneous state of a model is characterized by the distribution of vacancy concentrations and the location of dislocations. Dislocation motion is driven by the generalized Peach-Koehler force. Vacancy diffusion is powered by so-called osmotic force, which is generated from a gradient of chemical potential μ. Both can be derived

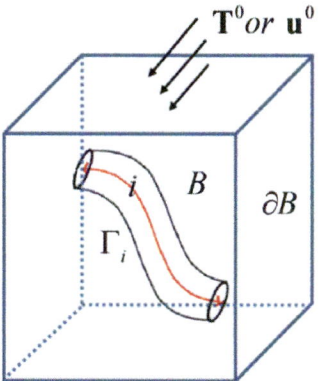

Figure 12.15

Three-dimensional model of crystal region including arbitrarily shaped dislocations. *Reprinted from Liu, F.X., Liu, Z.L., Pei, X.Y., Hu, J.Q., Zhuang, Z., 2017b. Modeling high temperature anneal hardening in Au submicron pillar by developing coupled dislocation glide-climb model. International Journal of Plasticity 99, 102—119. Copyright 2017, with permission from Elsevier.*

from the derivatives of Gibbs free energy. Simplifying assumptions are made here to enable the computations to be feasible: (1) elastic interaction energy is neglected between vacancies and dislocations; (2) the vacancy flux reaches a steady state at each time step, which is a good assumption when the climb rate is slow enough compared with the vacancy diffusion rate; (3) the diffusion coefficients are independent of stress; and (4) the dislocations act as perfect sources or sinks of vacancies as long as vacancy emission is rapid compared with the long-range vacancy diffusion rate. Then, the governing equations for vacancy diffusion in Domain B can be expressed (Nabarro, 1979)

$$\dot{c} = -\nabla \cdot \mathbf{J} \tag{12.4.1}$$

$$\mathbf{J} = -D\nabla\mu \tag{12.4.2}$$

$$\mu = \frac{kT}{\Omega}\left(\ln\frac{c}{1-c} + \frac{E_f}{kT} - \frac{P\Delta V}{kT}\right) \tag{12.4.3}$$

Eq. (12.4.1) is the continuity equation because matter is conserved during diffusion, where c denotes the vacancy concentration and $\mathbf{J}$ is the concentration flux. Eq. (12.4.2) follows Fick's diffusion law, in which $D = D_0\exp(-E_m/kT)/kT$ represents the diffusivity, D_0 denotes the diffusion coefficient, k is Boltzmann's constant, T is the absolute temperature, and E_m is the vacancy migration energy. Unless otherwise noted, units of terms in the current work are all given in the International System of Units. In Eq. (12.4.3), E_f is the vacancy formation energy, Ω is the atomic volume, ΔV indicates the relaxed volume resulting from the nucleation of a vacancy, and P is hydrostatic pressure. In most practical cases where the local stress is not high, the third term on the right-hand side of Eq. (12.4.3) can be neglected (Geers et al., 2014; Lothe, 1967). Under the assumption that each segment is a perfect sink, vacancies are assumed to be at equilibrium with the dislocations. Thus, the corresponding boundary condition on the ith internal boundary Γ_i can be expressed as

$$\mu = -f_c^i \big/ b \sin \theta^i \quad \text{on } \Gamma_i \tag{12.4.4}$$

where b is the magnitude of the Burgers vector; θ^i is an angle between the ith dislocation line direction unit vector ζ^i and its Burgers vector $\mathbf{b}^i$, which describes the dislocation character; and f_c^i is the climb component of total force acting on the ith dislocation line, including the climb component of Peach-Koehler force, line tension, and osmotic force resulting from the change in free energy caused by to the evolution of the local pressure field during dislocation climb (Bullough and Newman, 1970; Danas and Deshpande, 2013; Geers et al., 2014). By substituting Eq. (12.4.4) into Eq. (12.4.3), the equilibrium vacancy concentration on the internal boundary Γ_i can be derived as

$$c_{eq}^i = \exp\left(-\frac{E_f}{kT} - \frac{f_c^i \Omega}{b \sin \theta^i \, kT}\right) = c^0 \exp\left(-\frac{f_c^i \Omega}{b \sin \theta^i kT}\right) \quad \text{on } \Gamma_i \tag{12.4.5}$$

where $c_0 = \exp(-E_f/kT)$ is the equilibrium vacancy concentration expressed by the Arrhenius law with no external stress applied.

In addition, the corresponding boundary condition on the external boundary ∂B is

$$\mu = \mathbf{T}^0 \cdot \mathbf{n} \quad \text{on} \;\; \partial B \tag{12.4.6}$$

Eq. (12.4.6) is for the force boundary condition case in which $\mathbf{T}^0$ denotes external traction on the outer boundary and $\mathbf{n}$ represents an outward unit vector normal. Similar laws are valid for displacement boundary conditions.

Besides vacancy concentrations near dislocation cores, a vacancy concentrations in Domain B is assumed to be equal to the mean vacancy concentration of the whole computational cell, given as $c_{\text{ave}} = N_{\text{vacancy}} \cdot \Omega / V$, where N_{vacancy} is the total number of vacancies in the computational cell, Ω is the volume of a vacancy, and V is the volume of computational cells. c_{ave} depends on the spatial distribution of vacancy sources or sinks. For thin foils or small bulk materials, surfaces are perfect vacancy sinks; thus, the mean concentration remains constant, as $c_{\text{ave}} = c_0$. For bulk materials, vacancies are formed from the climb of dislocations at much more quickly than they move to the surface. The surface-interface effect can be neglected. The total vacancy number is then conserved. Thus, climbing of the dislocation line is the only way to generate or consume vacancies for simplicity (Mordehai et al., 2008). The evolution of c_{ave} can be determined by the swept area of dislocation climb:

$$\frac{dc_{\text{ave}}}{dt} = \frac{b}{V} \frac{dS}{dt} \tag{12.4.7}$$

where S is the swept area. If the length of the dislocation line is l, the area swept by dislocations during the climb in corresponding time step Δt can be expressed as $\Delta S = v_c \cdot l \cdot \sin\theta \cdot \Delta t$. It follows that the sign of climb velocity v_c determines the sign of swept area ΔS. Therefore, it determines whether dislocation segments are absorbing vacancies, as $\Delta S > 0$, or emitting vacancies, as $\Delta S < 0$.

As mentioned previously, dislocation climbs result from the diffusion of vacancies either from or into dislocation cores. We specify a positive direction for climb velocity when an edge dislocation climbs upward after emitting vacancies, and vice versa. The volume change of vacancies in the ith dislocation core resulting from dislocation climb per unit length can be denoted as $v_c^i b \sin \theta^i$, where $b \sin \theta^i$ is the length of the edge component of the dislocation line. As the matter conservation law requires, with the steady-state diffusion assumption, this volume change $v_c^i b \sin \theta^i$ should be equal to the volumetric vacancy flux ϕ_i of the ith dislocation (Liu et al., 2017a):

$$v_c^i b \sin \theta^i = \phi_i \tag{12.4.8}$$

where $\phi_i = \oint_{\Gamma_i} \mathbf{J} \cdot \mathbf{n} ds$. $\mathbf{n}$ denotes the inward unit vector normal and $\mathbf{J}$ is the vacancy flux across internal boundary Γ_i. Then, the corresponding climb rate is

$$v_c^i = \frac{1}{b \sin \theta^i} \oint_{\Gamma_i} \mathbf{J} \cdot \mathbf{n} ds \qquad (12.4.9)$$

Eq. (12.4.9) is the diffusion-controlled climb law, which is especially true when the climb rate is slow compared with the vacancy diffusion rate. It demonstrates that vacancy diffusion flux and the dislocation climb velocity are coupled by this climb law. Climb direction $\mathbf{n}_c^i$ of the ith dislocation line can be determined by Burgers vector $\mathbf{b}^i$ and dislocation line direction unit vector ζ^i:

$$\mathbf{n}_c^i = \frac{\mathbf{b}^i \times \zeta^i}{\left| \mathbf{b}^i \times \zeta^i \right|} \qquad (12.4.10)$$

These equations, including governing Eqs. (12.4.1)–(12.4.3) and boundary conditions (12.4.6), (12.4.4), and (12.4.5), as well as the climb law in Eq. (12.4.9), are sufficient to characterize the evolution of both dislocations and vacancies in the model. As long as the vacancy distribution is determined, the climb velocity can be derived from the vacancy flux resulting from the vacancy concentration gradient. However, it is complex and time-consuming to solve these equations, especially with a complicated dislocation network, because there are as many internal conditions (12.4.4) as the number of dislocations in the computational cell. These dislocations, each of which possesses a core region encircled by internal boundary Γ, are always in motion and the adaptively remeshing domain is required to track the moving boundaries. This makes large-scale numerical simulation troublesome and will cause much more difficulty when it is carries out in 3D. In the following work, a unified framework that couples 3D-DDD and FEM is developed to solve this problem and circumvent the complexity.

Fig. 12.16 mainly contains the following DDD-FEM transfer procedures: (1) Forces acting on dislocation lines and deriving equilibrium vacancy concentrations on internal boundaries $\Gamma_1 \ldots \Gamma_n$ are calculated according to Eq. (12.4.5) by the DDD method. Then, the boundary conditions and mean vacancy concentration c_{ave} are localized to the continuum material points in FEM, which is a crucial procedure for the whole calculation. (2) Diffusion Eqs. (12.4.1)–(12.4.3) are solved by FEM associated with the internal and external boundary conditions in a unified continuum mechanic framework to obtain the volumetric vacancy flux crossing the internal boundary of the dislocation cores. It will be transferred back to the DDD cell to derive dislocation climb rates according to Eq. (12.4.9), and then to update the dislocation configurations. This bidirectional, real-time coupling process will be conducted for each time step. Regularization methods are important to ensure the accuracy and efficiency of this process, including how to localize the discrete vacancy concentrations properly in the DDD model to continuum material

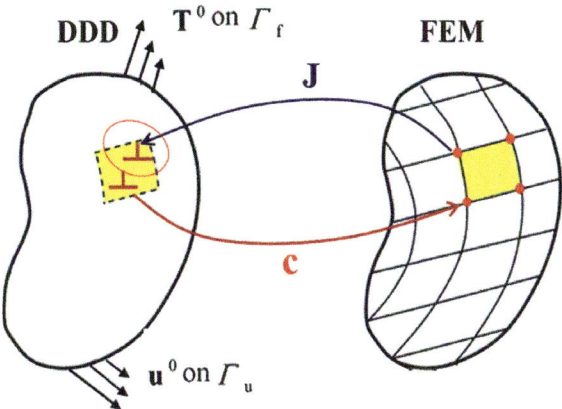

Figure 12.16
Schematic of variable-transferring procedure in the modified discrete-continuous method. *DDD,*
discrete dislocation dynamics; *FEM,* finite element method. *Reprinted from Liu, F.X., Liu, Z.L., Pei,*
X.Y., Hu, J.Q., Zhuang, Z., 2017b. Modeling high temperature anneal hardening in Au submicron pillar by
developing coupled dislocation glide-climb model. International Journal of Plasticity 99, 102–119. Copyright
2017, with permission from Elsevier.

points in FEM model, and how to transfer the continuum vacancy flux in FEM back to the
discrete dislocation segments to derive the climb rate.

12.4.2 Localize Vacancy Concentration Field of Discrete Dislocation Dynamics Segments to Finite Element Method Nodes

In the computational cell, the evolution of a vacancy field is computed by solving Eqs.
(12.4.1)–(12.4.3) using FEM. Dirichlet boundary conditions are imposed for vacancy
diffusion corresponding to equilibrium concentrations at the dislocation cores, including
internal boundaries as Eq. (12.4.5) and external boundaries as Eq. (12.4.6), according to
imposed traction $\mathbf{T}^0$. Initial mean vacancy concentration c_{ave}^0 is specified to be equal to c_0.
Then, the initial values and boundary conditions of the vacancy field are transferred to
FEM nodes, as shown in Fig. 12.17A to solve diffusion Eqs. (12.4.1)–(12.4.3). The FEM
element is designed to be eight-node linear brick.

To localize the continuum vacancy concentration field properly to discrete FEM nodes, a
distance-related weight function is employed:

$$w(r) = \frac{\ln(r/r_d)}{\ln(r_d/r_c)} \tag{12.4.11}$$

where, as shown in Fig. 12.17B, r_c and r_d are the inner and outer cutoff radii of the control
volume of the dislocation line, respectively, which are defined to specify the scope of the
vacancy concentration field affected by the dislocations. Considering that determining
different values of r_c and r_d for different dislocations needs plenty of searches for
dislocation positions, it is time-consuming and expensive. Thus, r_c is supposed to be equal

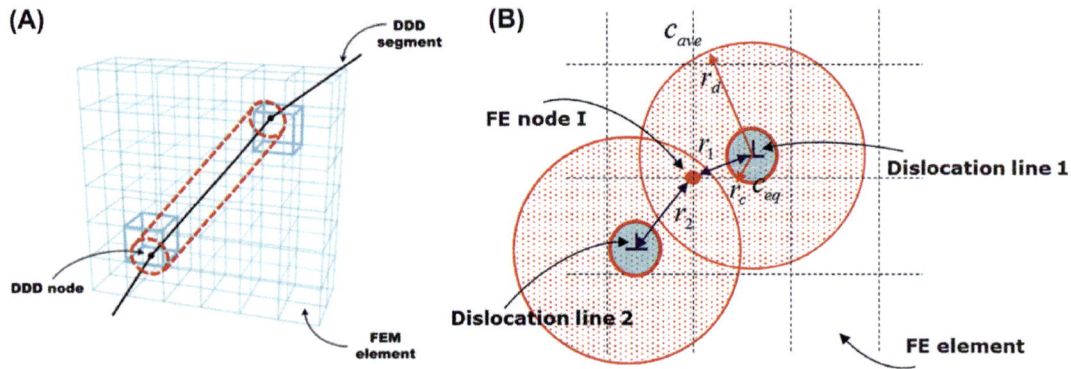

Figure 12.17

Localizing discrete vacancy concentrations on discrete dislocation dynamics (DDD) segments to finite element method (FEM) nodes. *FE*, finite element. *Reprinted from Liu, F.X., Liu, Z.L., Pei, X.Y., Hu, J.Q., Zhuang, Z., 2017b. Modeling high temperature anneal hardening in Au submicron pillar by developing coupled dislocation glide-climb model. International Journal of Plasticity 99, 102–119. Copyright 2017, with permission from Elsevier.*

to the cutoff radius of the dislocation core and r_d is assumed to be equal to mean dislocation spacing: that is, $r_d = 1/2\sqrt{\rho}$, where ρ is the dislocation density. Although the value of r_d is not accurate enough when there are large fluctuations in local dislocation density, an inverse distance weighted interpolation method is used in Eq. (12.4.14) to deal with the local problem to reduce error caused by this assumption. The weight function in Eq. (12.4.11) originates from the analytical solution for the cylindrical diffusion of a single straight dislocation line in an infinite field at steady state (Caillard and Martin, 2003):

$$c(r) = c_\infty + (c_\infty - c_c)\frac{\ln(r/r_\infty)}{\ln(r_\infty/r_c)} \tag{12.4.12}$$

with the following boundary conditions:

$$\begin{cases} c(r) = c_\infty, & r = r_\infty \\ c(r) = c_c, & r = r_c \end{cases} \tag{12.4.13}$$

where r is a distance away from the dislocation core and $c(r)$ is the vacancy concentration at the corresponding location. r_∞ is the outer cutoff for the distance away from the dislocation, which is supposed to be infinite. c_c and c_∞ are the concentrations corresponding to r_c and r_∞.

The area between the inner and outer cutoff boundaries r_c and r_d is defined as the affected area of corresponding dislocation. The cross-section of the affected area, which is the red dashed area around the dislocation core, is shown in Fig. 12.17B. Then, the concentration of FEM nodes can be divided into three categories according to their locations, as given in Fig. 12.17B: (1) When node I is in the inner boundary Γ of any dislocation i, $c^I = c^i_{eq}$, shown as the blue area inside the innermost circle; (2) when node I is in neither the inner

boundary Γ nor the affected area of any dislocation, $c^J = c_{ave}$, shown as the white area outside the sphere of radius r_d; and (3) for node I in the red dashed area, which is the affected area, the vacancy concentration can be obtained as

$$c^I = \frac{\sum_{i=1}^{n} \left[c_{ave} + \left(c_{eq}^i - c_{ave} \right) w(r_i) \right] \cdot w(r_i)}{\sum_{i=1}^{n} w(r_i)} \tag{12.4.14}$$

where the superscript in uppercase is the node number of FEM, whereas lowercase stands for the segment number of DDD. n is the total number of all DDD segments whose affected area contains FEM node I. r_i is a vertical distance from FEM node I to DDD segment i. Eq. (12.4.14) satisfies the boundary conditions in Eq. (12.4.13) and would degenerate into the analytical solution in Eq. (12.4.12) when there is only one dislocation segment in the computational cell. Thus, it has a clear physical meaning in that r_d is much larger than r_c.

12.4.3 Transferring Vacancy Flux From Finite Element Method Back to Discrete Dislocation Dynamics Segments

Eqs. (12.4.1)–(12.4.3) can be solved by FEM as long as the initial vacancy concentrations and Dirichlet boundary conditions are obtained. Both pipe and bulk diffusion are taken into consideration because diffusion happens wherever there is a concentration gradient. The diffusion coefficients of pipe and bulk diffusion are unified for simplification. It is practicable because bulk diffusion is the dominant mode in a high-temperature situation when the jogs are dense enough to guarantee a larger mean free path than jog spacing (Caillard and Martin, 2003). Once concentration flux field $\mathbf{J}$ is obtained for the discrete FEM nodes, it is necessary to transfer it to the DDD segments. Then, volumetric fluxes across the dislocation cores can be derived to obtain the climb rate of the DDD segments. According to the divergence theorem, the volumetric concentration flux across any closed surface ϕ_i that only contains specified dislocation segment i can be calculated. Thus, we can choose a proper envelope surface and calculate the vacancy flux across the surface, as shown in Fig. 12.18. The climb rate can be derived as

$$v_c^i = \frac{1}{b \sin \theta^i} \cdot 2\pi r \cdot \frac{\sum_{I=1}^{N} (\mathbf{J}_I \cdot \mathbf{n}_I)}{N} \tag{12.4.15}$$

where r is a radius of the envelope cylinder and N is the total number of FEM nodes on the cylinder surface.

Finally, the climb rates can be transferred back to the dislocation segments to update the dislocation configuration. Then, it is necessary to update the mean vacancy concentration.

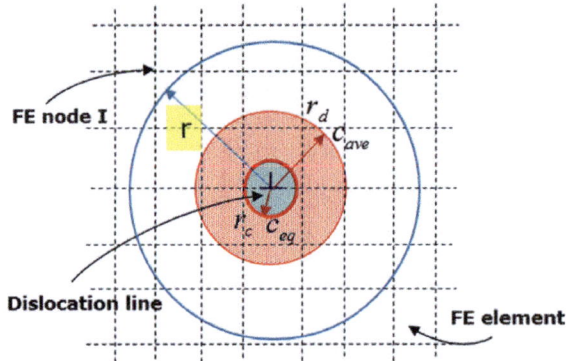

Figure 12.18

Transforming vacancy flux at the finite element (FE) method node to the discrete dislocation dynamics segment. *Reprinted from Liu, F.X., Liu, Z.L., Pei, X.Y., Hu, J.Q., Zhuang, Z., 2017b. Modeling high temperature anneal hardening in Au submicron pillar by developing coupled dislocation glide-climb model. International Journal of Plasticity 99, 102—119. Copyright 2017, with permission from Elsevier.*

In bulk crystals, according to Eq. (12.4.7), mean dislocation concentration c_{ave} can be derived from the climb velocity as

$$c_{\text{ave}}(t + \Delta t) = c_{\text{ave}}(t) + \frac{\sum b \cdot \left(v_c^i(t) \cdot l^i \cdot \sin \theta^i \cdot \Delta t \right)}{V} \tag{12.4.16}$$

where $l_i \sin\theta$ is a length of the edge component of dislocation segment i and V is the volume of the computational cell. In a thin foil, vacancies can be annihilated from the surface at a much faster rate than they are produced from the dislocations, and the mean concentration remains constant.

12.4.3.1 Case Study: Evolution of Prismatic Loop in a Bulk Single Crystal

Now, we focus on the evolution of prismatic loops resulting from vacancy aggregation after irradiation or rapid cooling from high temperatures. In a bulk single crystal without external stress, this prismatic loop will be driven by the competition between line tension and osmotic force. Mordehai et al. (2008) derived an approximate analytical expression for the climb velocity of a straight edge dislocation in a prescribed uniform vacancy field with a steady-state and bulk diffusion assumption:

$$v_c = \frac{1}{b \cdot \sin \theta} \cdot \frac{2\pi D \cdot (c_{\text{eq}} - c_d)}{\ln(r_d/r_c)} \tag{12.4.17}$$

where c_{eq} denotes the equilibrium concentration in a dislocation core. This analytical expression is in quantitative agreement with the atomistic simulation result (Clouet, 2011). Gu et al. (2015) proved that it is also applicable for the evolution of a single prismatic loop. A numerical simulation of the evolution of a prismatic loop in bulk single crystals is carried out and compared with the analytical solution in the following section to validate the capability and accuracy of DCM in capturing the evolution of dislocation configuration.

(A)

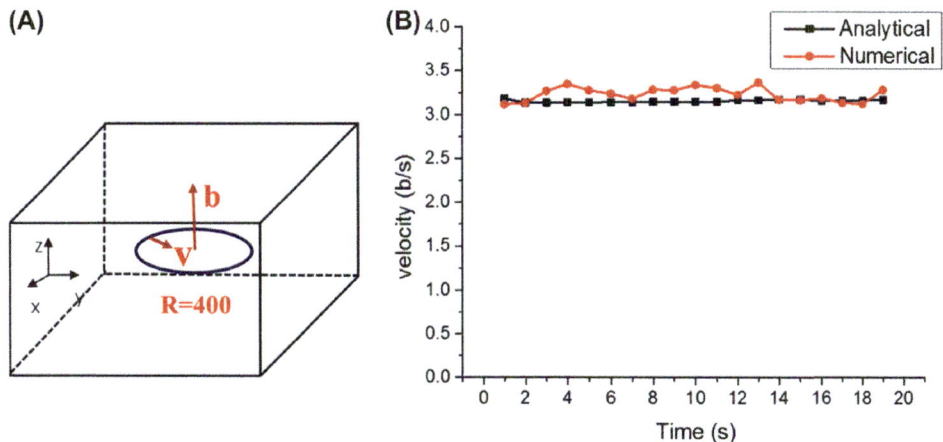

Figure 12.19
(A) Schematic of a prismatic loop in an aluminum single crystal; (B) Numerical simulation results compared with analytical solutions. *Reprinted from Liu, F.X., Liu, Z.L., Pei, X.Y., Hu, J.Q., Zhuang, Z., 2017b. Modeling high temperature anneal hardening in Au submicron pillar by developing coupled dislocation glide-climb model. International Journal of Plasticity 99, 102–119. Copyright 2017, with permission from Elsevier.*

A 36-segment, discretized, circular, prismatic, vacancy-type loop 400 nm in radius is introduced as an initial configuration in a computational cell 4000 nm × 4000 nm × 4000 nm in size, as shown in Fig. 12.19A. This loop has the usual plane in the [011] direction and a Burgers vector in the <011> direction, which can change size only by climbing onto the loop plane. The initial mean concentration is set to be $c_{ave} = c_0$. Since the prismatic loop contract homogeneously, there is a climb rates at nodes along the loop. Climb rate within a certain period of time is demonstrated in Fig. 12.19B. The black line with solid black squares represents the analytical solutions derived from Eq. (12.4.17), whereas the red line with solid red dots denotes the numerical results obtained from the modified DCM. The numerical results agree well with the analytical solutions, except for a slight fluctuation in numerical results caused by inevitable stochastic error during variable transmissions between the DDD and FEM. Moreover, the numerical results are slightly higher than the analytical solutions because only bulk diffusion is taken into account in the analytical solutions. In our numerical model, both pipe and bulk diffusions are considered in a natural manner although a unified diffusion coefficient is used, which results in a larger climb rate. Thus, the computational model is more in line with the actual situation and much more applicable.

Aside from applicability and accuracy, the sensitivity and independence of the element size and system parameters are important criteria for evaluating a numerical method. To this end, the climb rates at different finite element sizes are plotted in Fig. 12.20A when a radius of envelope cylinder r is set to be equal to element size a. The total number of elements used for a mesh with element size $a = 25, 50, 80,$ and 100 nm is $4,096,000,$ $521,000, 125,000,$ and $64,000,$ respectively, and the corresponding computation times are

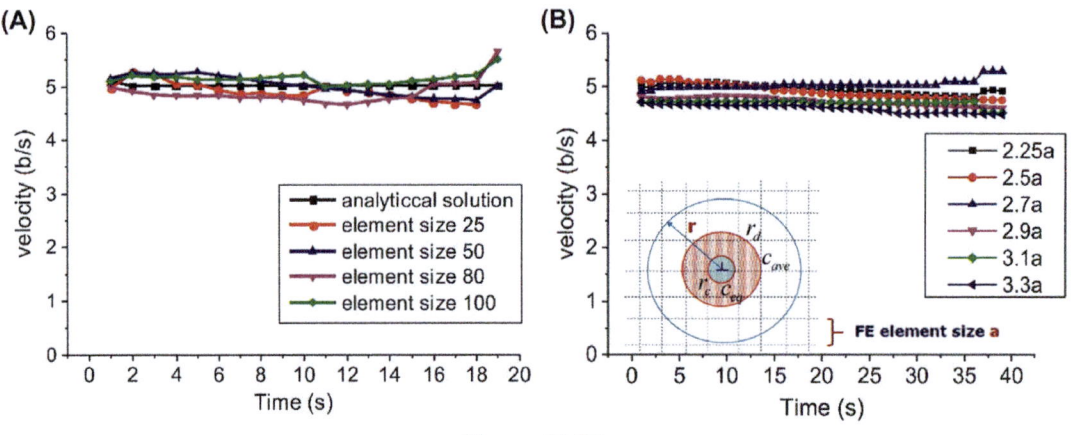

Figure 12.20

(A) Climb rates at different element sizes; (B) Climb rates with different radius *r* of envelope cylinder (*a* = 50 nm). *Reprinted from Liu, F.X., Liu, Z.L., Pei, X.Y., Hu, J.Q., Zhuang, Z., 2017b. Modeling high temperature anneal hardening in Au submicron pillar by developing coupled dislocation glide-climb model. International Journal of Plasticity 99, 102—119. Copyright 2017, with permission from Elsevier.*

65,040, 3840, 1620, and 1380 s. In fact, the most of the computation time is spent on transferring data between the DDD and FEM models. Climb rates with different *r*s at element size *a* = 50 nm are demonstrated in Fig. 12.20B. These simulation results prove that our computational model is independent of element size *a* and radius *r*, which is further evidence of the efficiency of the computational model.

12.4.4 Coupled Dislocation Glide-Climb Model

In this section, conventional dislocation glide is incorporated into the modified DCM to couple dislocation glide and climb into a unified model.

12.4.4.1 Dislocation Glide

Different from diffusion-controlled dislocation climb, in formulating the DDD approach, the motion of a glide dislocation of unit length is assumed to be governed by

$$m\dot{v}_g + B_g v_g = f_g \tag{12.4.18}$$

where v_g is the dislocation glide velocity and m is the effective mass. Inertia force $m\dot{v}_g$ is important in the case of very high strain rate, such as for shock or impact loadings. In most cases, this inertia effect can be neglected as long as the glide velocity is smaller than the speed of sound (Bulatov and Cai, 2006; Davoudi, 2017), because the glide velocity reaches a steady-state value in a time that is smaller than the typical time increment in DDD calculations. B_g is the drag coefficient, which is a function of temperature and pressure (Zbib et al., 2002). Davoudi (2017) proved that for an intermediate (that is, $\theta_D/3$, where θ_D is the Debye temperature) to high temperature, B_g varies linearly with T/θ_D.

Thus, in the current work, we assume that $B_g = B_g^0 T/T^0$, where $B_g^0 = 3kT/b^2 c_s$, with c_s denoting the speed of shear waves in the material (Hirth and Lothe, 1982). f_g is the glide component of nodal force, including Peach-Koehler force f_{PK}, image force f_{image}, and line tension f_{self}:

$$f_g = (f_{PK} + f_{image} + f_{self}) \times (n \times \zeta)$$
$$= [(\sigma \times b) \times \zeta + f_{image} + f_{self}] \times (n \times \zeta) \quad (12.4.19)$$

where **n** is the slip plane normal. $(\sigma \times b) \times \zeta$ denotes the Peach-Koehler force caused by externally applied stress and the long-range interactions of other dislocation segments. Line tension f_{self} is defined as the increase in energy per unit in the length of the dislocation line (Hull and Bacon, 2001), which is analogous to the surface tension of a soap bubble and tends to straighten the line and reduce the total energy. Image force f_{image} has an important role in attracting dislocations to the free surface because the material is more compliant there and the dislocation energy is lower (Hull and Bacon, 2001). An analytical solution for force acting on a straight dislocation segment near the free surface is adopted from Hirth and Lothe (1982) to add to the infinite-body stress field to capture the image force in the near-surface region. Details regarding image force in the DDD code were described in our previous publications (Cui et al., 2014; Gao et al., 2011a; Liu et al., 2009). Thus, glide velocity v_g^i of dislocation i is assumed to follow a viscous phonon-drag law:

$$v_g^i = f_g^i / B_g \quad (12.4.20)$$

12.4.4.2 Coupling Glide and Climb in a Unified Framework

Now that the nodal velocities of climb v_c^i and glide v_g^i are obtained from Eqs. (12.4.15) and (12.4.20), the remaining challenge of coupling glide and climb in a unified framework is how to tackle the timescale disparity at several orders of magnitude between glide and climb.

Two viable time-stepping schemes were proposed in pioneering work. Keralavarma et al. (2012) put forward an adaptive time-stepping scheme in which dislocation glide or climb occurs separately at different time steps. The dislocations glide with a small time step, Δt_g, at the very beginning. Dislocation climb is activated when jammed or quasiequilibrium dislocation states are detected. A much larger time step, Δt_c, is used to deal with the climb-related problem, whose value relies heavily on the temperature. Once the dislocations climb to a new slip plane, which enables further glide, glide motion occurs again and the time step will revert to Δt_g. This adaptive method has proved to be efficient for processes lasting a long time, such as creep (Keralavarma and Benzerga, 2015). Liu et al. (2017) proposed a sequentially coupled time-stepping scheme in which dislocations naturally glide and climb with a smaller time step at 0.1 ns. Because the climb distance

per time step is small compared with the glide distance, the small increments of dislocation climb per time step will accumulate and be superimposed onto the total slip distance until a climb step reaches at least one glide plane distance. This sequentially coupled scheme has proved to be much more accurate in the case of fast climb but it is inefficient compared with the adaptive one when the climb rate is too low. In the current work, the difference between the climb and glide rates is up to six orders of magnitude, and thus the adaptive time-stepping scheme is adopted.

Computational procedures in the modified DCM that couple both glide and climb are sorted out in this section. Main steps of this procedure are plotted in Fig. 12.21. Only the DDD code is employed in the glide-only steps. Once dislocation climb is activated, the FEM code is invoked to solve the diffusion equation. Climb time step Δt_c is set to guarantee that the slip distance of each dislocation segment is nearly one interplanar spacing at every 1000 step. The glide time step is fixed as $\Delta t_g = 2 \times 10^{-11}$ s. The iterative strategy is detailed subsequently.

1. Only dislocation glide is considered at the very beginning and the time step is initialized as glide time step Δt_g.
2. An initial microstructure is constructed with randomly distributed dislocations according to their specified densities. Then, glide simulations at 0 stress are performed to

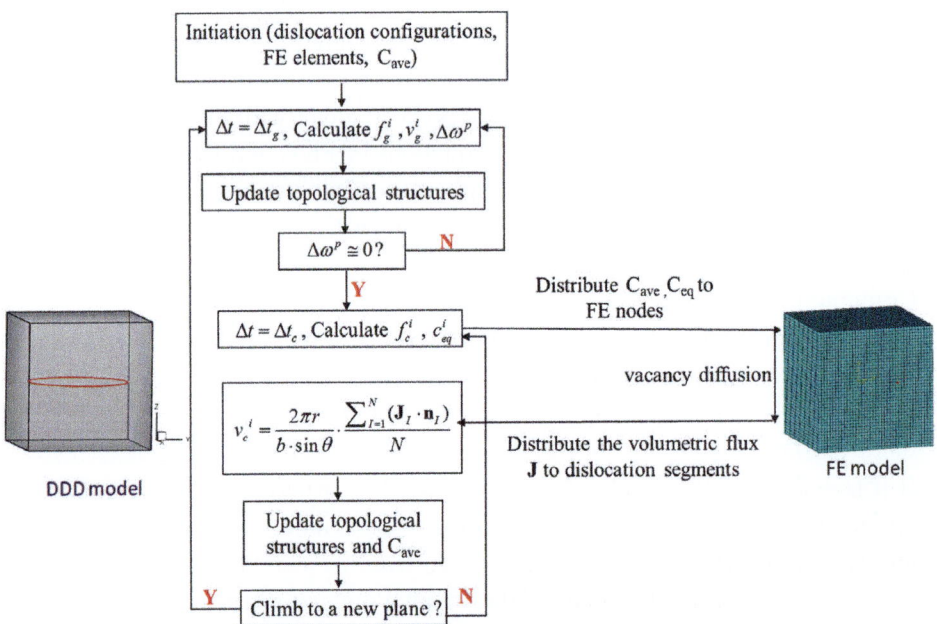

Figure 12.21

Computation procedure of coupled glide-climb model. *DDD*, discrete dislocation dynamics; *FE*, finite element. *Reprinted from Liu, F.X., Liu, Z.L., Pei, X.Y., Hu, J.Q., Zhuang, Z., 2017b. Modeling high temperature anneal hardening in Au submicron pillar by developing coupled dislocation glide-climb model. International Journal of Plasticity 99, 102—119. Copyright 2017, with permission from Elsevier.*

relax the initial microstructure so that the dislocation configurations reach a local equilibrium condition, which is taken to be the initial dislocation configuration.

3. Forces on the dislocation segments are computed using Eq. (12.4.19). The dislocation positions are updated according to the glide velocity obtained from Eq. (12.4.20). Once the displacement of dislocation segments is obtained, plastic strain rate $\Delta\varepsilon^p$ and plastic spin rate $\Delta\omega^p$ induced by their slip during time increment $\Delta t = \Delta t_g$ can be obtained according to Orowan's formula:

$$\Delta\boldsymbol{\varepsilon}^p = \frac{1}{2V\Delta t}(\mathbf{b}\otimes\mathbf{A} + \mathbf{A}\otimes\mathbf{b})$$

$$\Delta\boldsymbol{\omega}^p = \frac{1}{2V\Delta t}(\mathbf{b}\otimes\mathbf{A} - \mathbf{A}\otimes\mathbf{b})$$

(12.4.21)

where $\mathbf{A}$ is the slipped area and V is the volume of the computational cell.

4. Step 2 is repeated until $\Delta\varepsilon^p \cong 0$, which means a quasiequilibrium state is reached. The glide-related processes are stopped and the dislocations are frozen at their current positions. The climb-related processes including vacancy diffusion and climb motion are activated. Time step Δt_g is switched to a much larger climb time step, $\Delta t_c = 10^{-3}t$, where t is the time taken for a dislocation to climb a distance equal to the slip plane spacing.

5. At time $t + \Delta t_c$, the vacancy concentrations on the internal and external boundaries are localized to the FEM nodes to solve diffusion governing Eqs. (12.4.1)–(12.4.3) and vacancy flux J is obtained. Then, it is transferred back to the DDD code to derive climb rate v_c^i. Mean vacancy concentration c_{ave} is updated along with the evolution of the dislocation configuration. The governing equations of FEM and DDD are both solved by an implicit integration scheme. The implicit time integration in FEM is implemented in the commercial FEM software ABAQUS, whereas time integration in the DDD code is implemented thus:

   a. Compute predictor dislocation positions at the current time step $\mathbf{r}'(t + \Delta t) = \mathbf{r}(t) + \mathbf{v}(t)\cdot\Delta t$, where Δt is the time step and $\mathbf{r}(t)$ and $\mathbf{v}(t)$ are the dislocation positions and velocities, respectively, at previous time step t;

   b. Compute dislocation velocity $\mathbf{v}'(t + \Delta t)$ at the current time step according to information from the current dislocation positions, $\mathbf{r}'(t + \Delta t)$;

   c. If $|\mathbf{r}'(t + \Delta t) - \mathbf{r}(t) - (\mathbf{v}'(t + \Delta t) + \mathbf{v}(t))\cdot\Delta t/2| > \varepsilon$, where ε is the desired accuracy, replace predictor dislocation positions $\mathbf{r}'(t + \Delta t)$ with $\mathbf{r}(t) + (\mathbf{v}'(t + \Delta t) + \mathbf{v}(t))\cdot\Delta t/2$ and return to Step 2.

   d. Continue to the next time step unless the total number of time steps is reached.

6. If any dislocation climbs to a new glide plane and breaks the quasiequilibrium state, the time step will revert to Δt_g. Step 2 is repeated until a new steady state is reached.

12.5 High-Temperature Annealing Hardening

Annealing is recognized as an effective method for softening metals that are too hard, such as after forging. A proper combination of annealing time and temperature is useful for balancing the strength and ductility of bulk metallic materials. This annealing softening effect at a macroscopic level can be ascribed to the relaxation of residual stress, the refinement of the grain size, and a reduction in dislocation densities. In contrast to the typical behavior at a macro level, experiments at a submicron scale have shown that metals may harden rather than soften during high-temperature annealing. For example, Lee et al. (2009) observed that after annealing, the compressive strength of 35% prestrained Au micropillars, which exhibit a strain softening character, recovered comparable to that of pristine ones. This was attributed to the reduction in defects, which results in higher stress for dislocation nucleation from the free surface. Huang et al. (2006b) reported that after annealing at 150°Cfor 30 min, nanostructured aluminum is hardened, because heat treatment reduces the generation and interaction of dislocations, leading to a decrease in ductility and an increase in strength. Thermal annealing has been widely employed to eliminate unwanted ion bombardment-induced defects during focused ion beam (FIB) milling (Kiener et al., 2012; Lee et al., 2016), which is a crucial technique for exploring the effect of mechanical size and other deformation behaviors of submicron metallic pillars (Bei et al., 2007; El-Awady et al., 2009b; Kiener et al., 2007; Shim et al., 2009). In addition, it is used to achieve an ideal strength (Iskandarov et al., 2011; Lowry et al., 2010) or to reduce the threading dislocation density in Ge/Si epitaxial films (Liu et al., 2015b). Although extensive studies have shown that this unusual annealing-induced hardening is closely related to a reduction in internal defects or an increase in the fractal dimension of dislocation networks (Gu and Ngan, 2013), the basic thermal process leading to the annihilation of defects and the detailed evolution of these defects during annealing are poorly documented (Kiener et al., 2012; Lee et al., 2016).

During high-temperature annealing, point defects are activated to diffuse much more quickly to absorb into or emit from the dislocation core, leading to a nonconservative dislocation movement called dislocation climb (Caillard and Martin, 2003). In climb, dislocations can be translated perpendicular to their slip planes, which has a significant role in relaxing dislocation pileups and promoting dislocation annihilation between different slip planes. Thus, the dislocation climb motion cannot be ignored in thermally activated events. However, the rate of this diffusion-controlled dislocation climb is fairly low because diffusion occurs over a long timescale, whereas the glide rate can be up to a shear wave velocity for the material, leading to a timescale separation of several orders of magnitude between the glide and climb rate (Huang et al., 2014a; Keralavarma et al., 2012). Thus, high-temperature annealing is a multifield and multitimescale process that

creates a great challenge for researchers. Attempts have been made to develop numerical simulations to couple dislocation glide and climb into a unified framework (Ayas et al., 2014; Bakó et al., 2011; Geers et al., 2014; Keralavarma and Benzerga, 2015; Keralavarma et al., 2012). Remaining challenges for developing such a coupled glide-climb model involve physical coupling between vacancy diffusion and dislocation climb, and bridging the huge timescale separation between glide and climb.

12.5.1 Brief Description of the Experiment

Based on the coupled glide-climb model given in the previous section, the intrinsic mechanism of high-temperature anneal hardening at the submicron scale is investigated in this section. The behavior of high-temperature annealing was observed in a uniaxial compression experiment conducted by Lee et al. (2009) in Au single-crystal micron pillars. Details about the sample preparations are illustrated schematically in Fig. 12.22, which is reference from Lee et al. (2009). At first, all pristine pillars deform to 35% strain. Some are annealed at $\sim 260°C$ for 7 min in air for thermal recovery experiments. These pillars are subsequently milled down to $D \sim 300$ nm by FIB machining before compression tests. Thus, the surface structures of these pillars are the same, which enables us to assess the effects of internal microstructures without considering FIB-induced damage. The orientation of the pillars is [001].

Atypical results of softening after prestraining and hardening after annealing are observed in the experiment. A highly dense dislocation network with a density of $\sim 10^{15}$ m^{-2} is created by prestraining. It seems that prestraining significantly reduces the flow strength of pillars whereas annealing restores the strength to pristine levels. The numerical simulations are described in the following section.

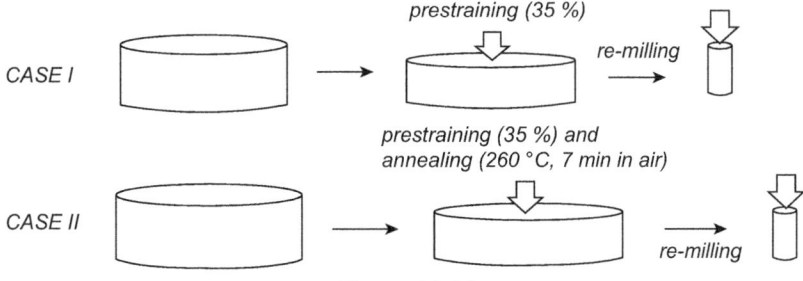

Figure 12.22

Case I: Prestraining experiments of pillars; Case II: prestraining and annealing experiments of pillars. All pillars were prestrained at 35% and had $D \sim 300$ nm after remilling (Lee et al., 2009). *Reprinted from Liu, F.X., Liu, Z.L., Pei, X.Y., Hu, J.Q., Zhuang, Z., 2017b. Modeling high temperature anneal hardening in Au submicron pillar by developing coupled dislocation glide-climb model. International Journal of Plasticity 99, 102–119. Copyright 2017, with permission from Elsevier.*

12.5.2 Simulation Procedures

Numerical simulations based on the modified DCM are carried out to investigate the intrinsic mechanisms for the prestraining softening and annealing hardening effects. The size of the pillar is 300 nm × 300 nm× 650 nm. The element type in FEM is a 3D eight-node brick 10 nm in characteristic element length, which is smaller than the minimum length of dislocation segments. The total element number is 58,500 and the node number is 63,426. A 35% prestrained Au micropillar is adopted as the initial configuration with free surfaces, as demonstrated in Fig. 12.23A, which possesses a high dislocation density of about 10^{15} m^{-2} owing to prestraining, as reported by Lee et al. (2009). The following properties of Au crystal are adopted: shear modulus $G = 27.6$ GPa, Poisson's ratio $v = 0.42$, magnitude of the Burgers vector $b = 0.29$ nm, density $\rho_{Au} = 19.26$ g/cm^3, and static viscous drag coefficient $B_g^0 = 1.22 \times 10^{-4}$ Pa/s at room temperature 298K. The pillars are fixed on the bottom and loaded on the top. The remaining side surfaces are free. Image forces are enforced on all six faces of the rectangular pillar, including the top and bottom surfaces.

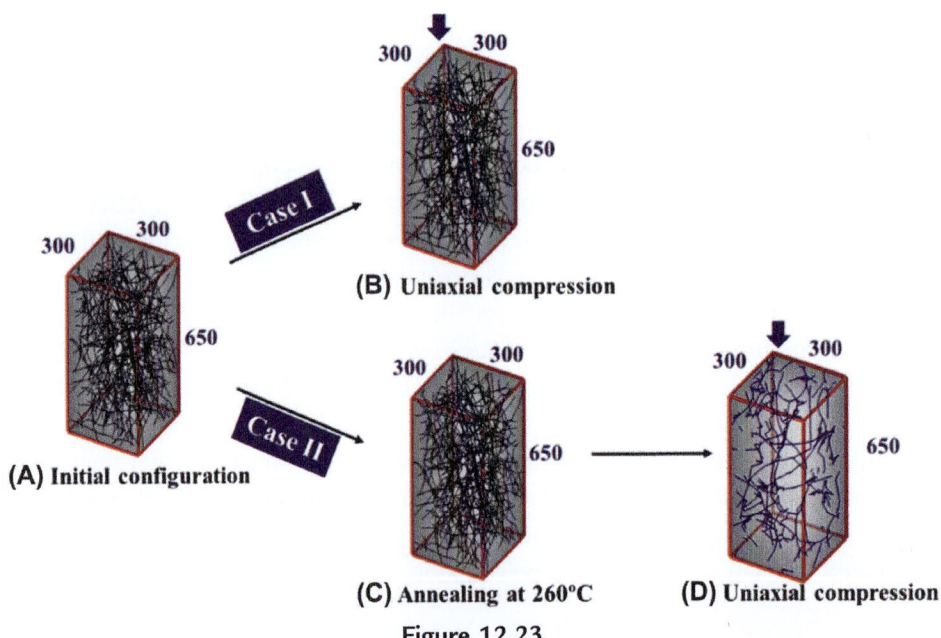

Figure 12.23

Simulation steps: (A) Generate an initial configuration; (B) Uniaxial compression of prestrained pillar; (c) Annealing process; (d) Uniaxial compression simulation of prestrained and annealed pillar. *Reprinted from Liu, F.X., Liu, Z.L., Pei, X.Y., Hu, J.Q., Zhuang, Z., 2017b. Modeling high temperature anneal hardening in Au submicron pillar by developing coupled dislocation glide-climb model. International Journal of Plasticity 99, 102−119. Copyright 2017, with permission from Elsevier.*

To generate such an initial configuration, a stress relaxation procedure is executed before the simulation, during which only dislocation glide is considered and all 12 slip systems of FCC crystals are taken into account. The configuration evolves without external loading until a global energy equilibrium state is approached. For each initial configuration, there are two cases of numerical tests. Case I is for simulating a 35% prestrained pillar, as schematically shown in Fig. 12.23B and Case II is for simulating the prestrained and annealed pillar, as shown in Fig. 12.23C and D.

The conventional 3D-DDD approach can be used for all simulation processes except annealing because the dislocation climb rate is too low to make a difference. During annealing, in which the specimen is annealed at 260°C, dislocation climb has an important role because of thermal activation. Thus, the coupled glide-climb model developed in Section 12.2 is used to simulate this process and a similar relaxation process is executed. After annealing, a uniaxial compression test is conducted, as shown in Fig. 12.23D. Stress-controlled compression is employed on the top of pillar for convenience instead of displacement-controlled compression, which is used in the experiment, because in annealing, dislocation climb and glide occur irregularly, inducing fluctuations in the slip distance at different steps and making it difficult to control the strain rate. It is fixed at the bottom of pillar.

In addition, the cross-slip of screw dislocations is believed to be a mechanism in the hardening response. Using the thermal activation theory, cross-slip can be implemented in the modified DCM. A search algorithm is employed to detect the screw-like dislocation segments. In the current simulations, dislocation segments are considered to be like screws if they lie within an angle of 15 degrees from the Burgers vector direction. After identifying the character of the dislocation, all screw segments are subjected to three additional rules to check whether cross-slip is favorable (Hussein et al., 2015): (1) The magnitude of the resolved shear stress in cross-slip plane RSS_c should be greater than, or at least equal to, that in initial glide plane RSS, i.e., $RSS_c \geq RSS$. (2) The resolved shear stress on the cross-slip plane must be greater than or equal to $Gb/10L$, i.e., $RSS_c \geq Gb/10L$, where G is the shear modulus and b is the magnitude of the Burgers vector. (3) The frequency of cross-slip events during simulated time step P_{step} must be greater than 1, i.e., $P_{\text{step}} \geq 1$. The first two rules are athermal, which guarantees that the dislocation will glide away after it cross-slips, whereas, the thermally activated cross-slip probability per time step can be computed through an Arrhenius-type equation as developed by Kubin et al. (1992):

$$P_{\text{step}} = \beta \frac{L}{L_0} \Delta t \cdot \exp\left(-\frac{(\tau_{\text{III}} - \tau)V}{k_B T}\right) \tag{12.5.1}$$

where β is a scaling factor, L is the length of the screw dislocation, and L_0 is a reference length of 1 μm, as defined in Kubin et al. (1992). Δt is the time step, which has already

Table 12.2: Parameters for Face-*Centered Cubic* Au Used in the Simulation

Symbol	Description	Value
b^a	Value of Burgers vector	0.29 nm
a^b	Lattice constant	0.408
Ω^c	Atomic volume	16.89×10^{-30} m^3
D_v^{0d}	Diffusion coefficient preexponential	1.18×10^{-5} m^2/s
E_v^{fc}	Vacancy formation energy	0.9 eV
E_v^{mc}	Vacancy migration energy	0.8 eV

[a]Data from Lee et al. (2009).
[b]Data from Foiles et al. (1986).
[c]Data from Lee et al. (2003).
[d]Data from Gao et al. (2013).

been defined, τ_{III} is the resolved shear stress at the onset of stage III hardening for bulk crystals, and V is the activation volume associated with cross-slip. Scaling factor β is defined to ensure that $P_{step} = 1$ when a screw dislocation of length $L = L_0$ subjected to a resolved shear stress $\tau = \tau_{III}$. According to the measurements of Bonneville et al, (1988), $V = 300b^3$ and $\tau_{III} = 28$ MPa for cross-slip in copper at room temperature. Because of the lack of available data, the same values are taken for gold. Cross-slip is activated once all three rules are met, and the slip plane switches from the original slip plane to the cross-slip one. Notably, during annealing, no external stress is applied; the resolved shear stress is too small to active cross-slip. Thus, the cross-slip events will be activated mostly in the compression process.

The whole simulation in Fig. 12.23A−D should be repeated with different initial configurations to eliminate random errors. The parameters used in the simulation are given in Table 12.2.

12.5.3 Simulation Results and Analyses

The simulation results of stress-strain curves for Au micropillars in different cases are given in Fig. 12.24, together with experimental data from Lee et al. (2009). However, data cited from the experiment are true stress-strain curves obtained from the assumption of volume conservation at the unloaded state whereas the stress-strain curves obtained from the current simulation are all engineering stress-strain curves. Stress at a given strain value on the engineering stress-strain curve should be a little bit higher than that on the true stress-strain curves, owing to the expansion of the compressed cross-sectional area. Considering the small deformation of about 5% strain, the engineering stress-strain curves are used in our simulation.

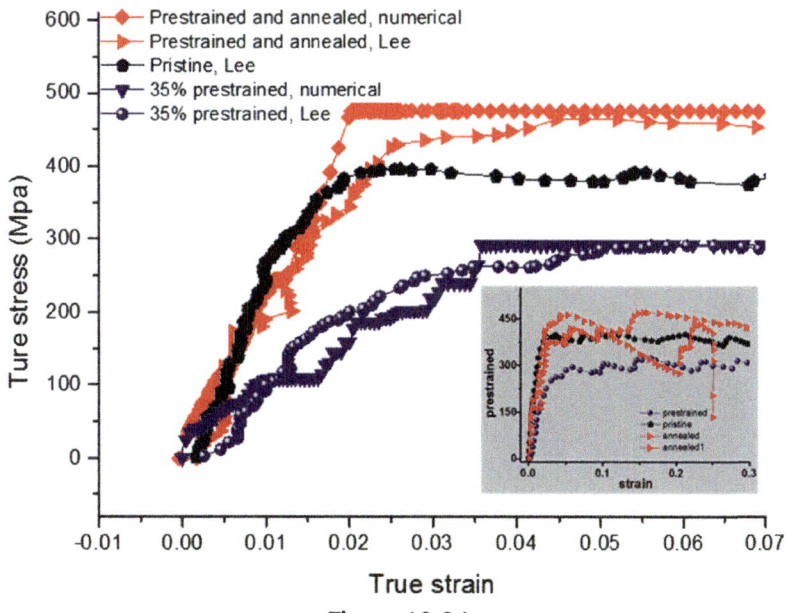

Figure 12.24

Stress-strain curves of 35% prestrained pillars before and after annealing. Results are from both experimental data from Lee et al. (2009) and our simulation. *Reprinted from Liu, F.X., Liu, Z.L., Pei, X.Y., Hu, J.Q., Zhuang, Z., 2017b. Modeling high temperature anneal hardening in Au submicron pillar by developing coupled dislocation glide-climb model. International Journal of Plasticity 99, 102—119. Copyright 2017, with permission from Elsevier.*

The black stress-strain curve in Fig. 12.24 is given by the pristine pillar with a low initial dislocation density of about 10^{13} m^{-2}. The inset shows the experimental stress-strain data up to 30% strain. The blue curve stands for prestrained pillars with a high initial dislocation density of about 10^{15} m^{-2}. The curve with circles is experimental data and the curve with triangles is the numerical result. The red lines are the results of prestrained and annealed pillars, in which the curve with solid squares is the experimental data and the curve with solid diamonds is the numerical results. The flow stresses computed in our simulation agree well with the experimental data, which turn out to be further evidence of the applicability of a modified DCM. The remarkable prestrain softening and annealing hardening effects are also well-captured, which result from the change in internal defect structures during thermal annealing.

Theoretically, during high-temperature annealing, dislocation glide, climb, surface nucleation, and cross-slip are all thermally activated and exhibit notable discrepancies with those at room temperature. Because no external stress is applied during annealing, the resolved shear stress is too small to active cross-slip.

For dislocation glide, thermal activation is demonstrated by temperature-dependent drag coefficient $B_g(T)$, which increases linearly with the temperature, resulting in a lower dislocation glide rate. For dislocation climb, thermal activation affects the vacancy

concentration and diffusivity, accelerating the climb rate remarkably. For dislocation surface nucleation, however, the surface nucleation rate can be expressed as the function of temperature T and stress σ (Ryu et al., 2011; Zhu et al., 2008b):

$$v = N_{\text{site}} v_0 \exp\left(- \frac{Q(\sigma, T)}{k_B T} \right) \tag{12.5.2}$$

where N_{site} is the number of effective surface nucleation sites, assumed to be equal to S/b^2, where S is the surface area, v_0 is the attempt frequency, and $k_B T$ is the thermal energy. Q is the activation free energy, which can be taken as $Q(\sigma, T) = (1 - T/T_m)Q_0(\sigma)$ (Zhu et al., 2008), where T_m is the surface-disordering temperature ranging from 400 to 500°C for Au (Wolf et al., 1978). Thus, the nucleation rate increases exponentially with the temperature. In the current work, the nucleation probability at the ith time step is

$$P_i = v \Delta t_i \tag{12.5.3}$$

where Δt_i is the time step. The number of the surface nucleated loops, N, is then derived by the time integration of nucleated probability P:

$$N = \begin{cases} \text{Integer}[P], & \text{if } \int_{t_m}^{t_n} P \geq 1 \\ \\ 0, & \text{if } \int_{t_m}^{t_n} P < 1 \end{cases} \tag{12.5.4}$$

Once integration $\int_{t_m}^{t_n} P$ is larger than that at a certain time step, surface nucleation will be activated. N dislocation loops will be added into the pillar near the surface. The nucleated loops are randomly located at any of 12 slip systems.

To distinguish among the specific influences of thermally activated glide, climb, and surface nucleation on flow stress, three sets of contrasting simulations of high-temperature annealing are performed: (I) glide only, (II) glide with consideration of surface nucleation, and (III) glide and climb with consideration of surface nucleation.

As shown in Fig. 12.25, the stress-strain curve with blue circles is the experimental data for a 35% prestrained pillar, the red curve with triangles is the experimental data for a 35% prestrained and annealed pillar, and the remaining curves are the numerical results for the 35% prestrained and annealed pillars: the orange curve with triangles is the numerical result of the glide-only case ([Case I]); the dark cyan curve with circles is the result of Case II; and the dark yellow curve with diamonds is the result of Case III. The thermal activation effects of glide and surface dislocation nucleation during annealing have only a small influence on flow strength. This is reasonable because the glide velocity is so large that a decrease in glide velocity caused by high temperature is negligible compared

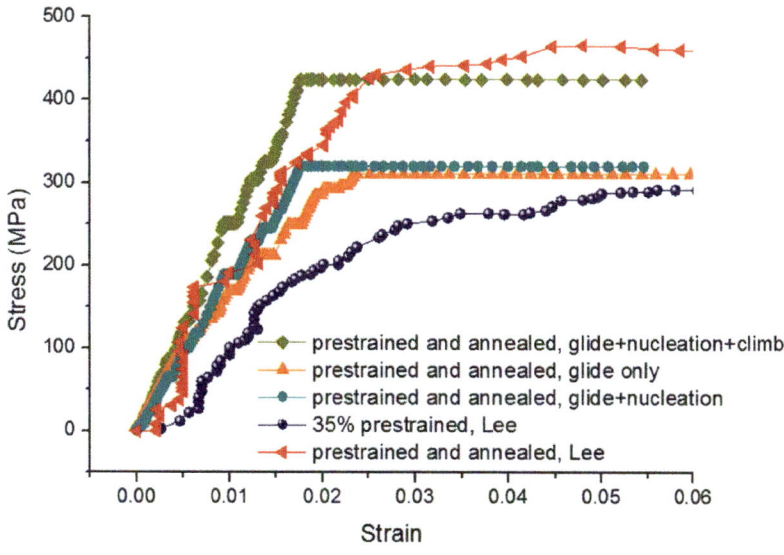

Figure 12.25

Stress-strain curves of 35% prestrained and annealed pillars considering different thermal activation mechanisms. *Reprinted from Liu, F.X., Liu, Z.L., Pei, X.Y., Hu, J.Q., Zhuang, Z., 2017b. Modeling high temperature anneal hardening in Au submicron pillar by developing coupled dislocation glide-climb model. International Journal of Plasticity 99, 102—119. Copyright 2017, with permission from Elsevier.*

with that of the climb velocity. Similarly, the increase in dislocation density during annealing caused by surface nucleation is also insignificant compared with the extremely high initial dislocation density. Only when the dislocation climb is accounted for can the flow strength in the numerical simulation match with that of the experimental data, which convincingly demonstrates that dislocation climb has a leading role during high-temperature annealing. A detailed microstructural analysis is needed to elucidate how dislocation climb induces the hardening of flow strength and why the flow strength of the prestrained or annealed pillar eventually approached a strength level similar to that of the pristine one after a large amount of plastic deformation, as shown in the inset of Fig. 12.24.

12.5.4 Microstructural Analysis

As reported in the experiment, the softening effect on the 35% prestrained pillar is ascribed to the existence of high dislocation density, which is about 10^{-15} m^{-2}. It allows the pillar to deform at a lower stress level. In contrast, the hardening effect on the annealed pillar results from a reduction in the dislocation density. In annealing, although there is no externally applied stress, the temperature is as high as 533K, which can activate dislocations to climb in a direction perpendicular to their original slip plane. It may facilitate dislocation interactions between different slip planes and promote dislocation annihilation, resulting in a reduction in dislocation density. Fig. 12.26 shows images of dislocation configurations before and after annealing.

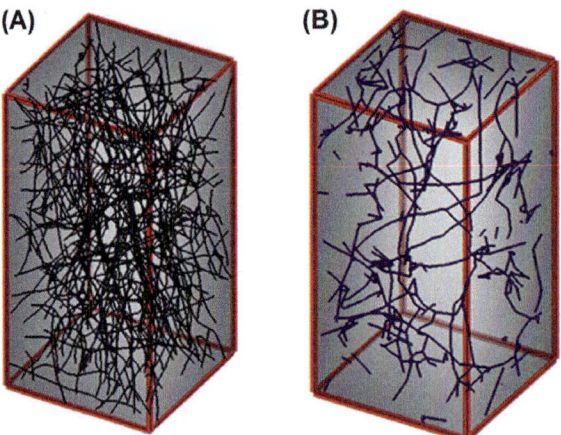

Figure 12.26

Dislocation configurations: (A) Before annealing; (B) After annealing. *Reprinted from Liu, F.X., Liu, Z.L., Pei, X.Y., Hu, J.Q., Zhuang, Z., 2017b. Modeling high temperature anneal hardening in Au submicron pillar by developing coupled dislocation glide-climb model. International Journal of Plasticity 99, 102–119. Copyright 2017, with permission from Elsevier.*

Unsurprisingly, a significant decrease in dislocation density ρ is observed after annealing. A determination of the density shows that it decreases from 1.14×10^{15} m^{-2} before annealing to 2.26×10^{14} m^{-2} afterward. However, with such a dramatic reduction, the density is still an order of magnitude higher than that of the pristine pillars, which is $\sim 10^{13}$ m^{-2}, but the flow strength has recovered almost to that of the pristine one. It seems that the dislocation density is not the only factor influencing flow strength.

To elucidate this issue, dislocation configurations are shown during the compression process, as demonstrated in Fig. 12.27. The most remarkable change observed is the existence of plenty of dislocation jogs, which are step-shaped with at least one segment of the dislocation line not on its original slip plane. Typical dislocation jogs in Fig. 12.27 are highlighted in blue and numbered from 1 to 4. The evolution of these four jogs at different time steps is shown from $t_1 = 11.2$ ns in Fig. 12.27A to $t_2 = 301$ ns in Fig. 12.27B, and then to $t_3 = 2887$ ns in Fig. 12.27C. Detailed configurations showing Jogs 1 and 2 in the blue-green box are magnified in Fig. 12.27A and B. These jogs are produced by dislocation climb during annealing and exist even in well-annealed crystals, because a dislocation line cannot climb as a whole at once, but instead move in stages, leaving plenty of jogs in the crystal. There is a thermodynamic equilibrium-determined number of jogs per unit length of dislocation (Hirth and Lothe, 1982), expressed as

$$n_j = n_0 \exp(-E_j/kT) \tag{12.5.5}$$

where E_j is the character energy resulting from the increase in dislocation length, n_0 is the number of atom sites per unit length of dislocation, and n_j/n_0 has an order of 10^{-5} at 1000K and 10^{-17} at 300K, respectively.

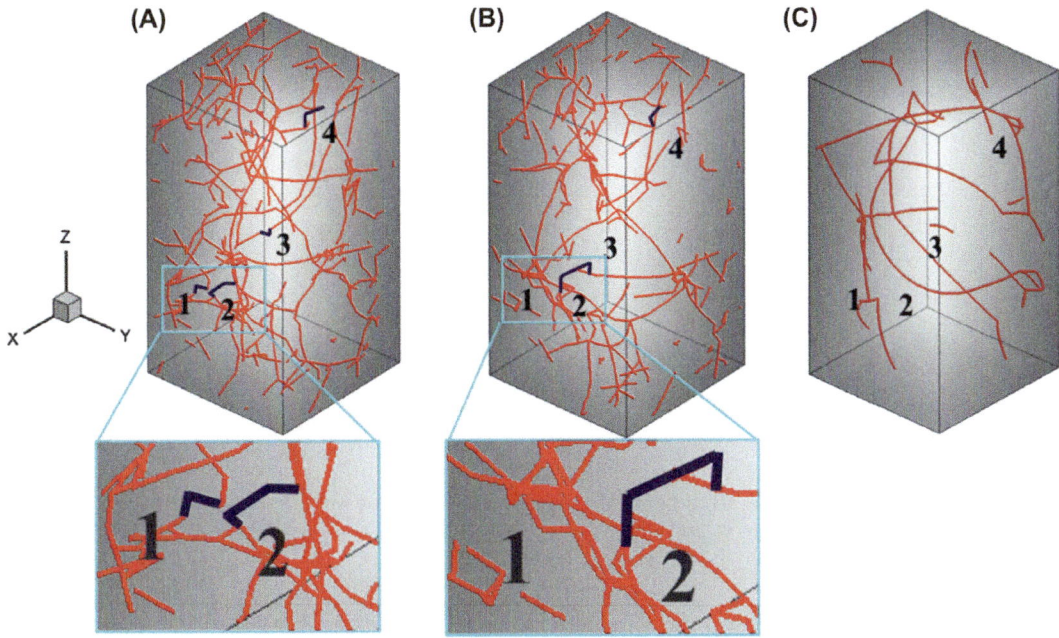

Figure 12.27
Dislocation configurations during compression: (A) at $t_1 = 11.2$ ns; (B) at $t_2 = 301$ ns; (C) at $t_3 = 2887$ ns. Dislocation lines are in *red* whereas jogs are highlighted in *blue*. *Reprinted from Liu, F.X., Liu, Z.L., Pei, X.Y., Hu, J.Q., Zhuang, Z., 2017b. Modeling high temperature anneal hardening in Au submicron pillar by developing coupled dislocation glide-climb model. International Journal of Plasticity 99, 102—119. Copyright 2017, with permission from Elsevier.*

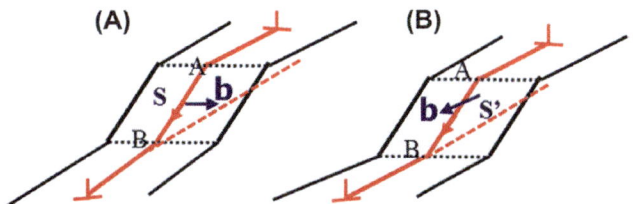

Figure 12.28
Schematics of different types of jogs: (A) On an edge dislocation; (B) On a screw dislocation. *Reprinted from Liu, F.X., Liu, Z.L., Pei, X.Y., Hu, J.Q., Zhuang, Z., 2017b. Modeling high temperature anneal hardening in Au submicron pillar by developing coupled dislocation glide-climb model. International Journal of Plasticity 99, 102—119. Copyright 2017, with permission from Elsevier.*

The jogs on an edge dislocation have very weak mobility. As illustrated in Fig. 12.28A, Jog AB must slip on Plane S to keep pace with the whole dislocation line, but Plane S may not be in the slip system. In other words, it is may not be in the close-packed {111} planes in an FCC crystal, which may cause much larger slip resistance for the jog to slip together with the whole dislocation line. On the other hand, the jog on a screw dislocation, as plotted in Fig. 12.28B, is sessile because the direction of the motion is in the plane normal direction, not in-plane motion. As the dislocation advances, Jog AB must climb to

slip along with the whole dislocation line. For unit jogs in which the normal component extends over interplanar spacing, the jogs can be dragged to move together and leave vacancy sites behind them, whereas superjogs are generally sessile and can be moved only by climbing. These can be verified by the dislocation configurations in Fig. 12.27A at time $t_1 = 11.2$ ns and Fig. 12.27B at time $t_2 = 301$ ns, respectively. It can be seen from Fig. 12.27A and B that as the configuration evolves from t_1 to t_2, small jogs fade away, such as Jogs 1 and 3. The next somewhat large jogs, such as Jog 4, remain more stable and change little. Much larger jogs, such as Jog 2, remain unchanged during the evolution. These jogs with low mobility or that are absolutely sessile may act as obstacles to dislocation motion and lead to higher flow strength. According to this analysis, the hardening effect after thermal annealing can be ascribed to (1) dislocation annihilation during annealing, which significantly reduces the dislocation density; and (2) the formation of dislocation jogs during annealing, which impedes dislocation motion at the early stage of loading, resulting in lower dislocation mobility and higher flow strength compared with those of the pristine pillars.

However, all of these jogs vanish as plastic deformation increases a certain amount, as shown in Fig. 12.27C at time $t_3 = 2887$ ns, because these climb-induced jogs are mostly edge-type with a small jog length. They are not stable enough to supply pinning points or serve as active dislocation sources and will be unlocked when the stress is large enough. Therefore, unlike the superjogs formed by multiple cross-slip, these climb-induced jogs have an important role in decelerating rapid dislocation annihilation at the early stage of loading rather than promoting more dislocation. This is why the flow strength of the prestrained or annealed pillar eventually approaches a level of strength similar to that of the pristine one after a large amount of plastic deformation, as reported in the experiment (Lee et al., 2009). For the prestrained micropillar, the dislocations are annihilated rapidly from the free surface with increasing compression and eventually experience mobile dislocation starvation after mass deformation, resulting in a stress level similar to that of the pristine pillar. For the annealed pillar, however, although the dislocation density is dramatically reduced by annealing, the density is still an order of magnitude larger than for the pristine one. The jogs block dislocation motion and lead to an increase in flow strength, which will recover to a stress level similar to that of the pristine pillar when plastic deformation is large enough to unlock the jogs and release the dislocations.

12.6 Summary

In this chapter, the coupled dislocation glide-climb model is developed to study the intrinsic formation mechanism of helical dislocations and high-temperature annealing hardening in submicron pillars. Both thermally activated dislocation glide and climb are dealt with in the framework of 3D-DDD. The climb rate is determined by vacancy volumetric flux across the dislocation core, which is obtained by solving vacancy diffusion equations analytically with an assumption of steady-state diffusion or using the FEM.

By adopting a sequentially coupled scheme to bridge the timescales, a coupled glide-climb model is established to study the formation of helical dislocation in pure aluminum micropillars. The supersaturated vacancy concentration promotes dislocation climb, which further promotes mixed dislocation to wind into helixes, whereas prismatic glide ensures spaces between spirals. In addition, the computational results are compared with the theoretical solution to figure out the underlying mechanism of the multiple helical turns; it turns out that evolution velocity breaks the stable equilibrium state and drives the helix into a configuration with multiple turns, which again validates the applicability of our coupled model.

A modified discrete-continuous model (DCM) is also proposed by coupling 3D dislocation dynamics with the FEM to deal with thermally activated dislocation motion. The climb rate is determined by the vacancy volumetric flux across the dislocation core, which is obtained by solving the vacancy diffusion equation using FEM code. The vacancy concentrations are updated as the dislocation configuration evolves and localizes from DDD segments to FEM nodes by a distance-related weight function originating from the classic analytical expression of cylindrical diffusion. It is a nonlocal model that breaks through the local limitation of analytical solution by taking the long-range contribution to the dislocation climb associated with vacancy diffusion into account. Moreover, both pipe and bulk diffusion are naturally considered in this model, which makes it much more appropriate for the realistic situation.

A coupled glide-climb model is further developed by incorporating dislocation glide into this model to shed light on the intrinsic mechanism of the hardening effect after-high temperature annealing, in which the timescale of separation between glide and climb is addressed by adopting an adaptive time-stepping scheme. The flow stresses computed in our simulations agree well with the experimental data. The remarkable prestrain softening and annealing hardening effects are also well-captured. These can be ascribed to two major aspects based on the microstructural analysis: (1) The climb of dislocation during high-temperature annealing promotes dislocation annihilation, leading to a decrease in dislocation density; and (2) the jogs, which are nucleated during annealing owing to the high-temperature climb, have weak mobility and act as obstacles to dislocation glide motion, resulting in a decrease in mobile dislocation density. The combined effect of these two aspects significantly decreases dislocation mobility and results in a higher level of flow strength compared with the prestrained pillars.

In general, the coupled dislocation glide-climb model developed in this chapter provides a useful method for studying the evolution of dislocation microstructures considering both the gliding and climbing of dislocations.

Single-Crystal Material Model and Pole Figures

A1.1 Introduction

The aim of this appendix is to present single-crystal material formulations that accurately characterize the micromechanics of crystals. Constructing appropriate homogenization techniques for polycrystals (Van Houtte, 1987; Van Houtte et al., 2005) to obtain polycrystalline aggregate responses would be a natural follow-up. However, additional complications arise to modeling the homogenized response; for example, with polycrystals it is more challenging to determine active slip directions (Boyce et al., 1989). Therefore, the focus is on computational modeling single crystals in a representative volume element.

The appendix is structured as follows. First, an extended discussion of crystals and their structures, Burgers vectors, and slip systems, is presented in Sections A1.2–A1.4. The discussion aims to provide a fairly detailed outline of the necessary crystalline properties that serve as input data for single-crystal plasticity models. In particular, that not all crystals possess orthogonal bases is cause for some additional preprocessing of slip-related data, as typically presented in literature, such as Miller indices of slip systems.

A1.2 Crystallographic Description of Cubic and Noncubic Crystals

This section introduces crystals and their different structures modeled by crystal plasticity and presents a method for identifying planes and directions in crystals, which is to lay the foundation for the definition of slip systems.

Crystalline materials are characterized by the fact that their atoms tend to be ordered with definiteness to form a space lattice (Azaroff, 1984; Kelly et al., 2000; Giacovazzo et al., 2002): that is, an infinite array of atoms stretching out in three-dimensional (3D) space. Each repeat unit within the space lattice is termed a unit cell. By applying discrete translations to these unit cells in 3D space the lattice is generated. It has been shown that only seven lattice systems exist (Azaroff, 1984), defined by the shape of the unit cells, which are illustrated in Fig. A1.1. The shape of a unit cell is defined by three axes, **a**, **b**, and **c**, and the three angles between them, α (between **b** and **c**), β (between **a** and **c**), and γ (between **a** and **b**).

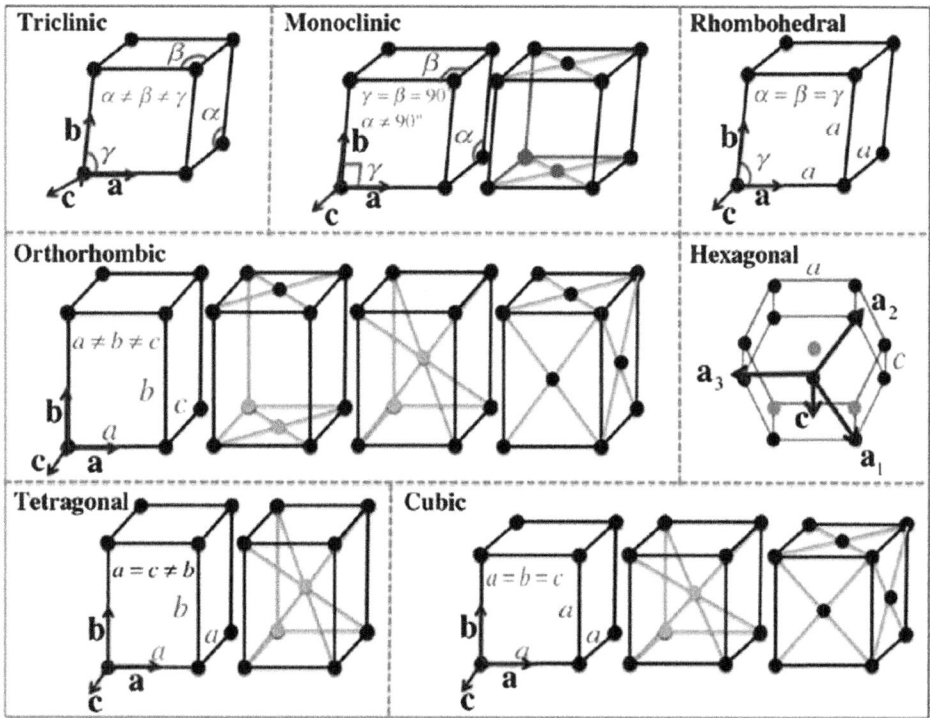

Figure A1.1
Possible unit cells in crystals.

Distributed within these seven lattice systems are 14 possible Bravais lattices, defined by whether the unit cell is: (1) primitive (P), that is, having only corner lattice points; (2) side-centered (C, B, or A) having lattice points centered on a pair of faces along an axis (**c**, **b**, or **a**, respectively) in addition to the corner points; (3) body-centered (I), having a lattice point in the center of the unit cell in addition to the corner points; or (4) face centered (F), having a lattice point centered on each face in addition to the corner points. Not every lattice system can exhibit P, C, I, and F unit cells, so those that remain possible are the 14 Bravais lattices shown in Fig. A1.1 (Belytschko et al., 2014). In metals and alloys, all points shown in the unit cells (whether on the vertices or faces or in the cell interior) correspond to atomic locations; however, in more general materials, these points would simply correspond to point groups (Azaroff, 1984). To label directions and planes of interest in these unit cells, Miller indices are introduced (Kelly et al., 2000).

A1.2.1 Specifying Directions

Miller indices for directions in the Bravais lattice can be obtained from the fractional coordinates of the first atom (more generally, lattice point) to intercept the desired direction vector. The coordinates of the atom are measured via a position vector pointing

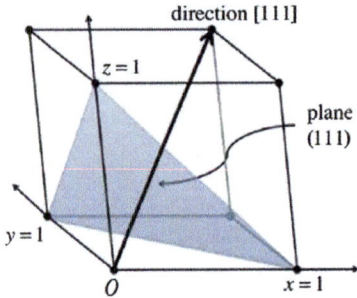

Figure A1.2
Planes and directions of a unit cell.

from an origin fixed at some corner of the unit cell (Fig. A1.2). The position vector may be written in terms of the Bravais lattice basis, a set of three primitive vectors (call them [**x**, **y**, **z**]) that define crystallographic axes of a unit cell:

$$\mathbf{r}_{uvw} = u\mathbf{x} + v\mathbf{y} + u\mathbf{z} \tag{A1.2.1}$$

where u, v, and w are respectively the **x**, **y**, and **z** fractional coordinates of the atom whose position is sought. Once the position vector is defined, its coefficients are grouped in square brackets, that is, $[uvw]$, and any fractions are cleared to yield the Miller indices for that direction, written $[hkl]$, where h, k, and l are integers. For example, a point of coordinates $(-1, 3/4, 1/2)$ will generate a direction with Miller indices $\left[\overline{4}32\right]$. By convention, negative numbers are written with an overbar instead of a minus sign.

A1.2.2 Specifying Planes

Before introducing Miller indices for planes, it is beneficial first to outline the concept of a reciprocal lattice (Azaroff, 1984; Kelly et al., 2000; Giacovazzo et al., 2002). Having chosen for the Bravais lattice a set of three primitive vectors as a basis, call them $[\mathbf{a}_1, \mathbf{a}_2, \mathbf{a}_3]$. Then, it is possible to define a reciprocal lattice with basis $[\mathbf{b}^1, \mathbf{b}^2, \mathbf{b}^3]$:

$$\left[\mathbf{b}^1, \mathbf{b}^2, \mathbf{b}^3\right]^T = [\mathbf{a}_1, \mathbf{a}_2, \mathbf{a}_3]^{-1} \tag{A1.2.2}$$

where each row of the inverse matrix corresponds to a basis vector in the reciprocal lattice. More explicitly,

$$\mathbf{b}^1 = \frac{\mathbf{a}_2 \times \mathbf{a}_3}{det[\mathbf{a}_1, \mathbf{a}_2, \mathbf{a}_3]}, \quad \mathbf{b}^2 = \frac{\mathbf{a}_3 \times \mathbf{a}_1}{det[\mathbf{a}_1, \mathbf{a}_2, \mathbf{a}_3]}, \quad \mathbf{b}^3 = \frac{\mathbf{a}_1 \times \mathbf{a}_2}{det[\mathbf{a}_1, \mathbf{a}_2, \mathbf{a}_3]} \tag{A1.2.3}$$

It follows from the definition of the cross-product that each vector $\mathbf{b}^i$ in the reciprocal lattice is normal to a plane defined by two corresponding basis vectors $(\mathbf{a}_j, \mathbf{a}_k)$ from the Bravais lattice. Hence, the reciprocal lattice basis and Bravais lattice basis are duals, and

$$\mathbf{b}^i \cdot \mathbf{a}_j = \delta^i_j, \quad 1 \leq i, \; j \leq 3 \tag{A1.2.4}$$

where δ^i_j is the Kronecker delta.

Any vector in the reciprocal lattice is but a linear combination of its basis vectors $\mathbf{b}^i$. Because crystal planes may be identified by their normal, a natural method with which to label planes of atoms is to associate with a vector from the reciprocal lattice a corresponding plane in the Bravais lattice, as shown in Fig. A1.3. Specifically, a plane of Miller index (hkl), where h, k, and l are integers, will be defined as the plane whose normal is the reciprocal lattice vector $\mathbf{g}^{hkl}$, given by

$$\mathbf{g}^{hkl} = h\mathbf{b}^1 + k\mathbf{b}^2 + l\mathbf{b}^3 \tag{A1.2.5}$$

To visualize the corresponding plane inside the Bravais lattice, note that any vector (say, $\mathbf{r}$) contained in plane (hkl) must satisfy the relation

$$\mathbf{r}\cdot\mathbf{g}^{hkl} = hr_1 + kr_2 + lr_3 = 0 \tag{A1.2.6}$$

This equation can be interpreted as a plane passing through the origin of the Bravais lattice. The orientation of the plane can be found from its intercepts with the crystal axes if the plane is shifted parallel to itself to set the right-hand side of Eq. (A1.2.6) to unity: that is, when

$$hr_1 + kr_2 + lr_3 = 1 \tag{A1.2.7}$$

Then, it is immediate that the plane of Miller indices (hkl) is only one whose intercepts are $(1/h + 1/k + 1/l)$ in the Bravais lattice, as shown in Fig. A1.2 for the plane (111). Conversely, given a plane of intercepts $(1/h + 1/k + 1/l)$ in a unit cell, its Miller index would be (hkl) and the corresponding unit normal is simply

$$\mathbf{n}^{hkl} = \left|\mathbf{g}^{hkl}\right|^{-1}\mathbf{g}^{hkl} \tag{A1.2.8}$$

It is easy to verify that for cubic crystals the normal to a plane (hkl) points in the direction $[hkl]$. However, this observation does not hold for noncubic crystals, and calculating

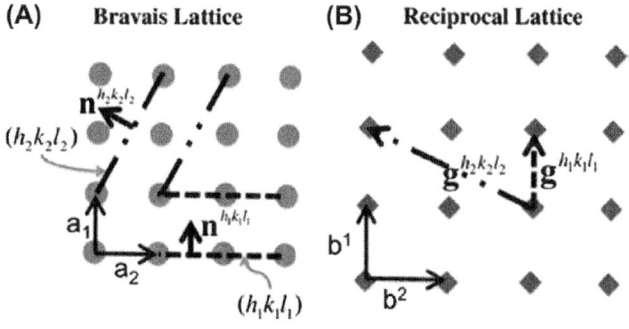

Figure A1.3
(A) Direct Bravais lattice; (B) Reciprocal lattice.

reciprocal lattice vectors as outlined earlier becomes necessary to implement single-crystal plasticity properly.

Correctly labeling planes and directions in crystals by Miller indices is especially important in materials engineering because of the directional dependency (anisotropy) of various material properties such as optical properties, reactivity, surface tension, and dislocation motion. Anisotropic motion of dislocations along preferential directions and planes is fundamental to the following developments of crystal plasticity (Rice and Asaro, 1977).

A1.3 Atomic Origins of Plasticity and the Burgers Vector in Single Crystals

This section presents a brief discussion of the atomic origins of plasticity in crystals with the aim of defining the Burgers vector, which is essential to formulating plastic shear strain in crystals. Plastic deformation in metals is isochoric, that is, volume-preserving or incompressible, so that plastic shear remains the mechanism of interest whose modeling lays the foundation for crystalline plasticity.

If shearing forces are applied on a perfect crystal lattice in which all atoms are in full registry, the shear stress needed to slide a layer of atoms over another as a function of displacement x is given by (Hull and Bacon, 2006)

$$\tau = \frac{Gc}{2\pi a} \sin \frac{2\pi x}{c} \tag{A1.3.1}$$

where a is the vertical spacing between atoms, c is the horizontal spacing, and G is the shear modulus. Thus, it could be concluded that the theoretical shear strength (τ_{yield}) of a perfect crystal is on the order of $\frac{G}{2\pi}$. Better estimates set the theoretical strength of perfect crystals at $\frac{G}{30}$ (Hull and Bacon, 2006). However, experimental measurements find that the shear strength of a crystal is well below these estimates, usually in the range (Hull and Bacon, 2006)

$$1 \times 10^{-8} G \leq \tau_{\text{yield}} \leq 1 \times 10^{-4} G \tag{A1.3.2}$$

To explain this difference, the dislocation theory was put forward (Orowan, 1934; Taylor, 1934). A dislocation is a topological defect (Anthony and Azirhi, 1995; Bulatov and Cai, 2006): that is, a defect in the way certain atoms in a lattice connect to their neighbors. Specifically, a dislocation is a line defect. It may be thought of as the boundary between slipped and unslipped material. Two basic types of dislocations exist in a crystal: edge and screw (Hirth and Lothe, 1982), as illustrated in Fig. A1.4.

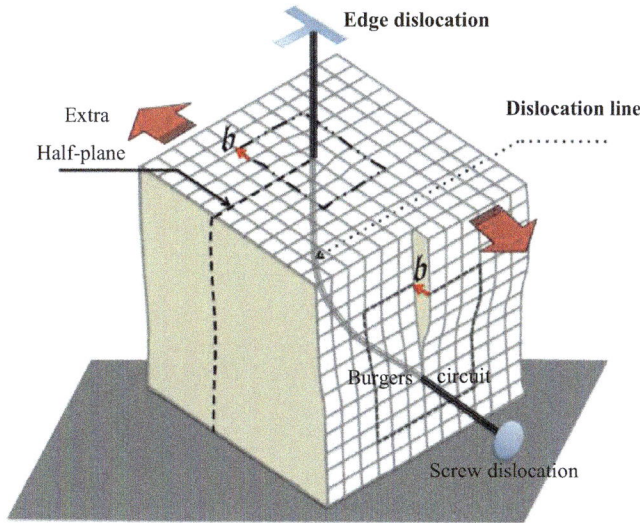

Figure A1.4
Dislocation line in a crystal.

An edge dislocation is caused by terminating a plane of atoms in the middle of a crystal. It can be visualized as the edge of an extra half-plane of atoms, as shown in Fig. A1.5C and on the top surface of Fig. A1.4. A screw dislocation may be visualized by slicing a crystal partway along some plane, say a vertical plane. Slide the right half against the left, such that the atoms on right come into full registry with new atoms on the left in their new positions (thus maintaining crystallinity). After this distortion, the boundary line at the tip of the slice defines the screw dislocation line. In a real crystal, dislocations are mixed in the sense that they could typically be resolved into edge and screw components. A mixed dislocation line can be visualized as a curved line that begins as a screw and ends as an edge, as shown in Fig. A1.4.

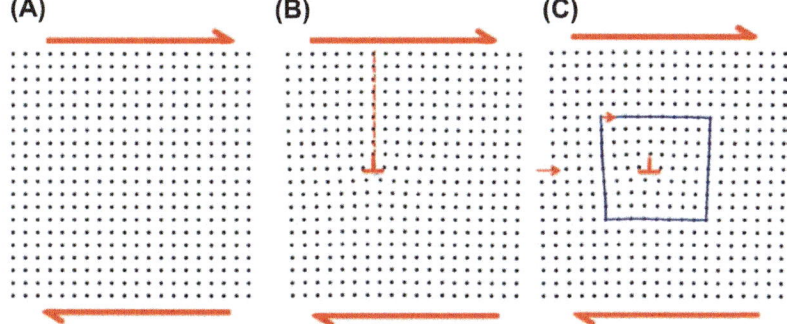

Figure A1.5
(A) Square lattice under shear; (B) Square lattice with dislocation under shear; (C) Dislocation moves.

Both screw and edge dislocations induce localized residual shear stresses around their cores. In addition to shear stress, edge dislocations induce normal stress because of their natural asymmetry. On the side of the extra half-plane compression arises, and on the side of the missing plane there is tension (Hull and Bacon, 2006). By virtue of the distortions and residual stresses they introduce in a lattice, dislocations make it easier for slip to occur in atomic layers containing them, and the energy barrier associated with their motion is low enough to explain the experimentally measured crystal shear strengths (Hull and Bacon, 2006).

Fig. A1.5A shows the side view of a square lattice subjected to shear forces. The edge of the extra half-plane (indicated by an upside-down "T" in the figure) corresponds with the side view of the dislocation line. The topological defect created by a dislocation may be inferred from Fig. A1.5B from the fact that atoms on a dislocation line have five nearest neighbors instead of six, as in a perfect simple cubic lattice in three dimensions.

It is customary to quantify the lattice defect that a dislocation introduces by means of a Burgers vector. A Burgers vector measures closure failure in any circuit drawn through the lattice when circumscribing a dislocation (Hull and Bacon, 2006). For instance, suppose a closed rectangular circuit is placed by marching through lattice points in a perfect portion of a crystal. Then, if the same circuit is placed around a dislocation, the start and end points of the circuit no longer coincide. Closure failure can arise from, for instance, the extra half-plane that the circuit needs to cover, as shown in Fig. A1.5C. To maintain crystallinity, the magnitude of a Burgers vector must correspond to a lattice vector, that is, the distance between two lattice points. The magnitude of a Burgers vector (b) is given by (De Graef and McHenry, 2007)

$$|b| = \sqrt{g_{ij}\Delta s^i \Delta s^j}, \quad \text{(sum on } i \text{ and } j\text{)} \qquad \text{(A1.3.3)}$$

where $|\triangle s| = |\mathbf{p}_m - \mathbf{p}_n|$ is the repeat distance along slip direction $\mathbf{s}$, that is, the distance between two consecutive lattice points m and n located at positions $\mathbf{p}_m$ and $\mathbf{p}_n$, respectively. g_{ij} are the components of a metric tensor, which defines distances in a crystal that may exhibit non-orthonormal bases and are given by the dot product of the lattice basis vectors: $g_{ij} = \mathbf{a}_i \cdot \mathbf{a}_j$. In its general form, and thus applicable to crystals of the lowest symmetry (triclinic) to those of highest symmetry (cubic), the metric tensor would be (De Graef and McHenry, 2007)

$$\mathbf{g} = \begin{bmatrix} a^2 & ab\cos\gamma & ac\cos\beta \\ ab\cos\gamma & b^2 & bc\cos\alpha \\ ac\cos\beta & bc\cos\alpha & c^2 \end{bmatrix} \qquad \text{(A1.3.4)}$$

where lengths a, b, c, and angles α, β, γ, are as defined in Fig. A1.1. Its inverse $\mathbf{G}$ represents the metric in reciprocal space and may be expressed as (De Graef and McHenry, 2007)

$$\mathbf{G} = \frac{1}{V^2} \begin{bmatrix} b^2 c^2 \sin^2 \alpha & abc^2 \Phi(\alpha, \beta, \gamma) & abc^2 \Phi(\gamma, \alpha, \beta) \\ abc^2 \Phi(\alpha, \beta, \gamma) & c^2 a^2 \sin^2 \beta & abc^2 \Phi(\beta, \gamma, \alpha) \\ abc^2 \Phi(\gamma, \alpha, \beta) & abc^2 \Phi(\beta, \gamma, \alpha) & a^2 b^2 \sin^2 \gamma \end{bmatrix} \tag{A1.3.5}$$

where $V^2 = a^2 b^2 c^2 \left(1 - \cos^2 \alpha - \cos^2 \beta - \cos^2 \gamma + 2 \cos \alpha \cos \beta \cos \gamma \right)$ is the determinant of $\mathbf{g}$, and the function of ordered angles $\Phi(A, B, C) = \cos A \cos B - \cos C$.

For an edge dislocation, the Burgers vector is perpendicular to the dislocation line, whereas for the screw dislocation, the Burgers vector is parallel to the dislocation line, as can be gathered from Fig. A1.4. In a mixed dislocation, the Burgers vector is neither normal nor parallel to the dislocation line. A dislocation retains its identity as it moves when a crystal deforms plastically, and in that sense, it may be thought of as a soliton (Bulatov and Cai, 2006), so that its Burgers vector is conserved.

In continuum mechanics, it is posited that the Burgers vector is defined by (Kroner, 1981; Bulatov and Cai, 2006; Gurtin, 2006)

$$b = \oint \mathbf{F}^e \mathrm{d}\mathbf{x} \tag{A1.3.6}$$

where $\mathbf{F}^e$ is the elastic part of the deformation gradient, obtainable from the multiplicative decomposition $\mathbf{F} = \mathbf{F}^e \mathbf{F}^p$ (see, Section 4.2). This formula yields what is called the Volterra dislocation (Bulatov and Cai, 2006), because of the continuous topology it assumes of a crystal in lieu of a discrete lattice structure.

When deformation is entirely elastic, $\mathbf{F}^e = \mathbf{F} = \nabla_0 \mathbf{u}$; that is, elastic deformation becomes the gradient of the displacement vector field. Elastic deformation may thus be seen as a conservative field whose potential is displacement. The closed-path integral of Eq. (A1.3.6) must therefore be 0, so that a 0 Burgers vector results. In the case of plasticity, $\mathbf{F}^e \neq \nabla_0 \mathbf{u}$, which is made clear by the multiplicative decomposition of $\mathbf{F} = \mathbf{F}^e \mathbf{F}^p$. It follows that elastic deformation is not a conservative field, and the closed-path integral does not vanish, so a Burgers vector results. In this sense, a Volterra dislocation quantifies the extent to which elastic deformation fails to be conservative, which ultimately relates to the evolving dislocation content of a deforming crystal.

A1.4 Defining Slip Planes and Directions in General Single Crystals

The slip planes and directions are defined in crystals, which are essential to defining the kinematics of single-crystal plasticity, and the discussion concludes with Table A1.1, which presents all preprocessing steps performed on crystallographic data to be fed into the single-crystal plasticity algorithm.

Table A1.1: Preprocessing: Determining Initial Slip Directions and Normal

1. Determine three basis vectors for the crystal to be modeled, $\{\mathbf{a}_1, \mathbf{a}_2, \mathbf{a}_3\}$

2. Normalize the basis vectors, $\mathbf{a}_i = \mathbf{a}_i \Big/ \sqrt{a_{1i}^2 + a_{2i}^2 + a_{3i}^2}, \quad i \in \{1, 2, 3\}$

3. Obtain the Miller indices for all slip planes $\{(hkl)^a, 1 \leq a \leq nss\}$ and slip directions $\{(uvw)^a, 1 \leq a \leq nss\}$ of the crystal. This information may be found from the literature or by selecting the densest planes and directions of the crystal in the absence of experimental data.

4. Define and normalize each of the slip direction vectors as:

$$\mathbf{s}^{uvw} = \left(u^2 + v^2 + w^2\right)^{-\frac{1}{2}}[u\mathbf{a}_1 + v\mathbf{a}_2 + w\mathbf{a}_3]$$

5. Pair each slip direction vector with the appropriate Burgers vector magnitude. Information on Burgers vectors for specific crystals may be found in the literature or by calculating the repeat distance of the lattice along the corresponding slip direction using

$$|\mathbf{b}| = \sqrt{g_{ij}\Delta s^j \Delta s^i}$$

where the metric $\mathbf{g}$ is defined by Eq. (A1.3.4) and $|\Delta s| = |\mathbf{p}_m - \mathbf{p}_n|$ is the repeat distance.

6. Transform Miller indices of planes to normal vectors using the reciprocal lattice construct:
   For cubic crystals: $\mathbf{n}^{hkl} = \left(h^2 + k^2 + l^2\right)^{-\frac{1}{2}}[h\mathbf{e}_1 + k\mathbf{e}_2 + l\mathbf{e}_3]$
   For noncubic crystals: $\mathbf{n}^{hkl} = \left|\mathbf{g}^{hkl}\right|^{-1}\mathbf{g}^{hkl}$, where $\mathbf{g}^{hkl}$ is defined by Eq. (A1.2.5).

7. Align the slip plane normal $\mathbf{n}^{hkl}$ and slip direction $\mathbf{s}^{uvw}$ of a crystal with respect to a predefined loading axis or according to a sample texture by applying the transformation

$$\left[s_t^\alpha, \mathbf{n}_t^\alpha\right] = \mathbf{T}_t^c\left[s_c^\alpha, \mathbf{n}_c^\alpha\right]$$

with $\mathbf{T}_t^c$ as defined in Eq. (A1.4.2).

The motion of dislocations typically characterizes plasticity in crystalline materials. As indicated by the arrows in Fig. A1.5C, applying shear activates the motion of a dislocation line and leaves behind a step of magnitude *b* along the direction of motion in the plane containing the dislocation, and defines a unit of slip. During the plastic deformation of metals and alloys, multitudes of dislocations move and interact; however, the crystalline material between them remains undistorted (except for typically negligible elastic lattice distortions).

In general, dislocations move within the most closely packed crystal planes and along the most closely packed crystal directions. Fig. A1.6 illustrates a face-centered cubic (FCC) crystal. The (001) and (111) planes have been singled out to illustrate the difference in atomic packing between them, where the (111) plane has a greater planar atomic density

(A)

(B) Planar density $2/a^2$ (atoms/Å^2)

$a = 2\sqrt{2}R$

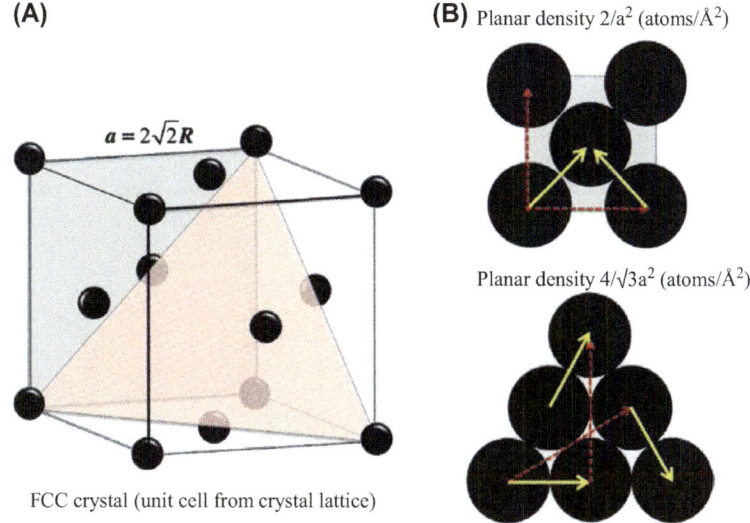

FCC crystal (unit cell from crystal lattice)

Planar density $4/\sqrt{3}a^2$ (atoms/Å^2)

Figure A1.6
(A) Face-centered cubic crystal; (B) (001) and (111) planes.

than the (001) plane. The figure also indicates in each plane by solid arrows possible slip directions (most closely packed), and by dotted arrows, directions that are not closely packed. Dislocation motion can occur: for example, in the (111) plane and along the $[0\bar{1}1]$, $[\bar{1}01]$, and $[\bar{1}10]$ directions (Kelly et al., 2000). These directions are members of a family. A family of directions is designated by angle brackets: for example, $\langle 1\bar{1}0 \rangle$.

Three other distinct planes crystallographically equivalent to (111) exist in an FCC crystal: $(\bar{1}11)$, $(\bar{1}\bar{1}1)$, and $(1\bar{1}1)$. These four planes also belong to a family. In general, a family of planes is designated by curly brackets, such as {111}. An example of families of planes and directions is shown in Fig. A1.7.

Table A1.2 gives examples of slip planes and directions in three basic crystals: FCC, body-centered cubic, and hexagonal close-packed (HCP) (Kelly et al., 2000). Materials scientists find it convenient to use four indices to define planes and directions in HCP crystals, in which three indices are used to describe the hexagonal base (along $\mathbf{a}_1$, $\mathbf{a}_2$, and $\mathbf{a}_3$ in Fig. A1.1) and one index defines the height of the unit cell (along the **c**-axis). To recover the three-index notation, the first and third index should be added: $(a_1, a_2, a_3, d) \rightarrow (a_2, (a_1 + a_3), d)$ in HCP crystals. This three-index notation is better suited to the crystal-plasticity algorithm developed here.

In general, a pair of vectors is always needed to characterize dislocation motion: slip plane normal **n** and slip direction **s**. This pair of vectors defines a slip system (Table A1.2) and ultimately defines plasticity in crystalline materials. The pair of vectors must satisfy the

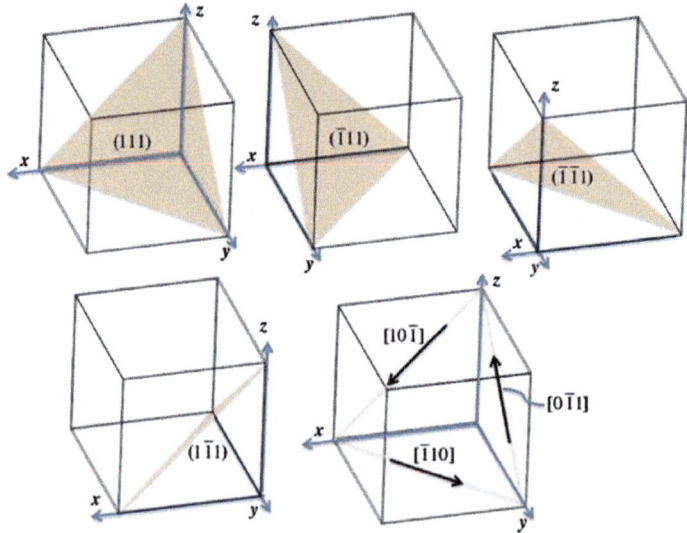

Figure A1.7
{111} family of planes in a cubic crystal, and $\langle 1\overline{1}0 \rangle$ family of directions.

Table A1.2: Slip Systems

Crystal	Slip Systems	Bravais Lattice
Face-centered cubic metals: Al, Cu ...	$\langle 1\overline{1}0 \rangle\{11\}$	F (Face-centered)
Body-centered cubic metals: Fe, W, Na ...	$\langle 1\overline{1}1 \rangle\{110\}$	I (Body-centered)
Hexagonal close-packed metals: Zn, Mg, Cd ...	$\langle 11\overline{2}0 \rangle(0001), \langle 11\overline{2}0 \rangle\{10\overline{1}1\}$	P (Primitive)
	$\langle 11\overline{2}0 \rangle\{10\overline{1}0\}, \langle 11\overline{2}0 \rangle\{11\overline{2}2\}$	

relation $\mathbf{n} \cdot \mathbf{s} = 0$ to form a slip system. Thus, [110] and (111) do not form a slip system because the dot product has a value of 2. On the other hand, the pair $\left[1\overline{1}0 \right]$ and (111), $\left[\overline{1}10 \right]$ and (111), or [110] and $\left(\overline{1}11 \right)$, and so on, can form a slip system because the dot product is 0.

In many applications, the slip systems of a crystal have a predetermined orientation relation with respect to the loading axis. For example, the cold rolling of metal billets may often induce the preferential crystallographic orientation of grains, which is referred to as texture (Van Houtte et al., 2005).

When loading a cold-rolled specimen, therefore, crystal orientations will have specific distributions with respect to the loading axis, a phenomenon that greatly contributes to anisotropic plastic deformation (Roters et al., 2010). This orientation effect can easily be captured in single-crystal plasticity by applying the following transformation to the slip

directions and normals of slip system a of the crystal $\left(\mathbf{s}_c^\alpha, \ \mathbf{n}_c^\alpha\right)$ to map them to the laboratory frame $\left(\mathbf{s}_l^\alpha, \ \mathbf{n}_l^\alpha\right)$:

$$\left[\mathbf{s}_l^\alpha, \mathbf{n}_l^\alpha\right] = \mathbf{T}_l^c\left[\mathbf{s}_c^\alpha, \mathbf{n}_c^\alpha\right] \qquad (A1.4.1)$$

where $\mathbf{T}_l^c$ is a 3D rotation matrix defined by Euler angle triplets $(\phi, \ \theta, \ \psi)$. Multiple conventions determine Euler angles, depending on the sequence of rotations. A common choice for crystallographers is the Bunge convention (Engler and Randle, 2009). A 3D rotation in this convention is assumed to arise from the sequence shown in Fig. A1.8: (1) a rotation ϕ about the Z-axis, which will change the orientation of vectors in the X−Y plane; (2) a rotation θ about the X-axis, which is now in a new position owing to the application of (1); and (3) a rotation ψ about the Z-axis in its new position after the application of (1) and (2). As such, the transformation matrix for a 3D rotation may be given by

$$\boldsymbol{T}_c^l = \begin{bmatrix} c_1 c_3 - c_2 s_1 s_3 & -c_1 s_3 - c_2 c_3 s_1 & s_1 s_2 \\ c_3 s_1 - c_1 c_2 s_3 & c_1 c_2 c_3 - s_1 s_2 & -c_1 s_2 \\ s_2 s_3 & c_3 s_2 & c_2 \end{bmatrix} \qquad (A1.4.2)$$

where $c_i = \cos(\delta_i), s_i = \sin(\delta_i)$, and d_i take the value of ϕ, θ, or ψ when $i = 1, 2,$ or 3, respectively. Euler angles may be read off experimentally obtained pole figures (Randle and Engler, 2009). Pole figures are 2D plots of orientation distributions of crystal planes in a material sample; they are briefly discussed in Section A1.5.

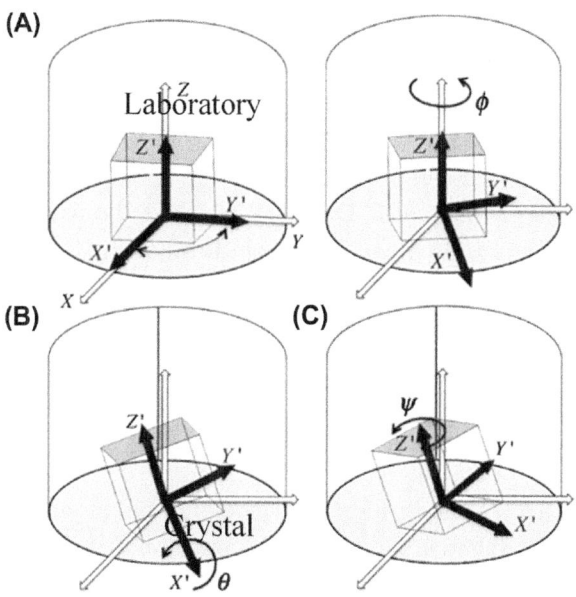

Figure A1.8
Euler angles that specify crystal orientation.

Table A1.1 lists the main equations in Sections A1.2—A1.4. The table generates slip-related input data for single-crystal plasticity material subroutines. The plasticity algorithm using these input data is developed in the remainder of this appendix.

There are situations in which a crystal's spatial orientation must follow that of its surrounding material. For example, precipitate crystals may nucleate and grow along energetically favorable planes (habit planes) and in specific directions of the surrounding matrix crystal, resulting in what is known as rational orientation relations (Wang and Starink, 2005). Such a case can also be modeled by single-crystal plasticity, provided the right sequence of plane and direction transformations is first applied (Elkhodary et al., 2011).

Example A1.1. Determine the slip planes and normals of an FCC crystal with Euler angles (0, 60, and 0 degrees) with respect to the sample axes.

Solution: From Table A1.3, the family of directions and planes of interest in FCC crystals is the $\langle 1\bar{1}0 \rangle \{111\}$. Only pairs satisfying the relation $\mathbf{n} \cdot \mathbf{s} = 0$ need be considered. There are only 12 independent such pairs if we identify positive slip and negative slip directions with the same system. FCC slip systems can be represented by the Miller indices in Table A1.3.

Table A1.3: Face-Centered Cubic Directions and Normals, Unrotated and With Euler Angles (0, 60, and 0 Degrees)

Slip System Number	Direction			Plane		
	Miller Indices	Vector s_c	Transformed Vector (s_t)	Miller Indices	Vector n_c	Transformed Vector (n_t)
1	$[\bar{1}01]$	$\frac{1}{\sqrt{2}}[-1,0,1]$	[−0.7071 −0.6124 0.3536]	(111)	$\frac{1}{\sqrt{3}}[1,1,1]$	[0.5774 −0.2113 0.7887]
2	$[\bar{1}10]$	$\frac{1}{\sqrt{2}}[-1,1,0]$	[−0.7071 0.3536 0.6124]	(111)	$\frac{1}{\sqrt{3}}[1,1,1]$	[0.5774 −0.2113 0.7887]
3	$[0\bar{1}1]$	$\frac{1}{\sqrt{2}}[0,-1,1]$	[0 −0.9659 −0.2588]	(111)	$\frac{1}{\sqrt{3}}[1,1,1]$	[0.5774 −0.2113 0.7887]
4	$[011]$	$\frac{1}{\sqrt{2}}[0,1,1]$	[0 −0.2588 0.9659]	$(\bar{1}\bar{1}1)$	$\frac{1}{\sqrt{3}}[-1,-1,1]$	[−0.5774 −0.7887 −0.2113]
5	$[\bar{1}10]$	$\frac{1}{\sqrt{2}}[-1,1,0]$	[−0.7071 0.3536 0.6124]	$(\bar{1}\bar{1}1)$	$\frac{1}{\sqrt{3}}[-1,-1,1]$	[−0.5774 −0.7887 −0.2113]

Continued

Table A1.3: Face-Centered Cubic Directions and Normals, Unrotated and With Euler Angles (0, 60, and 0 Degrees)—cont'd

	Direction			Plane		
Slip System Number	Miller Indices	Vector s_c	Transformed Vector (s_t)	Miller Indices	Vector n_c	Transformed Vector (n_t)
6	$[101]$	$\frac{1}{\sqrt{2}}[1,0,1]$	$[0.7071$ -0.6124 $0.3536]$	$(\bar{1}\bar{1}1)$	$\frac{1}{\sqrt{3}}[-1,-1,1]$	$[-0.5774$ -0.7887 $-0.2113]$
7	$[101]$	$\frac{1}{\sqrt{2}}[1,0,1]$	$[0.7071$ -0.6124 $0.3536]$	$(\bar{1}11)$	$\frac{1}{\sqrt{3}}[-1,1,1]$	$[-0.5774$ -0.2113 $0.7887]$
8	$[110]$	$\frac{1}{\sqrt{2}}[1,1,0]$	$[0.7071$ 0.3536 $0.6124]$	$(\bar{1}11)$	$\frac{1}{\sqrt{3}}[-1,1,1]$	$[-0.5774$ -0.2113 $0.7887]$
9	$[0\bar{1}1]$	$\frac{1}{\sqrt{2}}[0,-1,1]$	$[0$ -0.9659 $-0.2588]$	$(\bar{1}11)$	$\frac{1}{\sqrt{3}}[-1,1,1]$	$[-0.5774$ -0.2113 $0.7887]$
10	$[011]$	$\frac{1}{\sqrt{2}}[0,1,1]$	$[0$ -0.2588 $0.9659]$	$(1\bar{1}1)$	$\frac{1}{\sqrt{3}}[1,-1,1]$	$[0.5774$ -0.7887 $-0.2113]$
11	$[110]$	$\frac{1}{\sqrt{2}}[1,1,0]$	$[0.7071$ 0.3536 $0.6124]$	$(1\bar{1}1)$	$\frac{1}{\sqrt{3}}[1,-1,1]$	$[0.5774$ -0.7887 $-0.2113]$
12	$[\bar{1}01]$	$\frac{1}{\sqrt{2}}[-1,0,1]$	$[-0.7071$ -0.6124 $0.3536]$	$(1\bar{1}1)$	$\frac{1}{\sqrt{3}}[1,-1,1]$	$[0.5774$ -0.7887 $-0.2113]$

The slip vectors (s_c) are obtained by simply normalizing the direction Miller indices to unity. Slip normals (n_c) in the table are obtained for FCC (that is, cubic) crystals by applying the equation: $\mathbf{n}^{hkl} = \left(h^2 + k^2 + l^2\right)^{-\frac{1}{2}}[h\mathbf{e}_1 + k\mathbf{e}_2 + l\mathbf{e}_3]$, according to Step 6 in Table A1.1. Using $\mathbf{T}_l^C$ as defined in Eq. (A1.4.2) with Euler angles (0, $\pi/3$, 0), we find

$$\mathbf{T}_l^c = \begin{bmatrix} 1 & 0 & 0 \\ 0 & 0.5 & -0.866 \\ 0 & 0.866 & 0.5 \end{bmatrix}$$

Finally, according to Step 7 in Table A1.1, s_l and n_l in Table A1.3 are obtained.

A1.5 Euler Angles From Pole Figures

In this section, we briefly outline the utility of pole figures and how to determine Euler angles from them. For a more extended discussion, see Kelly et al. (2000) and Cullity and Stock (2001).

For a polycrystalline aggregate, each grain possesses a different crystallographic orientation, as indicated in Fig. A1.9. To plot these different crystal orientations on a 2D plot, stereographic projection is used. To identify a given plane, say the (001) plane, in each crystal, draw a unit normal to that plane. Because the plane could be oriented along any direction in 3D space, its unit normal could intersect any point on a unit sphere, as shown in Fig. A1.9B–D. From that intersection point a line is drawn to the south pole of the sphere, designated by an "S" in the figure. Then, the intersection of that line with the equatorial plane, designated by a black spot, would represent the crystal's orientation on what becomes called a pole figure (i.e., the equatorial circle with the black spots in it) (Fig. A1.9E and F).

For normal (poles) that point beneath the equator, a line from the sphere's surface to the north pole (N) is drawn instead of the south pole, and thus a separate pole figure of the lower hemisphere is generated. The pole figure plotted in Fig. A1.9E includes (as gray circles) the (001) planes of Crystals 1, 2, and 3 shown in Fig. A1.9B–D. It further shows a uniform distribution of spots, indicating no preferential orientation in space for a

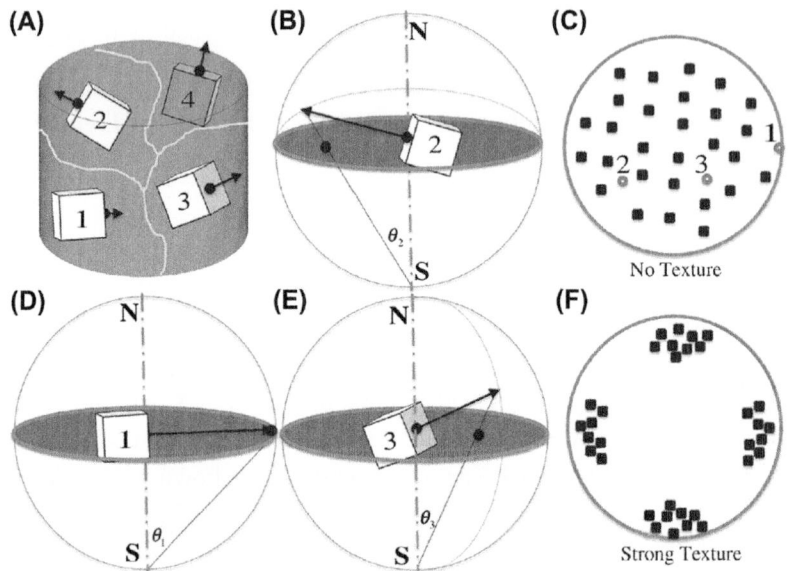

Figure A1.9
(A) Polycrystalline aggregate shown for grains; (B–D) Finding the stereographic projections of Grains 1–3; (E) A plot figure showing no texture; (F) A plot figure showing strong texture.

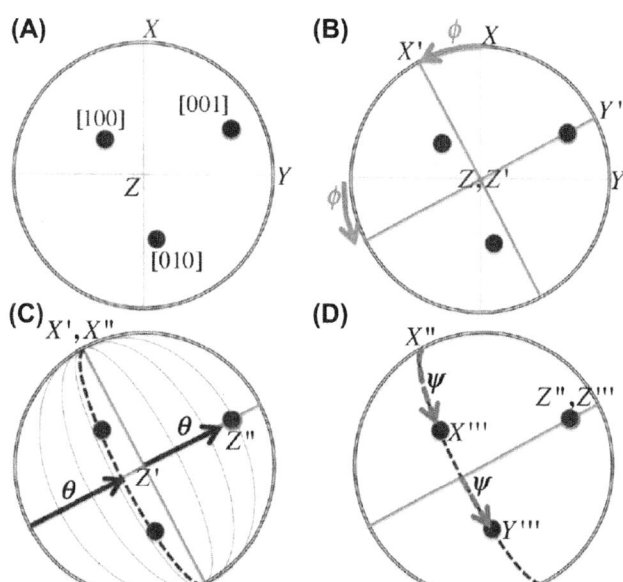

Figure A1.10
Finding Euler angles from a pole figure.

hypothetical polycrystal aggregate. Such a case is referred to as no texture. On the other hand, Fig. A1.9F shows clustering of spots in specific regions of the pole figure, which indicates a preferential orientation in space for crystals in a specimen: that is, texture. For example, the latter is common observed when metals are processed by rolling.

It is possible to extract the distribution of Euler angles for the corresponding planes from a pole figure, which is required in Section 4.2 in Chapter 4 to align slip directions and normal correctly in the single-crystal plasticity algorithm explained in Table A1.1. A representative procedure is shown in Fig. A1.10.

Crystallographically equivalent plane normal are identified, [001], [010] and [100], for this example. First, a rotation ϕ about the Z-axis is applied to make the Y-axis pass through the [001] pole, as shown in Fig. A1.10B, thereby creating new axes, X′, Y′, and Z′. Second, a rotation θ is applied about the X′-axis to bring the Z′-axis in coincidence with the [001] pole. The Z′-axis is here renamed Z″, and the X′-axis, X″, as shown in Fig. A1.10C. The plane whose normal is the Z″-axis is then identified with the major circle found from the outer diameter by the same angle θ, as shown in Fig. A1.10C. Finally, a rotation ψ is applied about the Z″-axis to bring the X″-axis in coincidence with the [100] pole, which is now labeled X‴. At this point, the crystal orientation is fully determined by three Euler angles measured from the given pole figure, and the [010] pole may be identified with Y‴-axis and the [001] pole with the Z‴-axis.

References

Abdolrahim, N., Zbib, H.M., Bahr, D.F., 2014. Multiscale modeling and simulation of deformation in nanoscale metallic multilayer systems. International Journal of Plasticity 52, 33−50.

Acharya, A., Bassani, J.L., 1995. On non-local flow theories that preserve the classical structure of incremental boundary value problem. In: IUTAM Symposium, Paris, 1995, Aug 29−Sept 1.

Acharya, A., 2001. A model of crystal plasticity based on the theory of continuously distributed dislocations. Journal of the Mechanics and Physics of Solids 49, 761−784.

Agnihotri, P.K., Van der Giessen, E., 2015. On the rate sensitivity in discrete dislocation plasticity. Mechanics of Materials 90, 37−46.

Aifantis, K.E., Ngan, A.H.W., 2007. Modeling dislocation—grain boundary interactions through gradient plasticity and nanoindentation. Materials Science and Engineering: A 459, 251−261.

Aifantis, E.C., 1984. On the microstructural origin of certain inelastic models. Transactions of ASME Journal of Engineering Materials and Technology 106, 326−330.

Aifantis, E.C., 1987. The physics of plastic deformation. International Journal of Plasticity 3, 211−247.

Akarapu, S., Zbib, H., Bahr, D., 2010. Analysis of heterogeneous deformation and dislocation dynamics in single crystal micropillars under compression. International Journal of Plasticity 26, 239−257.

Akasheh, F., Zbib, H.M., Ohashi, T., 2007. Multiscale modelling of size effect in fcc crystals: discrete dislocation dynamics and dislocation-based gradient plasticity. Philosophical Magazine 87, 1307−1326.

Allain, S., Bouaziz, O., 2010. A quantitative modeling of the unloading behavior of metals during a tensile test. International Journal of Materials Research 101, 1497−1502.

Alpay, S., Misirlioglu, I., Nagarajan, V., Ramesh, R., 2004. Can interface dislocations degrade ferroelectric properties? Applied Physics Letters 85, 2044−2046.

Amelinckx, S., Bontinck, W., Dekeyser, W., Seitz, F., 1957. On the formation and properties of helical dislocations. Philosophical Magazine 2, 355−378.

An, Q., et al., 2008. Melting of Cu under hydrostatic and shock wave loading to high pressures. Journal of Physics: Condensed Matter 20 (9), 1−8.

Argaman, N., Levy, O., Makov, G., 2001. Dislocation pattern formation—simulations of annealing in two dimensions. Le Journal de Physique IV 11. Pr5-61-Pr65-68.

Armstrong, R.W., Li, Q., 2015. Dislocation mechanics of high-rate deformations. Metallurgical and Materials Transactions A 46 (10), 4438−4453.

Armstrong, R., Walley, S., 2008. High strain rate properties of metals and alloys. International Materials Reviews 53, 105−128.

Armstrong, R.W., et al., 2007. Dislocation mechanics of shock-induced plasticity. Metallurgical and Materials Transactions A 38 (11), 2605−2610.

Arsenlis, A., Cai, W., Tang, M., Rhee, M., Oppelstrup, T., Hommes, G., Pierce, T.G., Bulatov, V.V., 2007. Enabling strain hardening simulations with dislocation dynamics. Modelling and Simulation in Materials Science and Engineering 15, 553.

Arsenlis, A., Rhee, M., Hommes, G., Cook, R., Marian, J., 2012. A dislocation dynamics study of the transition from homogeneous to heterogeneous deformation in irradiated body-centered cubic iron. Acta Materialia 60, 3748−3757.

Arzt, E., 1998. Size effects in materials due to microstructural and dimensional constraints: a comparative review. Acta Materialia 46, 5611−5626.

Asaro, R.J., Lubarda, V.A., 2006. Mechanics of Solids and Materials. Cambridge University Press.

Asaro, R.J., Needleman, A., 1985. Overview no. 42. Texture development and strain-hardening in rate dependent polycrystals. Acta Metallurgica 33, 923−953.

Asaro, R.J., Rice, J., 1977. Strain localization in ductile single crystals. Journal of the Mechanics and Physics of Solids 25, 309−338.

Asaro, R.J., 1983. Micromechanics of crystals and polycrystals. Advances in Applied Mechanics 23, 1−115.

Ashby, M.F., 1970. The deformation of plastically non-homogeneous materials. Philosophical Magazine 21, 399−424.

Atkinson, A., Leppington, F.G., 1977. The effect of couple stresses on the tip of a crack. International Journal of Solids and Structures 13, 1103−1122.

Aubry, S., Rhee, M., Hommes, G., Bulatov, V., Arsenlis, A., 2016. Dislocation dynamics in hexagonal close-packed crystals. Journal of the Mechanics and Physics of Solids 94, 105−126.

Austin, R.A., McDowell, D.L., 2011. A dislocation-based constitutive model for viscoplastic deformation of fcc metals at very high strain rates. International Journal of Plasticity 27 (1), 1−24.

Austin, R.A., McDowell, D.L., 2012. Parameterization of a rate-dependent model of shock-induced plasticity for copper, nickel, and aluminum. International Journal of Plasticity 32−33, 134−154.

Ayas, C., Deshpande, V., 2015. Climb enabled discrete dislocation plasticity of superalloys. Key Engineering Materials 651−653, 981−986.

Ayas, C., Deshpande, V.S., Geers, M.G.D., 2012. Tensile response of passivated films with climb-assisted dislocation glide. Journal of the Mechanics and Physics of Solids 60, 1626−1643.

Ayas, C., van Dommelen, J.A.W., Deshpande, V.S., 2014. Climb-enabled discrete dislocation plasticity. Journal of the Mechanics and Physics of Solids 62, 113−136.

Bakó, B., Groma, I., Györgyi, G., Zimányi, G., 2006. Dislocation patterning: the role of climb in meso-scale simulations. Computational Materials Science 38, 22−28.

Bakó, B., Clouet, E., Dupuy, L.M., Blétry, M., 2011. Dislocation dynamics simulations with climb: kinetics of dislocation loop coarsening controlled by bulk diffusion. Philosophical Magazine 91, 3173−3191.

Bardeen, J., Herring, C., Shockley, W., 1952. Imperfections in Nearly Perfect Crystals. Wiley, New York, p. 281.

Barnett, D., 1985. The displacement field of a triangular dislocation loop. Philosophical Magazine A 51, 383−387.

Bayley, C., Brekelmans, W., Geers, M., 2006. A comparison of dislocation induced back stress formulations in strain gradient crystal plasticity. International Journal of Solids and Structures 43, 7268−7286.

Beanland, R., 1995. Dislocation multiplication mechanisms in low-misfit strained epitaxial layers. Journal of Applied Physics 77, 6217−6222.

Begley, M.R., Hutchinson, J.W., 1997. The mechanics of size-dependent indentation. Journal of the Mechanics and Physics of Solids 46, 2049−2068.

Bei, H., Shim, S., Miller, M.K., Pharr, G.M., George, E.P., 2007. Effects of focused ion beam milling on the nanomechanical behavior of a molybdenum-alloy single crystal. Applied Physics Letters 91, 111915.

Bei, H., Shim, S., Pharr, G., George, E., 2008. Effects of pre-strain on the compressive stress-strain response of Mo-alloy single-crystal micropillars. Acta Materialia 56, 4762−4770.

Belytschko, T., Gracie, R., 2007. On XFEM applications to dislocations and interfaces. International Journal of Plasticity 23, 1721−1738.

Belytschko, T., Liu, W.K., Moran, B., Elkhodary, K.I., 2014. Nonlinear Finite Element for Continua and Structures, second ed. John Wiley & Sons, Ltd.

Bennett, S.E., 2010. Dislocations and their reduction in GaN. Materials Science and Technology 26, 1017−1028.

Benzerga, A., Brechet, Y., Needleman, A., Giessen, E., 2004. Incorporating three-dimensional mechanisms into two-dimensional dislocation dynamics. Modelling and Simulation in Materials Science and Engineering 12, 159.

Benzerga, A., 2008. An analysis of exhaustion hardening in micron-scale plasticity. International Journal of Plasticity 24, 1128−1157.

Benzerga, A.A., 2009. Micro-pillar plasticity: 2.5D mesoscopic simulations. Journal of the Mechanics and Physics of Solids 57, 1459—1469.

Bertin, N., Tomé, C., Beyerlein, I., Barnett, M., Capolungo, L., 2014. On the strength of dislocation interactions and their effect on latent hardening in pure magnesium. International Journal of Plasticity 62, 72—92.

Bittencourt, E., Needleman, A., Gurtin, M.E., Van der Giessen, E., 2003. A comparison of nonlocal continuum and discrete dislocation plasticity predictions. Journal of the Mechanics and Physics of Solids 51, 281—310.

Bonneville, J., Escaig, B., Martin, J., 1988. A study of cross-slip activation parameters in pure copper. Acta Metallurgica 36, 1989—2002.

Bontinck, W., Amelinckx, S., 1957. Observation of helicoidal dislocation lines in fluorite crystals. Philosophical Magazine 2, 94—96.

Borg, U., 2007. A strain gradient crystal plasticity analysis of grain size effects in polycrystals. European Journal of Mechanics - A/Solids 26, 313—324.

Brechet, Y., Canova, G., Kubin, L.P., 1996. Strain softening, slip localization and propagation: from simulations to continuum modelling. Acta Materialia 44, 4261—4271.

Bridgman, P., 1949. The Physics of High Pressure. Bell and Sons, London.

Brinckmann, S., Van der Giessen, E., 2004. A discrete dislocation dynamics study aiming at understanding fatigue crack initiation. Materials Science and Engineering: A 387, 461—464.

Brinckmann, S., Kim, J.-Y., Greer, J.R., 2008. Fundamental differences in mechanical behavior between two types of crystals at the nanoscale. Physical Review Letters 100, 155502.

Bringa, E.M., et al., 2006. Shock deformation of face-centred-cubic metals on subnanosecond timescales. Nature Materials 5 (10), 805—809.

Bulatov, V., Cai, W., 2006. Computer Simulations of Dislocations. Oxford University Press, New York.

Bulatov, V., 2002. Current developments and trends in dislocation dynamics. Journal of Computer-Aided Materials Design 9 (2), 133—144.

Bullough, R., Newman, R.C., 1970. The Kinetics of Migration of Point Defects to Dislocations.

Bushby, A.J., Dunstan, D.J., 2004. Plasticity size effects in nanoindentation. Journal of Materials Research 19, 137—142.

Cai, W., Arsenlis, A., Weinberger, C., Bulatov, V., 2006. A non-singular continuum theory of dislocations. Journal of the Mechanics and Physics of Solids 54, 561—587.

Cailard, D., Martin, J.L., 2003. Thermally Activated Mechanisms in Crystal Plasticity. Elsevier Science, Amsterdam.

Cao, B., et al., 2005. Effect of shock compression method on the defect substructure in monocrystalline copper. Materials Science and Engineering: A 409 (1—2), 270—281.

Cao, B., et al., 2007. Shock compression of monocrystalline copper: atomistic simulations. Metallurgical and Materials Transactions A 38 (11), 2681—2688.

Cao, B., et al., 2010. Shock compression of monocrystalline copper: experiments, characterization, and analysis. Materials Science and Engineering: A 527 (3), 424—434.

Caslavsky, J.L., Gazzara, C.P., 1971. The observation of helical dislocations in sapphire. Journal of Materials Science 6, 1139—1141.

Chaussidon, J., Robertson, C., Rodney, D., Fivel, M., 2008. Dislocation dynamics simulations of plasticity in Fe laths at low temperature. Acta Materialia 56, 5466—5476.

Chen, L.Q., Khachaturyan, A.G., 1991. Computer simulation of structural transformations during precipitation of an ordered intermetallic phase. Acta Metallurgica et Materialia 39, 2533—2551.

Chen, J.Y., Huang, Y., Ortiz, M., 1998. Fracture analysis of cellular materials: a strain gradient model. Journal of the Mechanics and Physics of Solids 46 (5), 789—828.

Chen, Y.J., Meyers, M.A., Nesterenko, V.F., 1999. Spontaneous and forced shear localization in high-strain-rate deformation of tantalum. Materials Science and Engineering: A 268, 70—82.

Chen, Z., et al., 2014. Effects of copper nanoparticle inclusions on pressure-induced fluid-polynanocrystalline structural transitions in krypton. Journal of Applied Physics 116 (23), 233506.

Chen, L.Y., Xu, J.Q., Choi, H., Pozuelo, M., Ma, X., Bhowmick, S., Yang, J.-M., Mathaudhu, S., Li, X.-C., 2015. Processing and properties of magnesium containing a dense uniform dispersion of nanoparticles. Nature 528, 539.

Cheng, G.J., Shehadeh, M.A., 2006. Multiscale dislocation dynamics analyses of laser shock peening in silicon single crystals. International Journal of Plasticity 22 (12), 2171–2194.

Chu, H.J., Wang, J., Zhou, C.Z., Beyerlein, I.J., 2011. Self-energy of elliptical dislocation loops in anisotropic crystals and its application for defect-free core/shell nanowires. Acta Materialia 59, 7114–7124.

Chu, H.J., Pan, E., Han, X., Wang, J., Beyerlein, I.J., 2012. Elastic fields of dislocation loops in three-dimensional anisotropic bimaterials. Journal of the Mechanics and Physics of Solids 60, 418–431.

Chu, H.J., Zhou, C.Z., Wang, J., Beyerlein, I.J., 2013. An analytical model for the critical shell thickness in core/shell nanowires based on crystallographic slip. Journal of the Mechanics and Physics of Solids 61, 2147–2160.

Clauset, A., Shalizi, C., Newman, M., 2009. Power-Law distributions in empirical data. SIAM Review 51, 661–703.

Cleveringa, H.H.M., Van Der Giessen, E., Needleman, A., 1997. Comparison of discrete dislocation and continuum plasticity predictions for a composite material. Acta Materialia 45, 3163–3179.

Cleveringa, H., Van der Giessen, E., Needleman, A., 2000. A discrete dislocation analysis of mode I crack growth. Journal of the Mechanics and Physics of Solids 48, 1133–1157.

Clouet, E., 2006. The vacancy–edge dislocation interaction in fcc metals: a comparison between atomic simulations and elasticity theory. Acta Materialia 54, 3543–3552.

Clouet, E., 2011. Predicting dislocation climb: classical modeling versus atomistic simulations. Physical Review B 84.

Cottrell, A.H., 1964. The Mechanical Properties of Materials. Wiley Interscience, New York.

Council, N.R., 2008. Integrated Computational Materials Engineering: A Transformational Discipline for Improved Competitiveness and National Security. National Academies Press.

Crone, J.C., Chung, P.W., Leiter, K.W., Knap, J., Aubry, S., Hommes, G., Arsenlis, A., 2014. A multiply parallel implementation of finite element-based discrete dislocation dynamics for arbitrary geometries. Modelling and Simulation in Materials Science and Engineering 22, 035014.

Crosby, T., Po, G., Ghoniem, N.M., 2014. Modeling plastic deformation of post-irradiated copper micro-pillars. Journal of Nuclear Materials 455, 126–129.

Csikor, F.F., Motz, C., Weygand, D., Zaiser, M., Zapperi, S., 2007. Dislocation avalanches, strain bursts, and the problem of plastic forming at the micrometer scale. Science 318, 251–254.

Cui, Y.N., Liu, Z.L., Zhuang, Z., 2013. Dislocation multiplication by single cross slip for FCC at submicron scales. Chinese Physics Letters 30, 046103.

Cui, Y., Lin, P., Liu, Z., Zhuang, Z., 2014. Theoretical and numerical investigations of single arm dislocation source controlled plastic flow in FCC micropillars. International Journal of Plasticity 55, 279–292.

Cui, Y.N., Liu, Z.L., Zhuang, Z., 2015a. Theoretical and numerical investigations on confined plasticity in micropillars. Journal of the Mechanics and Physics of Solids 76, 127–143.

Cui, Y., Liu, Z., Zhuang, Z., 2015b. Quantitative investigations on dislocation based discrete-continuous model of crystal plasticity at submicron scale. International Journal of Plasticity 69, 54–72.

Cui, Y.N., Liu, Z.L., Wang, Z.J., Zhuang, Z., 2016a. Mechanical annealing under low-amplitude cyclic loading in micropillars. Journal of the Mechanics and Physics of Solids 89, 1–15.

Cui, Y., Po, G., Ghoniem, N., 2016b. Controlling strain bursts and avalanches at the nano-to micrometer scale. Physical Review Letters 117, 155502.

Cui, Y., Po, G., Ghoniem, N., 2016c. Temperature insensitivity of the flow stress in body-centered cubic micropillar crystals. Acta Materialia 108, 128–137.

Cui, Y., Po, G., Ghoniem, N., 2017a. Does irradiation enhance or inhibit strain bursts at the submicron scale? Acta Materialia 132, 285–297.

Cui, Y., Po, G., Ghoniem, N., 2017b. Influence of loading control on strain bursts and dislocation avalanches at the nanometer and micrometer scale. Physical Review B 95, 064103.

Cui, Y., Po, G., Ghoniem, N.M., 2018. A coupled dislocation dynamics-continuum barrier field model with application to irradiated materials. International Journal of Plasticity 104, 54–67.

Cui, Y., 2016. The Investigation of Plastic Behavior by Discrete Dislocation Dynamics for Single Crystal Pillar at Submicron Scale. Springer.

Dahmen, K.A., Ben-Zion, Y., Uhl, J.T., 2009. Micromechanical model for deformation in solids with universal predictions for stress-strain curves and slip avalanches. Physical Review Letters 102, 175501.

Damian, C., Swift, C.R.R., 2006. Laser-Induced Mach waves for ultra-high-pressure experiments. AIP Conference Proceedings 845, 1297–1300.

Danas, K., Deshpande, V.S., 2013. Plane-strain discrete dislocation plasticity with climb-assisted glide motion of dislocations. Modelling and Simulation in Materials Science and Engineering 21, 045008.

Davoudi, K.M., Nicola, L., Vlassak, J.J., 2012. Dislocation climb in two-dimensional discrete dislocation dynamics. Journal of Applied Physics 111, 103522.

Davoudi, K., 2017. Temperature dependence of the yield strength of aluminum thin films: multiscale modeling approach. Scripta Materialia 131, 63–66.

Dayeh, S.A., Tang, W., Boioli, F., Kavanagh, K.L., Zheng, H., Wang, J., Mack, N.H., Swadener, G., Huang, J.Y., Miglio, L., Tu, K.-N., Picraux, S.T., 2013. Direct measurement of coherency limits for strain relaxation in heteroepitaxial core/shell nanowires. Nano Letters 13, 1869–1876.

De Guzman, M.S., Neubauer, G., Flinn, P., et al., 1993. The role of indentation depth on the measured hardness of materials. Materials Research Society Symposium Proceedings 308, 613–618.

de Wi, R., 1960. The Continuum Theory of Stationary Dislocations, Solid State Physics. Elsevier, pp. 249–292.

de Wit, R., 1959. Self-energy of a helical dislocation. Physical Review 116, 592–597.

Dekel, E., et al., 1998. Spallation model for the high strain rates range. Journal of Applied Physics 84 (9), 4851.

Déprés, C., Robertson, C., Fivel, M., 2004. Crack initiation in fatigue: experiments and three-dimensional dislocation simulations. Materials Science and Engineering: A 387, 288–291.

Déprés, C., Fivel, M., Tabourot, L., 2008. A dislocation-based model for low-amplitude fatigue behaviour of face-centred cubic single crystals. Scripta Materialia 58, 1086–1089.

Derlet, P.M., Maaß, R., 2013. Micro-plasticity and intermittent dislocation activity in a simplified micro-structural model. Modelling and Simulation in Materials Science and Engineering 21, 035007.

Derlet, P.M., Maaß, R., 2014. A probabilistic explanation for the size-effect in crystal plasticity. Philosophical Magazine 1–16.

Deshpande, V.S., Needleman, A., Van der Giessen, E., 2003. Finite strain discrete dislocation plasticity. Journal of the Mechanics and Physics of Solids 51, 2057–2083.

Deshpande, V.S., Needleman, A., Van der Giessen, E., 2005. Plasticity size effects in tension and compression of single crystals. Journal of the Mechanics and Physics of Solids 53, 2661–2691.

Devincre, B., Condat, M., 1992. Model validation of a 3D simulation of dislocation dynamics: discretization and line tension effects. Acta Metallurgica et Materialia 40, 2629–2637.

Devincre, B., Kubin, L., Hoc, T., 2006. Physical analyses of crystal plasticity by DD simulations. Scripta Materialia 54, 741–746.

Devincre, B., Hoc, T., Kubin, L., 2008. Dislocation mean free paths and strain hardening of crystals. Science 320, 1745–1748.

Devincre, B., Madec, R., Monnet, G., Queyreau, S., Gatti, R., Kubin, L., 2011. Modeling crystal plasticity with dislocation dynamics simulations: the 'microMegas' code. In: Mechanics of Nano-Objects, pp. 81–100.

Devincre, B., 1996. Meso-scale simulation of the dislocation dynamics. In: NATO ASI Series E Applied Sciences, vol. 308. Advanced Study Institute, pp. 309–324.

Diaz de la Rubia, T., Zbib, H.M., Khraishi, T.A., Wirth, B.D., Victoria, M., Caturla, M.J., 2000. Multiscale modelling of plastic flow localization in irradiated materials. Nature 406, 871–874.

Dimiduk, D.M., Uchic, M.D., Parthasarathy, T.A., 2005. Size-affected single-slip behavior of pure nickel microcrystals. Acta Materialia 53, 4065–4077.

Dimiduk, D.M., Woodward, C., LeSar, R., Uchic, M.D., 2006. Scale-free intermittent flow in crystal plasticity. Science 312, 1188–1190.

Dimiduk, D., Nadgorny, E., Woodward, C., Uchic, M., Shade, P., 2010. An experimental investigation of intermittent flow and strain burst scaling behavior in LiF crystals during microcompression testing. Philosophical Magazine 90, 3621–3649.

Duesbery, M., Louat, N., Sadananda, K., 1992. The numerical simulation of continuum dislocations. Philosophical Magazine A 65, 311–325.

Dunstan, D., Bushby, A., 2013. The scaling exponent in the size effect of small scale plastic deformation. International Journal of Plasticity 40, 152–162.

Dunstan, D., Bushby, A., 2014. Grain size dependence of the strength of metals: the Hall–Petch effect does not scale as the inverse square root of grain size. International Journal of Plasticity 53, 56–65.

Dupuy, L., Fivel, M., 2002. A study of dislocation junctions in FCC metals by an orientation dependent line tension model. Acta Materialia 50, 4873–4885.

Edington, J.W., 1969. The influence of strain rate on the mechanical properties and dislocation substructure in deformed copper single crystals. Philosophical Magazine 19, 1189–1206.

El-Awady, J.A., Bulent Biner, S., Ghoniem, N.M., 2008. A self-consistent boundary element, parametric dislocation dynamics formulation of plastic flow in finite volumes. Journal of the Mechanics and Physics of Solids 56, 2019–2035.

El-Awady, J.A., Wen, M., Ghoniem, N.M., 2009a. The role of the weakest-link mechanism in controlling the plasticity of micropillars. Journal of the Mechanics and Physics of Solids 57, 32–50.

El-Awady, J.A., Woodward, C., Dimiduk, D.M., Ghoniem, N.M., 2009b. Effects of focused ion beam induced damage on the plasticity of micropillars. Physical Review B 80, 104104.

El-Awady, J.A., Rao, S.I., Woodward, C., Dimiduk, D.M., Uchic, M.D., 2011. Trapping and escape of dislocations in micro-crystals with external and internal barriers. International Journal of Plasticity 27, 372–387.

El-Awady, J.A., Uchic, M.D., Shade, P.A., Kim, S.-L., Rao, S.I., Dimiduk, D.M., Woodward, C., 2013. Pre-straining effects on the power-law scaling of size-dependent strengthening in Ni single crystals. Scripta Materialia 68, 207–210.

Elkhodary, K., Lee, W., Suna, L.P., et al., 2011. Deformation mechanisms of an Ω precipitate in a high-strength aluminum alloy subjected to high strain rates. Journal of Materials Research 26 (4), 487–497.

El-Sherik, A.M., Erb, U., Palumbo, G., Aust, K.T., 1992. Deviations from hall-petch behaviour in as-prepared nanocrystalline nickel Scripta. Metallurgica et Materialia 27, 1185–1188.

Elssner, G., et al., 1994. The influence interface impurities on fracture energy of UHV diffusion bonded metal-ceramic bicrystals. Scripta Metallurgica et Materialia 31, 1037–1042.

Erel, C., Po, G., Ghoniem, N., 2017. Dependence of hardening and saturation stress in persistent slip bands on strain amplitude during cyclic fatigue loading. Philosophical Magazine 97, 2947–2970.

Espinosa, H., Panico, M., Berbenni, S., Schwarz, K., 2006. Discrete dislocation dynamics simulations to interpret plasticity size and surface effects in freestanding FCC thin films. International Journal of Plasticity 22, 2091–2117.

Estrin, Y., Kubin, L., 1986. Local strain hardening and nonuniformity of plastic deformation. Acta Metallurgica 34, 2455–2464.

Estrin, Y., Krausz, A., et al., 1996. Unified Constitutive Laws of Plastic Deformation. Academic Press, New York, p. 69.

Evans, A.G., Hutchinson, J.W., 2009. A critical assessment of theories of strain gradient plasticity. Acta Materialia 57, 1675–1688.

Evers, L., Brekelmans, W., Geers, M., 2004a. Scale dependent crystal plasticity framework with dislocation density and grain boundary effects. International Journal of Solids and Structures 41, 5209–5230.

Evers, L.P., Brekelmans, W.A.M., Geers, M.G.D., 2004b. Non-local crystal plasticity model with intrinsic SSD and GND effects. Journal of the Mechanics and Physics of Solids 52, 2379–2401.

Fan, H., Aubry, S., Arsenlis, A., El-Awady, J.A., 2015. The role of twinning deformation on the hardening response of polycrystalline magnesium from discrete dislocation dynamics simulations. Acta Materialia 92, 126–139.

Fan, H., Aubry, S., Arsenlis, A., El-Awady, J.A., 2016. Grain size effects on dislocation and twinning mediated plasticity in magnesium. Scripta Materialia 112, 50–53.

Fedelich, B., 2002. A microstructural model for the monotonic and the cyclic mechanical behavior of single crystals of superalloys at high temperatures. International Journal of Plasticity 18, 1–49.

Fertig, R.S., Baker, S.P., 2009. Simulation of dislocations and strength in thin films: a review. Progress in Materials Science 54, 874–908.

Fivel, M.C., Robertson, C.F., Canova, G.R., Boulanger, L., 1998. Three-Dimensional modeling of indent-induced plastic zone at a mesoscale. Acta Materialia 46 (17), 6183–6194.

Fleck, N.A., Hutchinson, J.W., 1993. A phenomenological theory for strain gradient effect in plasticity. Journal of the Mechanics and Physics of Solids 41, 1825–1857.

Fleck, N.A., Hutchinson, J.W., 1997. Strain gradient plasticity. In: Hutchinson, J.W., Wu, T.Y. (Eds.), Advanced Applied Mechanics, vol. 33. Academic Press, pp. 295–361.

Fleck, N.A., Shu, J., 1995. Microbuckle initiation in fiber composites: a finite element study. Journal of the Mechanics and Physics of Solids 43, 1887–1918.

Fleck, N.A., Muller, G.M., Ashby, M.F., Hutchinson, J.W., 1994. Strain gradient plasticity - theory and experiment. Acta Metallurgica et Materialia 42, 475–487.

Foiles, S.M., Baskes, M.I., Daw, M.S., 1986. Embedded-atom-method functions for the fcc metals Cu, Ag, Au, Ni, Pd, Pt, and their alloys. Physical Review B 33, 7983–7991.

Follansbee, P.S., Gray, G.T., 1991. Dynamic deformation of shock prestrained copper. Materials Science and Engineering: A 138 (1), 23–31.

Foreman, A.J.E., 1968. Junction reaction hardening by dislocation loops. Philosophical Magazine 17, 353–364.

Forest, S., 1998. Modeling slip, kink and shear banding in classical and generalized single crystal plasticity. Acta Materialia 46, 3265–3281.

Franciosi, P., Zaoui, A., 1982. Multislip tests on copper crystals: a junctions hardening effect. Acta Metallurgica 30, 2141–2151.

Frank, F., Van der Merwe, J., 1949. One-dimensional dislocations. II. Misfitting monolayers and oriented overgrowth. Proceedings of the Royal Society of London. Series A. Mathematical and Physical Sciences 198, 216–225.

Frank, F., 1958. Dislocation theory. Il Nuovo Cimento (1955–1965) 7, 386–413.

Freund, L.B., Suresh, S., 2003. Thin Film Materials. Cambridge University Press, Cambridge.

Freund, L.B., 2000. The mechanics of electronic materials. International Journal of Solids and Structures 37, 185–196.

Frick, C., Clark, B., Orso, S., Schneider, A., Arzt, E., 2008. Size effect on strength and strain hardening of small-scale [111] nickel compression pillars. Materials Science and Engineering: A 489, 319–329.

Friedman, N., Ito, S., Brinkman, B.A.W., Shimono, M., DeVille, R.E.L., Dahmen, K.A., Beggs, J.M., Butler, T.C., 2012a. Universal critical dynamics in high resolution neuronal avalanche data. Physical Review Letters 108, 208102.

Friedman, N., Jennings, A.T., Tsekenis, G., Kim, J.Y., Tao, M., Uhl, J.T., Greer, J.R., Dahmen, K.A., 2012b. Statistics of dislocation slip avalanches in nanosized single crystals show tuned critical behavior predicted by a simple mean field model. Physical Review Letters 109, 095507.

Fu, Y., Du, H., Huang, W., Zhang, S., Hu, M., 2004. TiNi-based thin films in MEMS applications: a review. Sensors and Actuators A: Physical 112, 395–408.

Gao, Y., Cocks, A.C.F., 2009. Thermodynamic variational approach for climb of an edge dislocation. Acta Mechanica Solida Sinica 22, 426–435.

Gao, H., Huang, Y., Nix, W.D., Hutchinson, J.W., 1999. Mechanism-based strain gradient plasticity-I. Theory. Journal of the Mechanics and Physics of Solids 47, 1239–1263.

Gao, Y., Liu, Z., You, X., Zhuang, Z., 2010. A hybrid multiscale computational framework of crystal plasticity at submicron scales. Computational Materials Science 49, 672–681.

Gao, Y., Zhuang, Z., Liu, Z.L., You, X.C., Zhao, X.C., Zhang, Z.H., 2011a. Investigations of pipe-diffusion-based dislocation climb by discrete dislocation dynamics. International Journal of Plasticity 27, 1055–1071.

Gao, Y., Zhuang, Z., You, X., 2011b. A hierarchical dislocation-grain boundary interaction model based on 3D discrete dislocation dynamics and molecular dynamics. Science China Physics, Mechanics and Astronomy 54, 625–632.

Gao, Y., Zhuang, Z., You, X., 2013. A study of dislocation climb model based on coupling the vacancy diffusion theory with 3D discrete dislocation dynamics. International Journal for Multiscale Computational Engineering 11, 59–69.

Gao, S., Fivel, M., Ma, A., Hartmaier, A., 2017. 3D discrete dislocation dynamics study of creep behavior in Ni-base single crystal superalloys by a combined dislocation climb and vacancy diffusion model. Journal of the Mechanics and Physics of Solids 102, 209–223.

García Ortega, M., Ramos de Debiaggi, S., Monti, A., 2002. Self-diffusion in FCC metals: static and dynamic simulations in aluminium and nickel. Physica Status Solidi (b) 234, 506–521.

Geers, M., Brekelmans, W., Bayley, C., 2007. Second-order crystal plasticity: internal stress effects and cyclic loading. Modelling and Simulation in Materials Science and Engineering 15, S133.

Geers, M.G.D., Cottura, M., Appolaire, B., Busso, E.P., Forest, S., Villani, A., 2014. Coupled glide-climb diffusion-enhanced crystal plasticity. Journal of the Mechanics and Physics of Solids 70, 136–153.

Geslin, P.-A., Appolaire, B., Finel, A., 2014. A phase field model for dislocation climb. Applied Physics Letters 104, 011903.

Geslin, P.A., Appolaire, B., Finel, A., 2015. Multiscale theory of dislocation climb. Physical Review Letters 115, 265501.

Ghoniem, N.M., Amodeo, R., 1988. Computer Simulation of Dislocation Pattern Formation. Trans Tech Publ.

Ghoniem, N.M., Sun, L., 1999. Fast-sum method for the elastic field of three-dimensional dislocation ensembles. Physical Review B 60, 128.

Ghoniem, N., Singh, B., Sun, L., de la Rubia, T.D., 2000a. Interaction and accumulation of glissile defect clusters near dislocations. Journal of Nuclear Materials 276, 166–177.

Ghoniem, N., Tong, S.-H., Sun, L., 2000b. Parametric dislocation dynamics: a thermodynamics-based approach to investigations of mesoscopic plastic deformation. Physical Review B 61, 913.

Ghoniem, N., Tong, S.-H., Singh, B., Sun, L., 2001. On dislocation interaction with radiation-induced defect clusters and plastic flow localization in fcc metals. Philosophical Magazine A 81, 2743–2764.

Giessen, E.V.D., Needleman, A., 1995. Discrete dislocation plasticity: a simple planar model. Modelling and Simulation in Materials Science and Engineering 3, 689–735.

Gil Sevillano, J., Ocana Arizcorreta, I., Kubin, L., 2001. Intrinsic size effects in plasticity by dislocation glide. Materials Science and Engineering: A 309, 393–405.

Gillis, P.P., Kratochvil, J., 1970. Dislocation acceleration. Philosophical Magazine 21 (170), 425–432.

Gillis, P.P., et al., 1969. Stress dependences of dislocation velocities. Philosophical Magazine 20 (164), 279–289.

Gilman, J.J., 1969. Micromechanics of Flow in Solids. McGraw-Hill.

Goldthorpe, I.A., Marshall, A.F., McIntyre, P.C., 2008. Synthesis and strain relaxation of Ge-core/Si-shell nanowire arrays. Nano Letters 8, 4081–4086.

Gracie, R., Ventura, G., Belytschko, T., 2007. A new fast finite element method for dislocations based on interior discontinuities. International Journal for Numerical Methods in Engineering 69, 423–441.

Gracie, R., Oswald, J., Belytschko, T., 2008. On a new extended finite element method for dislocations: core enrichment and nonlinear formulation. Journal of the Mechanics and Physics of Solids 56, 200–214.

Greer, J.R., De Hosson, J.T.M., 2011. Plasticity in small-sized metallic systems: intrinsic versus extrinsic size effect. Progress in Materials Science 56, 654–724.

Greer, J.R., Nix, W.D., 2006. Nanoscale gold pillars strengthened through dislocation starvation. Physical Review B 73 (24), 245410.

Greer, J.R., Oliver, W.C., Nix, W.D., 2005. Size dependence of mechanical properties of gold at the micron scale in the absence of strain gradients. Acta Materialia 53 (6), 1821–1830.

Greer, J., 2006. Bridging the gap between computational and experimental length scales: a review on nano-scale plasticity. Reviews on Advanced Materials Science 13, 59–70.

Greer, J.R., 2007. Effective use of focused ion beam (FIB) in investigating fundamental mechanical properties of metals at the sub-micron scale. In: Materials Research Society Symposium Proceedings. Cambridge University Press, pp. 0983-LL0908-0903.

Groh, S., Zbib, H., 2009. Advances in discrete dislocations dynamics and multiscale modeling. Journal of Engineering Materials and Technology 131, 041209.

Groh, S., Marin, E., Horstemeyer, M., Zbib, H., 2009. Multiscale modeling of the plasticity in an aluminum single crystal. International Journal of Plasticity 25, 1456–1473.

Groma, I., Csikor, F.F., Zaiser, M., 2003. Spatial correlations and higher-order gradient terms in a continuum description of dislocation dynamics. Acta Materialia 51, 1271–1281.

Groma, I., 1997. Link between the microscopic and mesoscopic length-scale description of the collective behavior of dislocations. Physical Review B 56, 5807–5813.

Gu, Y., El-Awady, J.A., 2018. Quantifying the effect of hydrogen on dislocation dynamics: a three-dimensional discrete dislocation dynamics framework. Journal of the Mechanics and Physics of Solids 112, 491–507.

Gu, R., Ngan, A., 2012. Effects of pre-straining and coating on plastic deformation of aluminum micropillars. Acta Materialia 60, 6102–6111.

Gu, R., Ngan, A.H.W., 2013. Dislocation arrangement in small crystal volumes determines power-law size dependence of yield strength. Journal of the Mechanics and Physics of Solids 61, 1531–1542.

Gu, Y., Xiang, Y., Quek, S.S., Srolovitz, D.J., 2015. Three-dimensional formulation of dislocation climb. Journal of the Mechanics and Physics of Solids 83, 319–337.

Gunkelmann, N., et al., 2016. Compaction and plasticity in nanofoams induced by shock waves: a molecular dynamics study. Computational Materials Science 119, 27–32.

Guo, Y., Zhuang, Z., Li, X., Chen, Z., 2007. An investigation of the combined size and rate effects on the mechanical responses of FCC metals. International Journal of Solids and Structures 44, 1180–1195.

Gurrutxaga-Lerma, B., et al., 2015a. The mechanisms governing the activation of dislocation sources in aluminum at different strain rates. Journal of the Mechanics and Physics of Solids 84, 273–292.

Gurson, A.L., 1977. Continuum theory of ductile rupture by void nucleation and growth: part I – yield criteria and flow rules for porous ductile media. Journal of Engineering Materials and Technology 99, 2–15.

Gurtin, M.E., Anand, L., Lele, S.P., 2007. Gradient single-crystal plasticity with free energy dependent on dislocation densities. Journal of the Mechanics and Physics of Solids 55, 1853–1878.

Gurtin, M.E., 2000. On the plasticity of single crystals: free energy, microforces, plastic-strain gradients. Journal of the Mechanics and Physics of Solids 48, 989–1036.

Gurtin, M.E., 2002. A gradient theory of single-crystal viscoplasticity that accounts for geometrically necessary dislocations. Journal of the Mechanics and Physics of Solids 50, 5–32.

Gurtin, M.E., 2006. The Burgers vector and the flow of screw and edge dislocations in finite-deformation single-crystal plasticity. Journal of the Mechanics and Physics of Solids 50 (1), 5–32.

Gurtin, M.E., 2008. A finite-deformation, gradient theory of single-crystal plasticity with free energy dependent on densities of geometrically necessary dislocations. International Journal of Plasticity 24, 702–725.

Gurtin, M.E., 2010. A finite-deformation, gradient theory of single-crystal plasticity with free energy dependent on the accumulation of geometrically necessary dislocations. International Journal of Plasticity 26, 1073–1096.

Guruprasad, P., Benzerga, A., 2008a. Size effects under homogeneous deformation of single crystals: a discrete dislocation analysis. Journal of the Mechanics and Physics of Solids 56, 132–156.

Guruprasad, P.J., Benzerga, A.A., 2008b. A phenomenological model of size-dependent hardening in crystal plasticity. Philosophical Magazine 88, 3585–3601.

Hafez Haghighat, S.M., Eggeler, G., Raabe, D., 2013. Effect of climb on dislocation mechanisms and creep rates in γ′-strengthened Ni base superalloy single crystals: a discrete dislocation dynamics study. Acta Materialia 61, 3709–3723.

Hall, E.O., 1951. The deformation and ageing of mild steel III: discussion of results. Proceedings of the Physical Society. Section B 64, 747–753.

Han, C.-S., Gao, H., Huang, Y., Nix, W.D., 2005a. Mechanism-based strain gradient crystal plasticity—I. Theory. Journal of the Mechanics and Physics of Solids 53, 1188—1203.

Han, C.-S., Gao, H., Huang, Y., Nix, W.D., 2005b. Mechanism-based strain gradient crystal plasticity—II. Analysis. Journal of the Mechanics and Physics of Solids 53, 1204—1222.

Hartmaier, A., Buehler, M.J., Gao, H., 2003. A Discrete Dislocation Plasticity Model of Creep in Polycrystalline Thin Films. Defect and Diffusion Forum. Trans Tech Publ, pp. 107—126.

Head, A., 1953a. The interaction of dislocations and boundaries. Philosophical Magazine 44, 92—94.

Head, A.K., 1953b. Edge dislocations in inhomogeneous media. Proceedings of the Physical Society. Section B 66, 793—801.

Hill, R., Rice, J.R., 1972. Constitutive analysis of elastic-plastic crystals at arbitrary strain. Journal of the Mechanics and Physics of Solids 20, 401—413.

Hill, R., 1950. The Mathematical Theory of Plasticity. Oxford University Press, Oxford.

Hiratani, M., Zbib, H.M., Khaleel, M.A., 2003. Modeling of thermally activated dislocation glide and plastic flow through local obstacles. International Journal of Plasticity 19, 1271—1296.

Hirth, J.P., Lothe, J., 1982. Theory of Dislocations. John Wiley and Sons, Inc., New York.

Hoivik, N.D., Elam, J.W., Linderman, R.J., Bright, V.M., George, S.M., Lee, Y., 2003. Atomic layer deposited protective coatings for micro-electromechanical systems. Sensors and Actuators A: Physical 103, 100—108.

Horstemeyer, M., Baskes, M., Plimpton, S., 2001. Length scale and time scale effects on the plastic flow of fcc metals. Acta Materialia 49, 4363—4374.

Hosemann, P., 2017. Small-scale mechanical testing on nuclear materials: bridging the experimental length-scale gap. Scripta Materialia 143, 161—168.

Hu, J.Q., Chen, Z., Liu, Z.L., Zhuang, Z., 2018. Pressure sensitivity of dislocation density in copper single crystals at submicron scale. Materials Research Express 5, 016504.

Hu, S.Y., Chen, L.Q., 2001. Solute segregation and coherent nucleation and growth near a dislocation - a phase-field model integrating defect and phase microstructures. Acta Materialia 49, 463—472.

Hu, J.Q., Liu, Z.L., Cui, Y.N., Wang, Z.J., Shan, Z.W., Zhuang, Z., 2014. Sensitive material behavior: theoretical model and experiment for compression collapse of gold particles at submicron scale. Journal of Applied Mechanics 81, 091007.

Hu, J.Q., Liu, Z.L., Van der Giessen, E., Zhuang, Z., 2017a. Strain rate effects on the plastic flow in submicron copper pillars: considering the influence of sample size and dislocation nucleation. Extreme Mechanics Letters 17, 33—37.

Hu, J.Q., Liu, Z.L., Chen, K.G., Zhuang, Z., 2017b. Investigations of shock-induced deformation and dislocation mechanism by a multiscale discrete dislocation plasticity model. Computational Materials Science 131, 78—85.

Huang, Y., Zhang, L., Guo, T.F., Hwang, K.C., 1997. Mixed mode near-tip fields for cracks in materials with strain gradient effects. Journal of the Mechanics and Physics of Solids 45, 439—465.

Huang, Y., Chen, J.Y., et al., 1999. Analytical and numerical studies on mode I and mode II fracture in elastic-plastic materials with strain gradient effects. International Journal of Fracture 100, 1—27.

Huang, Y., Gao, H., Nix, W.D., Hutchinson, J.W., 2000. Mechanism-based strain gradient plasticity—II. Analysis. Journal of the Mechanics and Physics of Solids 48, 99—128.

Huang, X., Hansen, N., Tsuji, N., 2006b. Hardening by annealing and softening by deformation in nanostructured metals. Science 312, 249—251.

Huang, M., Li, Z., Tong, J., 2014a. The influence of dislocation climb on the mechanical behavior of polycrystals and grain size effect at elevated temperature. International Journal of Plasticity 61, 112—127.

Huang, M., Tong, J., Li, Z., 2014b. A study of fatigue crack tip characteristics using discrete dislocation dynamics. International Journal of Plasticity 54, 229—246.

Hull, D., Bacon, D.J., 2001. Introduction to Dislocations. Butterworth-Heinemann.

Hull, D., Bacon, D.J., 2006. Introduction to Dislocations. Butterworth-Heinemann, Oxford.

Hull, D., Bacon, D.J., 2011. Introduction to Dislocations. Elsevier.

Hurtado, D.E., Ortiz, M., 2012. Surface effects and the size-dependent hardening and strengthening of nickel micropillars. Journal of the Mechanics and Physics of Solids 60, 1432—1446.

Lee, S.W., Han, S.M., Nix, W.D., 2009. Uniaxial compression of fcc Au nanopillars on an MgO substrate: the effects of prestraining and annealing. Acta Materialia 57, 4404–4415.

Lee, S.W., Jennings, A.T., Greer, J.R., 2013. Emergence of enhanced strengths and Bauschinger effect in conformally passivated copper nanopillars as revealed by dislocation dynamics. Acta Materialia 61, 1872–1885.

Lee, W.H., et al., 2015. High-velocity shock compression of SiC via molecular dynamics simulation. Computational Materials Science 98, 297–303.

Lee, S., Jeong, J., Kim, Y., Han, S.M., Kiener, D., Oh, S.H., 2016. FIB-induced dislocations in Al submicron pillars: annihilation by thermal annealing and effects on deformation behavior. Acta Materialia 110, 283–294.

Lee, E.H., 1969. Elastic-plastic deformation at finite strains. Journal of Applied Mechanics 36, 1–6.

Legros, M., Dehm, G., Arzt, E., Balk, T.J., 2008. Observation of giant diffusivity along dislocation cores. Science 319, 1646–1649.

Lei, L., Marin, J.L., Koslowski, M., 2013. Phase-field modeling of defect nucleation and propagation in domains with material inhomogeneities. Modelling and Simulation in Materials Science and Engineering 21.

Lemarchand, C., Devincre, B., Kubin, L.P., 2001. Homogenization method for a discrete-continuum simulation of dislocation dynamics. Journal of the Mechanics and Physics of Solids 49, 1969–1982.

Lepinoux, J., Kubin, L., 1987. The dynamic organization of dislocation structures: a simulation. Scripta Metallurgica 21, 833–838.

LeSar, R., 2014. Simulations of dislocation structure and response. Annual Review of Condensed Matter Physics 5, 375–407.

Li, X., Bhushan, B., 2003. Fatigue studies of nanoscale structures for MEMS/NEMS applications using nanoindentation techniques. Surface and Coatings Technology 163, 521–526.

Li, Z., Hou, C., Huang, M., Ouyang, C., 2009b. Strengthening mechanism in micro-polycrystals with penetrable grain boundaries by discrete dislocation dynamics simulation and Hall–Petch effect. Computational Materials Science 46, 1124–1134.

Li, B., et al., 2014. Shock response of He bubbles in single crystal Cu. Journal of Applied Physics 116 (21), 213506.

Li, J.C.M., 1964. Stress field of a dislocation segment. Philosophical Magazine 10, 1097–1098.

Liao, H., Wu, Y., Ding, K., 2013. Hardening response and precipitation behavior of Al–7%Si–0.3%Mg alloy in a pre-aging process. Materials Science and Engineering: A 560, 811–816.

Lin, E.Q., et al., 2012. Shock response of copper bicrystals with a $\sum 3$ asymmetric tilt grain boundary. Computational Materials Science 59, 94–100.

Lin, P., Liu, Z.L., Cui, Y.N., Zhang, Z., 2015. A stochastic crystal plasticity model with size-dependent and intermittent strain bursts characteristics at micron scale. International Journal of Solids and Structures 69–70, 267–276.

Lin, P., Liu, Z., Zhang, Z., 2016. Numerical study of the size-dependent deformation morphology in micropillar compressions by a dislocation-based crystal plasticity model. International Journal of Plasticity 87, 32–47.

Liu, X.Y., Biner, S.B., 2008. Molecular dynamics simulations of the interactions between screw dislocations and self-interstitial clusters in body-centered cubic Fe. Scripta Materialia 59, 51–54.

Liu, X.H., Schwarz, K.W., 2005. Modelling of dislocations intersecting a free surface. Modelling and Simulation in Materials Science and Engineering 13 (8), 1233–1247.

Liu, Z.L., et al., 2008. A mesoscale investigation of strain rate effect on dynamic deformation of single-crystal copper. International Journal of Solids and Structures 45 (13), 3674–3687.

Liu, Z.L., Liu, X., Zhuang, Z., You, X., 2009a. Atypical three-stage-hardening mechanical behavior of Cu single-crystal micropillars. Scripta Materialia 60, 594–597.

Liu, Z.L., Liu, X., Zhuang, Z., You, X., 2009b. A multi-scale computational model of crystal plasticity at submicron-to-nanometer scales. International Journal of Plasticity 25, 1436–1455.

Liu, Z.L., Zhuang, Z., Liu, X.M., Zhao, X.C., Zhang, Z.H., 2011. A dislocation-dynamics based higher-order crystal plasticity model and applications on confined thin-film plasticity. International Journal of Plasticity 27, 201–216.

Liu, D., He, Y., Dunstan, D.J., Zhang, B., Gan, Z., Hu, P., Ding, H., 2013a. Toward a further understanding of size effects in the torsion of thin metal wires: an experimental and theoretical assessment. International Journal of Plasticity 41, 30–52.

Liu, Y., Chen, Y., Yu, K.Y., Wang, H., Chen, J., Zhang, X., 2013b. Stacking fault and partial dislocation dominated strengthening mechanisms in highly textured Cu/Co multilayers. International Journal of Plasticity 49, 152–163.

Liu, C.M., et al., 2015a. Orientation-dependent responses of tungsten single crystal under shock compression via molecular dynamics simulations. Computational Materials Science 110, 359–367.

Liu, Z., Hao, X., Ho-Baillie, A., Tsao, C.-y., Green, M.A., 2015b. Cyclic thermal annealing on Ge/Si(100) epitaxial films grown by magnetron sputtering. Thin Solid Films 574, 99–102.

Liu, F.X., Liu, Z.L., Zhuang, Z., 2017a. Numerical investigations of helical dislocations based on coupled glide-climb model. International Journal of Plasticity 92, 2–18.

Liu, F.X., Liu, Z.L., Pei, X.Y., Hu, J.Q., Zhuang, Z., 2017b. Modeling high temperature anneal hardening in Au submicron pillar by developing coupled dislocation glide-climb model. International Journal of Plasticity 99, 102–119.

Lloyd, D.J., 1994. Particle reinforced aluminum and magnesium matrix composites. International Materials Reviews 39, 1–23.

Lothe, J., 1967. Dislocation climb forces. Journal of Applied Physics 38, 845.

Lowry, M.B., Kiener, D., LeBlanc, M.M., Chisholm, C., Florando, J.N., Morris, J.W., Minor, A.M., 2010. Achieving the ideal strength in annealed molybdenum nanopillars. Acta Materialia 58, 5160–5167.

Lubarda, V.A., 1997. Energy analysis of dislocation arrays near bimaterial interfaces. International Journal of Solids and Structures 34, 1053–1073.

Lubliner, L., 1990. Plasticity Theory. Macmillan, New York.

Ma, Q., Clarke, D.R. 1995. Size dependent hardness of silver single crystal. Journal of Materials Research 10, 853–863.

Maaß, R., Van Swygenhoven, H., Derlet, P.M., Volkert, C.A., Grolimund, D., 2007. Time-resolved Laue diffraction of deforming micropillars. Physical Review Letters 99, 145505.

Maaß, R., Derlet, P., 2018. Micro-plasticity and recent insights from intermittent and small-scale plasticity. Acta Materialia 143, 338–363.

Maaß, R., Uchic, M.D., 2012. In-situ characterization of the dislocation-structure evolution in Ni micro-pillars. Acta Materialia 60, 1027–1037.

Maaß, R., Van Petegem, S., Grolimund, D., Van Swygenhoven, H., Kiener, D., Dehm, G., 2008a. Crystal rotation in Cu single crystal micropillars: in situ Laue and electron backscatter diffraction. Applied Physics Letters 92, 071905.

Maaß, R., Van Petegem, S., Zimmermann, J., Borca, C.N., Van Swygenhoven, H., 2008b. On the initial microstructure of metallic micropillars. Scripta Materialia 59, 471–474.

Maaß, R., Van Petegem, S., Ma, D.C., Zimmermann, J., Roters, F., Van Swygenhoven, H., Raabe, D., 2009. Smaller is stronger: the effect of strain hardening. Acta Materialia 57, 5996–6005.

Maaß, R., Derlet, P.M., Greer, J.R., 2013. Small-scale plasticity: insights into dislocation avalanche velocities. Scripta Materialia 69, 586–589.

Maaß, R., Derlet, P.M., Greer, J.R., 2015. Independence of slip velocities on applied stress in small crystals. Small 11, 341–351.

Mackenchery, K., et al., 2016. Dislocation evolution and peak spall strengths in single crystal and nanocrystalline Cu. Journal of Applied Physics 119 (4), 044301.

Madec, R., Devincre, B., Kubin, L., 2002a. From dislocation junctions to forest hardening. Physical Review Letters 89, 255508.

Madec, R., Devincre, B., Kubin, L., 2002b. Simulation of dislocation patterns in multislip. Scripta Materialia 47, 689–695.

Malygin, G., 2010. Size effects under plastic deformation of microcrystals and nanocrystals. Physics of the Solid State 52, 49–57.

Malygin, G., 2012. Influence of the transverse size of samples with micro-and nano-grained structures on the yield and flow stresses. Physics of the Solid State 54, 559–567.

Marian, J., Cai, W., Bulatov, V.V., 2004. Dynamic transitions from smooth to rough to twinning in dislocation motion. Nature Materials 3, 158–163.

Martinez, E., Marian, J., Arsenlis, A., Victoria, M., Perlado, J., 2008. A dislocation dynamics study of the strength of stacking fault tetrahedra. Part I: interactions with screw dislocations. Philosophical Magazine 88, 809–840.

Matouš, K., Geers, M.G., Kouznetsova, V.G., Gillman, A., 2017. A review of predictive nonlinear theories for multiscale modeling of heterogeneous materials. Journal of Computational Physics 330, 192–220.

Mayeur, J.R., McDowell, D.L., 2013. An evaluation of higher-order single crystal strength models for constrained thin films subjected to simple shear. Journal of the Mechanics and Physics of Solids 61, 1935–1954.

McDowell, D., Ghosh, S., Kalidindi, S., 2011. Representation and computational structure-property relations of random media. Journal of the Minerals, Metals and Materials Society 63, 45–51.

McElhaney, K., Vlassak, J., Nix, W., 1998. Determination of indenter tip geometry and indentation contact area for depth-sensing indentation experiments. Journal of Materials Research 13, 1300–1306.

Meyers, M.A., et al., 2003. Laser-induced shock compression of monocrystalline copper: characterization and analysis. Acta Materialia 51 (5), 1211–1228.

Michael, Z., Paolo, M., 2005. Fluctuation phenomena in crystal plasticity—a continuum model. Journal of Statistical Mechanics: Theory and Experiment 2005, P08004.

Michalske, T., Houston, J., 1998. Dislocation nucleation at nano-scale mechanical contacts. Acta Materialia 46, 391–396.

Milne, R.J., Lockwood, A.J., Inkson, B.J., 2011. Size-dependent deformation mechanisms of Al nanopillars. Journal of Physics D Applied Physics 44, 485301.

Mindlin, R.D., 1965. Second gradient of strain and surface tension in linear elasticity. International Journal of Solids and Structures 1, 417–438.

Mirzadeh, H., Roostaei, M., Parsa, M.H., Mahmudi, R., 2015. Rate controlling mechanisms during hot deformation of Mg–3Gd–1Zn magnesium alloy: dislocation glide and climb, dynamic recrystallization, and mechanical twinning. Materials and Design 68, 228–231.

Mishin, Y., et al., 2001. Structural stability and lattice defects in copper:Ab initio, tight-binding, and embedded-atom calculations. Physical Review B 63 (22), 224106.

Mizuno, K., Morikawa, K., Okamoto, H., Hashimoto, E., 2014. Row of dislocation loops as a vacancy source in ultrahigh-purity aluminum single crystals with a low dislocation density. In: Transactions of the Materials Research Society of Japan, vol. 39, pp. 169–172.

Mogilevskii, M.A., Bushnev, L.S., 1990. Deformation structure development in Al and Cu single crystal on shock-wave loading up to 50 and 100 GPa. Combustion, Explosion and Shock Waves 26, 215–220.

Monnet, G., Devincre, B., Kubin, L., 2004. Dislocation study of prismatic slip systems and their interactions in hexagonal close packed metals: application to zirconium. Acta Materialia 52, 4317–4328.

Monnet, G., 2015. New insights into radiation hardening in face-centered cubic alloys. Scripta Materialia 100, 24–27.

Montemayor, L.C., Meza, L.R., Greer, J.R., 2014. Design and fabrication of hollow rigid nanolattices via two-photon lithography. Advanced Engineering Materials 16, 184–189.

Mook, W., Niederberger, C., Bechelany, M., Philippe, L., Michler, J., 2010. Compression of freestanding gold nanostructures: from stochastic yield to predictable flow. Nanotechnology 21, 055701.

Mordehai, D., Clouet, E., Fivel, M., Verdier, M., 2008. Introducing dislocation climb by bulk diffusion in discrete dislocation dynamics. Philosophical Magazine 88, 899–925.

Mordehai, D., Clouet, E., Fivel, M., Verdier, M., 2009. Annealing of dislocation loops in dislocation dynamics simulations. IOP Conference Series: Materials Science and Engineering 3, 012001.

Moshe, E., et al., 1998. An increase of the spall strength in aluminum, copper, and Metglas at strain rates larger than 10^7 s^{-1}. Journal of Applied Physics 83 (8), 4004.

Motz, C., Weygand, D., Senger, J., Gumbsch, P., 2009. Initial dislocation structures in 3-D discrete dislocation dynamics and their influence on microscale plasticity. Acta Materialia 57, 1744–1754.

Mughrabi, H., 2009. Cyclic slip irreversibilities and the evolution of fatigue damage. Metallurgical and Materials Transactions A 40, 1257–1279.

Mughrabi, H., 2010. Fatigue, an everlasting materials problem-still en vogue. Procedia Engineering 2, 3–26.

Muhlhaus, H.B., Aifantis, E.C., 1991. A variational principle for gradient plasticity. International Journal of Solids and Structures 28, 845–857.

Muhlstein, C.L., Brown, S.B., Ritchie, R.O., 2001. High-cycle fatigue of single-crystal silicon thin films. Journal of Microelectromechanical Systems 10, 593–600.

Munday, L.B., Crone, J.C., Knap, J., 2016. Prismatic and helical dislocation loop generation from defects. Acta Materialia 103, 217–228.

Munro, M., 1997. Evaluated material properties for a sintered alpha-alumina. Journal of the American Ceramic Society 80, 1919–1928.

Mura, T., 1987. Micromechanics of Defects in Solids. Springer.

Naamane, S., Monnet, G., Devincre, B., 2010. Low temperature deformation in iron studied with dislocation dynamics simulations. International Journal of Plasticity 26, 84–92.

Nabarro, F.R., Duesbery, M.S., 2002. Dislocations in Solids. Elsevier.

Nabarro, F.R.N., 1951. The synthesis of elastic dislocation fields. Philosophical Magazine 42, 1224–1231.

Nabarro, F.R.N., 1979. Dislocations in Solids: Dislocations in Metallurgy. North-Holland.

Nadgornyi, E., 1998. Progress in Materials Science. Dislocation Dynamics and Mechanical Properties. Pergamon Press, Oxford.

Nagtegaal, J.D., DeJong, J.E., 1981. Some computational aspects of elastic-plastic large strain analysis. International Journal for Numerical Methods in Engineering 17, 15–41.

Namazu, T., Isono, Y., 2009. Fatigue life prediction criterion for micro–nanoscale single-crystal silicon structures. Journal of Microelectromechanical Systems 18, 129–137.

Narasimhan, R., Rosakis, A.J., Moran, B., 1992. A three-dimensional numerical investigation of fracture initiation by ductile failure mechanisms in 4340 steels. International Journal of Fracture 56, 1–24.

Needleman, A., 2000. Computational mechanics at the mesoscale. Acta Materialia 48, 105–124.

Newberger, M., 1971. Handbook of Electronic Materials Group IV Semiconducting Materials, Handbook of Electronic Materials. IFI/Plenum, New York.

Ng, K.S., Ngan, A.H.W., 2008a. Breakdown of Schmid's law in micropillars. Scripta Materialia 59, 796–799.

Ng, K.S., Ngan, A.H.W., 2008b. Stochastic nature of plasticity of aluminum micro-pillars. Acta Materialia 56, 1712–1720.

Ng, K.S., Ngan, A.H.W., 2008c. Stochastic theory for jerky deformation in small crystal volumes with pre-existing dislocations. Philosophical Magazine 88, 677–688.

Ng, K., Ngan, A., 2009a. Effects of trapping dislocations within small crystals on their deformation behavior. Acta Materialia 57, 4902–4910.

Ng, K.S., Ngan, A.H.W., 2009b. Deformation of micron-sized aluminium bi-crystal pillars. Philosophical Magazine 89, 3013–3026.

Ngan, A., Ng, K., 2010. Transition from deterministic to stochastic deformation. Philosophical Magazine 90, 1937–1954.

Ni, X., Papanikolaou, S., Vajente, G., Adhikari, R.X., Greer, J.R., 2017. Probing microplasticity in small-scale fcc crystals via dynamic mechanical analysis. Physical Review Letters 118, 155501.

Nicola, L., Van der Giessen, E., Needleman, A., 2003. Discrete dislocation analysis of size effects in thin films. Journal of Applied Physics 93, 5920–5928.

Nicola, L., Van der Giessen, E., Gurtin, M.E., 2005a. Effect of defect energy on strain-gradient predictions of confined single-crystal plasticity. Journal of the Mechanics and Physics of Solids 53, 1280–1294.

Nicola, L., Van der Giessen, E., Needleman, A., 2005b. Two hardening mechanisms in single crystal thin films studied by discrete dislocation plasticity. Philosophical Magazine 85, 1507–1518.

Nicola, L., Xiang, Y., Vlassak, J.J., Van der Giessen, E., Needleman, A., 2006. Plastic deformation of freestanding thin films: experiments and modeling. Journal of the Mechanics and Physics of Solids 54, 2089–2110.

Nordson, C.F., Tvergaard, V., 2005. Instabilities in power law gradient hardening materials. International Journal of Solids and Structures 42, 2559–2573.

Nix, W.D., Gao, H., 1998. Indentation size effects in crystalline materials: a law for strain gradient plasticity. Journal of the Mechanics and Physics of Solids 46, 411–425.

Nix, W.D., Gibeling, 1985. Mechanisms of Time-Dependent Flow and Fracture of Metals. Metals/Materials Technology Series 8313-004. ASM, Metals Park, OH.

Nix, W., Lee, S., 2010. Micro-pillar plasticity controlled by dislocation nucleation at surfaces. Philosophical Magazine 91, 1084–1096.

Nix, W., 1989. Mechanical properties of thin films. Metallurgical Transactions A 20, 2217–2245.

Nogaret, T., Rodney, D., Fivel, M., Robertson, C., 2008. Clear band formation simulated by dislocation dynamics: role of helical turns and pile-ups. Journal of Nuclear Materials 380, 22–29.

Norfleet, D., Dimiduk, D., Polasik, S., Uchic, M., Mills, M., 2008. Dislocation structures and their relationship to strength in deformed nickel microcrystals. Acta Materialia 56, 2988–3001.

O'day, M., Curtin, W., 2005. Bimaterial interface fracture: a discrete dislocation model. Journal of the Mechanics and Physics of Solids 53, 359–382.

Oh, S.H., Legros, M., Kiener, D., Dehm, G., 2009. In situ observation of dislocation nucleation and escape in a submicrometre aluminium single crystal. Nature Materials 8, 95–100.

Ohashi, T., Kawamukai, M., Zbib, H., 2007. A multiscale approach for modeling scale-dependent yield stress in polycrystalline metals. International Journal of Plasticity 23, 897–914.

Ohno, N., Okumura, D., 2007. Higher-order stress and grain size effects due to self-energy of geometrically necessary dislocations. Journal of the Mechanics and Physics of Solids 55, 1879–1898.

Osetsky, Y.N., Stoller, R.E., Rodney, D., Bacon, D.J., 2005. Atomic-scale details of dislocation–stacking fault tetrahedra interaction. Materials Science and Engineering: A 400, 370–373.

Osetsky, Y.N., Rodney, D., Bacon, D.J., 2006. Atomic-scale study of dislocation–stacking fault tetrahedron interactions. Part I: mechanisms. Philosophical Magazine 86, 2295–2313.

Oswald, J., Gracie, R., Khare, R., Belytschko, T., 2009. An extended finite element method for dislocations in complex geometries: thin films and nanotubes. Computer Methods in Applied Mechanics and Engineering 198, 1872–1886.

Ouyang, C.J., Li, Z.H., Huang, M.S., Hu, L.L., Hou, C.T., 2009. Combined influences of micro-pillar geometry and substrate constraint on microplastic behavior of compressed single-crystal micro-pillar: two-dimensional discrete dislocation dynamics modeling. Materials Science and Engineering: A 526, 235–243.

Pan, J., Saje, M., Needleman, A., 1983. Localization of deformation in rate-sensitive porous plastic solids. International Journal of Fracture 21, 261–278.

Panda, D., Tseng, T.-Y., 2013. Growth, dielectric properties, and memory device applications of ZrO_2 thin films. Thin Solid Films 531, 1–20.

Pantleon, W., 2004. Stage IV work-hardening related to disorientations in dislocation structures. Materials Science and Engineering: A 387, 257–261.

Papanikolaou, S., Cui, Y., Ghoniem, N., 2017a. Avalanches and plastic flow in crystal plasticity: an overview. Modelling and Simulation in Materials Science and Engineering 26, 013001.

Papanikolaou, S., Song, H., Van der Giessen, E., 2017b. Obstacles and sources in dislocation dynamics: strengthening and statistics of abrupt plastic events in nanopillar compression. Journal of the Mechanics and Physics of Solids 102, 17–29.

Parthasarathy, T.A., Rao, S.I., Dimiduk, D.M., Uchic, M.D., Trinkle, D.R., 2007. Contribution to size effect of yield strength from the stochastics of dislocation source lengths in finite samples. Scripta Materialia 56, 313–316.

Patra, A., McDowell, D.L., 2016. Crystal plasticity investigation of the microstructural factors influencing dislocation channeling in a model irradiated bcc material. Acta Materialia 110, 364–376.

Peirce, D., Asaro, R.J., Needleman, A., 1983. Material rate dependence and localized deformation in crystalline solids. Acta Metallurgica 31, 1951–1976.

Peirce, D., Shih, C.F., Needleman, A., 1984. A tangent modulus method for rate dependent solids. Computers and Structures 18, 875–887.

Perzyna, P., 1971. Thermodynamic theory of viscoplasticity. Advances in Applied Mechanics 11. Academic Press, New York.

Petch, N.J., 1953. The cleavage strength of polycrystals. Journal of the Iron and Steel Institute 174, 25–28.

Picu, R., Soare, M., 2010. Asymmetric dislocation junctions exhibit a broad range of strengths. Scripta Materialia 62, 508–511.

Plimpton, S., 1995. Fast parallel algorithms for short-range molecular dynamics. Journal of Computational Physics 117 (1), 1–19.

Po, G., Ghoniem, N., 2014. A variational formulation of constrained dislocation dynamics coupled with heat and vacancy diffusion. Journal of the Mechanics and Physics of Solids 66, 103–116.

Po, G., Ghoniem, N., 2015. Mechanics of Defect Evolution Library, MODEL. https://bitbucket.org/model/model/wiki/home.

Po, G., Mohamed, M., Crosby, T., Erel, C., El-Azab, A., Ghoniem, N., 2014. Recent progress in discrete dislocation dynamics and its applications to micro plasticity. Journal of the Minerals, Metals and Materials Society 66, 2108–2120.

Po, G., Cui, Y., Rivera, D., Cereceda, D., Swinburne, T.D., Marian, J., Ghoniem, N., 2016. A phenomenological dislocation mobility law for bcc metals. Acta Materialia 119, 123–135.

Polmear, I.J., 2006. Light Alloys: From Traditional Alloys to Nanocrystals. Elsevier/Butterworth-Heinemann, Burlington, MA.

Pontes, J., Walgraef, D., Aifantis, E., 2006. On dislocation patterning: multiple slip effects in the rate equation approach. International Journal of Plasticity 22, 1486–1505.

Poole, W.J., Ashby, F.M., Fleck, N.A., 1996. Micro-hardness of annealed and work-hardened copper polycrystals. Scripta Metallurgica et Materialia 34, 559–564.

Prager, W., 1945. Strain hardening under combined stress. Journal of Applied Physics 16, 837–840.

Prager, W., 1961. Introduction to Mechanics of Continua. Ginn & Co., New York.

Prakash, A., Weygand, D., Bitzek, E., 2017. Influence of grain boundary structure and topology on the plastic deformation of nanocrystalline aluminum as studied by atomistic simulations. International Journal of Plasticity 97, 107–125.

Puri, S., Das, A., Acharya, A., 2011. Mechanical response of multicrystalline thin films in mesoscale field dislocation mechanics. Journal of the Mechanics and Physics of Solids 59, 2400–2417.

Püschl, W., 2002. Models for dislocation cross-slip in close-packed crystal structures: a critical review. Progress in Materials Science 47, 415–461.

Qu, S., Huang, Y., Jiang, H., Liu, C., Wu, P.D., Hwang, K.C., 2004. Fracture analysis in the conventional theory of mechanism-based strain gradient (CMSG) plasticity. International Journal of Fracture 129, 199–220.

Quek, S.S., Choi, Z.H., Wu, Z., Zhang, Y.W., Srolovitz, D.J., 2016. The inverse hall–petch relation in nanocrystalline metals: a discrete dislocation dynamics analysis. Journal of the Mechanics and Physics of Solids 88, 252–266.

Rajagopalan, J., Han, J.H., Saif, M.T.A., 2007. Plastic deformation recovery in freestanding nanocrystalline aluminum and gold thin films. Science 315, 1831–1834.

Ramirez, B.R., Ghoniem, N., Po, G., 2012. Ab initio continuum model for the influence of local stress on cross-slip of screw dislocations in fcc metals. Physical Review B 86, 094115.

Rao, S., Dimiduk, D., Tang, M., Uchic, M., Parthasarathy, T., Woodward, C., 2007. Estimating the strength of single-ended dislocation sources in micro-sized single crystals. Philosophical Magazine 87, 4777–4794.

Rao, S.I., Dimiduk, D., Parthasarathy, T.A., Uchic, M., Tang, M., Woodward, C., 2008. Athermal mechanisms of size-dependent crystal flow gleaned from three-dimensional discrete dislocation simulations. Acta Materialia 56 (13), 3245–3259.

Rao, S., Dimiduk, D., El-Awady, J., Parthasarathy, T., Uchic, M., Woodward, C., 2011. Calculations of intersection cross-slip activation energies in fcc metals using nudged elastic band method. Acta Materialia 59, 7135–7144.

Revanian, O., Zikry, M.A., Rajendran, A.M., 2007. Statistically necessary and grain boundary dislocation densities: microstructural representation and modelling. Proceedings of the Royal Society 463, 2833–2853.

Rhee, M., Zbib, H., Hirth, J., Huang, H., Rubia, T., 1998. Models for long-/short-range interactions and cross slip in 3D dislocation simulation of BCC single crystals. Modelling and Simulation in Materials Science and Engineering 6, 467–492.

Rice, J.R., Asaro, R., 1977. Strain localization in ductile single crystals. Journal of Physics and Solids 25, 309–338.

Rice, J.R., 1971. Inelastic constitutive relations for solids - an internal-variable theory and its application to metal plasticity. Journal of the Mechanics and Physics of Solids 19, 433–455.

Rivas, J.M., Quinones, S.A., Murr, L.E., 1995. Hypervelocity impact cratering microstructural characterization. Scripta Metallurgica et Materialia 33, 101–107.

Rodney, D., Phillips, R., 1999. Structure and strength of dislocation junctions: an atomic level analysis. Physical Review Letters 82, 1704–1707.

Rodney, D., Martin, G., Bréchet, Y., 2001. Irradiation hardening by interstitial loops: atomistic study and micromechanical model. Materials Science and Engineering: A 309, 198–202.

Rodney, D., Le Bouar, Y., Finel, A., 2003. Phase field methods and dislocations. Acta Materialia 51, 17–30.

Rodney, D., 2004. Molecular dynamics simulation of screw dislocations interacting with interstitial frank loops in a model FCC crystal. Acta Materialia 52, 607–614.

Roos, A., et al., 2001a. A two-dimensional computational methodology for high-speed dislocations in high strain-rate deformation. Computational Materials Science 20 (1), 1–18.

Roos, A., et al., 2001b. High-speed dislocations in high strain-rate deformations. Computational Materials Science 20 (1), 19–27.

Roters, F., Raabe, D., Gottstein, G., 1996. Calculation of stress—strain curves by using 2 dimensional dislocation dynamics. Computational Materials Science 7, 56–62.

Roters, F., Eisenlohr, P., et al., 2010. Overview of constitutive laws, kinematics, homogenization and multiscale methods in crystal plasticity finite-element modeling: theory, experiments, applications. Acta Materialia 58 (4), 1152–1211.

Roy, A., Acharya, A., 2005. Finite element approximation of field dislocation mechanics. Journal of the Mechanics and Physics of Solids 53, 143–170.

Ryu, S., Kang, K., Cai, W., 2011. Predicting the dislocation nucleation rate as a function of temperature and stress. Journal of Materials Research 26 (18), 2335–2354.

Ryu, I., Cai, W., Nix, W.D., Gao, H., 2015. Stochastic behaviors in plastic deformation of face-centered cubic micropillars governed by surface nucleation and truncated source operation. Acta Materialia 95, 176–183.

Schneider, M.S., et al., 2005. Laser shock compression of copper and copper—aluminum alloys. International Journal of Impact Engineering 32 (1–4), 473–507.

Schuh, C.A., Nieh, T.G., Yamasaki, T., 2002. Hall–Petch breakdown manifested in abrasive wear resistance of nanocrystalline nickel. Scripta Materialia 46, 735–740.

Schwarz, K., 1999a. Simulation of dislocations on the mesoscopic scale. I. Methods and examples. Journal of Applied Physics 85, 108–119.

Schwarz, K.W., 1999b. Simulation of dislocations on the mesoscopic scale. II. Application to strained-layer relaxation. Journal of Applied Physics 85, 120–129.

Semenov, A., Woo, C., 2003. Theory of Frank loop nucleation at elevated temperatures. Philosophical Magazine 83, 3765–3782.

Serrano, G.D., Pelegrina, J.L., Condó, A.M., Ahlers, M., 2006. Helical dislocations as vacancy sinks in β phase Cu–Zn–Al–Ni alloys. Materials Science and Engineering: A 433, 149–154.

Shade, P.A., Wheeler, R., Choi, Y.S., Uchic, M.D., Dimiduk, D.M., Fraser, H.L., 2009. A combined experimental and simulation study to examine lateral constraint effects on microcompression of single-slip oriented single crystals. Acta Materialia 57, 4580–4587.

Shan, Z.W., Mishra, R.K., Syed Asif, S.A., Warren, O.L., Minor, A.M., 2008. Mechanical annealing and source-limited deformation in submicrometre-diameter Ni crystals. Nature Materials 7 (2), 115–119.

Shanthraj, P., Zikry, M., 2011. Dislocation density evolution and interactions in crystalline materials. Acta Materialia 59 (20), 7695–7702.

Shao, S., Abdolrahim, N., Bahr, D.F., Lin, G., Zbib, H.M., 2013. Stochastic effects in plasticity in small volumes. International Journal of Plasticity, https://doi.org/10.1016/j.ijplas.2013.09.005.

Shehadeh, M.A., et al., 2005b. Multiscale dislocation dynamics simulations of shock compression in copper single crystal. International Journal of Plasticity 21 (12), 2369–2390.

Shehadeh, M.A., et al., 2006. Simulation of shock-induced plasticity including homogeneous and heterogeneous dislocation nucleations. Applied Physics Letters 89 (17), 171918.

Shehadeh, M.A., 2012. Multiscale dislocation dynamics simulations of shock-induced plasticity in small volumes. Philosophical Magazine 92 (10), 1173–1197.

Shen, C., Wang, Y., 2003. Phase field model of dislocation networks. Acta Materialia 51, 2595–2610.

Shen, Y.L., 2008. Externally constrained plastic flow in miniaturized metallic structures: a continuum-based approach to thin films, lines, and joints. Progress in Materials Science 53, 838–891.

Shi, J., Zikry, M.A., 2009. Grain-boundary interactions and orientation effects on crack behavior in polycrystalline aggregates. International Journal of Solids and Structures 46, 3914–3925.

Shi, X., Dupuy, L., Devincre, B., Terentyev, D., Vincent, L., 2015. Interaction of <1 0 0> dislocation loops with dislocations studied by dislocation dynamics in α-iron. Journal of Nuclear Materials 460, 37–43.

Shim, S., Bei, H., Miller, M.K., Pharr, G.M., George, E.P., 2009. Effects of focused ion beam milling on the compressive behavior of directionally solidified micropillars and the nanoindentation response of an electropolished surface. Acta Materialia 57, 503–510.

Shin, C., Fivel, M., Verdier, M., Robertson, C., 2005. Dislocation dynamics simulations of fatigue of precipitation-hardened materials. Materials Science and Engineering: A 400, 166–169.

Shu, J.Y., Fleck, N.A., 1998. The prediction of a size-effect in micro-indentation. International Journal of Solids and Structures 35, 1363–1383.

Shu, J.Y., Fleck, N.A., Van der Giessen, E., Needleman, A., 2001. Boundary layers in constrained plastic flow: comparison of nonlocal and discrete dislocation plasticity. Journal of the Mechanics and Physics of Solids 49, 1361–1395.

Silcox, J., Whelan, M., 1960. Direct observations of the annealing of prismatic dislocation loops and of climb of dislocations in quenched aluminium. Philosophical Magazine 5, 1–23.

Sills, R., Cai, W., 2016. Solute drag on perfect and extended dislocations. Philosophical Magazine 96, 895–921.

Sills, R.B., Kuykendall, W.P., Aghaei, A., Cai, W., 2016. Fundamentals of Dislocation Dynamics Simulations. Multiscale Materials Modeling for Nanomechanics. Springer, pp. 53–87.

Sluys, L.J., Estrin, Y., 2000. The analysis of shear banding with a dislocation based gradient plasticity model. International Journal of Solids and Structures 37, 7127–7142.

Smyshlyaev, V.P., Fleck, N.A., 1996. The role of strain gradients in the grain size effect for polycrystals. Journal of the Mechanics and Physics of Solids 44, 465–495.

Sobie, C., Bertin, N., Capolungo, L., 2015. Analysis of obstacle hardening models using dislocation dynamics: application to irradiation-induced defects. Metallurgical and Materials Transactions A 46, 3761–3772.

Stelmashenko, N.A., Walls, A.G., Brown, L.M., et al., 1993. Micro indentation on W and Wo oriented single crystals: an STM study. Acta Metallurgica et Materialia 41, 2855–2865.

Sternberg, E., Muki, R., 1967. The effect of couple-stresses on the stress concentration around a crack. International Journal of Solids and Structures 3, 69–95.

Stölken, J., Evans, A., 1998. A microbend test method for measuring the plasticity length scale. Acta Materialia 46, 5109–5115.

Stukowski, A., Albe, K., 2010. Extracting dislocations and non-dislocation crystal defects from atomistic simulation data. Modelling and Simulation in Materials Science and Engineering 18 (8), 085001.

Stukowski, A., 2010. Visualization and analysis of atomistic simulation data with OVITO–the open visualization tool. Modelling and Simulation in Materials Science and Engineering 18 (1), 015012.

Tang, M., Marian, J., 2014. Temperature and high strain rate dependence of tensile deformation behavior in single-crystal iron from dislocation dynamics simulations. Acta Materialia 70, 123–129.

Tang, M., Kubin, L.P., Canova, G.R., 1998. Dislocation mobility and the mechanical response of b.c.c. single crystals: a mesoscopic approach. Acta Materialia 46, 3221–3235.

Tang, X., Lagerlöf, K.P.D., Heuer, A.H., 2003. Determination of pipe diffusion coefficients in undoped and magnesia-doped sapphire (α-Al$_2$O$_3$): a study based on annihilation of dislocation dipoles. Journal of the American Ceramic Society 86, 560–565.

Tang, M., Cai, W., Xu, G., Bulatov, V.V., 2006. A hybrid method for computing forces on curved dislocations intersecting free surfaces in three-dimensional dislocation dynamics. Modelling and Simulation in Materials Science and Engineering 14, 1139.

Tang, H., Schwarz, K.W., Espinosa, H.D., 2007. Dislocation escape-related size effects in single-crystal micropillars under uniaxial compression. Acta Materialia 55, 1607–1616.

Tang, H., Schwarz, K., Espinosa, H., 2008. Dislocation-source shutdown and the plastic behavior of single-crystal micropillars. Physical Review Letters 100, 185503.

Taylor, G.I., 1934. The mechanism of plastic deformation of crystals. Part I. Theoretical. In: Proceedings of the Royal Society of London. Series A, Containing Papers of a Mathematical and Physical Character. pp. 362–387.

Taylor, G.T., 1938. Plastic strain in metals. Journal of the Institute of Metals 62, 307–324.

Terentyev, D., Monnet, G., Grigorev, P., 2013. Transfer of molecular dynamics data to dislocation dynamics to assess dislocation–dislocation loop interaction in iron. Scripta Materialia 69, 578–581.

Torrents Abad, O., Wheeler, J.M., Michler, J., Schneider, A.S., Arzt, E., 2016. Temperature-dependent size effects on the strength of Ta and W micropillars. Acta Materialia 103, 483–494.

Toupin, R.A., 1962. Elastic materials with couple stresses. Archive for Rational Mechanics and Analysis 11, 385–414.

Tu, K.N., Rosenberg, R., 1982. Treatise on Materials Science & Technology, Preparation and Properties of Thin Films. Academic Press, New York.

Tvergaard, V., Needleman, A., 1984. Analysis of the cup-cone fracture in a round tensile bar. Acta Metallurgica 32, 157–169.

Tvergaard, V., 1981. Influence of voids on shear band instabilities under plane strain conditions. International Journal of Fracture 17, 389–407.

Uchic, M.D., Dimiduk, D.M., Florando, J.N., Nix, W.D., 2004. Sample dimensions influence strength and crystal plasticity. Science 305, 986–989.

Uchic, M.D., Shade, P., Dimiduk, D., 2009a. Micro-compression testing of fcc metals: a selected overview of experiments and simulations. Journal of the Minerals, Metals and Materials Society 61, 36–41.

Uchic, M.D., Shade, P., Dimiduk, D., 2009b. Plasticity of micrometer-scale single crystals in compression. Annual Review of Materials Research 39, 361–386.

Uhl, J.T., Pathak, S., Schorlemmer, D., Liu, X., Swindeman, R., Brinkman, B.A., LeBlanc, M., Tsekenis, G., Friedman, N., Behringer, R., 2015. Universal quake statistics: from compressed nanocrystals to earthquakes. Scientific Reports 5, 16493.

Ungar, T., Li, L., Tichy, G., Choo, H., Liaw, P.K., 2011. Work softening in nanocrystalline materials induced by dislocation annihilation. Scripta Materialia 64, 876–879.

Van der Giessen, E., Needleman, A., 1995. Discrete dislocation plasticity: a simple planar model. Modelling and Simulation in Materials Science and Engineering 3 (5), 689–735.

Van Houtte, P., Li, S., et al., 2005. Deformation texture prediction: from the Taylor model to the advanced Lamel model. International Journal of Plasticity 21 (3), 589–624.

Van Houtte, P., 1987. The Taylor and the relaxed Taylor theory. Textures and Microstructures 7, 29–72.

Vattré, A., Devincre, B., Feyel, F., Gatti, R., Groh, S., Jamond, O., Roos, A., 2013. Modelling crystal plasticity by 3D dislocation dynamics and the finite element method: the Discrete-Continuous Model revisited. Journal of the Mechanics and Physics of Solids 63, 491–505.

Verdier, M., Fivel, M., Groma, I., 1998. Mesoscopic scale simulation of dislocation dynamics in fcc metals: principles and applications. Modelling and Simulation in Materials Science and Engineering 6, 755–770.

Volkert, C.A., Lilleodden, E.T., 2006. Size effects in the deformation of sub-micron Au columns. Philosophical Magazine 86, 5567–5579.

Wagner, G., Gottschalch, V., 2006. Helical dislocations in Sn-doped GaP epitaxial layers and their characterization by transmission electron microscopy. Philosophical Magazine A 52, 395–406.

Wallin, M., et al., 2008. Multi-scale plasticity modeling: coupled discrete dislocation and continuum crystal plasticity. Journal of the Mechanics and Physics of Solids 56 (11), 3167–3180.

Wang, Z.Q., Beyerlein, I.J., 2011. An atomistically-informed dislocation dynamics model for the plastic anisotropy and tension–compression asymmetry of BCC metals. International Journal of Plasticity 27, 1471–1484.

Wang, Y.U., Chen, L.Q., Khachaturyan, A.G., 1993. Kinetics of strain-induced morphological transformation in cubic alloys with a miscibility gap. Acta Metallurgica et Materialia 41, 279–296.

Wang, Y.U., Jin, Y.M., Cuitino, A.M., Khachaturyan, A.G., 2001. Nanoscale phase field microelasticity theory of dislocations: model and 3D simulations. Acta Materialia 49, 1847–1857.

Wang, Y.U., Jin, Y.M., Khachaturyan, A.G., 2002. Phase field microelasticity theory and modeling of elastically and structurally inhomogeneous solid. Journal of Applied Physics 92, 1351.

Wang, Y.U., Jin, Y.M., Khachaturyan, A.G., 2003. Phase field microelasticity modeling of dislocation dynamics near free surface and in heteroepitaxial thin films. Acta Materialia 51, 4209–4223.

Wang, Y.U., Jin, Y.M., Khachaturyan, A.G., 2004. Phase field microelasticity modeling of surface instability of heteroepitaxial thin films. Acta Materialia 52, 81–92.

Wang, Y.U., Jin, Y.M., Khachaturyan, A.G., 2005. Mesoscale modelling of mobile crystal defects-dislocations, cracks and surface roughening: phase field microelasticity approach. Philosophical Magazine 85, 261–277.

Wang, Z., Beyerlein, I., LeSar, R., 2007a. The importance of cross-slip in high-rate deformation. Modelling and Simulation in Materials Science and Engineering 15, 675–690.

Wang, Z.Q., et al., 2007b. Dislocation motion in high strain-rate deformation. Philosophical Magazine 87 (16), 2263–2279.

Wang, D., Volkert, C., Kraft, O., 2008a. Effect of length scale on fatigue life and damage formation in thin Cu films. Materials Science and Engineering: A 493, 267–273.

Wang, Z.Q., et al., 2008b. Slip band formation and mobile dislocation density generation in high rate deformation of single fcc crystals. Philosophical Magazine 88 (9), 1321–1343.

Wang, Z.J., Li, Q.J., Shan, Z.W., Li, J., Sun, J., Ma, E., 2012a. Sample size effects on the large strain bursts in submicron aluminum pillars. Applied Physics Letters 100, 071906.

Wang, Z.J., Shan, Z.W., Li, J., Sun, J., Ma, E., 2012b. Pristine-to-pristine regime of plastic deformation in submicron-sized single crystal gold particles. Acta Materialia 60, 1368–1377.

Wang, Z., Shan, Z., Li, J., Sun, J., Ma, E., 2014. An index for deformation controllability of small-volume materials. Science China Technological Sciences 57, 663–670.

Wang, Z.J., Li, Q.J., Cui, Y.N., Liu, Z.L., Ma, E., Li, J., Sun, J., Zhuang, Z., Dao, M., Shan, Z.W., Suresh, S., 2015. Cyclic deformation leads to defect healing and strengthening of small-volume metal single crystals. Proceedings of the National Academy of Sciences of the United States of America 112 (4), 13502–13507.

Wang, L.Y., Liu, Z.L., Zhuang, Z., 2016. Developing micro-scale crystal plasticity model based on phase field theory for modeling dislocations in heteroepitaxial structures, 17, 267–283.

Watanabe, O., Zbib, H.M., Takenouchi, E., 1998. Crystal plasticity: micro-shear banding in polycrystals using voronoi tessellation. International Journal of Plasticity 14, 771–788.

Watling, J.R., Paul, D.J., 2011. A study of the impact of dislocations on the thermoelectric properties of quantum wells in the Si/SiGe materials system. Journal of Applied Physics 110, 114508.

Weeks, R., Dundurs, J., Stippes, M., 1968. Exact analysis of an edge dislocation near a surface layer. International Journal of Engineering Science 6, 365–372.

Weertman, J. 1957. Helical dislocations. Physical Review 107, 1259–1261.

Wei, Y., Hutchinson, J.W. 1997. Steady-state crack growth and work of fracture for solids characterized by strain gradient plasticity. Journal of the Mechanics and Physics of Solids 45, 1253–1273.

Wei, Hutchinson, 1998. Interface strength, work of adhesion and plasticity in the peel test. Recent Advances in Fracture Mechanics 93, 315–333.

Wei, H., Wei, Y., 2012. Interaction between a screw dislocation and stacking faults in FCC metals. Materials Science and Engineering: A 541, 38–44.

Wei, Y., 2006. A new finite element method for strain gradient theories and applications to fracture analyses. European Journal of Mechanics - A/Solids 25, 897–913.

Weinberger, C.R., Cai, W., 2007. Computing image stress in an elastic cylinder. Journal of the Mechanics and Physics of Solids 55, 2027–2054.

Wen, P., et al., 2016. A molecular dynamics study of the shock-induced defect microstructure in single crystal Cu. Computational Materials Science 124, 304–310.

Wolf, D., Jagodzinski, H., Moritz, W., 1978. Diffuse LEED intensities of disordered crystal surfaces. Surface Science 77, 265–282.

Wright, T.W., 2002. The Physics and Mathematics of Adiabatic Shear Bands. Cambridge University Press, Cambridge.

Wulfinghoff, S., Forest, S., Bohlke, T., 2015. Strain gradient plasticity modeling of the cyclic behavior of laminate microstructures. Journal of the Mechanics and Physics of Solids 79, 1–20.

Wzorek, M., et al., 2007. Hydrostatic pressure effect on dislocation evolution in self-implanted Si investigated by electron microscopy methods. Vacuum 81 (10), 1229–1232.

Xia, Z.C., Hutchinson, J.W., 1996. Crack tip fields in strain gradient plasticity. Journal of the Mechanics and Physics of Solids 44, 1621–1648.

Xiang, Y., Srolovitz, D.J., 2006. Dislocation climb effects on particle bypass mechanisms. Philosophical Magazine 86, 3937–3957.

Xiang, Y., Vlassak, J.J., 2006. Bauschinger and size effects in thin-film plasticity. Acta Materialia 54, 5449–5460.

Xu, S., Guo, Y.F., Ngan, A.H.W., 2013. A molecular dynamics study on the orientation, size, and dislocation confinement effects on the plastic deformation of Al nanopillars. International Journal of Plasticity 43, 116–127.

Ye, C., et al., 2011. Bimodal nanocrystallization of NiTi shape memory alloy by laser shock peening and post-deformation annealing. Acta Materialia 59 (19), 7219–7227.

Yefimov, S., Groma, I., van der Giessen, E., 2004. A comparison of a statistical-mechanics based plasticity model with discrete dislocation plasticity calculations. Journal of the Mechanics and Physics of Solids 52, 279–300.

Yoffe, E.H., 1961. A dislocation at a free surface. Philosophical Magazine 6, 1147–1155.

Yoo, M., Morris, J., Ho, K., Agnew, S., 2002. Nonbasal deformation modes of hcp metals and alloys: role of dislocation source and mobility. Metallurgical and Materials Transactions A 33, 813–822.

Yuan, F., Wu, X., 2014. Hydrostatic pressure effects on deformation mechanisms of nanocrystalline fcc metals. Computational Materials Science 85, 8–15.

Zaiser, M., Aifantis, E.C., 2006. Randomness and slip avalanches in gradient plasticity. International Journal of Plasticity 22, 1432–1455.

Zaiser, M., 2006. Scale invariance in plastic flow of crystalline solids. Advances in Physics 55, 185–245.

Zapperi, S., Lauritsen, K.B., Stanley, H.E., 1995. Self-organized branching processes: mean-field theory for avalanches. Physical Review Letters 75, 4071–4074.

Zbib, H., Aifantis, E.C., 1989. On the localization and postlocalization behavior of plastic deformation, part 1, on the initiation of shear bands; part II, on the evolution and thickness of shear bands; part III, on the structure and velocity pf portevin-le chatelier bands. Res Mechanica 261, 279–292, 293–305.

Zbib, H.M., Diaz de la Rubia, T., 2002. A multiscale model of plasticity. International Journal of Plasticity 18 (9), 1133–1163.

Zbib, H.M., Rhee, M., Hirth, J.P., 1998. On plastic deformation and the dynamics of 3D dislocations. International Journal of Mechanical Sciences 40, 113–127.

Zbib, H.M., de la Rubia, T.D., Bulatov, V., 2002. A multiscale model of plasticity based on discrete dislocation dynamics. Journal of Engineering Materials and Technology 124, 78–87.

Zbib, H., Shehadeh, M., Khan, S., Karami, G., 2003. Multiscale dislocation dynamics plasticity. International Journal for Multiscale Computational Engineering 1.

Zbib, H.M., Overman, C.T., Akasheh, F., Bahr, D., 2011. Analysis of plastic deformation in nanoscale metallic multilayers with coherent and incoherent interfaces. International Journal of Plasticity 27, 1618–1639.

Zhang, X., Aifantis, K.E., 2011. Interpreting strain bursts and size effects in micropillars using gradient plasticity. Materials Science and Engineering: A 528, 5036–5043.

Zhang, X., Shang, F., 2014. A continuum model for intermittent deformation of single crystal micropillars. International Journal of Solids and Structures 51, 1859–1871.

Zhang, L., Huang, Y., Chen, J.Y., Hwang, K.C., 1998. The mode III full-field solution in elastic materials with strain gradient effects. International Journal of Fracture (92), 325–348.

Zhang, G., Schwaiger, R., Volkert, C., Kraft, O., 2003. Effect of film thickness and grain size on fatigue-induced dislocation structures in Cu thin films. Philosophical Magazine Letters 83, 477–483.

Zhang, G., Volkert, C., Schwaiger, R., Wellner, P., Arzt, E., Kraft, O., 2006. Length-scale-controlled fatigue mechanisms in thin copper films. Acta Materialia 54, 3127–3139.

Zhang, F., Saha, R., Huang, Y., Nix, W.D., Hwang, K.C., Qu, S., Li, M., 2007. Indentation of a hard film on a soft substrate: strain gradient hardening effects. International Journal of Plasticity 23 (1), 25–43.

Zhang, X., Pan, B., Shang, F., 2012. Scale-free behavior of displacement bursts: lower limit and scaling exponent. Europhysics Letters 100, 16005.

Zhang, H., et al., 2013a. A unified physically based crystal plasticity model for FCC metals over a wide range of temperatures and strain rates. Materials Science and Engineering: A 564, 431–441.

Zhang, J., Liu, G., Sun, J., 2013b. Strain rate effects on the mechanical response in multi-and single-crystalline Cu micropillars: grain boundary effects. International Journal of Plasticity 50, 1–17.

Zhang, X., Shang, F., Yu, Y., Yan, Y., 2014. A stochastic model for the temporal aspects of flow intermittency in micropillar compression. International Journal of Solids and Structures 51, 4519–4530.

Zhao, S., et al., 2016. Amorphization and nanocrystallization of silicon under shock compression. Acta Materialia 103, 519–533.

Zhou, C., LeSar, R., 2012. Dislocation dynamics simulations of plasticity in polycrystalline thin films. International Journal of Plasticity 30, 185–201.

Zhou, S.J., Preston, D.L., Lomdahl, P.S., Beazley, D.M., 1998. Large-scale molecular dynamics simulations of dislocation intersection in copper. Science 279, 1525–1527.

Zhou, C., Biner, S., LeSar, R., 2010a. Discrete dislocation dynamics simulations of plasticity at small scales. Acta Materialia 58, 1565–1577.

Zhou, C., Biner, S., LeSar, R., 2010b. Simulations of the effect of surface coatings on plasticity at small scales. Scripta Materialia 63, 1096–1099.

Zhou, C., Beyerlein, I.J., LeSar, R., 2011. Plastic deformation mechanisms of fcc single crystals at small scales. Acta Materialia 59, 7673–7682.

Zhu, T., Bushby, A., Dunstan, D., 2008a. Materials mechanical size effects: a review. Materials Technology 23, 193–209.

Zhu, T., Li, J., Samanta, A., Leach, A., Gall, K., 2008b. Temperature and strain-rate dependence of surface dislocation nucleation. Physical Review Letters 100, 025502.

Zhuang, Y.X., Hansen, O., Knieling, T., Wang, C., Rombach, P., Lang, W., Benecke, W., Kehlenbeck, M., Koblitz, J., 2006. Thermal stability of vapor phase deposited self-assembled monolayers for MEMS anti-stiction. Journal of Micromechanics and Microengineering 16, 2259–2264.

Ziegler, H., 1959. A modification of Prager's hardening rule. Quarterly of Applied Mathematics 17, 55–65.

Zuo, L., Ngan, A., 2006. Molecular dynamics study on compressive yield strength in Ni_3Al micro-pillars. Philosophical Magazine Letters 86, 355–365.

Further Reading

Amodeo, R.J., Ghoniem, N.M., 1990. Dislocation dynamics. I. A proposed methodology for deformation micromechanics. Physical Review B 41, 6958–6967.

Fivel, M.C., Canova, G.R., 1999. Developing rigorous boundary conditions to simulations of discrete dislocation dynamics. Modelling and Simulation in Materials Science and Engineering 7, 753.

Groh, S., Devincre, B., Kubin, L.P., Roos, A., Feyel, F., Chaboche, J.L., 2003. Dislocations and elastic anisotropy in heteroepitaxial metallic thin films. Philosophical Magazine Letters 83, 303–313.

Guo, Y., Huang, Y., Gao, H., Zhuang, Z., Hwang, K.C., 2001. Taylor-based nonlocal theory of plasticity: numerical studies of micro-indentation experiments and crack tip fields. International Journal of Solids and Structures 38 (42–43), 7447–7460.

Gurrutxaga-Lerma, B., et al., 2015b. The role of homogeneous nucleation in planar dynamic discrete dislocation plasticity. Journal of Applied Mechanics 82 (7), 071008.

Han, X., Ghoniem, N.M., 2005. Stress field and interaction forces of dislocations in anisotropic multilayer thin films. Philosophical Magazine 85, 1205–1225.

Huang, Y., Hu, K.X., 1995. A generalized self-consistent mechanics method for solid containing elliptical inclusions. ASME Journal of Applied Mechanics 62, 566–572.

Huang, Y., Zhang, F., Hwang, K.C., Nix, W.D., Pharr, G.M., Feng, G., 2006a. A model of size effects in nano-indentation. Journal of the Mechanics and Physics of Solids 54 (8), 1668–1686.

Hwang, K.C., Huang, Y.G., 1999. Solid Constitutive Relation. Tsinghua University Press (in Chinese).

Hwang, K.C., Jiang, H., Huang, Y., Gao, H.J., Hu, N., 2002. A finite deformation theory of strain gradient plasticity. Journal of the Mechanics and Physics of Solids 50 (1), 81–99.

Hwang, K.C., Jiang, H., Huang, Y., Gao, H.J., 2003. Finite deformation analysis of mechanism-based strain gradient plasticity: torsion and crack tip field. International Journal of Plasticity 19 (2), 235–251.

Hwang, K.C., Guo, Y., Jiang, H., Huang, Y., Zhuang, Z., 2004. The finite deformation theory of Taylor-based non-local plasticity. International Journal of Plasticity 20 (4–5), 831–839.

Kroner, E., 2006. Continuum theory of defects. In: Balian, R., Kleman, M., Poirier, J.P. (Eds.), Les Houches, Session XXXV, 1980-Physics of Defects. North-Holland, Amsterdam, pp. 219–315.

Li, H., Lin, J., Dean, T., Wen, S., Bannister, A., 2009a. Modelling mechanical property recovery of a linepipe steel in annealing process. International Journal of Plasticity 25, 1049–1065.

Ng, K., Ngan, A., 2008d. A Monte Carlo model for the intermittent plasticity of micro-pillars. Modelling and Simulation in Materials Science and Engineering 16, 055004.

Ouyang, C., Li, Z., Huang, M., Fan, H., 2010. Cylindrical nano-indentation on metal film/elastic substrate system with discrete dislocation plasticity analysis: a simple model for nano-indentation size effect. International Journal of Solids and Structures 3103–3114.

Shehadeh, M.A., et al., 2005a. Modelling the dynamic deformation and patterning in FCC single crystals at high strain rates: dislocation dynamics plasticity analysis. Philosophical Magazine 85 (15), 1667–1685.

Shi, M.X., Huang, Y., Gao, H., Hwang, K.C., 2000. Non-existence of separable crack tip field in mechanism-based strain gradient plasticity. International Journal of Solids and Structures 37 (41), 5995–6010.

Takahashi, A., Ghoniem, N., 2008. A computational method for dislocation—precipitate interaction. Journal of the Mechanics and Physics of Solids 56, 1534–1553.

Wen, J., Huang, Y., Hwang, K.C., Liu, C., Li, M., 2005. The modified Gurson model accounting for the void size effect. International Journal of Plasticity 21 (2), 381–395.

Weygand, D., Friedman, L.H., van der Giessen, E., Needleman, A., 2001. Discrete dislocation modeling in three-dimensional confined volumes. Materials Science and Engineering: A 309, 420–424.

Index

Note: 'Page numbers followed by "f" indicate figures, "t" indicate tables.'

CPI Antony Rowe
Eastbourne, UK
April 25, 2019